Quantum Dots 2010

Journal of Physics: Conference Series
Volume 245

Nottingham, United Kingdom
26-30 April 2010

Editor:

Robert A. Taylor

ISBN: 978-1-61738-933-7
ISSN: 1742-6588

Printed from e-media with permission by:

Curran Associates, Inc.
57 Morehouse Lane
Red Hook, NY 12571

CURRAN ASSOCIATES INC. proceedings.com

Some format issues inherent in the e-media version may also appear in this print version.

Printed by Curran Associates, Inc. (2010)

Institute of Physics
Dirac House, Temple Back
Bristol BS1 6BE UK

Phone: 44 1 17 929 7481
Fax: 44 1 17 920 0979

techtracking@iop.org

Additional copies of this publication are available from:

Curran Associates, Inc.
57 Morehouse Lane
Red Hook, NY 12571 USA
Phone: 845-758-0400
Fax: 845-758-2634
Email: curran@proceedings.com
Web: www.proceedings.com

Quantum Dots 2010

Journal of Physics: Conference Series
Volume 245

Nottingham, United Kingdom
26-30 April 2010

ISBN: 978-1-61738-933-7
ISSN: 1742-6588

IOP Institute of Physics

TABLE OF CONTENTS

PREFACE

These conference proceedings contain the written papers of the contributions presented at Quantum Dot 2010 (QD2010). The conference was held in Nottingham, UK, on 26th-30th April, 2010. The conference addressed topics in research on:

1. Epitaxial quantum dots (including self-assembled and interface structures, dots defined by electrostatic gates etc):

    optical properties and electron transport
    quantum coherence effects
    spin phenomena
    optics of dots in cavities
    interaction with surface plasmons in metal/semiconductor structures
    opto-electronics applications

2. Novel QD structures:

    Fabrication and physics of graphene dots, dots in nano-wires etc.

3. Colloidal quantum dots:

    growth (shape control and hybrid nanocrystals such as metal/semiconductor, magnetic/semiconductor)
    assembly and surface functionalisation
    optical properties and spin dynamics
    electrical and magnetic properties
    applications (light emitting devices and solar cells, biological and medical applications, data storage, assemblers)

The Editors

Acknowledgements

Conference Organising Committee:

Maurice Skolnick (Chair)
Alexander Tartakovskii (Programme Chair)
Pavlos Lagoudakis (Programme Chair)
Max Migliorato (Conference Secretary)
Paola Borri (Publicity)
Robert Taylor (Proceedings)
Manus Hayne (Treasurer)
Ray Murray (Sponsorship)
Mohamed Henini (Local Organiser)

International Advisory Committee:

Yasuhiko Arakawa (Tokyo University, Japan)
Manfred Bayer (Dortmund University, Germany)
Sergey Gaponenko (Stepanov Institute of Physics, Minsk, Belarus)
Pawel Hawrylak (NRC, Ottawa, Canada)
Fritz Henneberger (Institute for Physics, Berlin, Germany)
Atac Imamoglu (ETH, Zurich, Switzerland)
Paul Koenraad (TU Eindhoven, Nethehrlands)
Guglielmo Lanzani (Politecnico di Milano, Italy)
Jungil Lee (Korea Institute of Science and Technology, Korea)
Henri Mariette (CNRS-CEA, Grenoble, France)
Lu Jeu Sham (San Diego, USA)
Andrew Shields (Toshiba Research Europe, Cambridge, UK)
Yoshihisa Yamamoto (Stanford University, USA)
Artur Zrenner (Paderborn University, Germany)

International Programme Committee:

Alexander Eychmüller (TU Dresden, Germany)
Jonathan Finley (TU Munich, Germany)
Dan Gammon (NRL, Washington, USA)
Alexander Govorov (Ohio University, USA)
Neil Greenham (Cavendish Laboratory, UK)
Vladimir Korenev (Ioffe Institute, Russia)
Leo Kouwenhoven (TU Delft, Netherlands)
Wolfgang Langbein (Cardiff University, UK)
Xavier Marie (CNRS Toulouse, France)
David Ritchie (Cambridge, UK)
Andrew Sachrajda (IMS, Ottawa, Canada)
Katerina Soulantica (University of Toulouse, France)
Seigo Tarucha (University of Tokyo, Japan)
Carlos Tejedor (UAM, Madrid, Spain)
Euijoon Yoon (Seoul National University, Korea)
Ulrike Woggon (Tu Berlin, Germany)

Proceedings edited and compiled by Prof. Robert A. Taylor, University of Oxford

Coherent spin preparation, manipulation and read-out with light and microwaves in a quantum well and dot

H. Kosaka[1,2], H. Shigyou[1], T. Inagaki[1,2], Y. Mitsumori[1,2], K. Edamatsu[1], T. Kutsuwa[2], M. Kuwahara[2], K. Ono[3,2], Y. Rikitake[4,2], N. Yokoshi[5,2], H. Imamura[5,2]

[1] Research Institute of Electrical Communication, Tohoku University, Sendai 980-8577, Japan
[2] CREST-JST, 4-1-8 Honcho, Kawaguchi, Saitama 332-0012, Japan
[3] Low Temperature Physics Laboratory, RIKEN, Saitama 351-0198, Japan
[4] Department of Information Engineering, Sendai National College of Technology, Sendai 989-3128, Japan
[5] Nanotechnology Research Institute, AIST, Tsukuba 305-8568, Japan

E-mail: kosaka@riec.tohoku.ac.jp

Abstract. Spin is a quantum property of electrons. For spin-based quantum information technology, preparation and read-out of the electron spin state should be spin coherent. We demonstrate that the polarization coherence of light can be transferred to the spin coherence of electrons in a semiconductor quantum nanostructure [1], and the prepared coherence of the electron spin can also be read out with light by the developed tomographic Kerr rotation method [2]. We also demonstrate that a single photon is efficiently converted (~27%) into a single electron trapped in a gate-defined quantum dot, where the g-factor of electrons is tuned to zero, and the charge state is detected with an adjacent quantum point contact without destructing the spin state [3]. We further demonstrate that the spin coherence of a single electron trapped in one of double quantum dots is electrically manipulated with a microwave applied to the gate and read out via the Pauli spin blockade phenomenon [4]. These demonstrations were carried out in a condition where the up/down spin basis states of electrons remained degenerated under an in-plain magnetic field. As this condition ensures the energy conservation between photons and electrons, the entire Poincare sphere representing polarization states of photons can be mapped onto the Bloch sphere representing spin polarization states of electrons. We theoretically showed that relative spin coherence of two electrons can be also measured with the help of spin-flip tunneling of electrons between the dots [5]. Full Bell state measurement is also possible by the single-spin manipulation and Pauli spin blockade [6]. All of these functions are needed to build all semiconductor quantum repeaters and distributed quantum computers.

1. Quantum media conversion

None of the physical implementations of quantum bits or qubits is suitable for all aspects of quantum information technology. Quantum information interfaces connecting those qubits are thus needed to place the right qubit at the right position. Photons are the natural messenger qubits for fast and reliable communication, whereas electron spins in a semiconductor are suitable processor qubits for stable and scalable computation. Quantum media conversion (QMC) between photons and electrons[7] will drastically extend the potential of quantum information and communication technologies. The QMC from a photon to an electron might look similar to the photo-electric effect as explained by Albert Einstein with the assumption of a light quantum (photon) ; however, the photo-electric effect utilizes only one aspect of quantum nature as the energy quantum, since only energy of a photon is converted into that of an electron through the photo absorption. The QMC , on the other hand, utilizes another aspect of quantum nature as the information quantum, which is essential for both the quantum mechanics and quantum information technology.

2. Coherent spin preparation and read out with light

As electron spin coherence originates in the quantum coherence between up ($\uparrow_e$) /down ($\downarrow_e$) spin states, light polarization coherence originates in the quantum coherence between right (σ^+_{ph})/left (σ^-_{ph}) circular polarization states. Electron spin states can be prepared by injecting circularly polarized photons or by optical pumping, and read-out can occur by measuring the Kerr (Faraday) rotation angle of reflected (transmitted) light. However, we cannot directly prepare and read out a coherent superposition of up and down electron spin states using the conventional optical spin injection and Kerr rotation schemes. Conventionally, the mixture of up spin and down spin results in a completely incoherent mixture of those spin states likewise the classical states. We have overcome this obstacle by developing new methods for coherent spin state transfer and coherent spin state tomography based on the V-shaped band structure in a semiconductor quantum well. The combination of the two spin states now results in the coherent superposition of those spin states likewise the quantum states.

Figure 1 shows operating principles of the spin coherence transfer and the tomographic Kerr rotation. By applying an in-plane magnetic field B_x to configure the V-shaped three-level system, the Kramers degeneracy of the light hole (LH) is lifted and reconfigures the eigenstates as $|\pm x\rangle_{lh} = |\downarrow\rangle_{lh} \pm |\uparrow\rangle_{lh}$ while keeping the degeneracy of the electron with the smaller g-factor (**Fig. 1a**). Via the resonant transition between the $|-x\rangle_{lh}$ state and the degenerated electron states, the optical selection rules lead to the transition from $\alpha|\sigma^+\rangle_{ph}+\beta|\sigma^-\rangle_{ph}$ to $(\alpha|\uparrow\rangle_e+\beta|\downarrow\rangle_e)\otimes|-x\rangle_{LH}$, resulting in the coherent transfer of light polarization to electron spins [1]. The same scheme applies the tomographic Kerr rotation as a virtual excitation (**Fig. 1b**). Provided that the prepared electron spin s_1 is in $|+y\rangle_e$, the probe light in $|+z\rangle_{ph} = (|+y\rangle_{ph}+|-y\rangle_{ph})/\sqrt{2}$ virtually creates another electron spin s_2 in $|+z\rangle_e = (|+y\rangle_e+|-y\rangle_e)/\sqrt{2}$ together with an LH in $|-x\rangle_{LH}$. Then, two components $|\pm y\rangle_{ph}$ in the scattered light experience different phase shifts $\phi^{\pm}$ due to the exchange interaction $s_1 \cdot s_2$. The Stokes vector thus rotates about s_1 by the Kerr rotation angle θ_K $\phi^+ - \phi^-$ in the Poincaré-Bloch sphere. The θ_K in effect measures the projection of the electron spin onto the projection basis $|\pm y\rangle_e$. The developed tomographic Kerr rotation method enables the density matrix tomography of coherent spin states of electrons (**Fig. 2**), even when the spin does not precess around the applied magnetic field, as if the spin is frozen [2]. As this condition ensures energy conservation between photons and electrons, the entire Poincare sphere representing the polarization states of photons can be ideally mapped onto a Bloch sphere representing spin polarization states of electrons. The frozen spin also faithfully maintains the

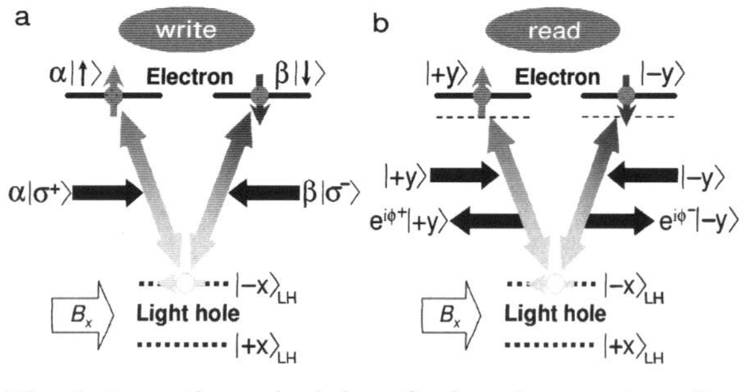

Fig. 1, Operating principles of spin coherence transfer (a: write) and tomographic Kerr rotation (b: read), based on the three-level V-shaped system.

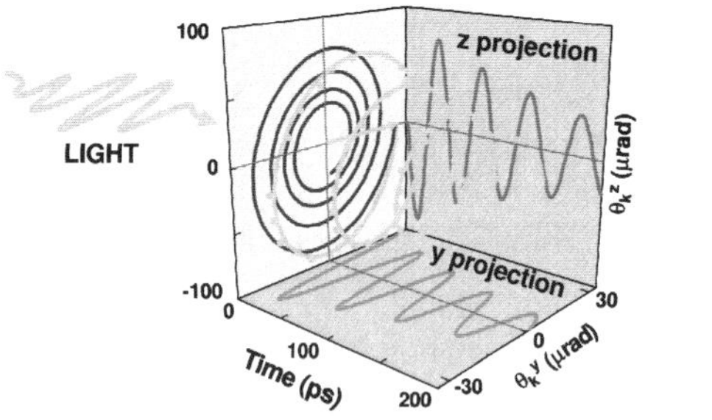

Fig. 2, Experimental demonstration of electron spin state tomography showing dynamic spin precession under an in-plain magnetic field Bx.

coherence transferred from photons encoded in their polarization.

The fundamental principle of TKR is the Pauli blockade (singlet state selection) between the target electron and the probe electron created virtually by the probe light. The TKR phenomenon is regarded as the reverse mapping of the electron spin states onto the photon states or the spin state swapping between a photon and an electron, through the resonant excitation of the spin trion with a light hole based on the degenerate Λ system. In other words, we can exchange the quantum information between the electron and the photon.

3. Single photon to single electron conversion in a single quantum dot

We demonstrated that a single photon creates a single electron in a semiconductor and detected by the charge [7,8]. We fabricated a gate-defined singe quantum dot consisting of a GaAs quantum well with multiple gates on the top surface to trap an electron (**Fig. 3**). The negatively biased gates confine only the negatively charged electron created by a single photon with removing the positively charged hole. The exceptional feature of this device is that the electron g-factor is designed to be zero, which enables the ideal spin state transfer from a photon to an electron. Another important feature is that a quantum point contact serves as an electron counter is implemented in the vicinity of the dot to count the number of electrons in the dot. This electron counter ensures the fidelity of the quantum media conversion by the post-selection of the conversion event. We succeeded to demonstrate that the device serves as a single photon to single electron converter with the conversion efficiency of 27%, excluding the coupling efficiency estimated by the area and the well thickness. The anti-bunching nature of the 1st captured electron and the 2nd captured electron (**Fig. 4**) and polarization selectivity (not shown) indicate that the photon polarization of a single electron is successfully transferred to the spin polarization of an single electron [3].

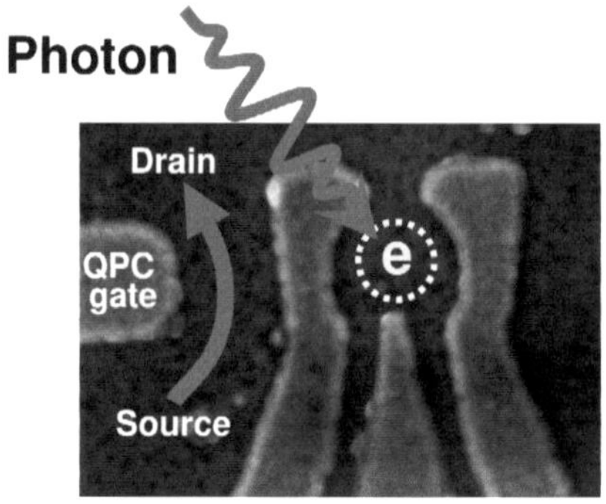

Fig. 3, SEM image of the device used for the demonstration of single photon to single electron conversion. QPC: quantum point contact.

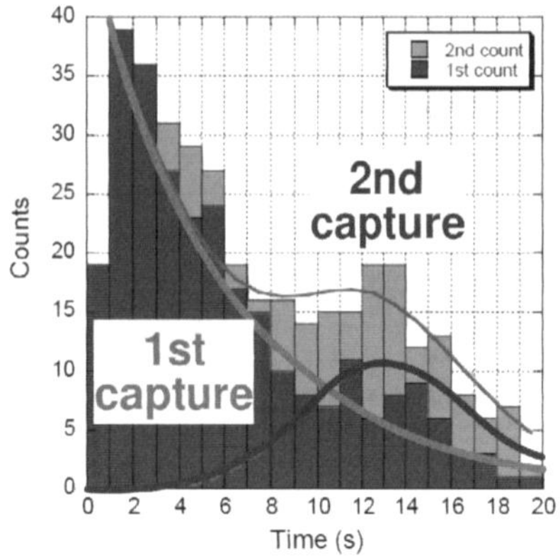

Fig. 4, Histogram of the electron-capture events detected by the quantum point contact as a function of light exposure time.

4. Coherent spin manipulation with microwave

Using a similar device with two coupled double quantum dots (**Fig. 5**), we succeeded in demonstrating single-electron spin manipulation by observing the Rabi oscillation in the dot current leaking through the dots as a function of microwave burst length (**Fig. 6**). The importance of this demonstration is that even an electron with nearly zero g-factor can be manipulated by microwaves using the electric-field component instead of the commonly used magnetic-field component through the spin-orbit interaction, and two-spin coherence is measured by the singlet detection through the Pauli spin blockade. The frozen spin can never be manipulated by a microwave via the magnetic field component. Instead, the electrical field component enables the spin manipulation through the modulation of the wavefunction of the electrons.

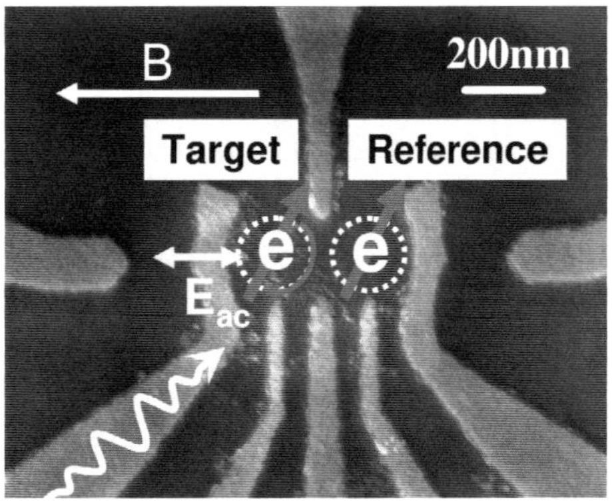

Fig. 5, SEM image of the coupled double quantum dots used for the demonstration of single electron spin manipulation.

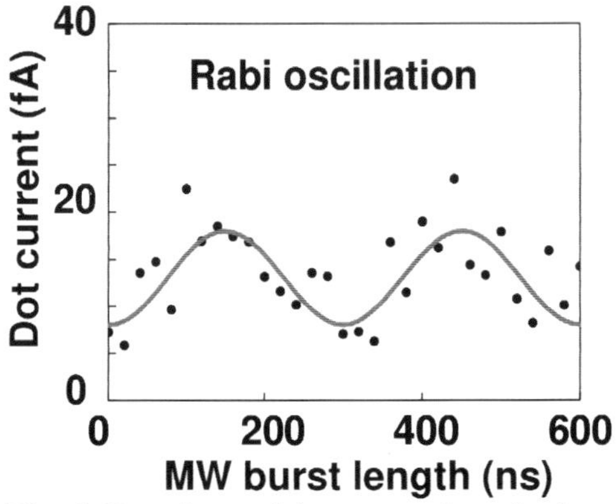

Fig. 6, Experimental demonstration of coherent spin manipulation of a single electron trapped in one of the coupled double quantum dots.

5. Full Bell state measurement

The above demonstrations of single electron spin resonance and singlet -state detection encourage us to perform the full Bell state measurement needed for the quantum repeater. Given that the g-factors of the electrons on both sides have the same magnitude but the opposite sign as shown in **Fig. 7**, we can swap any of the three triplet states for the single state and read out without disturbing the other states simply by the spin resonance in the x, y, and z directions and Pauli spin blockade, which enables full Bell state measurement [5,6].

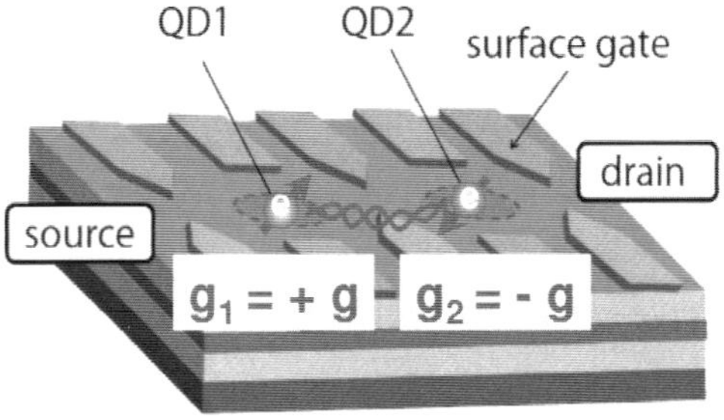

Fig. 7, Schematic structure of the proposed devise for the demonstration of full Bell state measurement.

6. Conclusions

We have demonstrated coherent spin preparation, manipulation and read-out with light and microwaves in a quantum well and dot. All of these functions are needed for building all semiconductor quantum repeaters and distributed quantum computers.

Acknowledgements: This work was supported by CREST-JST, SCOPE-MIC, MEXT and NEDO.

References

[1] H. Kosaka, H. Shigyou, Y. Mitsumori, Y. Rikitake, H. Imamura, T. Kutsuwa, K. Arai, K. Edamatsu, Phys. Rev. Lett. 100, 096602 (2008).
[2] H. Kosaka, T. Inagaki, Y. Rikitake, H. Imamura, Y. Mitsumori, K. Edamatsu, Nature 457, 702 (2009).
[3] M. Kuwahara, T. Kutsuwa, K. Ono, Hideo Kosaka, App. Phys. Lett., in press (2010).
[4] T. Kutsuwa, M. Kuwahara, K. Ono, H. Kosaka, Physica E, 42, 821 (2010).
[5] N. Yokoshi, H. Imamura, H. Kosaka, Phys. Rev. Lett. 103, 046806 (2009).
[6] N. Yokoshi, H. Imamura, H. Kosaka, Phys. Rev. B Rapid Comm., in press (2010).
[7] H. Kosaka, Y. Mitsumori, Y. Rikitake, H. Imamura, Appl. Phys. Lett. 90, 113511 (2007).
[8] H. Kosaka, D.S. Rao, H.D. Robinson, P. Bandaru, T, Sakamoto, E. Yablonovitch, Phys. Rev. B Rapid Communications, 65 (2002), R201307-1-4.
[9] H. Kosaka, D.S. Rao, H.D. Robinson, P. Bandaru, K. Makita, E. Yablonovitch, Phys. Rev. B 67, 045104 (2003).

Opportunities for single hole-spin control using delocalized states of quantum dot molecules

MF Doty[1], JI Climente[2], A Greilich[3], M Yakes[3], AS Bracker[3] and D Gammon[3]

[1] Dept. of Materials Science and Engineering, University of Delaware, Newark, DE 19716, USA [2]Dept. de Quimica Fisica i Analitica, Universitat Jaume I, 12080, Castello, Spain [3] Naval Research Laboratory, Washington, DC 20375, USA

E-mail: doty@udel.edu

Abstract. We present three unique spin properties of holes in the delocalized states of coupled pairs of InAs quantum dots: resonant changes in the hole g factor, antisymmetric hole molecular ground states, and hole spin mixing. These three properties suggest new opportunities for spin control that are unique to holes and further support holes as a competitive candidate for the storage and manipulation of quantum information.

1. Introduction

Holes confined in QDs have spin projections that could be used as bits in spin-based logic or quantum information processing devices. Hole spins were initially discounted as viable qubit candidates because the complex nature of valence band states was expected to drastically reduce the spin lifetimes and decoherence times. Recent work, however, has shown that hole spins in QDs can be initialized and stored with lifetimes comparable to those of electrons.[1, 2] Moreover, hole states have Bloch wavefunctions with p-like symmetry, which dramatically suppresses the hyperfine interaction with nuclear spins that is a primary source of decoherence for electron spins. Consequently, hole spins have emerged as a competitive choice for the fundamental unit of information in spin-based logic devices. In this work we present studies of coupled pairs of InAs QDs that illustrate how the formation of molecular states for holes leads to new opportunities for controlling single hole spin projections. These discoveries suggest that coupled pairs of quantum dots may provide an architecture that simplifies implementation of spin control protocols and further support holes as a competitive candidate for the storage and manipulation of spin-based quantum information.

2. Delocalized Molecular States

The Quantum Dot Molecules (QDMs) we describe here consist of two InAs QDs separated by a GaAs barrier and are grown by molecular beam epitaxy. The energy levels of the two dots are slightly different such that the electron and hole energy levels of the two dots are not degenerate, but the hole levels can be tuned into resonance with an applied electric field. On resonance, coherent hole tunneling between the QDs leads to the formation of molecular orbitals. The formation of molecular orbitals can be understood by considering only the lowest energy levels of each QD and temporarily ignoring spin. When there is no interaction between the states of

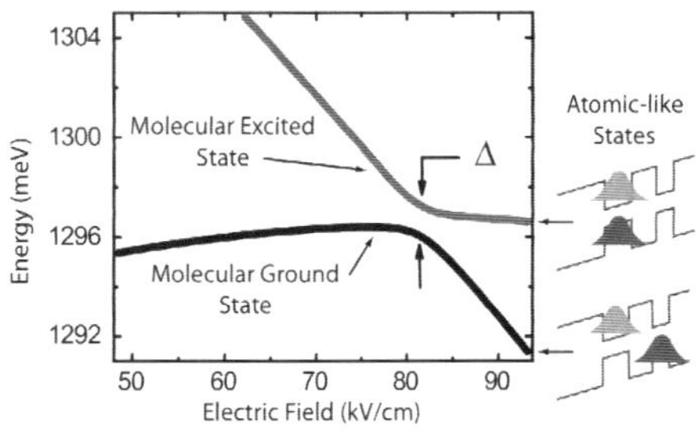

Figure 1. Measured PL energies of the neutral exciton in a QDM as a function of applied electric field. The diagrams schematically depict the wavefunction of the atomic-like basis states.

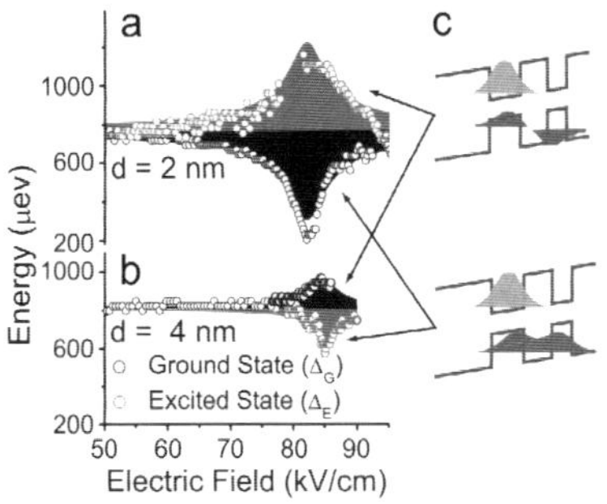

Figure 2. Zeeman splitting at 6 T of the molecular ground and excited states as a function of applied electric field for QDMs with (a) 2 nm and (b) 4 nm barrier. (c) Schematic depiction of the symmetric and antisymmetric molecular wavefunctions.

the individual QDs, there are two possible states for holes: the energetic ground states of the top and bottom dot. The QD asymmetry and applied electric field cause electrons in the top dot to rapidly relax into the bottom dot, so we only need to consider one possible state for the electron (the energetic ground state of the bottom QD). The two possible locations for holes create two atomic-like basis states with wavefunctions schematically depicted at the right side of Fig. 1.

Fig. 1 plots the energies of photoluminescence (PL) lines emitted from a QDM with 2 nm barrier as a function of the applied electric field. The two atomic-like basis states have PL energies that depend differently on the applied electric field, leading to the asymptotically horizontal and diagonal lines in Fig. 1. When the hole levels of the two dots come into resonance, coherent tunneling of the hole between the two dots leads to the formation of delocalized molecular states. The signature of the formation of molecular orbitals is the anticrossing gap (Δ) that appears in the spectra. On resonance, the molecular state can be described as the bonding (symmetric) or antibonding (antisymmetric) combination of the two basis states. The data have been plotted in black and red to illustrate that one line (black) is the molecular ground state on resonance and the other (red) is the molecular excited state.

3. g factor Resonance

Fig. 2a shows the measured Zeeman splitting for the sample presented in Fig. 1 when we apply a magnetic field in the Faraday configuration. The data in black show that the molecular ground state has a resonant decrease in Zeeman splitting at the electric field of resonance. In contrast, the data in red show that the molecular excited state has a resonant increase in Zeeman splitting at the same electric field. These changes in Zeeman splitting originate in a resonant change in the g factor for the hole due to the formation of hole states with molecular orbital character, as schematically depicted in Fig. 2c.[3, 4] The molecular ground state has a symmetric (bonding) molecular wavefunction with non-zero amplitude in the GaAs barrier between the two dots, which increases the contribution from the GaAs and leads to suppression of the hole g factor. The node in the wavefunction for the antisymmetric (antibonding) molecular excited state, on the other hand, suppresses the contribution of the GaAs and leads to the resonant increase in Zeeman splitting for the molecular excited state.

Electrically tunable g factors have previously been used to coherently modulate electron spin projections in static magnetic fields using a technique known as g Tensor Magnetic Resonance (gTMR).[5] However, the heterostructure used to demonstrate gTMR does not localize single spins and is inherently unable to provide the basis for a scalable spin-based logic device with individual localized, addressable spins. The QDM geometry, on the other hand, can localize a single hole spin in the QDM. Moreover, the magnitude of the resonant change in hole g factor observed in QDMs is more than an order of magnitude bigger than that observed in coupled quantum wells.[6] Hole states in QDMs are thus ideally suited for the application of gTMR to coherently control single spins.

4. Reversal of the molecular orbital character

The theory of molecular orbitals derived for natural diatomic molecules predicts that molecular ground states have symmetric (bonding) orbital character. The suppression of the Zeeman splitting for the molecular ground state, as shown by the black trace in Fig. 2a, is the experimental signature of the state's symmetric molecular orbital character. In Fig. 2b we present data on the Zeeman splitting of the molecular orbitals for a QDM in which the GaAs barrier has been increased from 2 nm to 4 nm. The data clearly demonstrate that the molecular ground state (black) now has the resonant *increase* in Zeeman splitting that is indicative of an antisymmetric (antibonding) molecular orbital. In contrast, the molecular excited state (red) now shows the resonant decrease in Zeeman splitting indicative of the symmetric (bonding) molecular orbital. This reversal of the molecular orbital character can not be explained by analogy with natural diatomic molecules and has surprising consequences.

The reversal of the molecular orbital character originates in spin orbit interactions between light and heavy hole spin components [7, 8] and has two important consequences for the control of hole spins and the use of QDMs for information storage and manipulation. First, it allows one to choose an energetic ground state with the desired resonant increase or resonant decrease in Zeeman splitting that is best suited to the overall device architecture and spin manipulation protocol. Second, if both electron and hole levels can be brought into resonance at the same value of the applied electric field the tunneling of the electron will form a symmetric molecular orbital state while the barrier thickness could be chosen to create an antisymmetric orbital state for holes. The opposite symmetry of the two states means that there is a zero matrix element for the dipole transition between these two states - i.e. there is no radiative recombination. The creation of an energetic excitonic ground state not subject to radiative recombination would be revolutionary for optoelectronic devices and lead to new opportunities for optically-driven information storage and manipulation. The creation of such a state is a unique opportunity provided by the molecular states of holes in QDMs.

5. Hole spin mixing

When the symmetry of the QDM is broken by lateral offset of one dot relative to the other, additional spin-orbit induced interactions become possible.[9] These additional mixing mechanisms couple states with orthogonal heavy-hole spin projections and opposite spatial locations, equivalent to a spin-flip tunneling processes. This novel hole spin mixing mechanism is mediated by states with significant light hole character and leads to new anticrossings in the experimental spectra, as indicated by the arrows in Fig. 3a.

The existence of a novel spin mixing mechanism suggest new opportunities for hole spin control. One example of a novel spin control mechanism using adiabatic fast passage techniques is depicted by the arrows in Fig. 3b. The hole is initialized in the spin down state within the bottom dot at an electric field well away from resonance so that the heavy-hole spin projections remain well isolated and viable states for the storage of quantum information.[9] The electric field is swept through the first anticrossing (arrow 1) at a rate that causes the hole state to 'jump' across the anticrossing and follow the asymptotic states. The electric field is then returned to the starting value at a rate that causes the hole state to follow the adiabatic line (arrow 2),

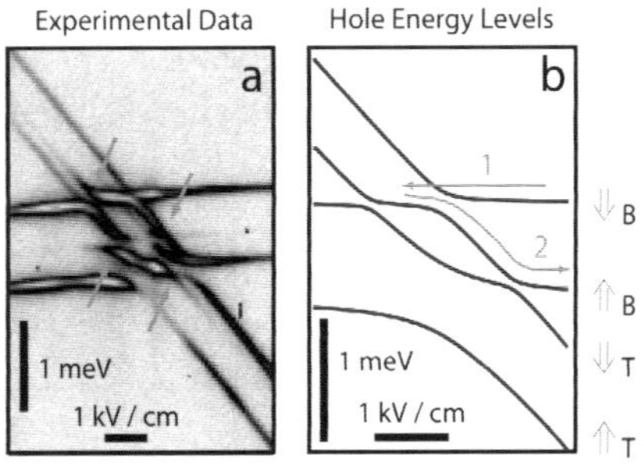

Figure 3. a) PL from the neutral exciton in a QDM with 4 nm barrier at 6 T. Arrows indicate the additional anticrossings arising from hole spin mixing. b) Energy level diagram for a single hole in a QDM calculated with representative values of the parameters. The arrows and subscripts describe the hole spin projection and spatial location (Bottom/Top dot) of the asymptotic energy levels. Arrows indicate a possible pathway for electrical manipulation of the hole spins.

ending as a spin up hole spin projection. This technique is similar to one recently used to control electron spins in lithographically defined QDs.[10] In the InAs QDMs proposed here, however, both optical and electrical spin manipulation methods can be utilized.

6. Summary
Coherent tunneling between neighboring InAs QDs leads to the formation of molecular states for holes. These molecular states have unique properties that are determined by the structure and symmetry of the QDM. The formation of molecular orbitals leads to resonant changes in g factors that could be the basis of gTMR control of single localized hole spins. Changing the spacing between the dots leads to the formation of antisymmetric molecular ground states for the holes that can control both the sign of the g factor resonance and the optical matrix elements for exciton states. Breaking the symmetry of the QDM by offsetting one QDM laterally leads to a novel spin mixing mechanism unique to holes that could enable electrically-driven coherent control of the hole spin projection. The combined results demonstrate that the molecular orbital states of holes provide a variety of novel spin control opportunities.

7. Acknowledgments
Acknowledgments
MFD acknowledges support from the University of Delaware Research Foundation and the NSF (0844747). Work at the Naval Research Lab was supported by NSA/ARO and ARO/MURI(W911NF0910406). JIC acknowledges support from the Ramon y Cajal program.

References
[1] Heiss D, Schaeck S, Huebl H, Bichler M, Abstreiter G, Finley J J, Bulaev D V and Loss D 2007 *Physical Review B* **76** 241306
[2] Gerardot B D, Brunner D, Dalgarno P A, Ohberg P, Seidl S, Kroner M, Karrai K, Stoltz N G, Petroff P M and Warburton R J 2008 *Nature* **451** 441–444
[3] Doty M F, Scheibner M, Ponomarev I V, Stinaff E A, Bracker A S, Korenev V L, Reinecke T L and Gammon D 2006 *Physical Review Letters* **97** 197202
[4] Doty M F, Scheibner M, Bracker A S and Gammon D 2008 *Physical Review B* **78** 115316
[5] Kato Y, Myers R C, Driscoll D C, Gossard A C, Levy J and Awschalom D D 2003 *Science* **299** 1201–1204
[6] Poggio M, Steeves G M, Myers R C, Stern N P, Gossard A C and Awschalom D D 2004 *Physical Review B* **70** 121305(R)
[7] Climente J I, Korkusinski M, Goldoni G and Hawrylak P 2008 *Phys. Rev. B* **78** 115323
[8] Doty M F, Climente J I, Korkusinski M, Scheibner M, Bracker A S, Hawrylak P and Gammon D 2009 *Physical Review Letters* **102** 047401
[9] Doty M F, Climente J I, Greilich A, Yakes M, Bracker A S and Gammon D 2010 *Physical Review B* **81** 035308
[10] Petta J R, Lu H and Gossard A C 2010 *Science* **327** 669–672 10.1126/science.1183628

In(Ga)As quantum dots grown on GaAs(111) substrates for entangled photons pairs

Irina A. Ostapenko[1], Erik Stock[1], Till Warming[1], Sven Rodt[1], Andrei Schliwa[1], Murat Öztürk[1], Jan Amaru Töfflinger[1], Anatol Lochmann[1], Dieter Bimberg[1], A. I. Toropov[2], S. A. Moshchenko[2], D. V. Dmitriev[2], V. A. Haisler[2]

[1]Institut für Festkörperphysik, Technische Universität Berlin, Hardenbergstrasse 36, Berlin 10623, Germany
[2]Institute of Semiconductor Physics, Lavrenteva av 13, Novosibirsk 630090, Russia

irina@sol.physik.tu-berlin.de

Abstract. Here we present the first successful growth and spectroscopic investigation of In(Ga)As/GaAs (111) quantum dots (QD). These QD were recently predicted as promising emitters of entangled photon pairs due to their intrinsic threefold degenerate symmetry resulting in zero fine structure splitting (FSS). Low spatial densities of the QDs were grown by droplet epitaxy in an MBE system. Emission of truly single photons is proved by correlation measurements. Using spatially-resolved power- and polarization- dependent experiments we discovered the characteristic excitonic luminescence fingerprints from a number of QDs. Excitonic FSS below 10 μeV, our resolution limit, is observed in agreement with our predictions.

1. Introduction

An easy-to-operate entangled-photon-pair source is required for efficient quantum key distribution in future secure data transmission [see e.g.1]. QDs have been proposed [2] and demonstrated [3, 4] as electrically-driven or optically-pumped sources of entangled photon pairs from a biexciton-exciton recombination cascade on condition that the FSS of the bright exciton state is less than the homogeneous linewidth of the transitions.

For In(Ga)As QDs on GaAs (001) even for highly symmetric QDs the lateral symmetry of the confining potential is lowered to C_{2v}. This originates from the lateral anisotropy of the confinement potential for the charge carriers due to elongation of the QD during the growth process, significant piezoelectric effects [5] and atomistic symmetry anisotropy [6]. The post-growth attempts to reduce the FSS to zero by annealing [7], external fields [4, 8], and stress [9] have to be adjusted for every single QD, therefore making it hardly applicable.

In contrast, for (111) substrates the piezoelectric field is directed along the growth direction leading to a confining potential of symmetry C_{3v}. Realistic calculations predicted recently zero FSS for this case [10, 11, 12].

2. Experiment

The self assembled QD samples were grown by solid-source molecular beam epitaxy using a Riber32P system on n-type doped GaAs (111) wafers. To increase the effective decomposition of the

metallorganic precursors and improve the GaAs buffer layer quality we used a miscut of 2° in the $[\bar{2}\bar{1}1]$ direction. A 500 nm-thick layer of the GaAs was followed by a 50 nm $Al_{0.6}Ga_{0.4}As$ diffusion barrier and then 90 nm of GaAs at 580 °C. The QDs layer was grown in a droplet epitaxy mode [13] at 510 °C. With closed As shutter nominally 1.5 ML of Ga were deposited, and then followed by nominally 2 ML of In to form In droplets on the GaAs surface, which in turn crystallized into InGaAs QDs under the As flux. The dots were than capped with 65 nm of GaAs, 50 nm of $Al_{0.4}Ga_{0.6}As$, followed finally by 10 nm GaAs. During the QD growth the wafer rotation was switched off, this allowed areas with low QD densities in the samples.
The samples were investigated with a microphotoluminescence set-up at 15 K with and excited with a frequency doubled Nd:YVO laser (532 nm) and a spot diameter of ~1 μm. The PL signal was dispersed by a triple monochromator and detected with a liquid-nitrogen cooled CCD array, resulting in a spectral resolution of 20 μeV. Using line shape analysis allows determine the energetic position of single line within an accuracy of 10 μeV. For the measurement of the second-order correlation function $g^{(2)}(\Delta t)$ a Hanbury-Brown-Twiss set-up (HBT) with 700 ps time- and 0.3 meV spectral resolution was used.

3. Results and discussion

Switching off the wafer rotation resulted in areas of the different QDs densities. Some of them show spectral sharp luminescence lines stemming from well separated luminescence centres, whereas we also observed areas with spectral broad luminescence. A typical spectrum of sharp emission lines, characteristic for the discrete QD energy levels is shown in Fig.1. The typical measured FWHM of the lines is up to 80 μeV, broadened by spectral diffusion. The spatial resolved investigation of the samples revealed optical active QD densities less than 10^9 cm^{-2} allowing single QD spectroscopy experiments without shadow masks or mesas.

True single photon emission was proven by the measurements of the second-order correlation function with a HBT. The inset of the Fig.1 shows the normalized $g^{(2)}(\Delta t)$ function of a single luminescence line with a measured value of $g^{(2)}(0)=0.3$, limited by the set-up time-resolution. The simulation with accounts for the time resolution [12] leads to true correlation function value $g^{(2)}(0)=0.1$. The cross-talk between the avalanche photo diodes of the HBT leads to the increased value of $g^{(2)}(\Delta t)$ at $\Delta t=\pm 5$ns.

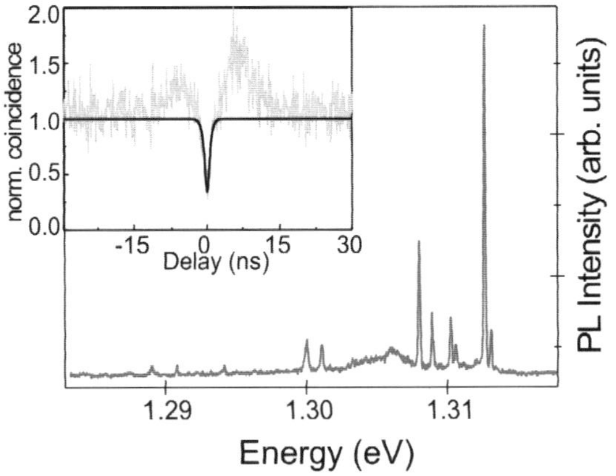

Figure 1. A typical microphotoluminescence spectrum of a sample showing single and sharp QD lines. The inset displays a second-order correlation measurement and a simulation for the measured in $g^{(2)}(\Delta t)$ (black line) [12]

Since the predicted FSS from In(Ga)As QDs grown on (111) GaAs substrates is zero we faced a challenge of investigation of the electronic origin of the emission lines. For this "novel" system we found some recurring luminescence line patterns in power-dependent measurements and performed polarization-dependent investigations.

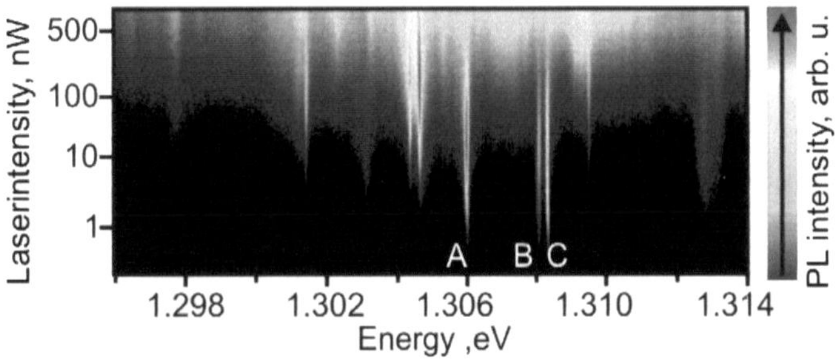

Figure 2. Excitation power dependence of the photoluminescence. The lines from excitonic complexes are labelled A, B, C

For several different QDs we found a similar pattern of excitonic emission behavior. One example is shown in Fig 2. Labeled with A, B and C the lines originate from the same QD and appear simultaneously with increasing excitation power. After saturation of these three lines (at about 100 nW excitation power) they fade out and further lines appear (presumably from multiexcitonic complexes). Moreover A, B and C show almost linear intensity dependence on excitation power. Since this is a characteristic excitonic behavior, we ascribe the lines to stem from single excitonic complexes of a single QD. Line A appears at 1.306 eV, whereas B and C at 1.308 eV with a 300 μeV spectral separation. The intensity ratio is found to be 4:1:2 for A:B:C respectively. Almost the same energetic separation, intensity ratio and power dependence between the lines was observed for five other QD in the same spectral luminescence range as a typical fingerprint of the In(Ga)As QD on the (111) GaAs substrate.

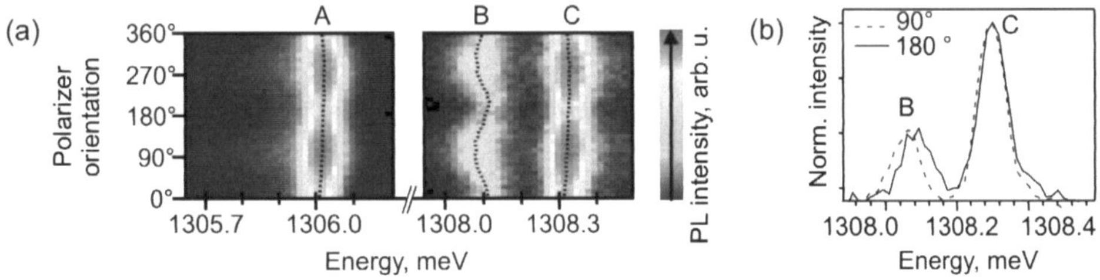

Figure 3. Polarization dependence of the lines A (a), B and C (b). The dotted lines show the center energy of the lines from a Gaussian fit. (c) Spectra of lines B and C for 90° and 180° polarizer position with a splitting of line B of 40μeV.

Fig. 3 presents the result of the polarization investigations of the excitonic lines (lines A, B and C from Fig.2). Lines A and C (Fig 3a, b) show clear polarization dependence of their intensity with no spectral position change. For an angle of 90° the intensity is 1.5 times larger, than that for 180°. Line B on the contrary shows the same intensity for all polarizer angles, but shifts in energetic position (Fig. 3b) up to 40 μeV with a periodicity of 180°. It is more visible in Fig 3c with spectra for 90° and 180° for B and C. In general, the FSS can be measured on uncharged X and a doubly charged exciton X++ [14]. Therefore lines A and C are possibly trions, whereas line B gives access to the FSS and stems either from X or X++.

For the other five QDs with the same luminescence fingerprint (like that described for Fig.2) we found the FSS of the line B ranging form 30 μeV down or below the determination limit (Fig.4 a, the dot shown on Fig. 2-3 corresponds to number 1). The polarization dependence of the lines B and C for the QD number 2 is shown on Fig. 4b. In contrast to the dot shown on Fig. 3 the excitonic line B exhibits an energetic shift on the order of the determination limit (10 μeV).

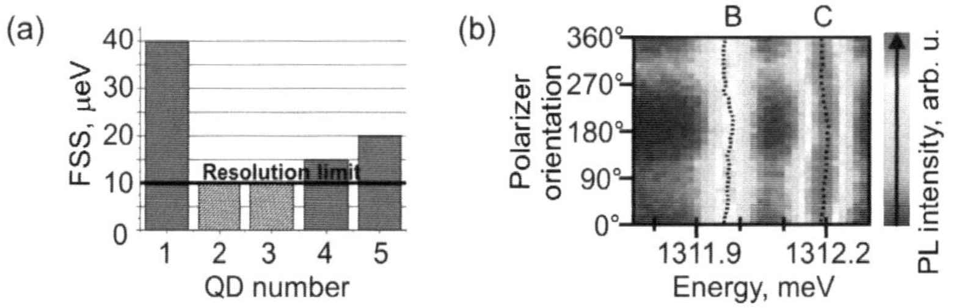

Figure 4. (a) FSS of excitonic line B, measured for five QDs with the same luminescence fingerprint. (b) Polarization dependence of lines B and C for QD number 2. The dotted lines show the centre energy of the lines from a Gaussian fit. Line B shows splitting below 10μeV.

We ascribe the nonzero FSS to two effects mainly. The miscut of 2° of the (111) substrate may lead to growth of elongated QDs on the steps on the (111) surface. Besides even the growth of GaAs material exhibits high densities of dislocations. This can hinder the three fold confining potential symmetry and lead to a preferential direction of the strain distribution, similar to (001) substrates.

4. Conclusion

We report on the realization of self-assembled In(Ga)As QDs on (111) GaAs substrates via droplet epitaxy. These novel types of QDs show sharp and intense lines. Densities lower than 10^9 cm^{-2} were achieved. True single photon emission was proven by antibunching measurements and $g^{(2)}(0)<0.3$ was observed. Power-dependent measurements reveal typical excitonic luminescence fingerprint for several QDs and one of the lines gives access to the FSS. The measured FSS is in some cases below our resolution limit of 10 μeV, precondition for entangled photon pairs.

5. Acknowledgements

This work was supported by Deutsche Forschungsgemeinschaft in the frame of SFB 787 and NATO SfP982735.

References

[1] N. Gisin, G. Ribordy, W. Tittel, and H. Zbinden, *Rev. Mod. Phys.* **74**, 145 (2002)
[2] O. Benson, C. Santori, M. Pelton, and Y. Yamamoto, *Phys. Rev. Lett.* **84**, 2513 (2000)
[3] N. Akopian et al. *Phys. Rev. Lett.* **96**, 130501 (2006)
[4] R. M. Stevenson, R. J. Young, P. Atkinson, K. Cooper, D. A. Ritchie, and A. J. Shields, *Nature* **439**, 179 (2006)
[5] R. Seguin et al. *Phys. Rev. Lett.* **95**, 257402 (2005)
[6] G. Bester, S. Nair, and A. Zunger, *Physical Review B* **67**, 161306 (2003).
[7] R. Seguin et al., *Appl. Phys. Lett.* **89**, 263109 (2006).
[8] K. Kowalik et al. *Appl. Phys. Lett.* **86**, 041907 (2005)
[9] S. Seidl et al. *Appl. Phys. Lett.* **88**, 203113 (2006)
[10] A. Schliwa, M. Winkelnkemper, A. Lochmann, E. Stock, and D. Bimberg, *Phys. Rev. B* **80**, 161307 (2009)
[11] R. Singh and G. Bester, *Phys. Rev. Lett.* **103**, 063601 (2009).
[12] D. Bimberg, E. Stock, A. Lochmann, A. Schliwa, J. Tofflinger, W. Unrau, M. Munnix, S. Rodt, V. Haisler, A. Toropov, *IEEE Photonics Journal* **1**, 58 (2009).
[13] T. Chikyow and N. Koguchi, *Jpn. J. Appl. Phys., Part 2* **29**, L2093 (1990)
[14] M. Ediger, G. Bester, B. D. Gerardot, A. Badolato, P. M. Petroff, K. Karrai, A. Zunger, and R. J. Warburton, *Phys. Rev. Lett.* **98**, 036808 (2007)

Ordered 1-D and 2-D InAs/InP quantum dot arrays at telecom wavelength

N Sritirawisarn[1], F W M van Otten[1] and R Nötzel[1]

[1] COBRA Research Institute on Communication Technology, Eindhoven University of Technology, 5600MB Eindhoven, The Netherlands

n.sritirawisarn@tue.nl

Abstract. Lateral one-dimensional (1-D) and two-dimensional (2-D) InAs/InP quantum dot (QD) arrangements are created by the concept of self-organized anisotropic strain engineering of InAs/InGaAsP superlattice (SL) templates on InP (100) and (311)B substrates by chemical-beam epitaxy (CBE). The SL templates comprise several-periods of an InAs QD layer plus a thin cap layer, post-growth annealing, and a separation layer. QDs order on top of the templates due to local strain recognition. Distinct preferential In adatom surface migration during annealing and substrate miscut lead to linear QD arrays along [001] for InP (100) substrates and a periodic square lattice aligned ±45° off [-233] for InP (311)B substrates. Optimization of the growth parameters balances In desorption and leads to well-separated and highly uniform QD arrays. Importantly, strong photoluminescence (PL) of defect-free InAs QD arrays is observed with the wavelength tuned into the 1.55-μm telecom region at room temperature through the insertion of GaAs interlayer beneath the QDs. Finally, the concept of self-organized anisotropic strain engineering for QD ordering is extended for formation of more complex architectures by combining it with step-engineering on shallow- and deep-patterned substrates. On the sidewall areas, the steps generated by the artificial patterns play the major role in determination of the In adatom surface migration during annealing, altering the QD arrays direction away from [001] on stripe-patterned InP (100) substrates. On the contrary, the sidewalls on patterned InP (311)B are faceted, not affecting the orientation of the 2-D InAs QD arrays.

1. Introduction

Lateral ordering of semiconductor quantum dots (QDs) is of prime importance for the exploitation of novel functionalities, e.g., coherent quantum state manipulation, quantum information processing, and quantum computing, which rely on the operation at the single and multiple electron and photon level with controlled quantum mechanical and electromagnetic interactions for future quantum functional devices. High quality QDs grown in the Stranski-Krastanow (S-K) growth mode are usually randomly distributed over the wafer surface. Therefore, efforts to position the QDs have gained prime attraction. In most of the case, artificially patterned substrates are prepared to control the QDs nucleation sites, however, often degrading the structural and optical qualities of the QDs due to lithographic imperfections and etching defects. We have previously introduced a concept for the lateral ordering of InAs QDs based on self-organized anisotropic strain engineering of InGaAs/GaAs superlattice (SL) templates in molecular-beam epitaxy (MBE) to produce well-ordered QD arrays with excellent optical properties [1,2]. To be compatible with telecom applications, the general concept has been transferred to the InAs/InP material system where the important 1.55-μm wavelength region at room temperature (RT) is covered [3,4].

InAs/InGaAsP SL template formation comprises InAs QD formation, thin InGaAsP capping, annealing, InGaAsP overgrowth, and stacking. This produces wire-like InAs nanostructures for InP (100) substrates and periodic shallow mound-like InAs nanoclusters for InP (311)B substrates due to anisotropic surface migration during annealing, together with lateral and vertical strain correlations and strain gradient driven In adatom migration during stacking. The lateral strain field modulations on the SL template surfaces govern InAs QD ordering due to local strain recognition. Excellent PL emission from the QD arrays is observed up to room temperature (RT) with the PL tuned into the 1.55-μm-wavelength region through the insertion of ultrathin GaAs interlayer beneath the InAs QD arrays. Finally, the self-organization concept is extended for formation of more complex architectures of lateral QD arrays by combining it with step engineering on artificially patterned shallow- and deep-etched substrates.

2. Experimental details

The samples were grown by CBE using pressure controlled trimethylindium (TMI), triethylgallium (TEG), AsH_3, and PH_3 as precursors. The AsH_3 and PH_3 gases were thermally decomposed in a high temperature injector at 900 °C. The planar and artificially patterned InP (100) substrates with miscut 2° toward (110) and singular InP (311)B substrates were mounted on Mo block. The mesa patterns were prepared by optical lithography and wet chemical etching in the 3 H_2SO_4 : 1 H_2O_2 : 1 H_2O solution. In this study, for InP (100) substrates, we concentrate on periodic mesa stripes with width and separation of 4 μm oriented along [010] and [001], and etched mesa depth of 50 and 200 nm. For InP (311)B substrates, periodic mesa stripes with width and separation of 2 μm oriented along [0-11], and etched mesa depth of 15 and 200 nm are presented. After etching, the samples were cleaned by oxygen plasma stripping followed by oxide removal in diluted 10 H_3PO_4 : 1 H_2O and rinsing in deionized water for three times before loading into the CBE growth chamber.

For InP (100) substrates, the growth started with 50 nm thick InP and 100 nm thick lattice-matched InGaAsP with RT bandgap at 1.27 μm (Q1.27), followed by a seven-periods InAs/InGaAsP SL template. Each of SL periods comprised 2.1 monolayers (ML) InAs, 10 s growth interruption under As flux, thin capping by 0.3 nm Q1.27, annealing for 2 min, and growth of a 15.3 nm Q1.27 spacer. On top of the SL template, 2.6-3.2 ML InAs were deposited for QD formation with a 0-0.8 ML GaAs interlayer inserted beneath the QD layer. For the PL studies, the samples were capped by 100 nm Q1.27 and 50 nm InP. The growth temperature was kept constant at 505 °C and the annealing temperature was at 520 °C. The growth rate of InAs was 0.23 ML/s, calibrated by high-resolution X-ray diffraction (HXRD).

For InP (311)B substrates, the growth commenced with 200 nm thick InP and 100 nm thick lattice-matched InGaAsP with RT bandgap at 1.18 μm (Q1.18). The nine-periods InAs/InGaAsP SL templates were grown on top of the buffer layers with each period consisting of 2.1 ML InAs, 10 s growth interruption under As flux, thin capping by 0.3-nm Q1.18, annealing for 2 min, and growth of a 5.5-nm Q1.18 spacer. On top of the SL template, a 3.2 ML InAs QD layer was grown with a 0-2.0 ML GaAs interlayer beneath the QD layer. The growth temperature was 505 °C throughout the entire structure and the annealing temperature was 514 °C. The growth rate of InAs was 0.18 ML/s, calibrated by HXRD.

These optimized growth parameters were chosen for well-separated and highly uniform QD arrays on both substrates. The surface morphology of uncapped samples was characterized by tapping mode atomic-force microscopy (AFM) in air. For the PL studies at RT and 4.8 K, a neodymium-doped yttrium aluminum garnet (Nd:YAG) laser (532 nm) with power density of 256 mW/cm^2 was used as excitation source and the PL signal was collected by a cooled InGaAs linear array detector and an InSb single channel detector.

3. Results and discussion

3.1. 1-D InAs QD arrays

The AFM image of the linear 1-D 2.6 ML InAs QD arrays on the optimized InAs/InGaAsP SL template on planar InP (100) is shown in the inset (a) of Figure 1. The QD arrays are aligned along the elastically soft [001] direction of the cubic crystal to minimize the total strain energy, selected by the substrate miscut generating steps in the same direction. Inset (b) shows the AFM image of the 2.6 ML InAs QD arrays with 0.8 ML GaAs interlayer inserted beneath the QD layer. The GaAs interlayer suppresses As/P exchange during InAs deposition which reduces the QD height from 7-11 to 4-5 nm and, consequently, the emission wavelength from far above 1.65 μm at RT into the 1.55-μm region, as shown in the PL spectra in Figure 1.

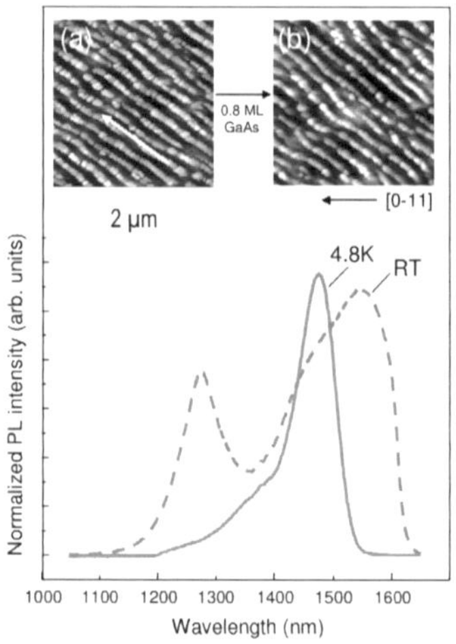

Figure 1. PL spectra of capped 2.6 ML InAs QD arrays with 0.8 ML GaAs interlayer taken at 4.8 K (solid line) and room temperature (dashed line). Inset: AFM image of uncapped 2.6 ML linear InAs QD arrays on the optimized SL template on InP (100) substrate (a) without GaAs interlayer and (b) with 0.8 ML GaAs interlayer beneath the InAs QD layer.

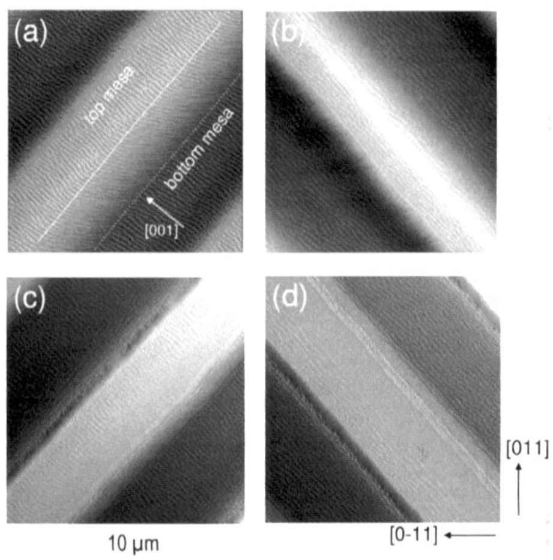

Figure 2. Deflection mode AFM images of uncapped 1-D 3.2 ML InAs QD arrays on the optimized SL template on 50-nm-deep (a) [010] and (b) [001], and 200-nm-deep (c) [010] and (d) [001] stripe-patterned InP (100) substrates with 4 μm stripe separation and width. The dashed lines in (a) are a guide for the eye following the mesa stripe.

Figure 2 shows the InAs QD arrays grown on the optimized SL template on the shallow- and deep-patterned InP (100) substrates with mesa stripes along (a), (c) [010] and (b), (d) [001]. For the vicinal sidewall planes towards (00-1) of the shallow [010]-oriented stripes with relatively large inclination [Figure 2 (a)], the linear QD arrays are rotated away from the natural [001] direction due to the sidewall step edges along [010] modifying the direction of adatom surface migration during annealing. On the opposite sidewall planes towards (001), the arrays orientation is rotated by 90° along [010], which is the other elastically soft direction, selected by the high density of steps along the [010] sidewall. On top of the mesa stripes, no change of the QD arrays orientation is observed due to the relatively small step density, maintaining the natural arrays direction along [001]. On the vicinal sidewall planes of [001]-oriented stripes, shown in Figure 2 (b), the sidewall step edges are in the same direction as those due to the substrate miscut having no influence on the orientation of the QD arrays. For the deep-patterned substrates the steep side facets (inclination more than 10 degrees) are QD free due to the slow-growing behavior. On the less steep vicinal sidewall planes similar trends of QD arrays formation as on the shallow-patterned substrates are observed.

3.2. 2-D InAs QD arrays

The AFM image of the 2-D InAs QD arrays on the optimized InAs/InGaAsP SL template on planar singular InP (311)B is shown in the inset (a) of Figure 3 without a GaAs interlayer and in the inset (b) with a 2.0 ML GaAs interlayer beneath the QDs. The QDs are aligned in a 2-D network along ±45° off [-233]. As the result of the GaAs interlayer, the QD height is reduced from 18-22 nm to 5-7 nm and the wavelength is shifted from far above 1.9 μm to the 1.55-μm region at RT, shown in the PL spectra in Figure 3.

Figure 4 shows the AFM images of the InAs QD arrays grown on the optimized SL template on the (a) shallow-patterned and (b) deep-patterned InP (311)B substrates with [0-11]-mesa stripes. For both shallow and deep patterns only QD free mesa side facets are developed. The 2-D QD arrays are not affected in accordance with the flat mesa top and bottom.

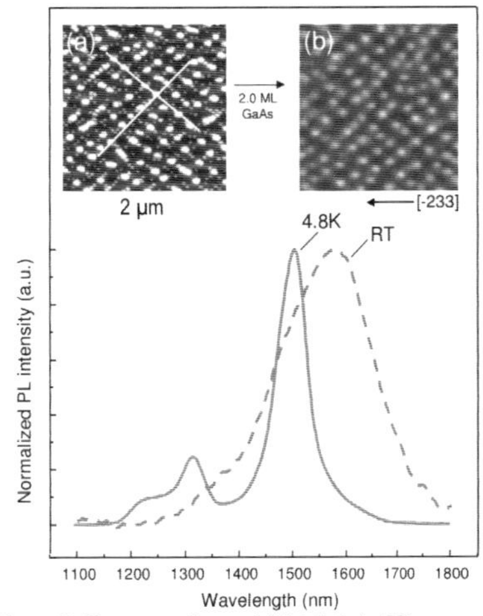

Figure 3. PL spectra of capped 3.2 ML InAs QD arrays with 2.0 ML GaAs interlayer taken at 4.8 K (solid line) and room temperature (dashed line). Inset: AFM image of uncapped 3.2 ML InAs QD arrays on the optimized SL template on InP (311)B substrate (a) without GaAs interlayer and (b) with 2.0 ML GaAs interlayer beneath the InAs QD layer.

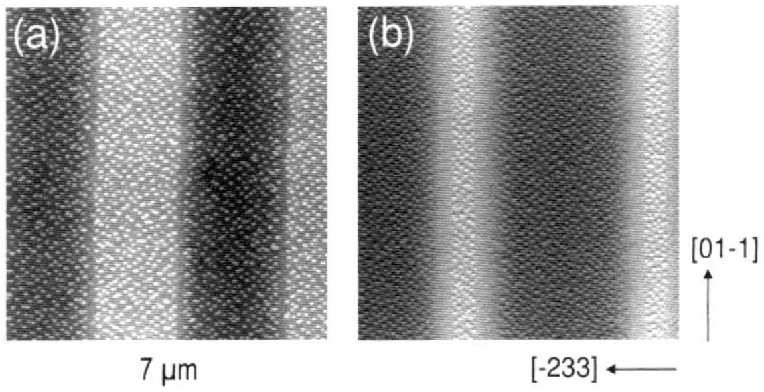

Figure 4. Deflection mode AFM images of uncapped 2-D 3.2 ML InAs QD arrays on the optimized SL template on (a) 15-nm-deep and (b) 200-nm-deep [0-11] stripe-patterned InP (311)B substrates with 2 μm stripe separation and width.

4. Conclusion

We have demonstrated the formation of ordered InAs QD arrays by self-organized anisotropic strain engineering of InAs/InGaAsP SL templates in CBE on planar and patterned InP (100) and (311)B substrates. The QD arrays emission wavelength at RT is tuned into the important 1.55-μm telecom wavelength region through the insertion of ultrathin GaAs interlayers. On shallow- and deep stripe-patterned InP (100) substrates, depending on the stripe geometry, the linear 1-D InAs QD arrays are rotated away from their natural [001] direction, selected by the substrate miscut, due to the presence of vicinal sidewall steps modifying the adatom surface migration during SL template formation. On shallow- and deep-patterned InP (311)B substrates only QD free sidewall facets are observed with flat top and bottom areas not affecting the formation of the natural 2-D InAs QD arrays. Hence, self-organized anisotropic strain engineering has shown the potential for the creation of laterally ordered QD arrays with excellent structural and optical properties compatible with telecom application. By combination with step engineering on artificially patterned substrates, complex QD arrays and networks have been realized for next generation quantum functional devices.

References

[1] Mano T, Nötzel R, Hamhuis G J, Eijkemans T J and Wolter J H 2001 *Appl. Phys. Lett.* **81** 1705
[2] van Lippen T, Nötzel R, Hamhuis G J and Wolter J H 2004 *Appl. Phys. Lett.* **85** 118
[3] Sritirawisarn N, van Otten F W M, Eijkemans T J and Nötzel R 2007 *J. Appl. Phys.* **102** 053514
[4] Sritirawisarn N, van Otten F W M, Soto Rodriguez P E D, Wera J L E and Nötzel R 2010 *J. Cryst. Growth* **312** 164

Highly efficient electrically triggered quantum dot micropillar single photon source

T Heindel, C Schneider, M Lermer, S Höfling, S Reitzenstein, L Worschech and A Forchel

Technische Physik, Wilhelm Conrad Röntgen Research Center for Complex Material Systems, University of Würzburg, D-97074 Würzburg, Germany

E-mail: tobias.heindel@physik.uni-wuerzburg.de

Abstract. We present electrically triggered single photon emission from low mode volume quantum dot-micropillar cavities at operating frequencies of up to 220 MHz. Due to an optimized layout of the doped planar microcavity and an advanced lateral current injection scheme, highly efficient single photon sources featuring $g^{(2)}(0)$-values well below 0.5 and a Purcell factor of 4 are realized. Single photon emission at a rate as high as (47.0±6.9) MHz with a corresponding overall efficiency of (21.4±3.1) % is demonstrated for a 2.0 μm-diameter device.

1. Introduction

Motivated by possible applications in quantum communication systems great efforts have been made in recent years with respect to the realization of efficient single photon sources (SPSs). Due to their discrete energy levels single quantum dots (QDs) can inherently act as single photon emitters and are therefore predestinated for solid state SPSs. However, QDs suffer from a poor outcoupling efficiency when embedded in a bulk semiconductor matrix since most of the isotropically emitted photons are trapped due to total internal reflection. The efficiency of a QD based SPS is strongly enhanced in combination with high quality microcavities [1, 2]. In this respect QD-micropillar cavities are very attractive systems due to a high extraction efficiency approaching 70%, as predicted by theory [3], and a highly directional emission normal to the sample surface. Moreover, in light of possible applications of SPSs, e.g. in quantum key distribution systems, and the associated device integration it is crucial to establish an electrical pumping scheme capable for high frequency operation. Up to now only a few approaches for electrically pumped SPS based on QDs embedded in oxide apertured pin-diodes with and without cavity effects and an efficiency of up to 14% have been demonstrated [4]. In this work we will demonstrate highly efficient single photon emission (SPE) from an electrically triggered QD-micropillar cavity as depicted schematically in Fig. 1. We observe overall efficiencies of up to 21% and record high SPE rates as high as 47 MHz, the latter even outperformes values reported in [5].

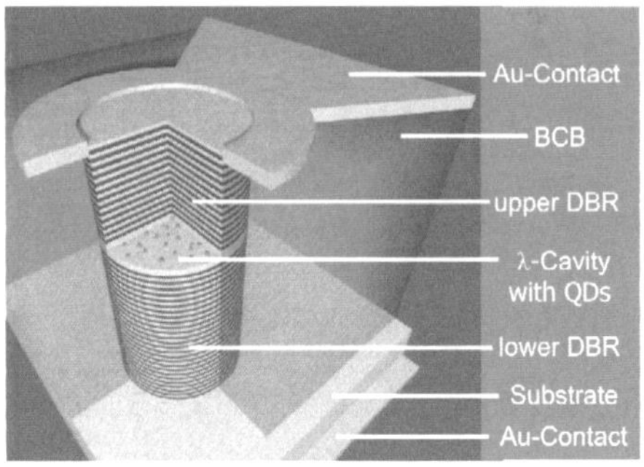

Figure 1. Schematic of an electrically contacted, quantum dot micropillar single photon source.

Journal of Physics: Conference Series **245** (2010) 012005 doi:10.1088/1742-6596/245/1/012005

2. Sample structure and measurement setup

The electrically contacted micropillar SPSs are based on a planar GaAs/AlAs microcavity sample with a n-doped lower and a p-doped upper distributed Bragg reflector (DBR) grown by molecular beam epitaxy on a n-doped GaAs substrate. The one-λ thick undoped GaAs cavity, which is embedded between the DBRs, contains a low density layer of InAs QDs. In the vicinity of the latter we introduced a n-type δ-doped layer to eliminate dark-state configurations that are known to reduce the efficiency of SPSs based on neutral QDs [2]. The asymmetric layout of the planar microcavity with 13 (26) AlAs/GaAs mirror pairs in the upper (lower) DBR was designed for highest outcoupling efficiencies using three-dimensional finite difference time-domain simulations. This optimization results in a moderate Q-factor of the planar cavity of about 3000. The lateral mode confinement is obtained by patterning pillars with a circular cross section with diameters d_C of 1-4 μm. The small number of mirror pairs in the upper DBR ensures a strongly directional emission normal to the sample surface so that photon losses due to sidewall imperfections of the micropillars are of minor importance. We refer to [6] for a more detailed discussion of the different loss channels. Furthermore, in our approach, the uncovered top facet of the micropillar guarantees efficient in- and outcoupling of light [7].

The device was investigated by means of high-resolution micro-electroluminescence (μEL) spectroscopy at low temperatures (10-70 K). Single photon emission was probed using a fiber coupled Hanbury-Brown and Twiss (HBT) setup with Si-based avalanche photon diodes (APDs) acting as detectors. The overall temporal resolution of the HBT setup is approximately 0.7 ns. Moreover, time resolved μEL-measurements were performed with a fast Si-APD enabling a temporal resolution of 40 ps. The SPS was triggered by an electrical pulse generator providing pulses with widths down to 200 ps (FWHM) at a repetition rate of up to 250 MHz. In addition, a DC offset could be applied to the sample.

3. Experimental results and discussion

In the following we present highly efficient, electrically triggered single photon emission of a micropillar with a diameter of d_C=2 μm.

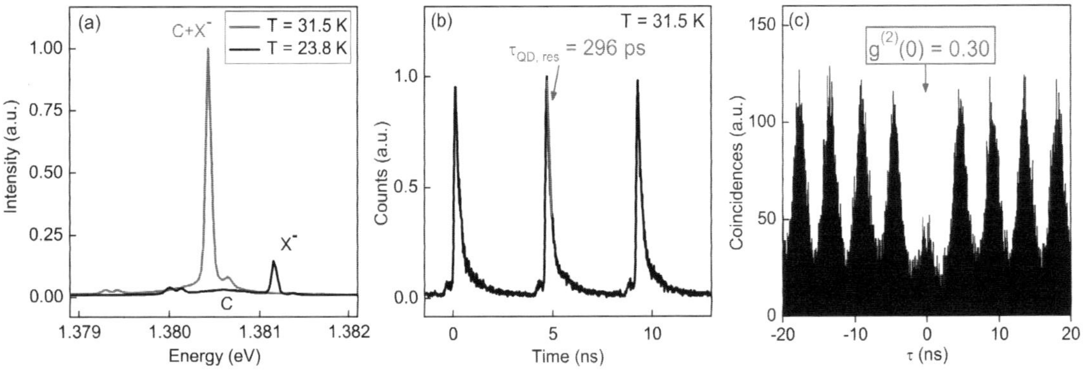

Figure 2. (a) μEL-spectra of a 2 μm diameter micropillar under pulsed excitation at 220 MHz (V_{DC}=1.65 V). Resonance of the cavity mode C and a single QD exciton X^- is observed at 31.5 K, where the intensity of the excitonic-line is strongly enhanced due to the Purcell effect. (b) Time resolved μEL-signal recorded for the resonance-condition in (a). (c) Photon autocorrelation measurement under on-resonance condition at V_{DC}=1.85 V.

3.1. Cavity enhanced single photon emission

Fig. 2(a) displays μEL-spectra of the micropillar for two different temperatures under pulsed excitation at 220 MHz. The device was biased with a DC voltage V_{DC} of 1.65 V, just below the onset of EL. In addition, pulses with an amplitude V_{AC} of 6.0 V were applied. At 23.8 K, a single QD exciton (X^-) as well as the very weak fundamental cavity mode (C) with a Q-factor of 2300 can be identified. The charged state of the exciton was thereby confirmed by polarization dependent measurements. By increasing the sample temperature to 31.5 K, the X^--line can be tuned on resonance with the cavity mode, which results in a strong enhancement of emission due to the Purcell effect [7].

The suitability of our devices with respect to the injection of short electrical pulses is demonstrated in Fig. 2(b) which shows time resolved μEL emitted at resonance (31.5 K). In fact, we observe a train of well shaped narrow pulses at a repetition rate of 220 MHz which allows us to estimate the Purcell factor for the resonant QD. The EL decay yields a recombination lifetime $\tau_{QD,res}$ = (296$\pm$8) ps of the resonant QD as illustrated in Fig. 2(b). The bulk QD lifetime $\tau_{QD,bulk}$=(1.20$\pm$0.05) ns was independently obtained by ensemble measurements on another sample from the same wafer whose top DBR was removed by plasma etching. From these numbers, we determine a Purcell factor of F_P=4.0$\pm$0.2 which is close to the maximal achievable value of $F_P = 3Q/(4\pi^2 V_{eff})(\lambda/n)^3 = 5$ for the present QD-micropillar system under spatial and spectral resonance [7]. It is worth mentioning that the fast response of the SPS should allow for repetition rates in the GHz range as reported in [8].

To test the emission of this resonantly coupled QD-cavity system with respect to its photon statistics, the μEL-signal was coupled to the HBT-setup. The corresponding photon autocorrelation function $g^{(2)}(\tau)$ was measured for V_{DC}=1.85 V (V_{AC}=6 V, f=220 MHz) and is depicted in Fig. 2(c). Non-classical emission of light is identified by the strongly reduced peak at τ=0 and the measured $g^{(2)}(0)$-value of (0.30$\pm$0.03)$<$ 0.5, which is an unambiguous proof of single photon emission from the QD-micropillar device. The deviation from the ideal value of $g^{(2)}(0)$=0 can be partly explained by taking uncorrelated background emissions from the cavity mode into account [9]. Further negative influence on $g^{(2)}(0)$ might be related to recapture processes and the associated multi-photon emission events.

3.2. Estimating the single photon emission rate

In order to evaluate the performance of the micropillar SPS in terms of the overall efficiency, i.e. the probability of emitting a single photon after an electrical input pulse, as well as the SPE rate under pulsed electrically current injection, we carefully calibrated our experimental HBT-setup to determine the photon emission rate $\dot{n}$. This detection rate needs to be corrected for multi photon emission events to obtain an estimate for the true SPE rate $\dot{n}_{SPE} = \dot{n}\sqrt{(1 - g^{(2)}(0))}$ [10]. Fig. 3 displays $\dot{n}_{SPE}$ and the corresponding $g^{(2)}(0)$-values as a function of V_{DC} (V_{AC}=6 V and f=220 MHz). Here, the DC bias allows for a precise adjustment of carrier capture rate into the QDs. At V_{DC}=1.85 V we obtain a SPE rate of $\dot{n}_{SPE}$=20.5 MHz. The latter increases up to $\dot{n}_{SPE}$=47.9 MHz at V_{DC}=2.10 V accompanied by successive saturation. This observation can be explained by QD saturation and an increase of uncorrelated background emission at high bias, which both limit the achievable SPE and result in an increase of $g^{(2)}(0)$ [2]. The highest

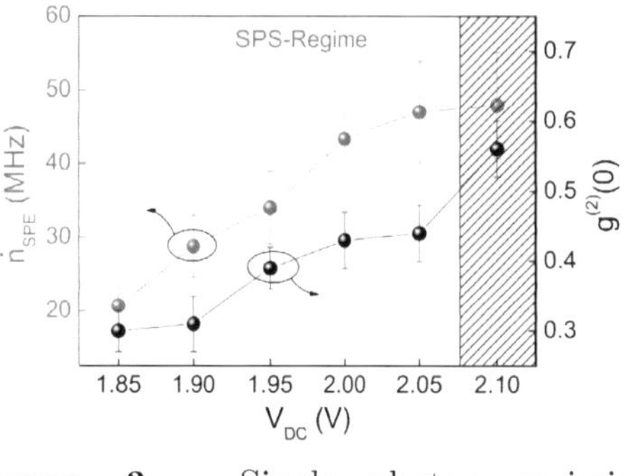

Figure 3. Single photon emission rate $\dot{n}_{SPE}$ (left y-axis) and corresponding $g^{(2)}(0)$-value (right y-axis) as a function of the dc offset V_{DC}.

$\dot{n}_{SPE}$-value in the SPS-regime ($g^{(2)}(0) < 0.5$) amounts to (47.0±6.9) MHz at V_{DC}=2.05 V. It is interesting to note that the achievable SPE rate of a given device depends sensitively on the particular settings of V_{DC}, V_{AC} and f, for the latter even a resonance like dependence has been observed. In the present case we therefore fine-tuned the pulse frequency f which allowed us to obtain even higher values of $\dot{n}_{SPE}$ compared to 27.5 MHz reported in [5] for a 2 μm diameter SPS. The value $\dot{n}_{SPE}$ =(47.0±6.9) MHz yields an overall SPS efficiency $\eta=\dot{n}_{SPE}/\text{f}$=(21.4±3.1)% when taking the repetition rate of f=220 MHz into account.

This finding can be classified by comparing the measured overall efficiency and the photonic outcoupling efficiency obtained indirectly by experimental cavity parameters. As derived in [3] the outcoupling efficiency η_{ext} of a SPS can be expressed by

$$\eta_{ext} = \frac{Q_{pillar}}{Q_{2D}} \cdot \frac{F_P}{F_P + 1}, \tag{1}$$

where F_P denotes the Purcell factor and Q_{2D} represents the Q-factor of the planar microcavity. Taking the experimental values of Q_{pillar}=2300, Q_{2D}=3000 and F_P=4 into account, we obtain an outcoupling efficiency of 61% close to the maximum (70%) derived in [3]. However, the overall efficiency of the SPS is strongly limited by a rather inefficient carrier injection into the QDs - a detrimental effect that can be attributed to non-radiative emission at the pillar sidewalls and the associated carrier losses [11, 12].

4. Summary

In summary we have presented electrically triggered SPE from a QD-micropillar with $g^{(2)}(0)$ as low as 0.30±0.03. We demonstrated record high SPE rates of up to (47.0±6.9) MHz corresponding to an overall efficiency of (21.4±3.1)% for a 2.0 μm-diameter device.

Acknowledgments

Acknowledgments This work was financially supported by the German Ministry of Education and Research project "QPENS". Expert sample preparation by M. Emmerling and A. Wolf is gratefully acknowledged.

References

[1] Michler P, Kiraz A, Becher C, Schoenfeld W V, Petroff P M, Zhang L, Hu E and Imamoğlu A 2000 *Science* **290** 2282–2285
[2] Strauf S, Stoltz N G, Rakher M T, Coldren L A, Petroff P M and Bouwmeester D 2007 *Nat. Photonics* **1** 704–708
[3] Barnes W, Björk G, Gérard J, Jonsson P, Wasey J, Worthing P and Zwiller V 2002 *Eur. Phys. J. D* **18** 197–210
[4] Ellis D J P, Bennett A J, Dewhurst S J, Nicoll C A, Ritchie D A and Shields A J 2008 *New J. Phys.* **10** 043035
[5] Heindel T, Schneider C, Lermer M, Kwon S H, Braun T, Reitzenstein S, Höfling S, Kamp M and Forchel A 2010 *Appl. Phys. Lett.* **96** 011107
[6] Reitzenstein S and Forchel A 2010 *J. Phys. D: Appl. Phys.* **43** 033001
[7] Böckler C, Reitzenstein S, Kistner C, Debusmann R, Löffler A, Kida T, Höfling S, Forchel A, Grenouillet L, Claudon J and Gérard J M 2008 *Appl. Phys. Lett.* **92** 091107
[8] Lochmann A, Stock E, Töfflinger J, Unrau W, Toropov A, Bakarov A, Haisler V and Bimberg D 2009 *Electron. Lett.* **45** 566–567
[9] Michler P, Imamoğlu A, Kiraz A, Becher C, Masona M, Carson P, Strouse G, Buratto S, Schoenfeld W and Petroff P 2002 *Phys. Status Solidi B* **229** 399–405
[10] Pelton M, Santori C, Vučković J, Zhang B, Solomon G S, Plant J and Yamamoto Y 2002 *Phys. Rev. Lett.* **89** 233602
[11] Englund D, Altug H and Vučković J 2007 *Appl. Phys. Lett.* **91** 071124
[12] Reitzenstein S, Heindel T, Kistner C, Rahimi-Iman A, Schneider C, Höfling S and Forchel A 2008 *Appl. Phys. Lett.* **93** 061104

Excitonic emission of colloidal nano-crystals embedded in Molecular Beam Epitaxy grown ZnSe

M Rashad (1), A Pawlis (1), D Schikora (1), K Lischka (1), O Schöps (2), M V Artemyev (2) and U Woggon (2)
(1) Department Physik, Universität Paderborn, Paderborn, Germany.
(2) Technische Universität Berlin, Berlin, Germany
E-mail: mahmed@mail.uni-paderborn.de

Abstract. We combine ZnSe layers grown by molecular beam epitaxy with colloidal CdSe core, CdSe/ZnSe and CdSe/ZnS core/shell nano-crystals (NCs) to achieve monolithic NC-semiconductor heterostructures. The NCs are prepared in solution and deposited by spray-coating on ZnSe buffer layers and subsequently overgrown by ZnSe. We find a blue shift of the photoluminescence of core/shell dots when they are overgrown by ZnSe. Rapid thermal annealing is used to improve the interface region between the NCs and the ZnSe matrix. The effect of different annealing temperatures on the optical properties of CdSe core, CdSe/ZnSe and CdSe/ZnS core/shell NCs overgrown with a cap layer of ZnSe is investigated. After annealing at 673 K the photoluminescence of these samples is red-shifted as compared to unprocessed samples. All photoluminescence results are explained by a model calculation with the following assumption about the 3D confining potential of the NCs: (i) the shell of core/shell NCs dissolves during ZnSe overgrowth, (ii) after overgrowth NCs are separated from the ZnSe matrix by an interface barrier, (iii) the height of this barrier is significantly reduced by annealing. For all three types of NCs we find an excellent quantitative agreement between the experimental and calculated NC transition energies. The absence of the barrier after annealing is further demonstrated by low temperature photoluminescence data of annealed samples which show enhanced diffusion of electron-hole pairs from ZnSe into the NCs .

1. Introduction

Wet chemically prepared colloidal nanoparticles of numerous shapes, sizes, and compositions have been reported over the past two decades and continue to exhibit exciting optical and electrical characteristics [1]. Further, such colloidal NCs exhibit an enormous potential regarding the tuning of their optical and structural properties [2]. As free parameters, the size, shape, material composition and NCs density can be varied and exactly matched to the requirements of specific devices. Alternatively, different types of NC which emit light at several wavelengths can be combined in the same device to achieve multi-color or white light emission from a single chip. In addition, wet-chemically prepared core and core-shell nanocrystals (NCs) can be deposited on different materials with very low concentrations. However, NCs show a relatively fast photodegradation and photoluminescence intermittency [3]. To overcome this effect, colloidal NCs can be capped with a larger band gap material and/or organic ligands to increase the confinement and passivate the surface states [4].
Our unique approach is the integration of the NCs in an epitaxial grown semiconductor matrix to stabilize their optical properties. In this paper we present the investigation of the photoluminescence

Journal of Physics: Conference Series **245** (2010) 012006 doi:10.1088/1742-6596/245/1/012006

(PL) of different types of colloidal NCs which were deposited on ZnSe/GaAs substrates and overgrown with ZnSe layers. Finally these samples were rapid thermal annealed at different temperatures. The peak energy of the NC-related luminescence of overgrown and annealed samples showed a pronounced shift relative to the emission of the NCs in solution. This shift is well reproduced by a model calculation which assumes a variation of the confining potential of the NCs due to overgrowth and annealing.

2. Experimental

Three types of NCs, CdSe core, CdSe (ZnSe) and CdSe (ZnS) core (shell) NCs with a shell thickness of 2 monolayers were synthesized by standard high-temperature reaction of organometallic precursors in strongly coordinating solvents, (see ref. [5]). We used 0.2 milliliters of the colloidal solution with a volume concentration of 10^{16} NCs/l. These colloidal solutions of NCs were sprayed on the surface of a 50 nm ZnSe buffer layer which was grown by molecular beam epitaxy on GaAs substrates. Then, the sample was transferred back to the MBE chamber and overgrown with a 30 nm thick ZnSe cap layer, for details see ref. [6]. The photoluminescence (PL) was measured using both the 325 nm emission line of a continuous wave He-Cd laser and the 405 nm emission line of a semiconductor laser diode in a μ-PL setup with a spectral resolution of 0.1 nm. The PL emission was detected with a N_2 cooled CCD detector. Annealing of the samples was performed in an AnnealSys AS-One rapid thermal annealing (RTA) system in N_2 gas. The RTA was accomplished isochronic with an annealing time of 30 sec.

3. Results and discussion

The room temperature PL spectra of the three different types of NCs mentioned in the experimental part were measured in solution, after overgrowth and after annealing at temperatures of 598 K, 623 K and 673 K for 30 sec. As an example fig. 1 shows the room temperature PL intensity versus photon energy of CdSe/ZnSe NCs. The black curve (a) is the PL spectrum of the NCs in solution, while the red curve (b) is referring to the PL of the NCs after overgrowth with ZnSe. The PL spectra after RTA treatment at three different temperatures are the curves (c). We find a blue shift between the PL emission of the NCs in solution and of the overgrown NCs. In contrast, a red shift is observed between the PL emission of the overgrown NCs and the annealed sample.

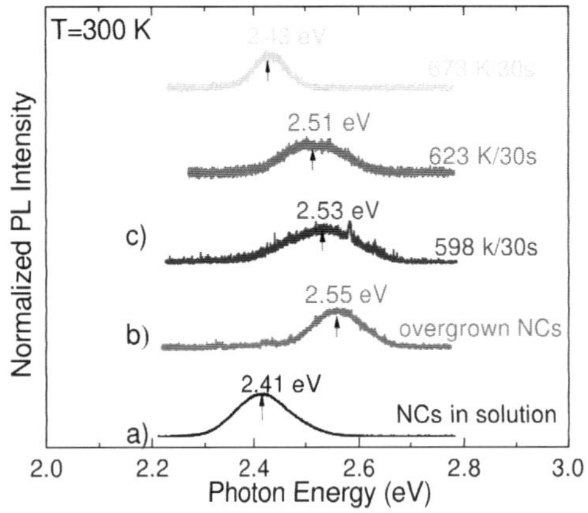

Figure 1. Room temperature PL spectra of CdSe/ZnSe NCs in solution (a), overgrown with a 30 nm ZnSe cap layer (b) and after rapid thermal annealing at 598 K, 623 K and 673 K for 30 s (c).

The blue shift was explained by a model presented in Ref. [6]. This model is based on two assumptions: (i) during the overgrowth process, the shell of the core/shell NCs is dissolved, and (ii) after overgrowth the NCs are separated from the ZnSe matrix by a thin barrier of yet unidentified origin which prevents excess electron-hole pairs generated in ZnSe to recombine within the NCs. The red shift after annealing may be explained also in the framework of this model, if we assumed that the barrier is reduced during the RTA process and after annealing at 673 K the barrier is completely extirpated. A similar trend regarding the red shift of the PL spectra after annealing was also observed for CdSe core and CdSe/ZnS core/shell NCs.

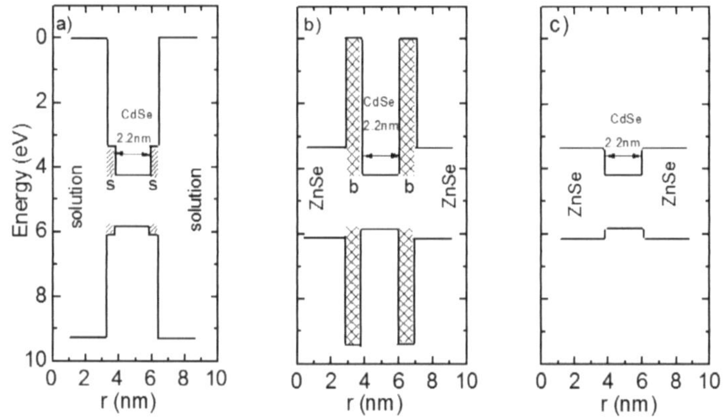

Figure 2. Confinement potential used in the calculation of excitonic energy of CdSe/ZnSe NCs in solution (a), overgrown in a ZnSe matrix (b) and after rapid thermal annealing at 673 K/30 sec (c).

Fig.2 illustrates the band structures used in our model calculations for ZnSe/CdSe core-shell NCs (a) in solution, (b) after overgrowth and (c) after annealing at 673K. We used an electron affinity of 3.92 eV [7] for CdSe and 3.5 eV for ZnSe [8]. The 0.5 nm thick ZnSe shell around the CdSe core is marked as **(s)** in fig. (2a). The barrier which is formed during the overgrowth process is marked **(b)** in fig. (2b). We set its width to 1 nm. Furthermore, we assume that during the annealing process the barrier is reduced. After annealing, the barrier height is zero and the ZnSe matrix is directly connected to the NCs shown in fig. (2c). Therefore, the carrier transfer between the ZnSe matrix and the NCs has high probability to occur.

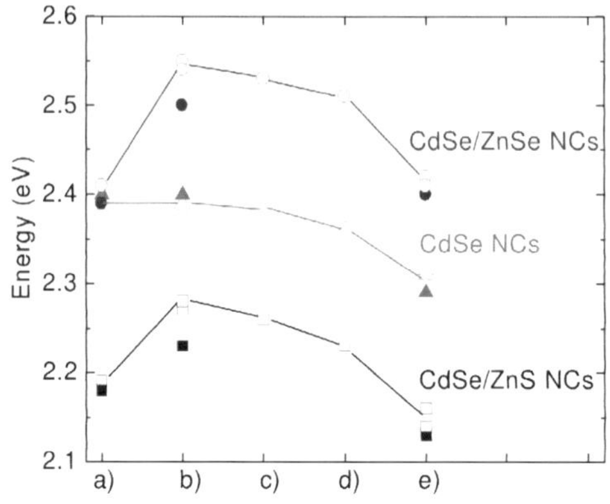

Figure 3. The PL peak energy (open symbols) and the calculated transition energies (filled symbols) for NCs in solution (a), overgrown in a ZnSe matrix (b) and after rapid thermal annealing at 673 K (e). Cases (c) and (d) indicated the experimental data annealed at 589 K and 623 K, respectively.

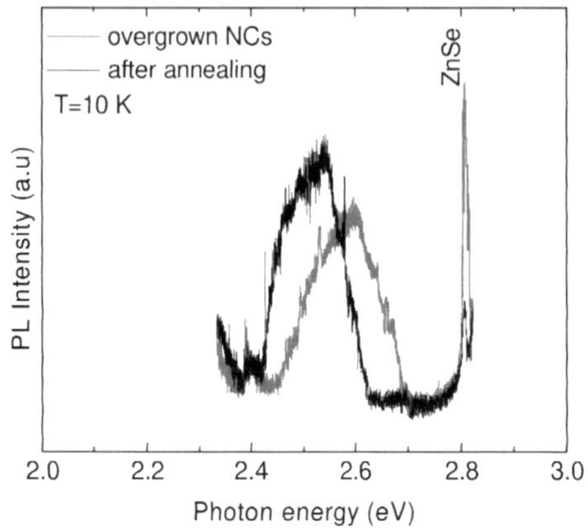

Figure 4. Low temperature PL spectra of CdSe/ZnSe NCs before and after 30sec rapid thermal annealing at 673 K.

The calculated exciton transition energies of three types of NCs (full symbols) together with PL peak energies (open marks) measured with NCs in solution, after overgrowth and after annealing at 673 K are shown Fig. 3. We find quantitative agreement between experimental and calculated data which we consider a strong support of the assumptions made, especially that the barrier between NCs and ZnSe matrix can be removed by RTA treatment.

Fig. 4 shows the low temperature PL spectra of CdSe/ZnSe NCs as overgrown (gray curve) and after annealing at 673 K (black curve). Two peaks can be identified in these spectra. The peak at 2.8 eV is due to recombination of excitons in the ZnSe matrix [9] and the emission band at about 2.5 eV is the PL of the CdSe/ZnSe NCs. The relative intensities of these emission lines are inverted after annealing of the sample. The spectrum before annealing (gray curve) exhibits a strong PL due to the recombination of excitons in ZnSe while the emission from the NCs is relatively week. In contrast, after annealing the ZnSe PL is strongly reduced while the emission from the NCs substantial enhanced. This behavior can be explained by an enhanced diffusion of excess e-h pairs from ZnSe into NCs when the barrier between NCs and ZnSe matrix is annihilated by annealing in excellent agreement with our model.

4. Conclusion

We have combined wet-chemical growth of colloidal NCs and molecular beam epitaxy to create complex semiconductor structures with NCs as the optical active material. After overgrowth of various core-shell NCs a distinct blue shift is observed which is not present in samples of CdSe core NCs overgrown with ZnSe. PL measurements indicate that the overgrown NCs are predominantly directly excited and a barrier layer prevents the diffusion of excited carriers from the ZnSe into NCs. Rapid thermal annealing of these heterostructures leads to a pronounced red-shift of the NC-related emission. Using a numerical model which contains distinctive assumptions about the confining potential of the NCs we can explain the measured PL peak energy of NCs in solution, after overgrowth and after annealing. One essential result of our investigation is that annealing reduces the barrier between NCs and the ZnSe matrix. This is supported by low temperature PL data which show an increase of the NC emission of annealed samples. Thus annealing of NCs-semiconductor heterostructures can pave the way towards applications of NCs in photonic and optoelectronic devices.

Acknowledgements

The Egyptian Mission Department and University of Paderborn for financial support.

References

[1] Murray B, Kagan C R and Bawendi M G 2000 *Rev. Mater. Sci.* **30** 545.
[2] Arens Ch, Roussau N, Schikora D, Lischka K, Schöps O, Herz E, Woggon U, Litvinov D ,Gerthsen D and Artemyev M V 2006 *Phys. Stat. Sol. (c)* **3** 861.
[3] Woggon U, Herz E, Schöps O, Artemyev M V, Arens Ch, Rousseau N, Schikora D, Lischka K, Litvinov D and Gerthsen D 2005 *Nano Lett.* **5** 483.
[4] Peng X, Schlamp M, Kadavanich A and Alivisatos A 1997 *J. Am. Chem. Soc.*, **119** 7019.
[5] Murray C B, Sun S, Gaschler W, Doyle H, Betley T A and Kagan C R 2001 *IBM J. Res. Dev.* **45** 47.
[6] Rashad M, Paluga M, Pawlis A, Lischka K, Schikora D, Artemyev M V and Woggon U 2010 *Phys. Stat. Sol.(c)* **7** 1523.
[7] Suzuki H 1966 *J. Appl. Phys.* **5** 1253.
[8] Chahboun A and Zorkani I 2001 *J. Condensed Matter* **4** 52.
[9] Theis, D 1977 *Phys. Stat. Sol. (b)* **79** 125.

Simulation of MeV electron energy deposition in CdS quantum dots absorbed in silicate glass for radiation dosimetry

R Baharin, P R Hobson and D R Smith

Centre for Sensors & Instrumentation, School of Engineering and Design, Brunel University, Uxbridge UB8 3PH, UK

E-mail: ruzalina.baharin@brunel.ac.uk

Abstract. We are currently developing 2D dosimeters with optical readout based on CdS or CdS/CdSe core-shell quantum-dots using commercially available materials. In order to understand the limitations on the measurement of a 2D radiation profile the 3D deposited energy profile of MeV energy electrons in CdS quantum-dot-doped silica glass have been studied by Monte Carlo simulation using the CASINO and PENELOPE codes. Profiles for silica glass and CdS quantum-dot-doped silica glass were then compared.

1. Introduction

Recent studies have investigated the change in the optical response of quantum-dots contained in glass matrices as a function of the absorbed ionizing radiation dose. A variety of different techniques have been exploited, for example change in optical absorption [1], change in thermoluminescence [2] and, for quantum-dots in a liquid environment, the change in the optical-stimulated fluorescent yield [3].

In order to investigate electron irradiation of quantum-dot-doped silica glass, Monte Carlo simulations using the CASINO [4] and PENELOPE [5] software packages were carried out. With these simulations, the electron interactions in material are grouped and referred to using a multiple scattering theory [6]. In CASINO, the electrons trajectory is moving from the entry point through the material until their energy is absorbed or they are backscattered out of the material. CASINO is however limited to planar layer structures and does not simulate bremsstrahlung emission of photons explicitly. Mathematical models are applied to calculate the probability of the scattered electrons, the scattering angle, distance between scattering events and the energy rate loss. The more sophisticated Monte Carlo code PENELOPE enables the simulation of electrons, positrons, and photons in any material, in complex geometries, over the energy range of a few hundred eV to 1 GeV. The cross-sections acquired from both numerical databases and analytical models were used in scattering models for the different types of interaction mechanisms [7].

The motivation for this work was the development of new 2D imaging radiation dosimeters for electrons and photons up to a biologically important absorbed dose of a dose level commonly used in radiation therapy to high dose rate for industrial usage. There is considerable interest in dosimetric techniques which have either 2D/3D spatial information or the ability to be rapidly and remotely interrogated during an irradiation or both. Optical absorbance and fluorescence are both techniques which can provide this capability. In this work, we report on simulations with bulk and surface absorbed quantum-dots in a silica matrix. A silica matrix was chosen as this material is highly radiation-tolerant to both electron and gamma irradiation and thus any changes in its bulk optical properties, critically absorption in our application, will be negligible at the doses we are using. Our initial aim is to investigate dosimeters, using near-UV excited fluorescence of the quantum dots, that are optically imaged onto a sensitive video camera to provide a two-dimensional dose map. The use of CdS-silica-CdS sandwich structures potentially allows some level of discrimination between electron and gamma irradiation as the dose-depth characteristics can be very different, different sized quantum dots could be used on the two surfaces to provide colour discrimination when the dosimeter is read out.

2. Monte Carlo Simulation

The two Monte Carlo simulation codes used in this work were CASINO and PENELOPE. Comparisons between silica glass only profiles with quantum-dot-doped silica glass were then made to investigate the depth dose distribution and the energy of backscattering electrons.

In both simulations, a 0.1cm slab of silica (SiO_2) glass was exposed to a beam of 1 MeV accelerating electrons. The source point is located at the sample surface and irradiates towards the sample. For the quantum-dot-doped silica glass simulation, the same geometries were considered with 1000 nm of CdS quantum-dots deposited on each face and a third geometry with the same ratio of CdS to SiO_2 but with the CdS uniformly distributed throughout the bulk silica matrix.

3. Results and Discussion

3.1. Depth-dose distribution

Figure 1 shows the dose distribution by PENELOPE over the sample's depth following irradiation by 1 MeV electrons. The dose grows a bit higher in CdS-Silica-CdS at the beginning and increases to a peak dose at 0.057 cm depth. In Silica, the dose is very slightly smaller than that in CdS-Silica-CdS with a peak also occurring at 0.057 cm depth. The scattering of primary electrons that lose energy along the path is responsible for the build-up zone.

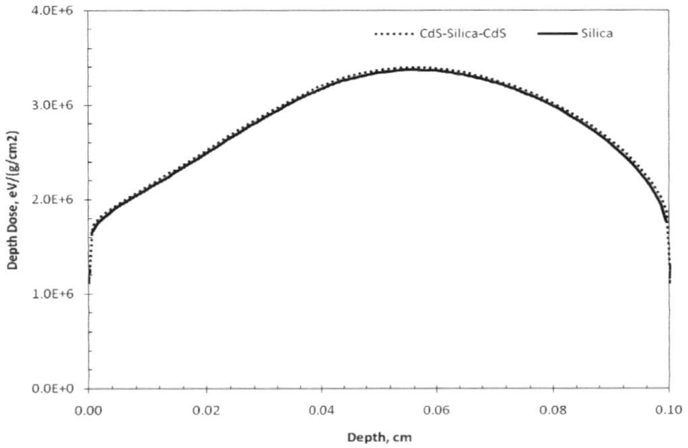

Figure 1. The depth-dose distribution in quantum-dots-doped silica glass (CdS-Silica-CdS) and silica glass only (Silica) from PENELOPE. The slabs were irradiated by 1 MeV accelerating voltage.

Figure 2 shows the distribution of penetration depth of backscattered electrons for quantum-dots-doped silica glass and silica glass only at 1 MeV by CASINO. 200000 electrons were used in each simulation to simulate the model. The backscattered electrons (BE) in CdS-Silica-CdS can travel through the sample further than in Silica. At a depth of 0.055 cm in Silica, the number of BE is starting to decrease and they have the potential to leave the sample surface. However, because of the fluctuations in the data, further analysis is required to interpret these data.

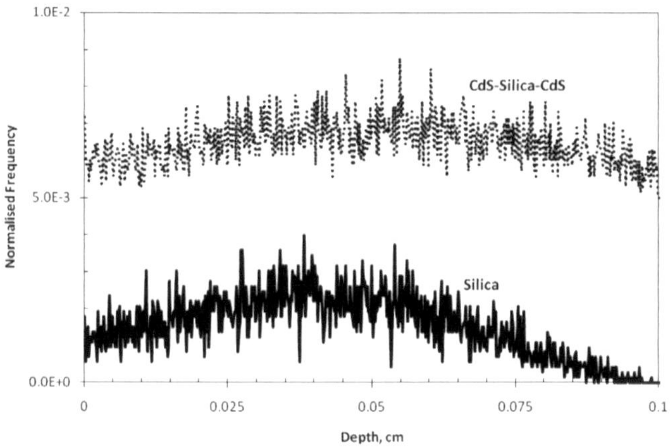

Figure 2. The maximum penetration depth of backscattered electrons in quantum-dots-doped silica glass (CdS-Silica-CdS) and silica glass only (Silica) by CASINO. The slabs were irradiated by 1 MeV accelerating voltage. The fluctuations reflect limited statistics. The CdS-Silica-CdS curve is multiplied by a factor of 5 for clarity.

Figure 3 shows the depth dose of porous Vycor glass with CdS quantum dots absorbed (Vycor-CdS) to compare to CdS-Silica-CdS. The thickness of the porous Vycor has been increased, compared to pure silica, because of its lower density.

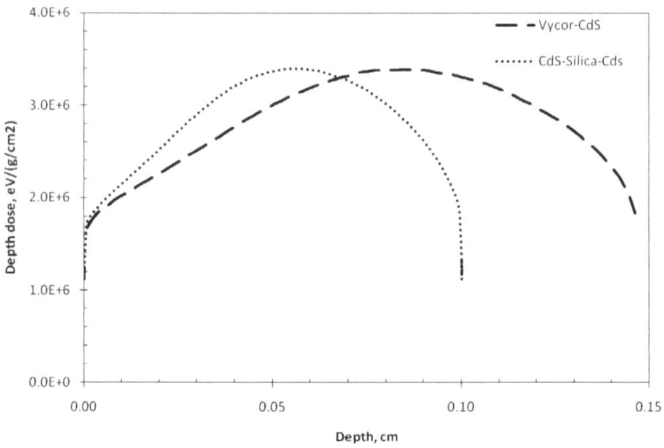

Figure 3. The depth-dose distribution in CdS quantum dots in Vycor glass matrix (Vycor-CdS) and quantum-dots-doped silica glass (CdS-Silica-CdS) from PENELOPE. The slabs were irradiated by electrons with 1 MeV accelerating voltage.

3.2. Backscattering electrons

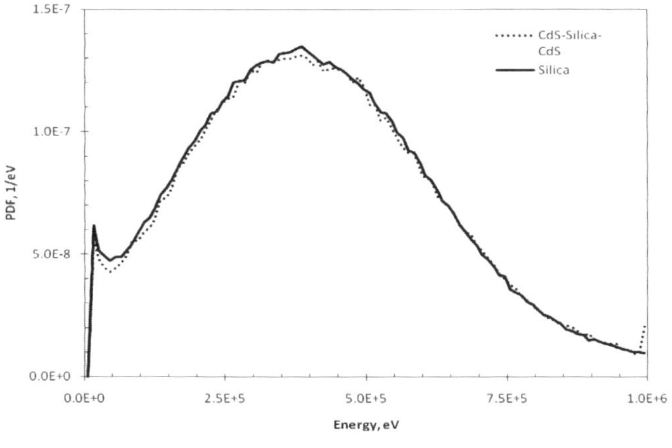

Figure 4. The energy distribution of backscattered electrons in quantum-dots-doped silica glass (CdS-Silica) and silica glass only (Silica) from PENELOPE. The slabs were irradiated by electrons with 1MeV accelerating voltage.

Figure 4 and 5 show the energy of backscattered electrons when leaving the surface sample. In CASINO, there is no difference between two samples. Meanwhile, PENELOPE shows a more complex profile since it correctly takes into account bremsstrahlung photons as well as electrons and a small enhanced full energy backscatter peak is predicted for the CdS-Silica-CdS sample.

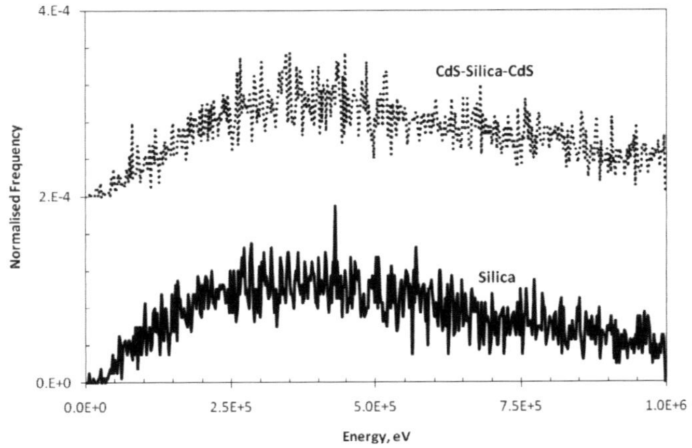

Figure 5. The energy of backscattered electrons in quantum-dots-doped silica glass (CdS-Silica-CdS) and silica glass only (Silica) from CASINO. The slabs were irradiated by electrons with 1 MeV accelerating voltage. The fluctuations reflect limited statistics. The CdS-Silica-CdS curve is multiplied by a factor of 2 for clarity.

Data from the CASINO simulations shown in Figures 2 and 5 are expressed in terms of 'normalised frequency' on the y-axis, the default output of CASINO, which provides measure of the frequency of occurrence of a particular process. In contrast, PENELOPE is able to provide output in terms of the specific energy deposited as a function of distance into a material, as in Figures 1 and 3, or as a probability distribution function, as in Figure 4.

4. Conclusion

Our preliminary results show that in terms of parameters such as backscattered electron spectra for 1MeV incident electrons CASINO and PENELOPE give similar, but not identical results. CASINO is not directly able to provide either a 1D or 2D profile of dose versus depth in contrast to PENELOPE. PENELOPE simulates processes involving photon emission and subsequent scattering and absorption which are absent from CASINO. This also enables PENELOPE to be used in applications where the incident radiation field consists of high energy gamma rays, such as those from ^{60}Co sources.

Our simulated results of layered CdS-Silica-CdS dosimeters have been compared with a possible dosimeter using quantum dots absorbed into a porous silica glass matrix, such as Vycor 7930 [8]. As expected, the differences in response are subtle but for some applications, lower energy incident electrons for example, or mixed radiation fields, they may be significant. The use of the PENELOPE Monte Carlo will enable us to model these effects more fully in our subsequent experimental work.

References

[1] Gomonnai A V, Azhniuk Yu M, Kranjcec M, Solomon A M, Lopushansky V V and Megela I G 2001 *Solid State Commun.* **119** 447
[2] Dekanozishvili G, Driaev D, Kalabegishvili T and Kvatchadze V 2009 *J. Lumin.* **129** 1154
[3] Stodilka R.Z., Carson J.J.L., Kui Y., Zaman M.B., Li C., Wilkinson D., 2009 *Journal of Physical Chemistry C* **113** 2580
[4] Drouin D, Couture A R, Joly D, Tastet X, Aimez V and Gauvin R 2007 *Scanning* **29** 92
[5] Salvat F and Fernandez-Varea JM 2009 *Metrologia* **46** S112
[6] Vilches M, Garjia-Pareja S, Guerrero R, Anguiano M and Lallena AM 2007 *Nuclear Instruments and Methods in Physics Research B* **254** 219
[7] Chica U, Anguiano M and Lallena AM 2009 *Physica Medica* **25** 51
[8] http://www.corning.com/docs/specialtymaterials/pisheets/Vycor%25207930.pdf

Manipulating fine structure splitting in semiconductor quantum dots

Ranber Singh and Gabriel Bester
Max-Planck-Institut für Festkörperforschung, Heisenbergstrasse 1, 70569 Stuttgart, Germany.
E-mail: G.bester@fkf.mpg.de

Abstract. We perform million-atom empirical pseudopotential calculations of fine structure splitting (FSS) in *realistic* InAsP/InP and InGaAs/GaAs QD structures. We find that in alloy In(Ga)As self-assembled QDs (001) there exist a lower bound to the tuning of FSS by applying uniaxial stress due to an anticrossing of the excitonic lines, while in pure InAs self-assembled QDs (001) the FSS goes through zero to negative values due to the crossing of latter. We further predict that semiconductor nanostructures grown along the [111] direction such as self-assembled quantum dots or nanowire quantum dots have a vanishing FSS on grounds of their symmetry and are therefore ideal candidates for the generation of entangled photon pairs.

1. Introduction

At the heart of the emerging fields of quantum teleportation [1], quantum cryptography [2] and quantum computations [3] is the generation and manipulation of entangled photon pairs. Semiconductor quantum dots (QDs) are investigated as a potential source of adequate photons. Indeed, in an idealized QD with degenerate intermediate exciton states the polarization of a photon stemming from a biexciton decay ($|XX\rangle \rightarrow |X\rangle$) is entangled with a photon stemming from the exciton recombination ($|X\rangle \rightarrow |0\rangle$) [4]. However, in real self-assembled semiconductor QDs grown along the [001] direction the existence of non-vanishing energetic difference between the intermediate exciton states, *so-called* fine structure splitting (FSS), destroys the polarization entanglement between the photons produced in the biexciton cascade process [5]. A schematic illustration of the biexciton cascade process generating a polarization-entangled photon pair in a QD is shown in Figure 1. When FSS is smaller than the radiative linewidth of exciton emission the two single exciton decay paths become indistinguishable and the radiative decay lose the "which path" information [6]. This is the ideal condition for the production of polarization-entangled photon pairs in a semiconductor QD. There is tremendous ongoing efforts to manipulate and reduce FSS below the radiative linewidth in order to generate polarization-entangled photon pairs using self assembled QDs grown along the [001] crystal direction. The ongoing efforts to manipulate and reduce the FSS include electric fields [7], magnetic field [8], strain [9], local annealing techniques [10], spectral filtering [6] and selection of QDs with low FSS [11]. An alternative approach to manipulate the FSS is the use of uniaxial stress [9]. But it is still an open fundamental question whether FSS can be tuned through zero by this mean.

Here, we investigated the manipulation of FSS in *realistic* and atomistically described self-assembled In(Ga)As/GaAs QD (001) structures by uniaxial stress applied along different crystal directions. The nanostructures grown along [001] and [111] directions with high structural

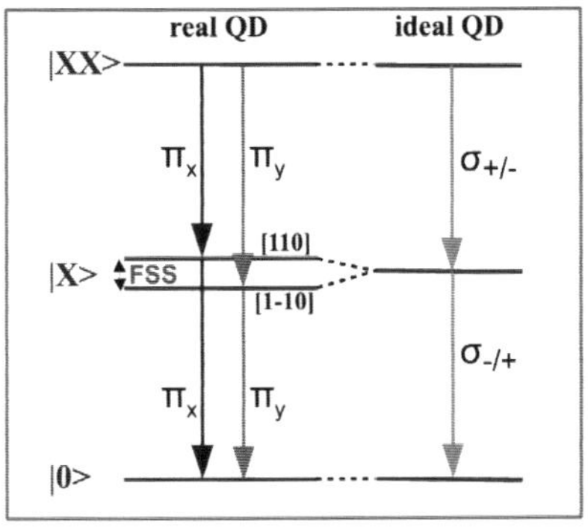

Figure 1. A schematic illustration of the biexciton cascade process in a QD. The π_x and π_y denotes the polarizations along [110] and [1$\bar{1}$0], respectively. σ_+ and σ_- denotes clockwise and anticlockwise circularly polarized photons. In the case where FSS≈0, the photons emitted in this biexciton cascade lose the "which-path information" and hence become "polarization-entangled".

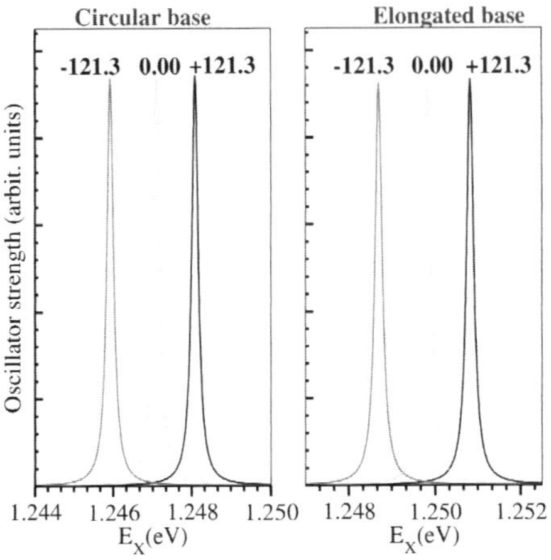

Figure 2. Oscillator strengths of $In_{0.6}Ga_{0.4}As$/GaAs QD's excitonic transition energies at different uniaxial stresses of 0.00 MPa, -121.3 MPa (tensile stress) and +121.3 MPa (compressive stress) applied along [110] crystallographic direction. The lineshape of excitonic lines is Lorentzian determined by radiative line broadening of 0.1 meV.

symmetries are also investigated for the QDs which intrinsically have zero FSSs. We predict that semiconductor nanostructures grown along the [111] direction such as self-assembled quantum dots or nanowire quantum dots have a vanishing FSS and therefore, are ideal candidates for the generation of entangled photon pairs [12].

2. Results and discussion

The QD structures are constructed with the information for shape, size and composition directly from the experimental reports [13, 14]. We relax the atomic positions of this QD structure to minimize the strain energy using the valence force field (VFF) method [15]. The single particle orbitals and energies of the QD are calculated by using the atomistic pseudopotential approach [16, 15], taking strain, band coupling, coupling between different parts of the Brillouin zone and spin-orbit coupling into account, retaining the atomistically resolved structure. The Coulomb and exchange integrals are calculated from the atomic wave functions [17] and the correlated excitonic states are calculated by the configuration interaction (CI) approach [18]. For the CI calculations we use all possible determinants constructed from the twelve lowest energy electron and twelve lowest hole states (spin included), thus accounting for correlations.

2.1. Manipulating FSS by uniaxial stress

In this work, we consider In(Ga)As/GaAs lens shaped QDs with (i) circular base (base diameter = 25.2 nm, height = 3.5 nm), (ii) elongated ellipsoidal base with long axis along the [110] direction [long (short) axis = 26.5 nm (23.9 nm), height = 3.5 nm] and truncated cone shaped QDs (base diameter = 24 nm, top diameter =18 nm) with heights 3.5 nm and 7 nm. We take different alloy compositions with constant 60% and 80% In content across the QD volume and a graded In composition profile according to the "model 5" reported by Mlinar *et al.* [19]. In this latter model, the In concentration profile was obtained by a combination of cross-section

STM, photoluminescence and theory. The In concentration varies linearly between the base center of the dot (55% In) the base edge (20% In), the top center (100% In) and the top edge (80% In). We investigate the effect of uniaxial stress along the [$1\bar{1}0$], [110] and [100] crystallographic directions. The uniaxial stress is created by changing the lattice parameters along the crystallographic direction of the applied stress. The stress values are calculated by using the relation, $S = Y[(a_0 - a)/a_0]$ where Y is the Young's modulus, a_0 and a are the equilibrium and distorted lattice parameters, respectively. The Young's modulus of zinc-blende cubic crystals is anisotropic. We take Y=85.3 GPa and 121.3 GPa [20, 21] as the Young's Modulus of GaAs along the [100] and [110] crystallographic directions, respectively. For each applied stress the atomic positions within the simulation cell, as well as the simulation cell along the [001] growth direction, are relaxed to the minimum strain energy using the VFF method. The unaxial stress applied here is much lower than the breaking limit of 1.0 GPa [9]. The

Table 1. The lower bound of FSS in *realistic* alloy In(Ga)As/GaAs QDs (001) under uniaxial stress along different crystal directions. Graded composition profiles of QDs are as defined in the text. Here, negative and positive values of uniaxial stress means tensile and compressive stresses, respectively.

structure	material	uniaxial stress [crystal direction] (MPa)	FSS (μeV)
	circular base		
Lens-shaped	$In_{0.6}Ga_{0.4}As$	21.3 [100]	2.86
Lens-shaped	$In_{0.6}Ga_{0.4}As$	-60.7 [$1\bar{1}0$]	2.74
Lens-shaped	$In_{0.6}Ga_{0.4}As$	60.7 [110]	2.74
Lens-shaped	$In_{0.8}Ga_{0.2}As$	30.3 [110]	5.25
Lens-shaped	Graded composition	-91.0[110]	5.60
Truncated-Cone	Graded composition	-91.0[110]	3.58
Long Truncated-Cone	Graded composition	-151.6 [110]	2.15
	elongated base		
Lens-shaped	$In_{0.6}Ga_{0.4}As$	151.6 [110]	2.98

uniaxial stress on InAs/GaAs QDs give rise to a red-shift of excitonic energy, while uniaxial stress on alloy QDs give rise to a blue-shift of excitonic energies. These results are in agreement with experimental results [9]. The detailed understanding of these shifts in excitonic energies by uniaxial stress will be presented in the subsequent work. Whereas QD's excitonic transition energy can be tuned substantially, the oscillator strength of transitions remain almost constant over the range of applied stresses [see Figure 2], in contrast to the use of electric fields. This property of the system is useful if the QD's emission wave length has to be tuned into resonance with a constant wavelength (light or cavity mode) in order to exploit the Purcell effect, and possibly use the QD as an efficient single photon source [22]. We find that in realistic alloy In(Ga)As self-assembled QDs (001) there exist a lower bound to the tuning of FSS by applying uniaxial stress due to anticrossing of excitonic lines, while in pure InAs self-assembled QDs (001) the FSS goes through zero to negative values due to crossing of excitonic lines [23]. The lower bound of FSS in alloy QDs under uniaxial stress along different crystal directions is given Table 1. We notice that for different alloy structures and stresses the lower bound to the tuning of FSS is about 2 μeV. The structural symmetry of pure InAs QDs [23] allow crossings of excitonic lines in pure QD structures under stress along the [110] or [$1\bar{1}0$] direction.

2.2. Manipulating FSS by growing high symmetric QD structures

On the basis of symmetry analysis using group theoretical arguments we predicted that QD structures with D_{2d} and C_{3v} symmetries intrinsically have zero FSSs. We have confirmed this by empirical pseudopotential calculations on *realistic* QD structures [12]. The FSS in [111] and [001] grown $InAs_xP_{1-x}/InP$ QDs with different structural symmetries are given in Table 2. Whereas QDs with D_{2d} symmetry are difficult to grow, the QDs with C_{3v} symmetry in nanowire hetrostructures [111] can be easily grown [24]. Thus, we predict that nanowire QDs are the ideal candidates for the generation of entangled photon pairs [12]. We further investigated how robust the results are against structural deformations. The simple in-plane elonagation of hetrostructure nanowire QDs redcues the C_{3v} symmetry to C_s and leads to nonzero FSSs.

Table 2. FSSs in InAsP/InP QDs with different shape, composition and structural symmetry.

QD material	Shape	Size (d, h) nm	FSS (μeV)
InAs [111]	disk (C_{3v})	33.0, 8.0	0.2
InAs [111]	hex. (C_{3v})	35.0, 8.0	0.0
$InAs_{0.25}P_{0.75}$ [111]	disk (C_{3v})	33.0, 8.0	0.0
InAs [111]	disk (C_{3v})	25.2, 3.5	0.1
InAs [111]	lens (C_{3v})	25.2, 3.5	0.1
InAs [001]	disk (D_{2d})	25.2, 3.5	0.0
InAs [001]	disk (C_{2v})	25.2, 3.5	12.9
		5%, 3,5	2.8
InAs [001]	elongated disk (C_s)	10%, 3.5	5.4
		15%, 3.5	8.2

References

[1] Jennewein T, Weihs G, Pan J W and Zeilinger A 2001 *Phys. Rev. Lett.* **88** 017903
[2] Gisin N, Ribordy G, Tittel W and Zbinden H 2002 *Rev. Mod. Phys.* **74** 145
[3] Chen J *et al* 2008 *Phys. Rev. Lett.* **100** 133603
[4] Benson O, Santori C, Pelton M and Yamamoto Y 2000 *Phys. Rev. Lett.* **84** 2513
[5] Bayer M *et al* 1999 *Phys. Rev. Lett.* **82** 1748
[6] Akopian N *et al* 2006 *Phys. Rev. Lett.* **96** 130501
[7] Kowalik K *et al* 2005 *Appl. Phys. Lett.* **86** 041907
[8] Stevenson R M, *et al* 2006 *Phys. Rev. B* **73** 033306
[9] Seidl S, Kroner M, Högele A, Karrai K, Warburton R, Badolato A and Petroff P 2006 *Appl. Phys. Lett.* **88**
[10] Langbein W, Borri P, Woggon U, Stavarache V, Reuter D and Wieck A D 2004 *Phys. Rev. B* **69** 161301
[11] Hafenbrak R, Ulrich S M, Michler P, Wang L, Rastelli A and Schmidt O G 2007 *New J. Phys.* **9** 315
[12] Singh R and Bester G 2009 *Phys. Rev. Lett.* **103** 063601
[13] Walther T, Cullis A G, Norris D J and Hopkinson M 2001 *Phys. Rev. Lett.* **86** 2381
[14] Kegel R *et al* 2001 *Phys. Rev. B* **63** 035318
[15] Williamson A J, Wang L W and Zunger A 2000 *Phys. Rev. B* **62** 12963
[16] Wang L W and Zunger A 1999 *Phys. Rev. B* **59** 15806
[17] Bester G, Nair S and Zunger A 2003 *Phys. Rev. B* **67** 161306
[18] Franceschetti A, Fu H, Wang L W and Zunger A 1999 *Phys. Rev. B* **60** 1819
[19] Mlinar V *et al* 2009 *Phys. Rev. B* **80** 165425
[20] Bateman T B, McSkimin H J and Whelan J M 1959 *J. Appl. Phys.* **30** 544
[21] Brantley W A 1973 *J. Appl. Phys.* **44** 534
[22] Pelton M *et al* 2002 *Phys. Rev. Lett.* **89** 233602
[23] Singh R and Bester G 2010 *Phys. Rev. Lett.* **in press**
[24] Minot E *et al* 2007 *Nano Lett.* **7** 367

MOVPE grown InGaAs quantum dots of high optical quality as seed layer for low-density InP quantum dots

D Richter, R Hafenbrak, K D Jöns, W-M Schulz, M Eichfelder, R Roßbach, M Jetter and P Michler

E-mail: d.richter@ihfg.uni-stuttgart.de
Institut für Halbleiteroptik und Funktionelle Grenzflächen, Universität Stuttgart, Allmandring 3, 70569 Stuttgart, Germany.

Abstract. To achieve a low densitiy of optically active InP-quantum dots we used InGaAs islands embedded in GaAs as a seed layer. First, the structural InGaAs quantum dot properties and the influence of the annealing technique was investigated by atomic force microscope measurements. High-resolution micro-photoluminescence spectra reveal narrow photoluminescence lines, with linewidths down to 11 μeV and fine structure splittings of 25 μeV. Furthermore, using these InGaAs quantum dots as seed layer reduces the InP quantum dot density of optically active quantum dots drastically. InP quantum dot excitonic photoluminescence emission with a linewidth of 140 μeV has been observed.

In recent years, investigations on semiconductor quantum dots (QDs) have been motivated by their potential application as single photon sources and as quantum bits in the field of quantum information processing [1, 2, 3, 4]. For this purpose it is essential to optically or electrically address individual QDs. However, the growth of self-assembled QDs formed by Stranski-Krastanov growth mode [5], often leads to high QD densities using MOVPE. Therefore the samples have to be structured into mesas or shadow masks have to be applied to isolate an individual QD. A solution to overcome this problem should be the growth of self-assembled QDs exhibiting a low density. A narrow linewidth of QD PL emission makes them suitable for research, e.g. for single spin measurement [6], two photon interference [7] and single-photon emitters, respectively. As current single-photon detectors have their highest photon detection efficiency in the red spectral range it is also preferable to fabricate single QDs emitting at such wavelengths [8]. For that reason we focused on the growth of InP QDs embedded in GaInP barrier material that is lattice-matched to GaAs [9]. This material system can emit over a wide spectral range in the visible. The growth of InP QDs in GaInP on such substrates produces QD densities of around $10^{10}\,\mathrm{cm}^{-2}$ which is orders of magnitude too high to study a single QD on an unstructured sample even when using micro-photoluminescence (μ-PL) techniques. To achieve a low density of optically active InP QDs we used InGaAs islands embedded in GaAs as a seed layer. Using the strain produced by the underlying InGaAs islands the InP QDs are expected to preferentially grow on top of the InGaAs seed QDs resulting in a low density of InP QD.

The sample structure was fabricated by MOVPE with standard sources at low pressure (100 mbar) on (100) GaAs substrates oriented 6° toward the [111]A direction. Before depositing a GaAs buffer layer the sample is heated up to 750°C to desorb surface oxides and reach the growth temperature for the following layers. The 50 nm thick $Al_{0.5}Ga_{0.5}As$ layer prevents from carrier diffusion towards the substrate during optical measurements, due to the higher bandgap compared to the subsequently grown GaAs barrier. The layer of self-assembled InGaAs QDs was grown using the Stranski-Krastanov growth mode by depositing 2 monolayers (ML) of InAs at 520°C and a growth rate of 0.63 ML/s on top of a 134 nm thick GaAs barrier layer. After the deposition of the QDs a growth interruption of 60 s was applied to allow QD ripening. This step was followed by an additional deposition of 3 nm GaAs on top of the InAs QDs at 520°C to reduce the Indium desorption for the following steps. Additionally, a combined temperature ramping to 600°C and annealing step was applied. During the 15 min lasting annealing step Indium-Gallium intermixing took place and therewith a blueshift of the QD PL emission was induced. PL samples were capped with 30 nm GaAs. (Fig.1(a) lower part)

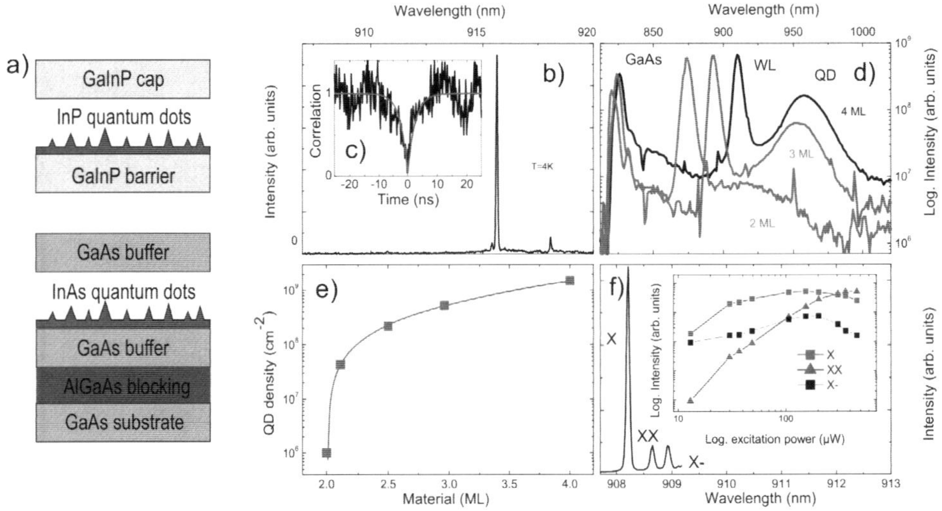

Figure 1. *(a) Reference sample structure. (b) µPL emission of the InGaAs QDs. (c) Autocorrelation measurement* $g^{(2)}(\tau)$ *for the investigated sample, which is nearly background free and dips to a value of* $g^{(2)}(\tau = 0) = 0.03$*. The red line represents a fit to the experimetal data. (d) Ensemble PL measurements for different InAs deposition amounts. (e) QD density as a function of the amount of deposited material. The red line is a guide to the eye. (f) shows a µ-PL spectra and the corresponding power dependence of the excitonic, biexcitonic and trionic labeled peaks.*

To gain control of the InGaAs QD growth, the amount of deposited material was varied. Using AFM and scanning electron microscopy (SEM) measurements the structural QD properties were examined for not overgrown samples. The QD density can be controllably adjusted over a range of three orders of magnitude between $10^9\,cm^{-2}$ and $10^6\,cm^{-2}$. The average QD height ranges from 10.2 nm (4 ML) to 4.2 nm (2 ML). This very pronounced influence on the structural QD properties should also be reflected in the optical properties. The samples were investigated by ensemble PL measurements (Fig.1(d)). The spectra show distinct peaks at around 826 nm originating from the GaAs barrier and a wetting layer (WL) emission around 885 nm. The relatively broad emission at higher wavelength originates from the QD ensemble. The QD ensemble emission blueshifts by reducing the InAs amount, corresponding to smaller sized QDs. Furthermore, the decrease in the integrated PL intensity can be attributed to the reduced QD density.

In order to perform µ-PL measurements a standard resolution PL [10, 11] and high resolution photoluminescence (HRPL) setups are used which are described elsewhere [10]. Investigations on two-photon correlations have been performed in terms of second-order $g^2(\tau)$ auto-correlation measurements on a Hanbury-Brown and Twiss (HBT)-type setup on a sample with 2 ML InAs deposition.

Due to the low QD density ($\sim 10^7$ cm^{-2}) we were able to adress single QDs optically without any need for mask processes. A typical QD emission is displayed in detail in the μ-PL spectra in Fig. 1(b). The emission energy is typically located around 920 nm. In Fig. 1(b) the signal to noise ratio of the PL peak at 916 nm is of about 120, signal to noise ratios of up to 140 were found. Since, there was no significant background observed, the use of the QDs as a single photon-source at high repetition rate is feasible. For single photon emission with high signal to noise ratios, no further background correction has to be applied. This is essential for high bit rate data communication.

In order to get a deeper insight into the emission properties of the grown QDs, we performed power-dependent μ-PL measurements to identify the exciton complexes responsible for the observed lines. The spectrum is dominated by the peak at 908.2 nm which we attributed to an exciton (X) emission (Fig. 1(f)). A second peak is observed at 908.6 nm and becomes more pronounced for higher excitation energies (Biexciton, (XX)). The third peak at 908.9 nm is of less intensity and maybe attributed to charged exciton emission. Many-body interactions like carrier exchange and Coulomb interactions give rise to a red shift of the charged excitonic states compared to the neutral states [1], therewith the redshifted, X* labeled PL peak is estimated to be a charged exciton. The assignment of X, XX and X*-emission to the PL peaks is based on analysis of the different integrated PL intensities (Inset in Fig. Fig. 1(f)). We extracted a exponent p_X of 1.1 for the excitonic PL peak and a value of p_{XX}=1.9 for the biexcitonic PL emission.

In order to prove the zero-dimensional state-densities of the investigated InAs QDs, we performed continuous-wave (cw) second-order auto-correlation measurements. The correlation function $g^2(\tau)$ is shown in Fig. 1(c). The $g^2(\tau)$ measurements were performed on an excitonic emission. For true single photon emitters a $g^2(\tau = 0)$ value of zero is expected. The measurement yields a value of $g^2(\tau = 0)$=0.03. Thus, the emitted photons depict a clear antibunching behavior. A deviation of $g^2(\tau = 0)$ from zero takes place due to background. The nearly background free $g^2(\tau)$ and signal to noise ratios of up to 140 already indicate a good optical quality. In order to determine the intrinsic linewidth, HRPL measurements were accomplished. The linewidth of the PL peak was determined in terms of full width of half maximum of a fitted Lorentzian function. For an excitation power density of 0.7 W/cm^2 a FWHM of 11 μeV was found. To further resolve the fine structure splitting ΔE_{FS}, HRPL measurements with a resolution of $\Delta E_{\mathrm{res}}^{\mathrm{HRPL}} = 0.3$ μeV have been performed. Under stepwise variation of the linear polarization detection angle a clear fine-structure splitting of $\Delta E_{FS}^{HRPL} = 25$ μeV becomes obvious.

In a next step InGaAs QDs were used as an seed for InP QDs. Therefore the sample structure was changed corresponding to Fig. 1(a) upper part. The InGaAs QDs were overgrown by 20 nm GaAs. A 10 nm lattice-matched $Ga_{0.51}In_{0.49}P$ layer followed at 720.C. The self-assembled InP QDs were grown by depositing 2 ML of InP at 710°C and a growth rate of 1.05 ML/s . After a growth interruption of 20 s to ripen of the QDs a 30 nm cap of GaInP followed.

In Fig. 2(b) the AFM height topography of some InP QDs is shown. A density of approximately 4×10^9 cm^{-2} could be observed. This is more than two orders of magnitude higher than expected. Also the heights of the InP QDs are much larger (up to 40 nm) than the heights without a seed layer (around 2 nm [9]). Two different island types can be distinguished: small truncated pyramid structures and large dome shaped islands that show highly indexed crystal planes. The strain produced by the underlaying InAs islands results in a different growth environment for the InP/GaInP QDs. μPL measurements of these InP QDs show that only a small percentage of the QDs are optically active. Therefore no structuring was needed to

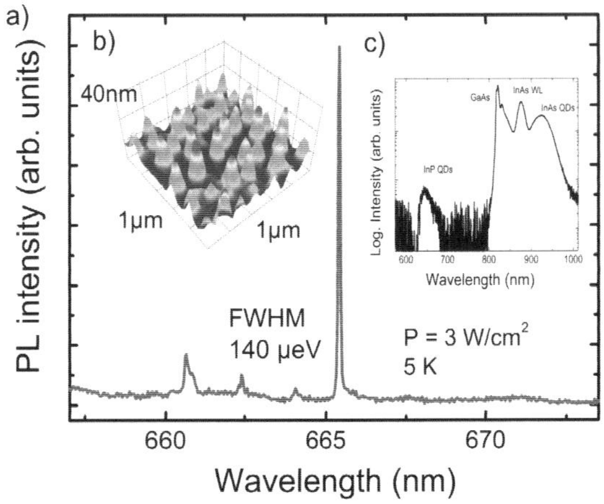

Figure 2. *(a) PL emission of a single InP QD exhibting a linewidth of 140 µeV. (b) AFM measurement of the InP QD top layer. (c) Ensemble PL spectrum exhibiting the InGaAs seed layer emission, the InP QD and the GaAs barrier emission.*

investigate single QDs. A typical emission linewidth of 140 μeV could be measured (Fig. 2(a)). Autocorrelation measurements reveal a $g^2(\tau = 0)$=0.4 for the PL emission in Fig. 2(a). In ensemble PL (2(c)) measurements the PL of the InGaAs seed layer (around 930 nm), the InP QDs (around 660 nm) as well as GaAs barrier are observed. Only the small QDs as shown in the AFM measurements seem to be optically active. The growth of InP QDs on such defective InAs clusters might induce also a defect in the overgrowing layer, and thus in the InP QDs on top. Therefore the large dome shaped InP islands seem to be incoherent and optically inactive. These structures may provide non-radiative decay channels to the carriers leading to decreased quantum efficiency. Only the small InP QDs which where grown on top of the small InAs islands or in between big InAs clusters are optically active.

In summary, we have shown that we are able to deposit low-density (In_xGa_{1-x})As QDs by MOVPE using partial capping and annealing steps for narrow linewidth single dot emission. HRPL measurements show a FWHM of the excitonic QD emission of 11 μeV. Polarization dependent measurements revealed a fine structure splitting of 25 μeV. We proved the zero-dimensional state-density of the InAs islands by performing correlation measurements ($g^{(2)}(\tau = 0) = 0.03$). The InGaAs QDs have been used as seed for InP QDs, resulting in a lower density of optically active InP QDs. This work was supported by the Deutsche Forschungsgemeinschaft (FOR730).

[1] Michler P 2009 *Single Semiconductor Quantum Dots* (Berlin: Springer)
[2] Robledo L, Elzerman J, Jundt G and Atat M
[3] Knill E, Laflamme R and Milburn G J 2001 *Nature* **409** 46
[4] Li X, Wu Y, Steel D, Gammon D, Stievater T H, Katzer D S, Park D, Piermarocchi C and Sham L J 2003 *Science* **301** 809
[5] Stranski I N and Krastanow L 1938 *Akad. Wiss. Wien Kl.IIb* **146** 797
[6] Berezovsky J, Mikkelsen M H, Gywat O, Stoltz N G, Coldren L A and Awschalo D D 2006 *Science* **314** 1916
[7] Hong C K, Ou Z Y and Mandel L 1987 *Phys. Rev. Lett.* **59** 2044
[8] Zwiller V, Blom H, Jonsson P, Panev N, Jeppesen S, Tsegaye T, Goobar E, Pistol M E, Samuelson L and Bjrk G 2001 *Appl. Phys. Lett.* **78** 2476
[9] Beirne G J, Reischle M, Robach R, Schulz W M, Jetter M, Seebeck J, Gartner P, Gies C, Jahnke F and Michler P 2007 *Phys. Rev. B* **75** 195302
[10] Hafenbrak R, Ulrich S M, Michler P, Wang L, Rastelli A and Schmidt O G 2007 *New J. Phys.* **9** 315
[11] Schulz W M, Robach R, Reischle M, Beirne G J, Bommer M, Jetter M and Michler P 2009 *Phys. Rev. B* **79** 035329

Inversion recovery measurements of exciton fine-structure beats in a single quantum dot

A. F. A. Khatab, A. J. Ramsay*, S. J. Boyle, A. M. Fox, and M. S. Skolnick

Dept. Physics and Astronomy, University of Sheffield, Sheffield S3 7RH, UK

E-mail: a.j.ramsay@shef.ac.uk

Abstract. We present a method for time-resolving the coherent precession of the spin of a single neutral exciton confined to an InAs/GaAs quantum dot. A pair of circularly polarized π pulses are used to pump and probe the exciton spin, rather than a pair of interferometrically stable $\pi/2$-pulses in a quantum interference type experiment, faster, technically simpler, measurements of the small 15 μeV exciton fine-structure can be made in the photocurrent regime. A small change in the fine-structure splitting with vertically applied electric-field is observed.

1. Introduction

Semiconductor quantum dots possess a small anisotropy due to a combination of bulk asymmetries [1], spin-orbit coupling [2], piezo-electric, and dot elongation [3]. This anisotropy breaks the spin degeneracy of the neutral exciton, resulting in a small fine-structure splitting in the energies of the bright exciton states.

The fine-structure splitting is a property of practical importance, being the main limitation on the fidelity of dot-based polarization entangled photon sources [4], and also spin preparation techniques based on exciton ionization [5]. This motivates strong efforts to suppress the fine-structure splitting using post growth annealing [6], lateral magnetic-fields [7], lateral electric-fields [8, 9, 10], and optical Stark shifts [11]. The fine-structure splitting can be measured using polarization resolved single dot cw spectroscopy techniques such as micro-PL [?, 7, 3, 12], transmission [11], or photocurrent [13]. The fine-structure splitting can be deduced from the exciton spin beats observed in ensembles of dots [14, 6, ?, 15], or at a single dot level using quantum interference [16, 13], and recently four-wave mixing techniques [17].

Here we employ the inversion recovery method [18] to observe the exciton spin precession of a single quantum dot using photocurrent detection. Compared to quantum interference based techniques for observing the fine-structure beats [16, 13], an inversion recovery method has a number of technical advantages. It is a simpler technique that does not require a stabilized interferometer, has a faster acquisition time, and the signal is less sensitive to the pulse-area, and therefore more robust against the mechanical instability of the setup.

2. Principle of experiment

The bright exciton is composed of an electron with $m_J = \mp 1/2$, and a heavy-hole of $m_J = \pm 3/2$, and is created using light of $\sigma_\pm$ circular polarization respectively. The electron-hole exchange interaction couples the exciton spin-states, to form linearly spin-polarized energy-eigenstates $|\alpha\rangle = (|\uparrow\rangle \pm |\downarrow\rangle)/\sqrt{2}$, where the label $\alpha = x, y$. These states are separated in energy by the fine-structure splitting $\hbar\omega_{\text{fs}}$, and have linearly polarized optical selection rules.

A pair of picosecond laser pulses resonantly excite the neutral exciton transition with a pulse-area of π. The pulses arrive with a time delay of τ between them. Since the time-duration of the laser pulses is short compared with the period of the fine-structure beat, the pulse drives a Rabi rotation between the crystal ground-state $|0\rangle$, and the exciton spin up/down state selected by the $\sigma_\pm$ circular polarization of the pulse respectively. For a pulse-area of π the pulse inverts the occupations of these two states. Initially the dot is in the crystal ground-state. At time $t = 0$, the first σ_+-polarized pulse prepares a spin-up exciton. This is a superposition of the x,y energy eigenstates, which beat due to their energy separation: $|\psi\rangle = \cos(\omega_{\text{fs}}t/2)\,|\uparrow\rangle + \sin(\omega_{\text{fs}}t/2)\,|\downarrow\rangle$. The exciton spin freely evolves due to the fine-structure beats and electron tunneling for a time τ.

The absorption of the second pulse at time $t = \tau$ depends on the population of the exciton spin state that is probed. The π-pulse inverts the occupations of the crystal-ground-state, and the exciton spin state selected by the polarization of the laser pulse. This results in a change in the final exciton population, and therefore a change in photocurrent that is proportional to the population inversion between the selected exciton spin and the crystal-ground state. Hence, a photocurrent signal that oscillates between the exciton spin up/down states due to the fine-structure beat is observed.

A calculation of the measured photocurrent per pulse-pair is: $PC = e\nu_{rep}\int_0^\infty \Gamma_t(t)\rho_{XX}dt$, where $\rho_{XX} = (\rho_{\uparrow\uparrow} + \rho_{\downarrow\downarrow})$ is the total exciton population. The density-matrix operator is defined as: $\hat{\rho}_{\alpha\beta} = |\alpha\rangle\langle\beta|$, with a corresponding expectation value of $\rho_{\alpha\beta}$. The probability that the state $(\alpha = 0, \uparrow, \downarrow)$ is occupied at time t is: $\rho_{\alpha\alpha}(t)$. Γ_t is the tunneling rate, and ν_{rep} is the repetition rate of the laser, which for this experiment gives $e\nu_{rep} = 12.18$ pA. The integral can be split into before, and after the probe pulse. The first integral is proportional to the change in the exciton population during $(0 < t < \tau)$: $PC_1(\tau) = e\int_0^\infty \Gamma_t(t)(\rho_{XX}(t)dt = e\nu_{rep}\eta(f - \rho_{XX}(\tau))$, where η is the photocurrent detector efficiency, and f is fidelity of the inversion operation performed by the laser pulse. The second integral is proportional to the exciton population after the α-polarized probe has been applied: $PC_2(\tau) = e\nu_{rep}\int_\tau^\infty \Gamma_t(t)\rho_{XX}(t)dt = e\nu_{rep}\eta(f\rho_{00}(\tau) + \rho_{\bar{\alpha}\bar{\alpha}}(\tau) + (1-f)\rho_{\alpha\alpha}(\tau))$. A comparison of the photocurrent signals measured for co and cross-circular excitation gives:

$$\begin{aligned} PC_{-+} - PC_{++} &= e\nu_{rep}\eta f(\rho_{\uparrow\uparrow}(\tau) - \rho_{\downarrow\downarrow}(\tau)) \\ &= Ae^{-\Gamma_s\tau}\cos(\omega_{\text{fs}}\tau) \end{aligned} \quad (1)$$

where Γ_e, Γ_h are the electron and hole tunneling rates respectively, and $\Gamma_s \approx \Gamma_e$ is the relaxation rate of the exciton spin. The exciton spin should exhibit a damped-cosine behavior due to the fine-structure beat [14].

3. Experiments and discussion

The experiments were performed on an n-i-Schottky diode with a single layer of low density InAs/GaAs quantum dots. The results presented are from a single dot emitting at 951 nm in the high energy tail of the dot-distribution. The two identical pulses have a Gaussian shape, with a Fourier transform limited full-width half-maximum of 0.2 meV. The photodiode sits in a cold-finger cryostat, at a temperature of about 15 K. Further details of both the device, and the setup can be found in ref. [18].

A set of polarization resolved inversion recovery measurements at various gate voltages is presented in figure 1. As discussed earlier, a pair of circularly polarized laser pulses excite the

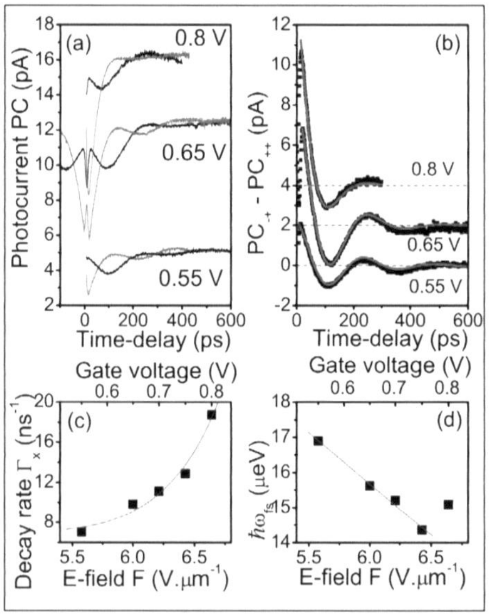

Figure 1. Inversion recovery measurements of exciton spin. (a) Photocurrent versus time-delay for on-resonant excitation of neutral exciton transition with two π pulses. The red (black) traces correspond to co (cross) circular polarized excitation schemes. (b) The difference in photocurrent ($PC_{-+} - PC_{++}$) between cross and co-circular excitation, which is proportional to the exciton spin, versus time-delay. The traces are calculated using data in (a), and the red-lines show fits to data using eq. (1). The fitting parameters are displayed in (c,d). (c) The decay rate Γ_x, due to electron tunneling versus gate voltage. (d) Small change in exciton fine-structure with vertically applied electric-field.

dot on resonance with the neutral exciton transition with a pulse-area of π, and the resulting photocurrent is measured as a function of the inter-pulse time-delay. For co (cross) circular excitation the data are shown as red (black) traces in fig. 1(a) respectively. The pulse-area is calibrated by measuring a Rabi rotation [18].

The time-resolved data are presented in figs. 1(a,b). Figure 1(a) shows a comparison of the photocurrent versus time-delay measured for a pair of co and cross-circularly polarized laser pulses that oscillate in anti-phase. The oscillation is due to the fine-structure beat as discussed above. Figure 1(b) shows the difference between the photocurrents measured for cross and co-circular excitation, and are calculated using the data in fig. 1(a). The signal is proportional to the net exciton spin: $(\rho_{\uparrow\uparrow} - \rho_{\downarrow\downarrow})$, and exhibits a damped-cosine behavior, due to the precession of the spin and a decay due to electron tunneling that increases with applied gate voltage. The red-lines shows the fits to eq. (1) that are used to extract the fine-structure and the electron tunneling rate, which are presented in figs. 1(c) and (d) respectively.

The data presented in fig. 1(a) is symmetric with respect to time-delay, because the two pulses are identical except for their polarization, and the spin-insensitive photocurrent detection only responds to the relative polarization of the laser pulses. For near zero time-delay, where the pulses are not temporally overlapped, the photocurrent signal is low for co-circular excitation, because the two π-pulses act as a single 2π-pulse and de-excite the exciton. Whereas for cross-circular excitation the second pulse does not interact with the spin-up exciton prepared by the first pulse, resulting in a net photocurrent equivalent to one exciton. As the electron tunnels from the dot, the exciton spin state relaxes to a hole-state, resulting in the convergence of the co/cross-circular photocurrent signals as seen in fig. 1(a). Both photocurrent signals saturate at a time-delay that is long compared with all tunneling times, since each pulse creates an exciton that subsequently tunnels resulting in a net photocurrent equivalent to two excitons per pulse-pair for both co and cross-circular excitation.

When the pulses are temporally overlapped there is a peak (dip) in the photocurrent for co (cross) circular excitation respectively, as seen in fig. 1(a). In the case of co-polarized excitation, the interference of the pulses results in an effective single pulse excitation with a pulse-area varying between $0 - 2\pi$, resulting in a peak of non-zero net-photocurrent. For cross-polarized excitation, the temporally overlapped pulses act as a single pulse with a pulse-area of $\pi\sqrt{2}$, and a linear polarization with an orientation angle equal to the inter-pulse phase. Ideally this would create $\sin^2(\pi\sqrt{2}/2)$ excitons per pulse-pair, versus one exciton per pulse-pair when the pulses are no longer overlapped, resulting in the observed dip in photocurrent.

Figure 1(d) shows a small decrease in the exciton fine-structure splitting with increasing vertically applied electric-field. A small change of about 3 μeV is observed for a change in gate-voltage from 0.5-0.8 V, with an electric-field gradient of $dE_{\mathrm{fs}}/dF = -2.9 \pm 0.2$ μeV.V^{-1}.μm. This is of a similar value to previous measurements using a micro-PL technique [19, 20]. We note that this change in the fine-structure splitting is small compared with the 80 μeV achieved using a lateral electric-field [8, 10]. We speculate that this may be due to a decrease in the overlap of the electron and hole wavefunction as the electric-field increases the electron-hole separation, resulting in a decrease in the electron-hole exchange energy.

4. Acknowledgments

The authors thank the EPSRC (UK) EP/G001642, and the QIPIRC UK for financial support. AFAK acknowledges the University of Cairo for financial support. We thank M. Hopkinson, and H.-Y. Liu for sample growth.

References

[1] H. Fu, L-W. Wang, and A. Zunger, Phys. Rev. B **59** 5568 (1999).
[2] G. Bester, S. Nair, and A. Zunger, Phys. Rev. B **69** 161306R (2003).
[3] R. Seguin, *et al*, Phys. Rev. Lett. **95** 257402 (2005).
[4] R. M. Stevenson, *et al*, Nature **439** 179 (2006); N. Akopian, *et al*, Phys. Rev. Lett. **96** 130501 (2006).
[5] M. Kroutvar, *et al*, Nature **432** 81 (2004); A. J. Ramsay, *et al* Phys. Rev. Lett. **100** 197401 (2008).
[6] A. I. Tartakovskii, *et al* Phys. Rev. B **70** 193303 (2004); W. Langbein, *et al*, Phys. Rev. B **69** 161301R (2004); R. J. Young, *et al*, Phys. Rev. B **72** 113305 (2005).
[7] R. M. Stevenson, *et al* Phys. Rev. B **73** 033306 (2006).
[8] K. Kowalik, *et al*, Appl. Phys. Lett. **86** 041907 (2005).
[9] K. Kowalik, *et al*, Appl. Phys. Lett. **91** 183104 (2007); M. M. Vogel, *et al*, Appl. Phys. Lett. **91** 051904 (2007).
[10] B. D. Gerardot, *et al* Appl. Phys. Lett. **90** 041101 (2007).
[11] G. Jundt, *et al*, Phys. Rev. Lett. **100** 177401 (2008).
[12] M. Bayer, *et al*, Phys. Rev. Lett. **82** 1748 (1999).
[13] S. Stufler, *et al* Phys. Rev. B **72** 121301R (2005).
[14] A. I. Tartakovskii, *et al* Phys. Rev. Lett. **93** 057401 (2004).
[15] I. A. Yugova, *et al*, Phys. Rev. B **75** 195325 (2007).
[16] N. H. Bonadeo, *et al*, Science **282** 1473 (1998).
[17] J. Kasprzak, and W. Langbein, Phys. Rev. B **78** 041103R (2008).
[18] R. S. Kolodka, *et al*, Phys. Rev. B **75** 193306 (2007).
[19] K. Kowalik, *et al*, Phys. Stat. Sol. (c) **3** 3890 (2006).
[20] S. Marcet, K. Ohtani, and H. Ohno, Appl. Phys. Lett. **96** 101117 (2010).

Negative transverse magnetoresistance of the two-dimensional electron gas in quantum well with parabolic confinement potential under an in-plane magnetic field

M A Nizametdinova[1], F M Hashimzade[2], Kh A Hasanov[2], M M Babayev[2] and B H Mehdiyev[2]

[1]Azerbaijan Architecture and Construction University, Baku, Azerbaijan
[2]Institute of Physics, National Academy of Sciences of Azerbaijan, AZ 1143, Baku, Azerbaijan

E-mail: firudin_hashimzade@yahoo.com

Abstract. We developed a quantitative theory of transverse magnetoresistance for the two-dimensional electron gas (2DEG) in a quantum well with parabolic confinement potential in the magnetic field in the plane of the free movement of electrons. Our theory is based on the solution of Boltzmann kinetic equation for the components of the tensors of kinetic coefficients, when the perturbing force is directed along the plane of the free movement of electrons, and Titeike's formula for the diagonal component of the conductivity along the direction of confinement, based on the idea of the drift of the centres of cyclotron orbits in the crossed electric and magnetic fields. Numerical calculations have been performed for the 2D electron gas in $GaAs/Al_xGa_{1-x}As$ QWs. It is shown, that the magnetoresistance is negative and equal about 50% in the weak magnetic field case, $\omega_c \ll \omega_0$ if the quantum limit $\hbar\omega_0 > k_0T$ is fulfilled.

1. Introduction

Much interest has been aroused recently in the behavior of two-dimensional electron systems in the presence of magnetic field parallel to the 2DEG layer. The electrical characteristics of these systems are of interest both for fundamental and applied physics. The study of magnetotransport in these structures makes it possible to understand better the mechanisms and parameters of electron scattering. The theory of the quantum galvanomagnetic effects in size-quantized systems was studied in [1-4]. In [1,2] the case of a strongly degenerate electronic gas was considered with the focus was on oscillation phenomena. In [3] the conductivity in QW structures has been calculated, and the size dependence of conductivity in the quantum limit for different mechanisms of electronic scattering has been considered. The authors used the kinetic equation method and the density matrix approach. In the latter case, the scattering was entered into the equation of motion for the density matrix through the lifetime of a quantum state. In [5] the magnetoresistance was studied for the case of electron scattering on a deformation potential of a three-dimensional crystal in a transverse magnetic field. It has been shown that for semiconductors with low mobility the magnetoresistance is negative in weak transverse magnetic fields. In the same paper it was noted that the negative magnetoresistance is not directly associated with weak localization.

In this paper we have calculated the conductivity tensor components for the current density in a QW for degenerate and non-degenerate electron gas. We considered electron scattering by point defects, with the magnetic field across the direction of confinement, i.e in the plane of a 2DEG. Thus, two cases for the relative arrangement of the current direction and the confinement direction are possible. In the case where the current is located in a plane of a 2DEG it is sufficient to consider the relaxation time approximation and to use the kinetic equation. In the case when the current is along the direction of confinement it is necessary to use the density matrix approach obtained in [6-7] for calculation of the diagonal conduction tensor components.

2. Theory

For practical purpose it is convenient to use the harmonic potential model which enables us to calculate various quantities analytically. We consider a quantum well, in which a 2DEG is confined in the x-direction and a homogenous static magnetic field $\vec{B}=(0,0,B)$, with the vector potential $\vec{A}=(0,xB,0)$ lies in the y-z plane of the layer of a 2DEG parallel to the z-axis. The confining potential in the x-direction is characterized by the parabolic potential of frequency ω_0, $U(x)=m^*\omega_0^2x^2/2$, where m^* is the effective mass of a conduction electron.

The eigenvalues and eigenfunctions of the Schrodinger equation are given by

$$\varepsilon_\alpha=\left(\frac{1}{2}+N\right)\omega\hbar+\left(\frac{\omega_0}{\omega}\right)^2\frac{\hbar^2k_y^2}{2m^*}+\frac{\hbar^2k_z^2}{2m^*} \tag{1}$$

$$\phi_\alpha(x,y,z)=\frac{1}{\sqrt{L_yL_z}}\frac{1}{\sqrt[4]{\pi}\sqrt{R}\sqrt{2^NN!}}\exp\left(-\left(\frac{x-x_\alpha}{\sqrt{2}R}\right)^2\right)H_N\left(\frac{x-x_\alpha}{R}\right)e^{i(k_yy+k_zz)} \tag{2}$$

where $\omega=\sqrt{\omega_0^2+\omega_c^2}$ - is the "hybrid" frequency, $\omega_c=eB/m^*c$ is the cyclotron frequency of electrons and N -is the oscillation quantum number, $x_\alpha=-\omega_cR^2k_y/\omega$ is the oscillator center, and $R=\sqrt{\hbar/m\omega}$, $H_N(\xi)$ -is the Hermite polynomial, $\alpha=(N,k_y,k_z)$ is a set of quantum numbers that determine the electron states in a magnetic field.

The transverse component of the magnetoresistance in a magnetic field is defined by

$$\rho_{yy}=\frac{\sigma_{xx}}{\sigma_{xx}\sigma_{yy}-\sigma_{xy}\sigma_{yx}} \tag{3}$$

For the calculation of kinetic coefficients ρ_{yy}, it is necessary to calculate both diagonal and non-diagonal components of conductivity tensor σ_{ik}. Note that in bulk semiconductors $\sigma_{yx}>>\sigma_{xx}$. It is related to the fact that a decrease in scattering potential results in the diagonal electric conductivity tensor components tending to zero, while the non-diagonal components stay finite [7, 8]. In QW, in contrast to the bulk crystal, the diagonal components of the conductivity tensor can exceed the non-diagonal ones [4].

The expression for non-diagonal components, which was obtained using the density matrix approach, is given in [9] as

$$\sigma_{yx}=-\sigma_{xy}=\frac{\omega_ce^2n}{m^*\omega^2} \tag{4}$$

where

$$n=\frac{k_0Tm^*}{\pi\hbar^2}\frac{\omega}{\omega_0}\sum_N\ln\left(1+e^{\eta-x_N}\right) \tag{5}$$

is the density of the two-dimensional electron gas, and

$$\eta = \frac{\zeta}{k_0 T}, \quad x_N = \frac{\hbar\omega}{k_0 T}\left(N + \frac{1}{2}\right). \tag{6}$$

Here ζ is the chemical potential of the electrons, and k_0 is the Boltzmann constant.

In the limit of strong magnetic fields, $\omega_0 << \omega_c$, or equivalently, in the bulk case, when $\omega_0 \to 0$, the energy spectrum (1) equals that of an electron in a magnetic field. In this case the expression for σ_{yx} in (5) coincides with that for the non-diagonal component of the conductivity tensor of the bulk semiconductor material.

For the calculation of the diagonal components of conductivity tensor σ_{xx} when the electric field is perpendicular to the plane of two-dimensional electron gas we will make use of the expressions obtained in [6] and [8], modified so as to allow for 2D character of the electron gas.

The scattering mechanisms considered in the present work are the elastic electron scattering on short-range potential.

Further we will focus on the extreme situation, namely, the quantum limit in which the scattering of electrons is confined within the $N = N' = 0$ level. For the quantum well in a magnetic field the quantum limit criterion is $\hbar\omega \gg k_0 T$. This case is the most interesting in practice for the transport phenomena.

The expression for σ_{xx} is given by

$$\sigma_{xx} = \frac{ne^2\omega_c^2}{m\tau_0\omega_0^4} A_1 \tag{7}$$

$$A_1 = \frac{1}{Log\left(1+e^{\tilde{\eta}}\right)} \int_0^\infty \left(-\frac{\partial f_0}{\partial x}\right) e^{-2ax} I_0(ax)^2 \left(1 - \frac{I_1(ax)}{I_0(ax)}\right) x dx \tag{8}$$

where

$$\tau_0 = \frac{L_y L_z}{N_d} \frac{\pi R^2 \hbar^3}{m V_0^2}, \; a = \frac{k_0 T \omega_c^2}{\omega \hbar \omega_0^2}, \; f_0 = \left(1 + \exp[x - \tilde{\eta}]\right)^{-1}, \; \tilde{\eta} = \eta - x_0 \tag{9}$$

Here $I(\xi)$, N_g and V_0 are the modified Bessel function of the first kind, two-dimensional concentration, and the strength of point defects, respectively.

For the case of the electric field and the temperature gradient directed along the plane of two-dimensional electron gas, we use the solution of the kinetic equation to calculate the diagonal components of tensors σ_{yy}

$$\sigma_{yy} = \frac{ne^2\tau_0\omega_0^4}{m\omega^4} A_2 \tag{10}$$

$$A_2 = \frac{1}{Log\left(1+e^{\tilde{\eta}}\right)} \int_0^\infty \left(-\frac{\partial f_0}{\partial x}\right) e^{2ax} \left(1 + \frac{I_1(ax)}{I_0(ax)}\right) x dx \tag{11}$$

For the relative change of the resistance in transverse magnetic field we obtain

$$\frac{\Delta\rho_{yy}(B)}{\rho(0)} = \left(\frac{\omega}{\omega_0}\right)^4 \frac{A_1}{A_1 A_2 + 1} - 1 \tag{12}$$

3. Results and discussion

We present the numerical calculations for a $GaAs / Al_x Ga_{1-x} As$ parabolic quantum well. We use the following set of physical parameters. The effective mass, $m^* = 0.066 m_0$, where m_0 is the free electron mass. The parameter of the parabolic potential is $\omega_0 = 1.4 \times 10^{13}\ s^{-1}$, and for the temperature we assume *T = 5 K*. Notice that in the calculations for the quantum limit criterion the Fermi level must be between the first and second subbands $\hbar\omega/2 \le \zeta < 3\hbar\omega/2$.

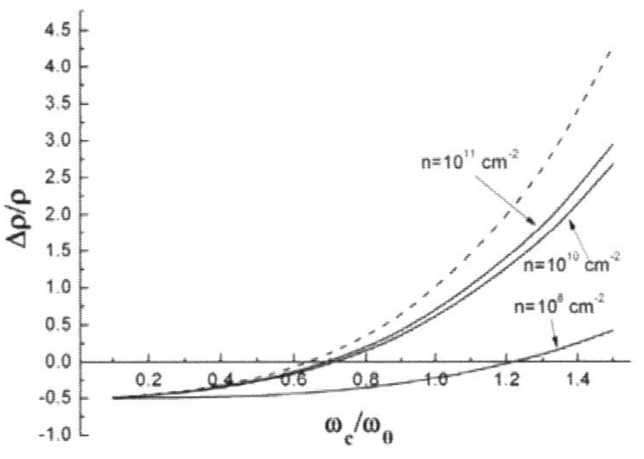

Fig.1 The dependence of the relative change of magnetoresistance as a function of dimensionless parameter ω_c / ω_0 for different 2DEG concentrations (T=5 K). The dashed line correspond to the case $A_1 = A_2 = 1$

In Fig.1 we plot the relative change of magnetoresistance in QW as a function of dimensionless parameter ω_c / ω_0 for different 2DEG concentration. In a weak magnetic field, when the frequency of the confinement potential exceeds the cyclotron frequency, the magnetoresistance of electron gas is negative. For any degree of degeneration of the electron gas the sign of the magnetoresistance changes as the magnetic field increases.

The analysis of the results for scattering on point defects shows that when the dependence of the form factor on the magnetic field in terms of the matrix element of scattering is neglected the dependence of the magnetoresistance on the magnetic field does not change significantly. With a weak filling of the conduction band this is quite justified, as it is evident from Fig.1. For the case of DA scattering mechanism on acoustic phonons and alloy disorder scattering, the scattering probability does not depend on the momentum transfer, and in this approximation the dependence of magnetoresistance on the magnetic field takes the form

$$\frac{\Delta\rho_{yy}(B)}{\rho(0)} = \frac{1}{2}\left(\frac{\omega}{\omega_0}\right)^{7/2} - 1$$

For other scattering mechanisms such a simple relationship cannot be obtained, but in a weak field, as seen from (7) and (10), when conditions $\omega_0 >> \omega_c$ and $\omega_c \tau >> 1$ are satisfied, we have $\sigma_{xx}\sigma_{yy} = \sigma_{xy}^2$, and magnetoresistance is negative.

References

[1] Simserides C D, 1999 *J. Phys.: Condens. Matter* **11**, 5131
[2] *Blokh M D, 1975 Sov. Phys. Solid State* ***17****, 567 (Engl.Transl).*
[3] Kubakaddi S S, Mulimani B G and Jali V M, 1986 *Phys. Stat.Sol. (b)* **137**, 683
[4] Sinyavskii E P, and Khamidullin R A,2002 *Semiconductors* **36**, 924
[5] Kaminskii V E, 2002 *Semiconductors* **36**, 1276
[6] Adams E N and Holstein T D, 1959 *J.Phys. Chem. Sol.* **10** , 254
[7] Anselm A I, and Askerov B M 1967 Sov. *Phys. Solid State* **9**, 22
[8] Askerov B M 1994 *Electron Transport Phenomena in Semiconductors (Singapore: World Scientific)*
[9] Hashimzade F M , Hasanov Kh A, Mehdiyev B H and Cakmak S 2010 *Phys. Scr.* **81**, 015701

Journal of Physics: Conference Series **245** (2010) 012012 doi:10.1088/1742-6596/245/1/012012

HYDROSTATIC PRESSURE EFFECTS ON THE STATE DENSITY AND OPTICAL TRANSITIONS IN QUANTUM DOTS.

G Galindez-Ramirez[1], S T Perez-Merchancano[1], H Paredes Gutierrez[2] and J D González[3]

[1] Departamento de Física, Universidad del Cauca, calle 5 4-70, Popayán, Colombia.

[2] Escuela de Física, Universidad Industrial de Santander, A. A. 678, Bucaramanga, Colombia.

[3] Grupo de Investigación en teoría de la Materia Condensada, Universidad del Magdalena, A.A. 731, Santa Marta, Colombia.

Abstract. Using the effective mass approximation and variational method we have computed the effects of hydrostatic pressure on the absorption and photoluminescence spectra in spherical quantum dot GaAs-(Ga, Al) As, considering a finite confinement potential of this particular work we show the optical transitions in quantum of various sizes in the presence of hydrogenic impurities and hydrostatic pressure effects. Our first result describes the spectrum of optical absorption of 500 A QD for different values of hydrostatic pressure P = 0, 20 and 40 Kbar. The absorption peaks are sensitive to the displacement of the impurity center to the edge of the quantum dot and even more when the hydrostatic pressure changes in both cases showing that to the extent that these two effects are stronger quantum dots respond more efficiently. Also this result can be seen in the study of the photoluminescence spectrum as in the case of acceptor impurities consider them more efficiently capture carriers or electrons that pass from the conduction band to the valence band. Density states with randomly distributed impurity show that the additional peaks in the curves of the density of impurity states appear due to the presence of the additional hydrostatic pressure effects.

E-mail: jdavid0831@gmail.com

1. Introduction

In the last few years research about hydrostatic pressure in optical properties has been increased [1]. The study of the optical properties in GaAs-(Ga,Al)As or QDs spherical quantum dots in presence of impurities and tunable external perturbations such as external fields and hydrostatic pressure is a very current research field for the nanosciences and nanotechnologies mainly because they possess an enormous potentiality for the application in production of fast re sponse electronic devices that are dominated low dimensionality quantum devices. It is well known that the hydrostatic pressure can be used to modify the band structure of semiconductors, for instance at hydrostatic pressures above 40 kbar the GaAs becomes an indirect gap semiconductor [2]. The use of low temperature photoluminescence measures has made possible to study the electronic and optical properties semiconductor microstructures under hydrostatic pressure. Raigoza and collaborators reported the studies in hydrostatic pressure effects on optical transitions of quantum wires [3]. In this work we will develop a theoretical study about the effects of hydrostatic pressure on ground energy and impurities associated effects in the absorption and photoluminescence spectra for shallow impurities (donors and acceptors) in a GaAs- (Ga 0.55, Al 0.45)As spherical quantum dot. We will use the effective mass approach and the variational method. The optic absorption spectrum will be calculated from the firs valence sub band to the acceptor impurities band.

2. Theory

In the effective mass approximation the Hamiltonian for a shallow hydrogenic impurity in a GaAs-(Ga $_{0.55}$, Al $_{0.45}$)As quantum dot under the effects of hydrostatic pressure (P) and low temperature (T) is given by:

$$H = -\frac{\hbar^2}{2}\nabla.\left(\frac{1}{m^*_{wc,bc}(P;T)}\nabla\right) - \frac{e^2}{\epsilon^*_{w,b}(P,T)r} + V(r,P) \quad (1)$$

Where r is the electron – impurity distance and the w and b subscript define the well and barrier regions [4]. In the well region the electron – donor effective mass and the dielectric constant had been used

$$\frac{m^*_e(P,T)}{m_0} = \left[1 + E^{\Gamma}_p\left(\frac{2}{E^{\Gamma}_g(P,T,0)} + \frac{1}{E^{\Gamma}_g(P,T,0)+\Delta_0}\right)\right]^{-1} \quad (2)$$

And

$$\epsilon(P,T) = \epsilon_0(T_0)e^{\partial_1(T-T_0)}e^{\partial_2 P} \quad (3)$$

Where m_0 is the electron effective mass of a free electron $E^{\Gamma}_p = E^{\Gamma}_p(P,T)$, shows the gap dependency of the GaAs with pressure and temperature, $E^{\Gamma}_p = 7.51\text{eV}$, $\Delta_0 = 0.341\text{eV}$, $\varepsilon_0 = 12.74$, $\partial_1 = 9.4\text{x}10^{-5}\text{K}^{-1}$, $\partial_2 = 1.67\text{x}10^{-3}\text{kbar}^{-1}$ and $T_0 = 75.6\text{K}$. In the Eq.1. $V(r,P)$ is the potential that confines the donor electrons to the well region and is given by:

$$V(r,P) = \begin{cases} 0, r < R \\ V_0(P), r > R \end{cases} \quad (4)$$

Where $V_0(P)$, is the height of the barrier that depends of the pressure. Taking into account the cross dependency $\Gamma - X$, with the pressure the phenomenological method suggested by Elabsy has been fallowed [4]. The trial hydrogenic wave function for the ground state has been chosen as [5]:

$$\Psi(r) = \begin{cases} N\dfrac{\sin(\varsigma_{10}r)}{r}e^{-\lambda r}, r \le R \\ N\dfrac{\sin(\varsigma_{10}R)}{r}e^{\chi_{10}(R-r)}e^{-\lambda r}, r \ge R \end{cases} \quad (5)$$

Where N is the normalization constant and λ is the variational parameter, whose eigenvalues associated with the trial function are given by the transcendental equation

$$-\varsigma_{10} = \chi_{10}\tan(\varsigma_{10}R) \quad (6)$$

Where the constants ς_{10} and χ_{10}, are given by,

$$\varsigma_{10} = \left(\frac{2mE_{10}}{\hbar^2}\right)^{1/2} \quad \text{and} \quad \chi_{10} = \left(\frac{2m(V-E_{10})}{\hbar^2}\right)^{1/2} \tag{7}$$

The bond energy dependence with the hydrostatic pressure can be written as,

$$E_b(R, r_0, P) = \frac{\hbar^2 \varsigma_{10}^2}{2m}(P) - \xi(R, r_0, P) \tag{8}$$

Where the second term of the right is the eigen values of the Hamiltonian Eq.1 minimizing from the variational parameter. The transition probability for time unit between the initial and final states, in the effective mass approach for the GaAs and using a dipole approximation is proportional to the square of the matrix element of the electron – photon interaction between the initial and final states wave functions and is determined.

$$P_{i-f} = \tfrac{2\pi}{\hbar}\left|\langle f \,|\, H_{int} \,|\, i\rangle\right|^2 \delta(E_f - E_i - \hbar\omega) \tag{9}$$

With H_{int} ~**e** · **p**, where $\vec{e}$ is the polarization vector in the direction of the electric field of the radiation and $\vec{p}$ is the momentum operator. For a homogeneous impurities distribution and assuming that the structure dimensions are larger than the lattice spacing, the total transition probability is obtained by the sum of the contributions that come from the position of all the impurities [7]. To calculate the density of the impurity states we assume that the circular cross section of the quantum dot is not too small in order that we might treat the impurity positions as a continuous random variable. In this case the density of the impurity states $g(E)$ is given by the following relation [4]:

$$g(E_b) = 4\pi\varsigma_r^{\,2}(E_b)\left|\frac{d\varsigma_r(E_b)}{dE_b}\right| = 4\pi\varsigma_r^{\,2}\left|\frac{dE_b(\varsigma_r)}{d\varsigma_r}\right|^{-1} \tag{10}$$

In Fig. 1 and 2 the graphics can be noticed that for any position of the impurity inside the quantum dot the bond energy is incremented with the pressure. The effects of the hydrostatic pressure are more noticeable for those impurities far away from the edge of the quantum dot because there the repulsion of the wavefunction of the potential barrier is stronger and this diminishes the confinement effect due to the hydrostatic pressure. The behavior of the donor binding energies is also reflected in the curves of density of impurity states in quantum dot.

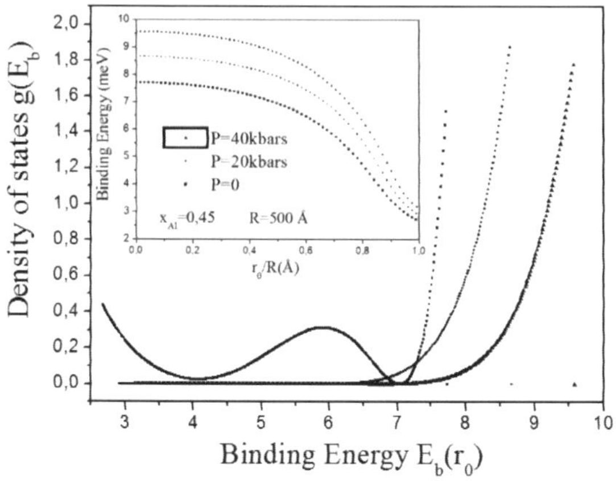

Fig.1 Density of the donor impurity states as a function of the binding energy for the conduction band in a QD. Considering three different pressure values.

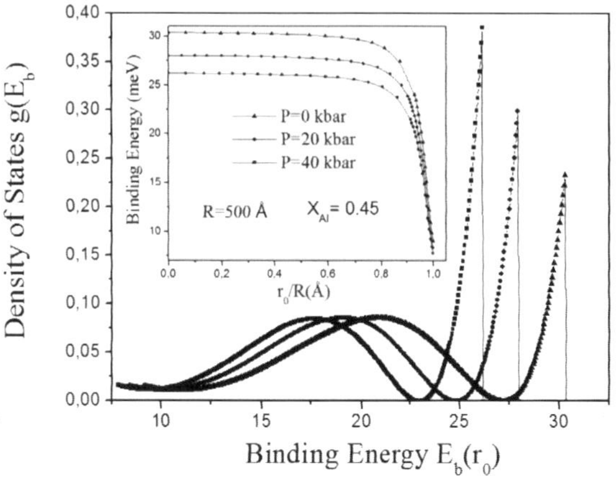

Fig.2 Density of the donor impurity states as a function of the binding energy for the acceptor states in a QD. Considering three different pressure values.

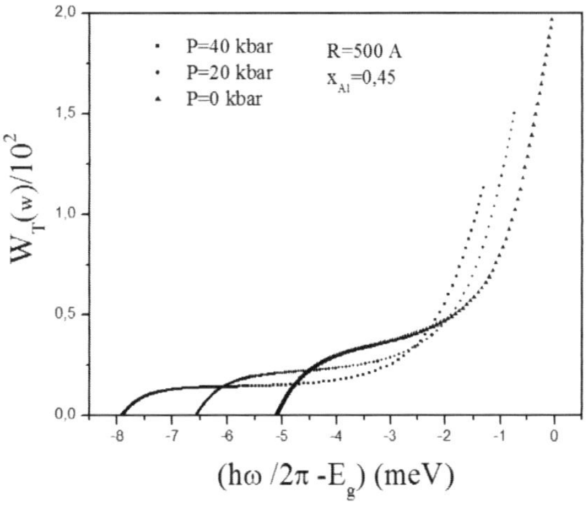

Fig.3 Total optical absorption spectrum, for a quantum dot of gallium arsenide of 500 Å radius and an aluminum concentration of 45%, for different values of hydrostatic pressure values.

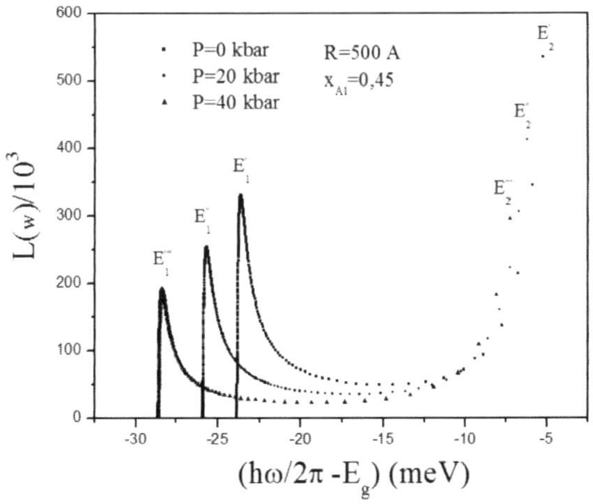

Fig.4. Photoluminescence spectrum for quantum dot with a 500 Å radius and aluminum concentration of 45%, for different hydrostatic pressure values.

The Fig. 3 shown that the energy is lower a new singular behavior change is present showing the presence of the singularity but now it is because there are no more linked states inside the point and we can´t detect the presence of levels in the quantum dot studied**.** The hydrostatic pressure increases the optic absorption decreases allowing a greater number of electron to be present in the conduction band because the bond energy is bigger when the pressure increases. The Fig. 4 shown that the photoluminescence spectrum as function of the system energy taking into account the shallow or hydrogen like impurities energy. Such figure shows that the van hove singularity occurs at lower energies as the hydrostatic pressure increases showing that an appreciable concentration of electrons comes down from the conduction band to the valence band as the hydrostatic pressure increases, this indicates that the system under study is more efficient when under the influence of greater pressures.

3. Conclusion

The theoretical behavior of the optic absorption and photoluminescence spectra affected by hydrostatic pressure is important due to the application in the development of optoelectronic devices such as electro absorption infrared modulators, wave length tunable laser and detectors where the properties of such devices can be adjusted by controlling the size of the quantum dot and the effect of the hydrostatic pressure upon theiŕ heterostructure.

Acknowledgement
This work was partially financed by the Universidad del Cauca, Universidad del Magdalena and Universidad Industrial de Santander.

4. References

[1] N. Raigoza, A.L. Morales, A. Montes, N. Porras- Montenegro,Phys. Rev. B **69**, 045323 (2004).
[2] M. Altarelli, U. Ekenberg, and A Fasolino, Phys. Rev. B **32**, 5138 (1985).
[3] N. Raigoza, C. A. Duque, N. Porras- Montenegro, L. E. Oliveira, Physica B 371 (2006) 153-157.
[4] A. M. Elabsy, J. Phys.Condes. Matter **6,** 10025 (1994).
[5] N. Porras-Montenegro, and S. T. Perez - Merchancano, Phys. Rev. B 46, 9780 (1992)
[7] L. E. Oliveira and R. Perez-Alvarez, Phys. Rev. B **40**, 10460 (1989).
[8] Silva-Valencia J, and Porras-Montenegro, _ *1997 J.* Appl. Phys. 81 901.

Modification of optical properties at bioconjugation of core-shell CdSe/ZnS quantum dots

T V Torchynska[1], A L Quintos Vazquez[2], R Peña Sierra[3], K Gazarian[4] and L Shcherbyna[5]
[1]ESFM – National Polytechnic Institute, México D. F. 07738, México
[2]ESIME – National Polytechnic Institute, México D. F. 07738, México
[3]CINVESTAV - IPN, México D.F. 07320, México
[4]Institute of Biomedical Investigations at UNAM, Mexico D.F. Mexico
[5]V. Lashkaryov Institute of Semiconductor Physics at NASU, Kyiv, 03028, Ukraine

E-mail: ttorch@esfm.ipn.mx

Abstract. This paper presents the results of photoluminescence study of core-shell CdSe/ZnS quantum dots (QDs) with radiative interface states in nonconjugated and bioconjugated conditions. PL spectra of nonconjugated QDs are characterized by a superposition of PL bands related to exciton emission in CdSe cores (2.36 eV) and to electron-hole emission via interface states (2.00, 2.75 and 3.04 eV) at the CdSe/ZnS or ZnS/polymer interfaces. The CdSe/ZnS QDs with emission at 525 nm have been conjugated with bio-molecules - mouse ovarian cancer (OC 125) antibodies. It is revealed that PL spectrum of bioconjugated QDs has changed dramatically with essential decreasing of the hot electron-hole recombination flow via interface states. This effect is explained on the base of the model deals with re-charging of QD interface states at the bioconjugation with OC125 antibodies. The nature of interface states and their changes at the QD aging process have been discussed as well.

1. Introduction

Semiconductor core/shell quantum dots (QDs) have unique optical properties such as QD-size tunable photoluminescence (PL) with narrow emission bandwidths and sufficiently high quantum efficiency. The most popular QDs are core/shell CdSe/ZnS QDs that is characterized by the high photoluminescence quantum yield. Their emission can be changed with the variation of CdSe core size due to the quantum confinement effect [1-3]. The ability to cover these QDs with biomolecules is a critical step in producing the efficient bio-luminescent markers [2,4,5]. Spectroscopic confirmation of bioconjugation with the detection not only the variation of PL intensity but also the spectral transformation of emission spectra, the changes in peak positions and in a band full width at half maximum (FWHM) is highly important. This paper presents the results of PL analysis of CdSe/ZnS core/shell QDs nonconjugated and bioconjugated to bio-molecules – the mouse ovarian cancer (OC 125) antibodies (mab).

2. Experimental details

Commercially available core-shell CdSe/ZnS QDs covered with polymer are used in a form of colloidal particles diluted in a phosphate buffer (PBS) with a 1:200 volumetric ratio. Studied QDs are characterized by the color emission with the maxima at 525 nm (2.36 eV). The part of CdSe/ZnS QDs has been bioconjugated (named 525P) to the mouse ovarian cancer (OC 125) antibodies using the commercially available 525 nm QD conjugation kit [6]. The conjugation reaction is based on the efficient coupling of thiols that present in reduced antibodies, to reactive maleimide groups which exist on the QD surface after SMCC activation. The part of CdSe/ZnS QDs (525N) has been left nonconjugated and serves as a reference object. Samples of QDs (bioconjugated and nonconjugated) in the form of a 5 mm size spot were dried on a polished surface of crystalline Si substrates as described earlier in [7-11]. PL spectra were measured at 300 K and at the excitation by a He-Cd laser with a wavelength of 325 nm and a beam power of 20 mW using a PL setup described in [9, 10].

3. Experimental results and discussion

Normalized PL spectra of pure nonconjugated CdSe/ZnS QDs measured at 300K demonstrate the broad PL band in the spectral range of 1.80-3.20 eV with a main maximum at 2.69 -2.75 eV and shoulders (or small peaks) (Fig.1a).

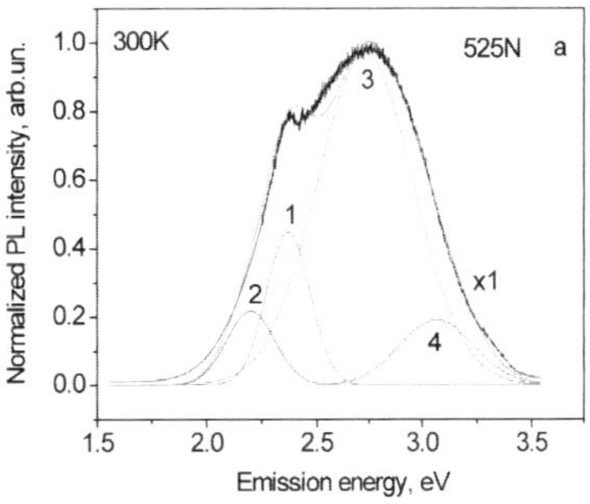

Figure 1a. Normalized PL spectra of nonconjugated 525N QDs.

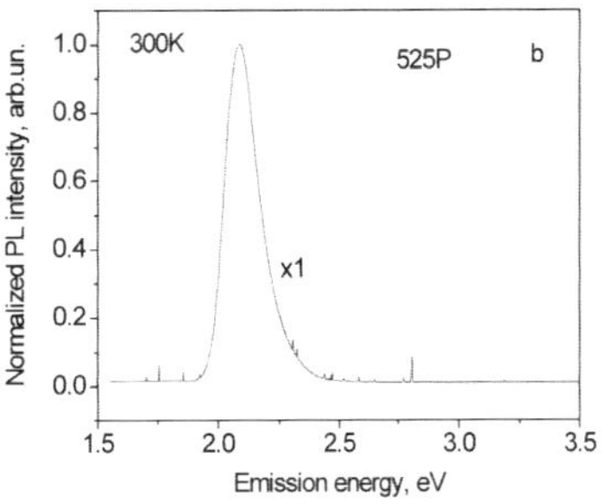

Figure 1b Normalized PL spectra of bioconjugated 525P QDs.

The deconvolution procedure has been applied to the PL spectrum permitting to represent it as a superposition of four elementary PL bands with the peaks at 2.20, 2.36, 2.73 and 3.06 eV (Fig.1a, curves 1-4). The PL band with a peak at 2.36 eV is related to the emission of ground state exciton in the CdSe core. Three other PL bands with the peaks at 2.20, 2.73 and 3.06 eV need to be discussed. High energy PL bands (2.73 and 3.06 eV) can be assigned to the recombination via: i) interface states at the CdSe/ZnS interface; ii) excited states in the CdSe core or iii) states at the ZnS/polymer interface. The PL spectrum of non-conjugated CdSe/ZnS QDs has been studied earlier at a low (10K) temperature with the aim to clarify the nature of high energy PL bands [12]. The PL band related to exciton emission in the CdSe core shifted toward high energy due to the increase of CdSe optical band gap with temperature decreasing [12]. At the same time the PL bands (2.20, 2.73 and 3.06 eV) did not change their peak positions versus temperature [12]. The last fact testifies that high energy PL bands do not connect with excited states in the CdSe core. Thus these PL bands (2.20, 2.73 and 3.06 eV) can be assigned to the interface states. This conclusion has been confirmed additionally by the same position of high energy PL bands in core/shell CdSe/ZnS QDs of different core sizes [10].

Earlier it was revealed the very fast (subpicosecond) electron and hole relaxation dynamics in CdSe QDs with the rate exceeding that due to phonon emission in the bulk semiconductors [13,14]. Taking into account very fast depopulation dynamics of electron and hole quantized states in CdSe QDs it is possible to suppose that radiative states responsible for the 2.20, 2.73 and 3.06 eV PL bands are localized at the CdSe/ZnS or at ZnS/polymer interfaces (Fig.2).

The PL spectrum of CdSe/ZnS QDs bioconjugated with the mouse ovarian cancer (OC 125) antibodies has been modified dramatically (Fig.1b). At the same time the PL intensity has not changed. The PL spectrum of bio-conjugated QDs is characterized by the one Gaussian shape PL band with a narrow full width at half maximum (90-150 meV) and the peak postion of 2.36 eV at 300K (Fig.1b) related to exciton emission in the CdSe core.

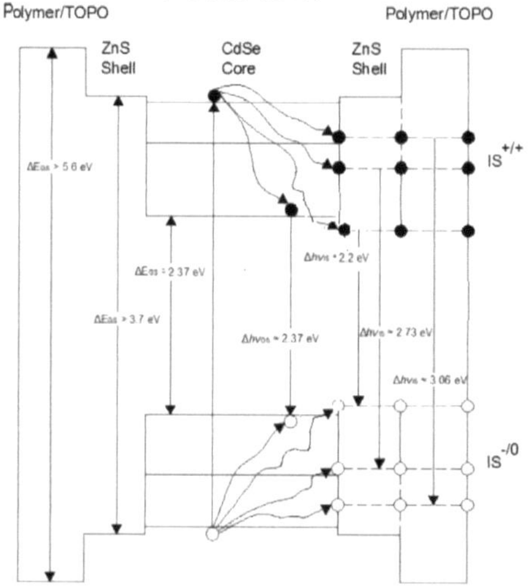

Figure 2 The energy diagram of CdSe/ZnS core/shell QDs covered by polymer. Symbols $IS^{+/+}$ and $IS^{-/0}$ present the charge of donor-like and acceptor-like interface states (IS), respectively, in nonconjugated (IS^{+} and IS^{-}) and bioconjugated (IS^{+} and IS^{0}) CdSe/ZnS QDs [11]. Dashed lines below IS symbols show the ways of carrier tunnelling from CdSe/ZnS interface states to the ZnS/polymer interface.

To explanation presented results we use the bioconjugation model proposed in [11]. Actually it is possible to suppose that the interface states (IS), responsible for the hole trapping in nonconjugated QDs, are negatively charged acceptor-like defects (IS^{-}) [15-17]. Simultaneously, the interface states, responsible for the electron trapping in non-conjugated QDs, are positively charged donor-like defects (IS^{+}) [15-17]. The negative charge of acceptor-like interface states is due to their compensation by the electrons from donor-like interface states in nonconjugated CdSe/ZnS QDs (Fig.2). The localization of defect states at the QD interface and their attractive potentials permit them to compete in the recombination process with exciton recombination inside the CdSe core (Fig.2) [11, 18-21]. To explain the PL spectrum transformation in bioconjugated QDs the interface state re-charging was proposed in [11]. It is known that some biomolecules have a positive charge on their surface (related to H^{+} ions) that can participate in electrostatic interaction during the bio-conjugation process with the negative charge on the QD surface. In this case follow the model presented in [11] in bioconjugated QDs the acceptor-like interface states are neutral (IS^{o}) and cannot compete with exciton recombination inside QDs. As a result the recombination flow via interface states decreases.

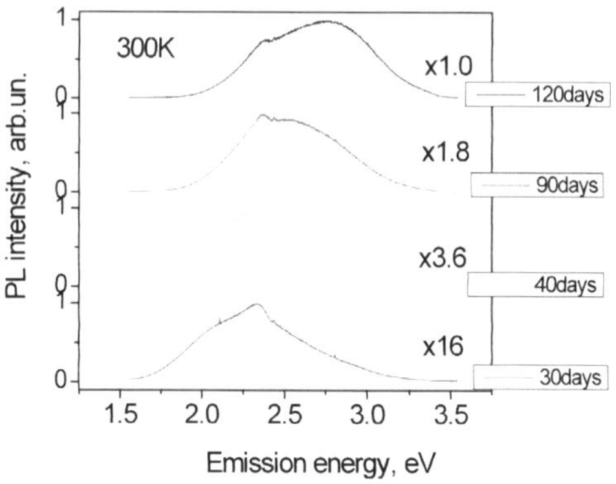

Figure 3. Normalized PL spectra of nonconjugated 525 nm QDs measured at 300 K in different moments of aging in air: 1-30 days, 2-40 days, 3-90 days, 4-120 days. Numbers at the curves indicate the multiplication coefficients (x1.0, x1.8, x3.6, x16) used at the normalization of experimental PL spectra.

Earlier we have seen such type of effects in the 605 nm CdSe/ZnS QDs bioconjugated with anti human Interleukin 10 antibodies [9,11]. The variation of PL spectra at the bioconjugation was similar, but the integrated PL intensity in bioconjugated QDs increased six- or tenfold. In the case of bioconjugation of 525 nm CdSe/ZnS QDs with the mouse ovarian cancer (OC 125) antibodies the integrated PL intensity of bioconjugated QDs did not change. The last effect can be explained by the re-absorption of light emitted by 525nm QDs in the mouse ovarian cancer (OC 125) antibodies.

The nature of interface states is not clear. It is supposed that the interface states appear at the ZnS/polymer interface in the process of polymer aging in air. To understand the interface state origin the variation of PL spectra of nonconjugated QDs at the aging in air has been studied. Fig. 3 presents the PL spectra of nonconjugated 525 nm QDs measured in the period of 30-120 days after the nonconjugated QD sample preparation (525N). As one can see two effects can be revealed from these PL spectra: i) the transformation of PL spectra and ii) PL intensity increasing. The first effect related to the intensity stimulation of 2.73 and 3.06 eV PL bands, apparently, due to appearing the corresponding interface states. It is assumed that the second effect appears owing the polymer transparency increasing for 525 nm QD emission at the aging in air of polymer covered QDs.

Acknowledgements. The work was partially supported by CONACYT Mexico (project 000000000131184), as well as by the SIP-IPN, Mexico.

[1] Kuno M, Fromm DP, Hamann HF, Gallagher A, Nesbitt DJ 2001 J. Chem. Phys. **115** 1028.
[2] Dybiec M, Chomokur G, Ostapenko S, Wolcott A, Zhang JZ, Zajac A, Phelan C, Sellers T, Gerion G 2007 Appl. Phys. Lett. **90** 263112.
[3] Mews A, Eychmuller A, Giersig M, Schoos D, Weller H 1994 J. Phys. Chem. **98** 934.
[4] Ebenstein Y, Mokari T, Banin U 2004 J. Phys. Chem. B **108** 93.
[5] Diaz Cano A, Jimenes Sandoval S, Vorobiev Y, Rodriguez Melgarejo F, Torchynska TV, 2010 Nanotechnology **21** 134016.
[6] www.invitrogen.com
[7] Torchynska TV, Douda J, Ostapenko S, Jimenez-Sandoval S, Phelan C, Zajac A, Zhukov T, Sellers T, 2008 J. of Non-Crystal. Solid. **354** 2885.
[8] Torchynska TV, Diaz Cano A, Dybic M, Ostapenko S, Morales Rodrigez M, Jimenes Sandoval S, Vorobiev Y, Phelan C, Zajac A, Zhukov T, Sellers T, 2007 phys. stat. sol. (c), **4** 241.
[9] Torchynska TV, Douda J, Calva PA, Ostapenko S and Peña Sierra R. 2009 J. Vac. Sci. &Technol. **27(2)**, 836.
[10] Torchynska TV, Douda J, and Peña Sierra R, 2009 phys. stat. sol. (c), **6** S143.
[11] Torchynska TV, 2009 Nanotechnology, **20** 095401.
[12] Vega Macotela LG, Douda J, Torchynska TV, Pena Sierra R and Shcherbyna L 2010 phys.stat.solid. (c) **7** 724.
[13] Klimov VI, McBranch DW, Leatherdale CA, Bawendi MG, 1999 Phys. Rev. B **60** 13740.
[14] Semiconductor and Metal Nanocrystals. Synthesis and Electronic and Optical Properties. Edited by Klimov VI, Marcel Dekker, Inc. NY, USA, 2004.
[15] Korsunskaya NE, Markevich IV, Torchinskaya TV and Sheinkman MK 1980 phys. stat. sol (a), **60** 565.
[16] Korsunskaya NE, Markevich IV, Torchinskaya TV and Sheinkman MK 1980 J. Physics C: Solid Sttate Physics, **13** 2975.
[17] Korsunskaya NE, Markevich IV, Torchinskaya TV and Sheinkman MK 1982 J. Phys. Chem. Solids, **43,** 475.
[18] Torchynska TV, Diaz Cano A, Morales Rodriguez M, Khomenkova L, 2003 Physica B, Conden, Matter. **340-342** 1113.
[19] Khomenkova L, Korsunska N, Sheinkman M, Stara T, Torchynska TV, Vivas Hernandez A 2005 J. Lumines. **115** 117.
[20] Torchynska T, Aguilar-Hernandez J, Diaz Cano A, Contreras-Puente G, Becerril Espinoza FG, Vorobiev Y, Goldstein Y, Shcherbyna LV 2001 Physica B: Condensed Matter, **308-310** 1108.
[21] Torchynska TV, Korsunska NE, Dzhumaev BR, Khomenkova L 2000 J. of Physics and Chemistry of Solids **61** (6) 937.

Journal of Physics: Conference Series **245** (2010) 012014 doi:10.1088/1742-6596/245/1/012014

Band structures of laterally coupled quantum dots, accounting for electromechanical effects

Sanjay Prabhakar[1], **Roderick Melnik**[1,2], **and Sunil Patil**[1]

[1]M^{2}NeT Laboratory, Wilfrid Laurier University, 75 University Avenue West, Waterloo, ON, Canada, N2L 3C5

[2]BCAM, Bizkaia Technology Park, 48160 Derio, Spain

E-mail: rmelnik@wlu.ca, http://www.m2netlab.wlu.ca

Abstract. In a series of recent papers we demonstrated that coupled electro-mechanical effects can lead to pronounced contributions in band structure calculations of low dimensional semiconductor nanostructures (LDSNs) such as quantum dots, wires, and even wells. Some such effects are essentially nonlinear. Both strain and piezoelectric effects have been used as tuning parameters for the optical response of LDSNs in photonics, band gap engineering and other applications. However, these effects have been largely neglected in literature while laterally coupled quantum dots (QDs) have been studied. The superposition of electron wave functions in these QDs become important in the design of optoelectronic devices as well in tayloring properties of QDs in other applications areas. At the same time, laterally grown QDs coupled with electric and mechanical fields are becoming increasingly important in many applications of LDSN-based systems, in particular where the tunneling of electron wave function through wetting layer (WL) becomes important and the distance between the dots is treated as a tuning parameter. Indeed, as electric and elastic effects are often significant in LDSNs, it is reasonable to expect that the separation between the QDs may also be used as a tuning parameter in the application of logic devices, for example, OR gates, AND gates and others. In this contribution, by using the fully coupled model of electroelasticity, we build on our previous results while analyzing the influence of these effects on optoelectronic properties of QDs. Results are reported for III-V type semiconductors with a major focus given to GaN/AlN based QD systems.

1. Introduction

Wide band gap semiconductor materials such as AlN/GaN QDs have attracted significant attention due to their current and potential applications in optical, optoelectronic and electronic devices used in nano- and bionano- technological applications. Strongly coupled self-assembled QDs are grown either in the same wetting layer or a vertically stack of closely spaced layers. Laterally coupled QDs in the same wetting layer are of special interest because of its potential applications in QDs lasers, light emitting diodes as qubits for quantum computation and other applications [1, 2, 3]. In this work, the coupled QDs have been considered as a wurtzite structure which can be easily grown on a sapphire substrate in c-direction (0001) by any conventional method such as MBE, MOCVD and others [4, 5]. Strain is induced due to the lattice mismatch at the interfaces between two different types of semiconductors which can be used as a tuning parameter in tailoring the electronic and optical properties of single and multiple self assembled quantum dots. Atomistic approach, pseudopotential approach and tight binding formalism have been applied to investigate the optical and electronic properties of such nano-objects [6, 7, 8].

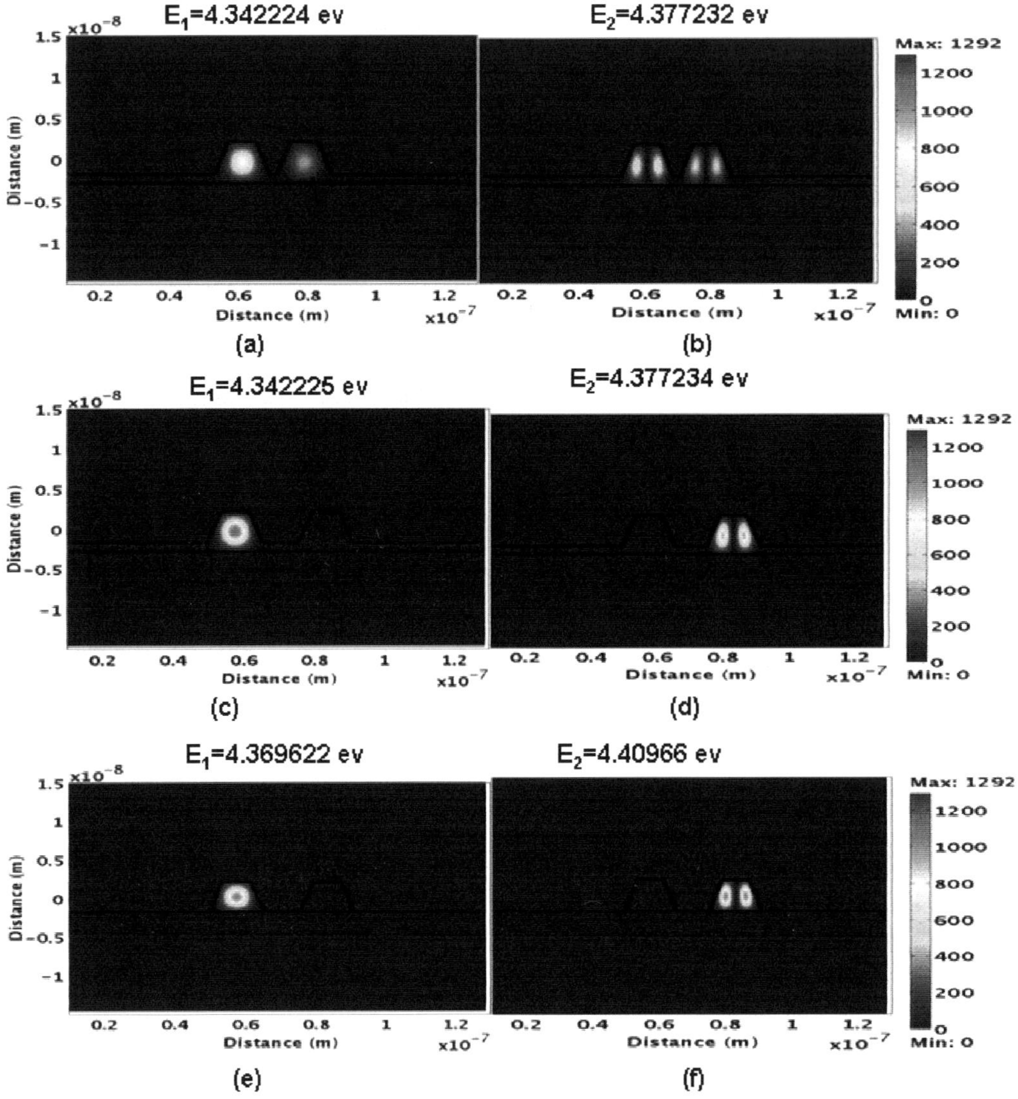

Figure 1. The influence of electromechanical effects on the band structure of two laterally coupled truncated pyramidal GaN/AlN QDs (a) ground state energy E_1 and (b) first excited state energy E_2 when the lateral QD separation is 2 nm with WL, (c) ground state energy E_1 and (d) first excited state energy E_2 when the lateral QD separation is 10 nm with WL, (e) ground state energy E_1 and (d) first excited state energy E_2 when the lateral QD separation is 10 nm without WL.

In this paper, we present a numerical analysis of the band structure of two coupled truncated GaN/AlN QDs under the influence of electroelasticity. By using the Finite Element Method (FEM), we study the effect of elasticity on the electronic properties of the two coupled truncated GaN/AlN QDs in the presence of WLs based on the 8×8 $\mathbf{k} \cdot \mathbf{p}$ Hamiltonian. Several parameters such as electromechanical fields, eigenvalues and electron wave functions in the two coupled truncated GaN/AlN QDs are compared to the previously reported results for single GaN/AlN truncated QDs [1].

2. Mathematical model for coupling classical and quantum mechanical parts

Based on boundary conditions and geometry of the system, we conclude that the electromechanical balance equations for wurtzite (WZ) structures are axisymmetric. Therefore, the original 3D problem can be reduced to a 2D problem. The coupled equations of WZ structure in the presence of electromechanical effects in cylindrical coordinates can be written as [10]

$$\frac{\partial \sigma_{rr}}{\partial r} + \frac{\partial \sigma_{rz}}{\partial z} + \frac{\sigma_{rr} - \sigma_{\theta\theta}}{r} = 0, \quad \frac{\partial \sigma_{rz}}{\partial r} + \frac{\partial \sigma_{zz}}{\partial z} + \frac{1}{r}\sigma_{rz} = 0, \quad \frac{\partial D_r}{\partial r} + \frac{\partial D_z}{\partial z} + \frac{1}{r} D_r = 0. \tag{1}$$

The stress tensor components and the electric displacement vector components can be written as

$$\sigma_{rr} = C_{11}\varepsilon_{rr} + C_{12}\varepsilon_{\theta\theta} + C_{13}\varepsilon_{zz} - e_{31}E_z, \; \sigma_{rz} = C_{44}\varepsilon_{rz} - e_{15}E_r, \tag{2}$$

$$\sigma_{zz} = C_{13}\varepsilon_{rr} + C_{13}\varepsilon_{\theta\theta} + C_{33}\varepsilon_{zz} - e_{33}E_z, \; D_r = e_{15}\varepsilon_{rz} + \epsilon_1 E_r. \tag{3}$$

The strain tensor components due to lattice mismatch in a semiconductor can be written as [10] $\varepsilon_{rr} = \frac{\partial u_r}{\partial r} - \varepsilon_a^\star$, $\varepsilon_{zz} = \frac{\partial u_z}{\partial z} - \varepsilon_c^\star$, $\varepsilon_{\theta\theta} = \frac{u_r}{r} - \varepsilon_a^\star$, $\varepsilon_{rz} = \frac{1}{2}\left(\frac{\partial u_r}{\partial z} + \frac{\partial u_z}{\partial r}\right)$. Here $\varepsilon_a^\star = \frac{a_m - a_{QD}}{c_{QD}}$ and $\varepsilon_c^\star = \frac{a_m - a_{QD}}{c_{QD}}$ are the local intrinsic strains along a and c directions, respectively. Also, a_m, c_m and a_{QD}, c_{QD} are the lattice constants of the matrix and the QD, respectively.

We analyze the band structures of QDs with an 8 band $\mathbf{k}\cdot\mathbf{p}$ model. Here we consider two conduction bands coupled with six valence bands including heavy hole, light hole and spin orbit bands [9, 11]. The 8×8 Hamiltonian can be written as

$$H = \begin{pmatrix} H_e^k + H_e^\varepsilon + E_c & & & 0 \\ & H_h^k + H_h^\varepsilon + E_v & & \\ & & H_e^k + H_e^\varepsilon + E_c & \\ 0 & & & H_h^k + H_h^\varepsilon + E_v \end{pmatrix}, \tag{4}$$

where H_e^k and H_e^ε are the kinetic and strain dependent parts of the electron Hamiltonian, respectively, whereas H_h^k and H_h^ε are the 3×3 matrix of the kinetic and strain dependent parts of the hole Hamiltonian respectively. E_c and E_v are the energies of unstrained conduction and valence band edges respectively. The physical parameters were chosen as in [10].

3. Results and Discussions

We have observed and quantified the electron wave functions overlap in the laterally grown coupled pyramidal shape quantum dots. The wave functions for the ground state and the first excited state for two laterally coupled truncated pyramidal shape of GaN/AlN QDs with and without WL have been calculated by applying the model described in section 2 (see [12] and [13] for more details on the influence of the WL. Here we include the coupled equations of electroelasticity (see (1)) into the 8-band **k.p** model to account for the influence in electron wave functions with and without wetting layers. These results are presented in Figure 1. Here we observe the ground state energy E_1 at 4.342224 eV in Fig. 1(a)and first excited state at 4.377232 eV in Fig. 1(b) when the lateral separation between the dots are 2 nm and $E_1 = 4.342225$ eV in Fig. 1(c), $E_2 = 4.377234$ eV in Fig. 1(d) when the lateral separation between the dots are 10 nm with WL. These reported values for ground and first excited states are consistent with the results published in [1]. However, here we consider the distance between the two coupled lateral QDs as a tuning parameter.

We observe that electron wave functions are localized in both QDs when the separation between them is approximately 4 nm and are restricted to a single quantum dot when the separation between the dots exceeds about 5 nm in presence of wetting layers. The electron wave functions are localized in one QD in the absence of wetting layers for any distance of lateral separation of the QDs. We compared the magnitude of the electron wave function to the previously reported results of [1] single QDs of identical geometry in both cases and found that the magnitude is approximately one order higher which is likely due to the electron wave functions overlap between the two QDs.

In addition, we also carried out a series of comparison studies of ground and first excited states eigenvalues with and without WL as shown in Figure 1. We found that the localization of electron wave functions is pushed down by approximately 3 meV.

4. Conclusions

Combined contributions of electromechanical effects in the band structure of two laterally coupled GaN/AlN QDs have been analyzed for the first time with and without wetting layer. The magnitude of the electron wave functions in ground and first excited states for two laterally coupled QDs is approximately one order higher than that of a single QD of identical geometry in both cases, pointing out to the evidence of electron wave functions overlap. Localization of electron wave functions changes significantly from one QD to the other QD or it spreads out into both QDs with the variation of lateral distance between the two QDs. Thus in addition to the electromechanical tuning, the distance between the QDs can also provide an additional tuning parameter in the design of QDs-based systems.

5. Acknowledgments

This work was supported by NSERC and CRC program.

References

[1] Patil, S. R., Melnik, R. V. N., 2009, Coupled electromechanical effects in II-VI group finite length semiconductor nanowires, Journal of Physics D - Applied Physics, 42 (14), Art. No. 145113.

[2] Wang, J., Mao, H., Yu, J., Zhao, Q., Zhang, H., Yang, P., Zhu, Z., and Chu, J., 2010, Electrically tunable electron g factors in coupled InAs/GaAs pyramid quantum dots, Appl. Phys. Lett. 96, 062108.

[3] Gong, W., Zheng, Y., Liu, Y., and Lu, T., 2006, Well-defined insulating band for electronic transport through a laterally coupled double-quantum-dot chain: Nonequilibrium Greens function calculations, Phys. Rev. B 73, 245329.

[4] Nakamura, S., Mukai, T., and Senoh, M., 1994, Candela-class high-brightness InGaN/AlGaN double-heterostructure blue-light-emitting diodes, Appl. Phys. Lett. 64, 1687.

[5] Yang, X. H., Schmidt, T. J., Shan, W., Song, J. J., and Goldenberg, B., 1995, Above room temperature near ultraviolet lasing from an optically pumped GaN film grown on sapphire, Appl. Phys. Lett. 66, 1.

[6] Jaskolski, W., Zielinski, M., Bryant, G. W., Aizpurua, J., 2006, Strain effects on the electronic structure of strongly coupled self-assembled InAs/GaAs quantum dots: Tight-binding approach, Phys. Rev. B 74, 195339

[7] Santoprete, R., Koiller, B., Capaz, R. B., Kratzer, P., Liu, Q. K. K., and M. Scheffler, 2003, Tight-binding study of the influence of the strain on the electronic properties of InAs/GaAs quantum dots, Phys. Rev. B 68, 235311.

[8] Bester, G., Shumway, J., Zunger, A., 2004, Theory of Excitonic Spectra and Entanglement Engineering in Dot Molecules, Phys. Rev. Lett. 93, 047401.

[9] Melnik, R. and Mahapatra, R., 2007, Coupled effects in quantum dot nanostructures with nonlinear strain and bridging modelling scales, Computers and Structures, 85 (11-14), 698-711.

[10] Patil, S. R. and Melnik, R. V. N., 2009, Thermoelectromechanical effects in quantum dots, Nanotechnology, 20 (12), Art. No. 125402.

[11] Chuang, S. L., and Chang, C. S., 1996, $\mathbf{k} \cdot \mathbf{p}$ method for strained wurtzite semiconductors, Phys. Rev. B 54, 2492.

[12] Melnik, R. V. N., and Willatzen, M., 2004, Bandstructures of conical quantum dots with wetting layers Nanotechnology 15, 1-8.

[13] Melnik, R. V. N., and Zotsenko, K. N., 2004, Modelling Simul. Mater. Sci. Eng. 12, 465-77.

Magnetothermoelectric Effects of 2D Electron Gas in Quantum Well with Parabolic Confinement Potential in-plane Magnetic Field

F M Hashimzade, M M Babayev, B H Mehdiyev and Kh A Hasanov

Institute of Physics, National Academy of Sciences of Azerbaijan, Baku, Azerbaijan

E-mail: firudin_hashimzade@yahoo.com

Abstract. We developed a quantitative theory of the phonon-drag thermopower and the Nernst-Ettingshausen effects for the 2D electron gas in a quantum well with parabolic confinement potential in the magnetic field in the plane of the free movement of electrons. Our theoretical results are in a good agreement with experimental data [1] in the range 1-10K. In this temperature interval the main contribution to the thermopower comes from phonon-drag. Our calculations show that the temperature dependence of both the longitudinal and transverse Nernst-Ettingshausen effects is non-monotonic. We find that phonon-drag Nernst-Ettingshausen effects change sign with the temperature.

1. Introduction

The thermomagnetic effects in quantum wells have been studied theoretically and experimentally in [1-4]. Here we have study the phonon-drag thermopower and the Nernst-Ettingshausen effects in a quantum well with parabolic confinement potential. The magnetic field is directed across the confinement direction, i.e. along the plane of a two-dimensional electron gas. Thus, two cases for the relative arrangement of the current direction and the confinement direction are possible. In the case where the current is directed along the plane of the two-dimensional electron gas it is sufficient to consider the relaxation time approximation and to use the Boltzmann equation. In the case when the current is along the direction of confinement it is necessary to use the density matrix approach [5] for calculation of the diagonal components of the conductivity tensor.

2. Theory

We consider a simple model for the quantum well, in which a two-dimensional electron gas is confined in the x-direction and a homogenous static magnetic field $\boldsymbol{B}$, with the vector-potential $\boldsymbol{A}(0,xB,0)$ in the Landau gauge, is parallel to the z-axis,. The parabolic confining potential can be written as $U_x = m\omega_0^2 x^2/2$, where m is the effective mass of a conduction electron, and ω_0 is the parameter of the parabolic potential. For this model, the one-particle Hamiltonian, its eigenvalues and eigenfunctions are given in [6].

Using the concept of generalized force given in [5] ,

$$\vec{\Phi} = -e\vec{E} - \frac{\varepsilon - \varsigma}{T}\nabla T - A_{ph} k_B \nabla T\,, \tag{1}$$

where the last term is related to the phonon-drag effect, we can obtain the non-diagonal, β_{xy}^{ph} , and diagonal, β_{yy}^{ph}, components of tensor β:

$$\beta_{xy}^{ph} = \frac{k_0 e\,\omega_c}{m\,\omega^2}\sum_{\alpha} f_0 A_{ph}\,, \qquad \beta_{yy}^{ph} = -e k_0 \sum_{\alpha} \tau_e \upsilon_y^2 \left(-\frac{\partial f_0}{\partial \varepsilon}\right) A_{ph}\,. \tag{2}$$

Here ε, $\vec{\upsilon}$, ζ, τ_e and $\omega_c = eB/mc$ are the energy, velocity, chemical potential, relaxation time and cyclotron frequency of electrons, respectively, $\omega = \left(\omega_0^2 + \omega_c^2\right)^{1/2}$, f_0 is the equilibrium electron distribution function, $\alpha = \left(N, k_y, k_z\right)$ is a set of quantum numbers that determine electronic states in a magnetic field, N is the oscillation quantum number, $\vec{k}\left(k_y, k_z\right)$ is the wave vector of electrons,

$$A_{ph} = -\frac{s^2}{k_0}\sum_{\vec{q}} 2\hbar\left(w_{da}(q) + w_{pa}(q)\right)\exp\left(-\frac{R^2 q_x^2}{2}\right)\tau_{ph}(q)\frac{(\vec{q}\nabla T)}{(\vec{\upsilon}\nabla T)}\frac{dN_q^0}{dT}\delta\left(\varepsilon_{\vec{k}+\vec{q}_{\|}} - \varepsilon_{\vec{k}}\right), \quad (3)$$

$$w_{da}(q) = \frac{\pi E_1^2 q^2}{\varepsilon\left(q_{\|}\right)^2 \rho\omega_q}, \qquad w_{pa}(q) = \frac{\pi e^2 \beta^2}{\varepsilon\left(q_{\|}\right)^2 \rho\omega_q}, \quad (4)$$

$\vec{q}$ and $\tau_{ph}(q)$ are the phonon wave vector and the relaxation time of phonons, respectively, N_q^0 is the occupation number (the Planck function) for phonons with frequency $\omega_q = sq$, s is the sound velocity, ρ is the density of the material, and $q_{\|}$ is the component of wave vector of phonon in the plane of the free movement of electrons. In (3) both the piezoelectric (pa) and the deformation (da) interaction of electrons with acoustic phonons are taken into account. Here E_1 is the deformation potential coefficient, $\beta = 0.89 e_{14}/\chi$, e_{14} is the piezoelectric constant, χ is the static dielectric constant, and $\varepsilon(q)$ is dielectric function of the electron gas. For the 2D electron gas we have

$$\varepsilon\left(q_{\|}\right) = 1 + \frac{2me^2}{\sqrt{\pi}\chi\hbar^2 q_{\|}} e^{\frac{1}{2}R^2 q_{\|}^2} erfc\left(\frac{Rq_{\|}}{\sqrt{2}}\right), \quad (5)$$

where $R = (\hbar / m\omega)^{1/2}$ is the magnetic length.

The scattering mechanisms of electrons explicitly considered in the present paper are the acoustic phonon scattering via deformation and piezoelectric couplings, and the impurity scattering arising from ionized impurities in the quantum well.

For the different mechanisms of relaxation of phonons we have

$$A_{ph}(y) = \frac{(e\beta)^2 s^2 m^2}{\pi^2 \rho\hbar (k_0 T)^2 R}\tau_{ph,\ell}\int_0^1 F_\ell(yx)\left(1 + \frac{me^2 R e^{2y^2x^2} erfc\left(\sqrt{2}\,yx\right)}{\sqrt{\pi}\chi\hbar^2 yx}\right)^{-2}\frac{x^2}{\sqrt{1-x^2}}dx, \quad (6)$$

where $y = Rk$,

$$F_\ell(yx) = \int_{-\infty}^{\infty} q^\ell\left(Sinh\frac{\hbar sq}{2k_0T}\right)^{-2}\left(1 + \left(\frac{E_1 q}{e\beta R}\right)^2\right)e^{-\frac{t^2}{2}}dt, \qquad q = \left(4R^2k^2x^2 + t^2\right)^{\frac{1}{2}}. \quad (7)$$

For Herring's ($\ell = 2$), Simons's ($\ell = 1$) and boundary scattering ($\ell = 0$) mechanisms, respectively,

$$\tau_{ph,2} = \frac{\rho \hbar^2 s^3 R^2}{(k_0 T)^3}\,;\ \tau_{ph,1} = \frac{\rho \hbar^3 s^4 R}{(k_0 T)^4}\,; \qquad \tau_{ph,0} = \frac{L}{s}, \tag{8}$$

where L is the phonon mean free path due to boundary scattering. For the calculation of thermopower, α_{yy}, and the Nernst-Ettingshausen coefficient, Q_{xy}, it is necessary to calculate both the diagonal and the non-diagonal components of σ_{ik} and β_{ik} [5]. β_{ik} consists of two components, the diffusion and the phonon-drag: $\beta_{ik} = \beta_{ik}^{dif} + \beta_{ik}^{ph}$.

For the degenerated electron gas we obtain

$$\sigma_{yx} = -\sigma_{xy} = \frac{\omega_c e^2 n}{m\omega^2},\ \sigma_{yy} = \frac{e^2 k_F^2 \tau_F}{2\pi m},\ \sigma_{xx} = \frac{2\pi e^2 n^2 \omega_c^2}{m\omega^4 k_F^2 \tau_F},\ \sigma_{xx}\sigma_{yy} = \sigma_{yx}^2, \tag{9}$$

$$\beta_{yy}^{ph} = -\frac{k_0}{e}\sigma_{yy} A_{ph}(y_F),\ y_F = R k_F;\ \beta_{xy}^{ph} = -\frac{k_0}{e}\sigma_{xy}\bar{A}_{ph},\ \bar{A}_{ph} = \frac{1}{\pi n R^2}\int_0^\infty A_{ph}(y) f_0(y)\, y dy, \tag{10}$$

where k_F and $n = k_F^2/2\pi$ are the Fermi's wave number and the real density of electrons.

The diffusion component of thermopower and the Nernst-Ettingshausen effects are given in [6], so here we consider only the phonon component. In the magnetic fields satisfying conditions $\omega_c << \omega_0$, $\omega_c \tau_F >> 1$ (here τ_F is the relaxation time of electrons at the Fermi surface), for the phonon component of the change in thermoelectric power in a magnetic field, $\Delta\alpha_{yy}^{ph}(B) = |\alpha_{yy}^{ph}(B)| - |\alpha_{yy}^{ph}(0)|$, and for the Nernst-Ettingshausen coefficient we find

$$\alpha_{yy}^{ph}(0) = -\frac{k_0}{e} A_{ph}(y_F),\ \Delta\alpha_{yy}^{ph}(B) = -\frac{k_0}{2e}\left[A_{ph}(y_F) - \bar{A}_{ph}\right], \tag{11}$$

$$Q_{xy}^{ph} = -\frac{k_0}{2eB}\left[A_{ph}(y_F) - \bar{A}_{ph}\right]\left(\frac{\omega_0}{\omega_c}\right)^2 (\omega_c \tau_F),\ \bar{A}_{ph} = \frac{1}{\pi n R^2}\int_0^\infty A_{ph}(y) f_0(y)\, y dy. \tag{12}$$

3. Results and discussion

In this section we apply our theoretical results to the 2D degenerate electron gas in $GaAs/Al_xGa_{1-x}As$ quantum wells. Numerical calculations for the thermopower have been performed for sample 1 in [1], with the electron density $n=1.78\cdot10^{15} m^{-2}$, the mobility $\mu=22.7\ m^2/Vs$, the mean free path of phonons $L=3\cdot10^{-4}\ m$, the width of the quantum well $L_x = 10^{-8} m$ [1], the mass density $\rho=5.3\cdot10^3\ kg/m^3$, the longitudinal sound velocity $s=5\cdot10^3\ m/s$, the effective mass $m = 0.067 m_0$ (m_0 is the free electron mass), the acoustic deformation potential $E_1=7.4\ eV$, and the piezoelectric constant $e_{14}=0.16\ C/m^2$ [7].

Our analysis shows that the situations considered in [1-3] satisfied the conditions of the quantum limit and the strong degeneracy of electron gas. Estimation shows that the dominant mechanism of scattering of the electrons is the scattering by ionized impurities, and for the phonons the dominant mechanism is the surface scattering.

We estimated parameter $\omega_0=7\cdot10^{13}\ s^{-1}$ of the parabolic potential using condition $R\sim L_x/2$. Our theoretical results for the variation of zero-field thermopower α_{yy} as a function of temperature T are shown in Fig. 1. The results are in a good agreement with experimental data in the range 1-$10\ K$. In this temperature interval the diffusion thermopower is by 3 to 8 times smaller than the phonon-drag thermopower. Thus, for example, at *T=5K*, $\alpha_{yy}^{ph} = 215\mu V/K$ and $\alpha_{yy}^{dif} = 36\mu V/K$. At temperatures below *1.5K* the main contribution to the thermopower comes from the piezoelectric interaction, and from the deformation interaction with acoustic phonons. The calculations indicate the importance of screening, so the value of the thermoelectric power in the range *1-10K*, without taking into account the screening, is 1.5-2 times higher.

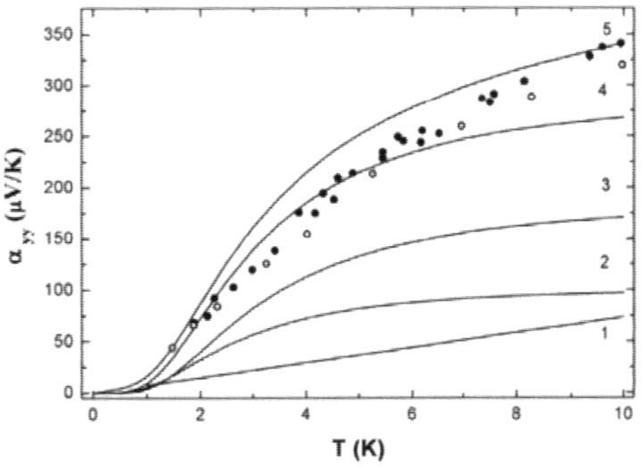

Figure 1. Variation of zero-field thermopower α_{yy} as a function of temperature *T*: the diffusion component (1), the phonon-drag thermopower (2: due to pa-coupling, 3: due to da-coupling, 4: the total phonon-drag thermopower), and the total thermopower (5). Open and solid points are the experimental data[1].

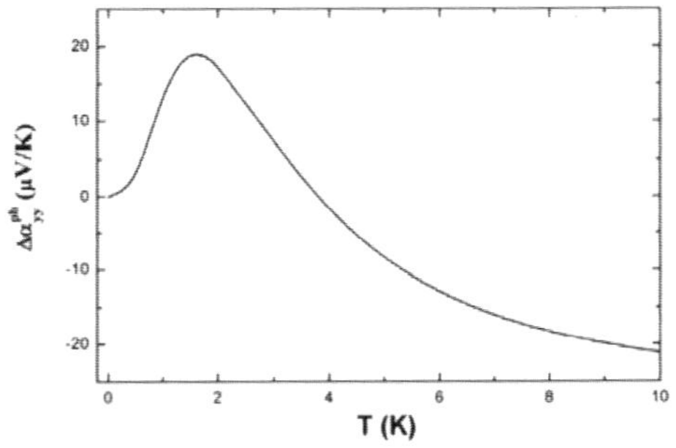

Figure 2. Phonon-drag magneto-thermopower $\Delta\alpha_{yy}^{ph}(B)$ as a function of temperature *T*.

Our calculations show that the temperature dependence of both the longitudinal and transverse Nernst-Ettingshausen effects is nonmonotonic (Fig. 2). Both effects change sign with temperature. Note that usually the change of sign of the Nernst-Ettingshausen effects with temperature arises due to the change in the mechanism of scattering, but in this case the mechanism of scattering remains unchanged.

It is also interesting to note that the transverse thermopower $\alpha_{xy}^{ph} = BQ_{xy}^{ph}$ exceeds the zero-field thermopower α_{yy}^{ph} by one order of magnitude. Thus, for example, at *T=5K*, *B=1T*, and $\omega_0 = 10^{13} sec^{-1}$, for the transverse thermopower we have $\alpha_{xy}^{ph} \approx 3mV/K$.

References

[1] Fletcher R, Maan J C, Weimann G 1985 *Phys. Rev.* B **32** 8477
[2] Fletcher R, Maan J C, Ploog K, Weimann G 1986 *Phys. Rev.* B **33** 7122
[3] Fletcher R, Harris J J, Foxon C T, Tsaousidou M and Butcher P N 1994 *Phys.Rev.* B **50** 14991
[4] Lyo S K 1988 *Phys. Rev.* B **38** 6345
[5] Askerov B M 1994 *Electron Transport Phenomena in Semiconductors* (Singapore: World Scientific)
[6] Hashimzade F M , Hasanov Kh A, Mehdiyev B H and Cakmak S 2010 *Phys. Scr.* **81**, 015701
[7] Gantmacher V F, Levinson I B 1987 *Carrier Scattering in Metals and Semiconductors* (Amsterdam, New-York: Elsevier)

Electron Spin Resonance in Triple Quantum Dots

Maria Busl[1], **Rafael Sanchez**[2], **Gloria Platero**[1]
[1] Instituto de Ciencia de Materiales Madrid, CSIC, Cantoblanco, Spain
[2] Universite de Geneve, Switzerland
E-mail: maria.busl@icmm.csic.es

Abstract. Electron spin resonance in triple quantum dots in series is analyzed by means of the density matrix formalism. We show that an ac magnetic field is able both to remove and to induce spin blockade.

1. Introduction

In solid state physics on the nanoscale one of the main protagonists is the electron spin. The two-level system formed by electron spin-up and down is a good candidate for a qubit and can easily be realized in quantum dots[1]. How to manipulate electron spins in quantum dots is one of the aims of recent experiments: Crossed DC and AC magnetic fields and also AC electric fields in combination with spin-orbit or hyperfine interaction are used to excite coherent spin rotations of one single electron in single and double quantum dots (DQD)[2, 3]. Electron spin resonance (ESR) has also been investigated quite recently theoretically for a DQD[4, 5].

Lately, a next step towards quantum dot arrays has been achieved: triple quantum dots (TQD), both in series[6] and triangular configurations[7]. Theoretical works in these systems concerned the analysis of eigenstates and stability diagrams, the effect of a magnetic field piercing the TQD area[8] and also strongly correlated electrons in the Kondo regime[9]. TQDs have also been proposed as spin entanglers[10] and as three-level systems where coherent electron trapping occurs[11, 12]. In the present paper, we analyze transport through a TQD driven by a time-dependent magnetic field $B_{\rm AC}$ in a serial configuration, coupled to two electric baths, cf. Fig. 1. We discuss the electron spin dynamics and the electronic current for the case of two extra electrons in the TQD. We show that in the spin blockade regime, an AC magnetic field is able to break spin blockade if its frequency matches the Zeeman splitting in the sample. However, modifying the frequency, we found an antiresonance in the current, in which the two electrons form spin-blockaded entangled states.

2. Model

We consider a system of three weakly coupled quantum dots, where the dots are coupled through tunnel barriers, and dots 1 and 3 are connected to source and drain contacts respectively, shown schematically in Fig. 1. The Hamiltonian of the system is given by $\hat{H}(t) = \hat{H}_{sys} + \hat{H}_{QD-leads} + \hat{H}_{leads}$. The first term, $\hat{H}_{sys} = \hat{H}_{QD} + \hat{H}_{tun} + \hat{H}_B$, includes the on-site energies of the uncoupled TQD, $\hat{H}_{QD} = \sum_{i\sigma} \epsilon_i \hat{c}^\dagger_{i\sigma}\hat{c}_{i\sigma}$, the coherent tunneling between the dots, $\hat{H}_{tun} = -\sum_{ij\sigma}(t_{ij}\hat{c}^\dagger_{i\sigma}\hat{c}_{j\sigma} + h.c.)$ and the magnetic field term (see below). The second term

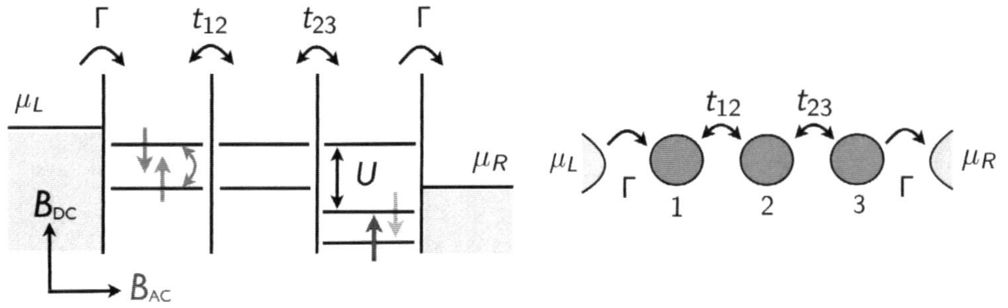

Figure 1. Schematic diagram of a TQD in the presence of crossed $B_{\rm DC}$ and $B_{\rm AC}$. The dots are coupled coherently to each other by tunneling amplitudes $t_{12,23}$ and incoherently to leads by rates $\Gamma_{L,R}$.

describes the coupling between the TQD and the leads, $\hat{H}_{QD-leads} = \sum_{l\in L,Rk\sigma}(\gamma_l \hat{d}^{\dagger}_{lk\sigma}\hat{c}_{l\sigma} + h.c.)$, and the last one describes the leads, $\hat{H}_{leads} = \sum_{lk\sigma}\epsilon_{lk}\hat{d}^{\dagger}_{lk\sigma}\hat{d}_{lk\sigma}$.

We introduce now a magnetic field with two components: a DC one along the z-axis that induces a Zeeman-splitting $\Delta_i = g_i B_{zi}$, and a circularly polarized AC component in the xy-plane that rotates the S_z-component when its frequency fulfills the ESR condition $\hbar\omega_i = \Delta_i$. The time-dependent Hamiltonian of the magnetic field is then

$$\hat{H}_B(t) = \sum_{i=1}^{3}[\Delta_i S_{zi} + B_{\rm AC}(\cos(\omega t)S_{xi} + \sin(\omega t)S_{yi})], \tag{1}$$

with the spin operators $S_i = \frac{1}{2}\sum_{\sigma\sigma'} c^{\dagger}_{i\sigma}\sigma_{\sigma\sigma'}c_{i\sigma'}$.

The dynamics of the system within the Born-Markov-approximation is given by the equation of motion of the reduced density matrix elements:

$$\begin{aligned}\dot{\rho}_{ln}(t) = & -i \ \ \langle l|[\hat{H}_{TQD} + \hat{H}_t + \hat{H}_B(t), \rho]|n\rangle \\ & + \ \sum_{k\neq n}(\Gamma_{nk}\rho_{kk} - \Gamma_{kn}\rho_{nn})\delta_{ln} - \Lambda_{ln}\rho_{ln}(1-\delta_{ln}).\end{aligned} \tag{2}$$

The commutator accounts for the coherent dynamics in the TQD, Γ_{ln} are the transition rates from state $|n\rangle$ to state $|l\rangle$ induced by the coupling to the leads - being $\Gamma_i = 2\pi|\gamma_i|^2$ when they occur through lead $i \in L, R$. Decoherence is considered by the terms $\Lambda_{ln} = \frac{1}{2}\sum_k(\Gamma_{kl} + \Gamma_{kn})$.

We consider a configuration where the dot coupled to the drain is permanently occupied by one electron (see Fig. 1), and only up to two electrons can be in the system. Double occupancy is only allowed in the drain dot. This is the case when the chemical potentials in the leads satisfy $\epsilon_3 + V < \mu_R < \epsilon_3 + U_3$ and $\mu_L < \epsilon_1 + 2V$. For resonant tunneling, $\epsilon_1 = \epsilon_2$ and $\epsilon_{1,2} + V = \epsilon_3 + U_3$. The system is described by the full two-electron TQD basis, out of which the relevant states for the dynamics are $|0,\sigma,\sigma\rangle$, $|0,\sigma,\sigma'\rangle$, $|\sigma,0,\sigma\rangle$,$|\sigma,0,\sigma'\rangle$ and $|0,0,\uparrow\downarrow\rangle$, where $\sigma = \{\uparrow,\downarrow\}$ and $\sigma' = \{\uparrow,\downarrow\}$. We consider a bias high enough to have unidirectional transport (from left to right), so the current is given by $I = \sum_\sigma \Gamma_R \rho_{|0,0,\uparrow\downarrow\rangle}$.

3. Results

3.1. $B_{AC} = 0$

The DC magnetic field produces a Zeeman splitting in the three dots, which we consider to be *inhomogeneous*, here especially we treat the case $\Delta_1 = \Delta_2 \neq \Delta_3$. In general, different Zeeman

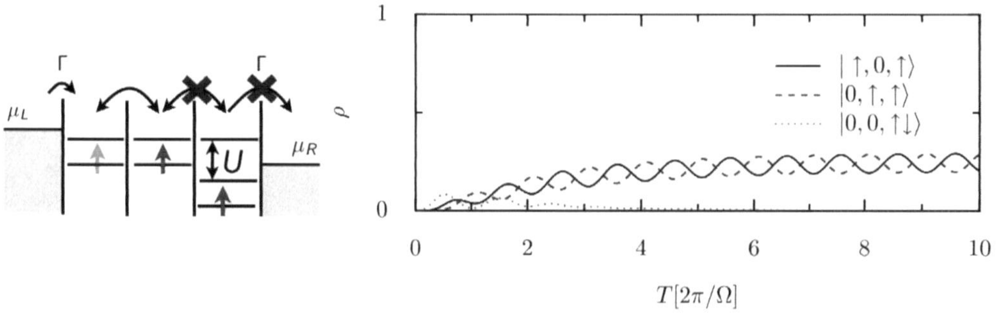

Figure 2. $\rho(t)$ for $B_{\mathrm{AC}} = 0$. Without magnetic field B_{AC} the two electrons end up in a triplet state (↑ or ↓, here only shown for ↑) and current ($\propto |0, 0, \uparrow\downarrow\rangle$) is blocked due to spin blockade, see red dotted line. Parameters: $\Gamma = 0.01$, $t = 0.005$, units in meV.

splittings occur due to the presence of hyperfine interaction, but can also be achieved by g-factor engineering[13]. The difference leads to a mixing of the two-electron triplet and singlet subspace. Without AC field applied to the system however, there is no *continuous* mixing between different spin channels: If the entering electron is in the same spin state as the confined electron in the third dot, Pauli principle forbids tunneling to the third dot. The occupation of the current blocking states increases and no current will flow to the drain, see Fig. 2.

3.2. $B_{AC} \neq 0$

Let us now apply a B_{AC}. The time-dependent Hamiltonian (1) for a circularly polarized magnetic field can be transformed into the rotating frame by means of $\hat{U}(t) = \exp\{-i\omega t \sum_{i=1}^{3} \hat{S}_{zi}\}$ and one ends up with a time-independent Hamiltonian[5, 12]:

$$\hat{H}'_{\mathrm{B}} = \sum_{i=1}^{3} [(\Delta_i - \hbar\omega)\hat{S}_{zi} + B_{\mathrm{ac}}\hat{S}_{xi}] \tag{3}$$

As a first case, we set the AC frequency to be resonant with the Zeeman splitting induced by B_{DC}, i.e. $\hbar\omega = \Delta_{1,2}$.

This induces the rotation of the electron spin. Due to the mixing of singlet and triplet states, the AC field is able to break the spin blockade by inducing transitions like $|0, \uparrow, \uparrow\rangle \Rightarrow |0, \downarrow, \uparrow\rangle \Rightarrow |0, 0, \uparrow\downarrow\rangle$, such that a finite current flows. For resonant transport, the current is shown as a function of B_{AC} and for different values of the Zeeman splitting difference $\delta = \Delta_{1,2} - \Delta_3$, see Fig. 3. As it is expected, the smaller δ, the smaller the current, since then the electrons stay mainly in the triplet subspace. The maximum current is shifted to higher intensities B_{AC}, the bigger δ.

If the resonance conditions $\omega = \Delta_{1,2}$ or $\omega = \Delta_3$ are not fulfilled though, the behavior of the current changes. In Fig. 4 we plot the current versus the AC frequency: For $\omega = \Delta_{1,2}$ and $\omega = \Delta_2$, the current reaches a maximum, whereas at $\omega_0 = \frac{\Delta_{1,2}+\Delta_3}{2}$, current is completely blocked. This can be explained by having a look at the eigenstates of (3). The Hamiltonian (3) can be diagonalized at ω_0, and two of its eigenstates are:

$$|\Psi_1\rangle = \frac{1}{2}(|\uparrow, 0, \uparrow\rangle - |\downarrow, 0, \downarrow\rangle - |0, \uparrow, \uparrow\rangle + |0, \downarrow, \downarrow\rangle) \tag{4}$$

$$|\Psi_2\rangle = \frac{1}{2}(-|\uparrow, 0, \uparrow\rangle + |\downarrow, 0, \downarrow\rangle - |0, \uparrow, \uparrow\rangle + |0, \downarrow, \downarrow\rangle) \tag{5}$$

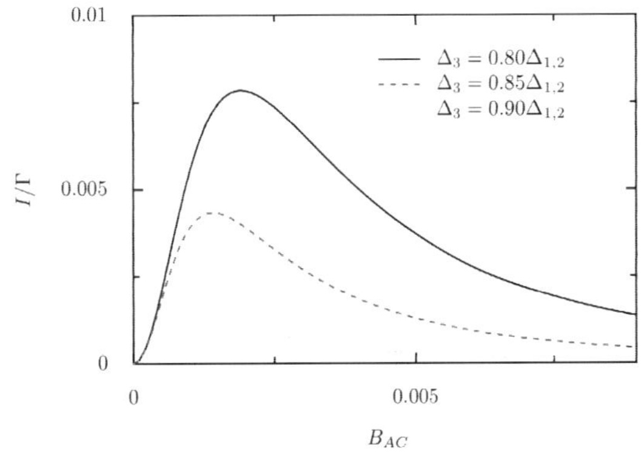

Figure 3. I-$B_{\rm AC}$ in a serial TQD exposed to crossed DC and AC magnetic fields, for $\omega = \Delta_{1,2} \neq \Delta_3$. For large $B_{\rm AC}$ current eventually vanishes due to a competition between tunneling oscillations and spin rotations: the larger $B_{\rm AC}$, the longer the electrons remain performing spin rotations and thus do not tunnel to the subsequent dot. The smaller the difference $\Delta_{1,2} - \Delta_3$, the smaller the current, due to less mixing of singlet and triplet subspace. Parameters: $\Gamma = 0.001$, $t = 0.01$, $\Delta_{1,2} = 0.013$, units in meV.

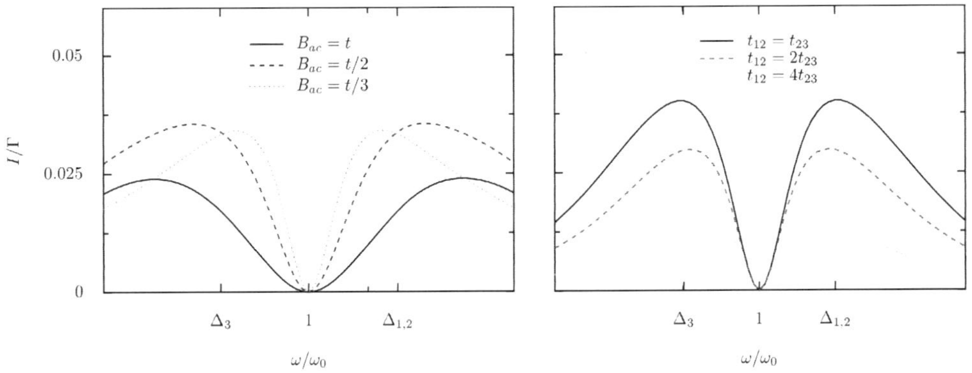

Figure 4. I-ω in a serial TQD exposed to crossed DC and AC magnetic fields. The current shows an antiresonant behavior once the AC frequency approaches $\hbar\omega = \frac{\Delta_1+\Delta_3}{2}$, independently of the system parameters t_{ij} and $B_{\rm AC}$. *Left*: I-ω for $t_{12} = t_{23} = t$ and different intensities $B_{\rm AC}$ of the AC magnetic field, *Right*: I-ω for different tunneling amplitudes t_{12}, t_{23}. Parameters: $\Gamma = t = 0.01$, units in meV.

These states only consist of *parallel* spin states, i.e. they are not coupled to the transport state $|0, 0, \uparrow\downarrow\rangle$. Thus, at ω_0, the magnetic field *forces the system back into spin blockade.* This antiresonant behavior is a general property of the AC magnetic field, as it occurs also in triangular TQDs and DQDs[12]. The width of the antiresonance depends mainly $B_{\rm AC}$, as one can appreciate in Fig. 4.

4. Conclusion

In summary, we present results for single electron transport through a specific configuration of a triangular TQD driven by crossed DC and AC magnetic fields. In the undriven case we calculated analytically the current as a function of both detuning and magnetic flux, showing the formation of a dark state as well as AB oscillations. The special configuration considered here leads to asymmetric current characteristics. ESR conditions allow to consider the TQD as a qubit where Rabi oscillations between two dark states can be controled. Both in the undriven and driven case, finite spin scattering destroys the coherent electron trapping and allow for a finite current to flow through the TQD.

Acknowledgments
We acknowledge financial support from grant MAT2008-02626/NAN. M. B. acknowledges JAE predoctoral grants program (CSIC) for support. R.S. acknowledges support from Swiss NSF and MaNEP.

[1] D. Loss *et al.*, Phys. Rev. A **57**, 120 (1998); D. Klauser *et al.*, Phys. Rev. B **73**, 205302 (2006).
[2] J.R. Petta *et al.*, Science **309**, 2180 (2005).
[3] F.H.L. Koppens *et al.*, Nature **442**, 766 (2006); K.C. Nowack *et al.*, Science **318**, 1430 (2007); E. A. Laird *et al.*, Phys. Rev. Lett. **99**, 246601 (2007); M. Pioro-Ladrière *et al.*, Nature Physics **4**, 776 (2008).
[4] R. Sánchez *et al.*, New J. Phys. 10 (2008) 115013.
[5] R. Sánchez *et al.*, Phys. Rev. B, 77 (2008) 165312; R. Sánchez *et al.*, Physica E 40 (2008) 1457.
[6] D. Schröer *et al.*, Phys. Rev. B 76 (2007) 075306.
[7] M. C. Rogge *et al.*, Phys. Rev. B 77 (2008) 193306; L. Gaudreau *et al.*, Phys. Rev. Lett. 97 (2006) 036807.
[8] M. Korkusinski *et al.*, Phys. Rev. B 75 (2007) 115301; F. Delgado *et al.*, Phys. Rev. B 76 (2007) 115332; F. Delgado *et al.*, Phys. Rev. Lett. 101 (2008) 226810.
[9] T. Kostyrko *et al.*, Phys. Rev. B 79 (2009) 075310; T. Kuzmenko *et al.*, Phys. Rev. Lett. 96 (2006) 046601.
[10] D. S. Saraga *et al.*, Phys. Rev. Lett. 90 (2003) 166803.
[11] B. Michaelis *et al.*, Europhys. Lett. 73 (2006) 677, C. Emary, Phys. Rev. B **76**, 245319 (2007), C. Pöltl *et al.*, Phys. Rev. B **80**, 235321 (2009).
[12] M. Busl *et al.*, Phys. Rev. B **81**, 121306(R) (2010), M. Busl *et al.*, Physica E **42**, 830 (2010).
[13] S.M. Huang *et al.*, Phys. Rev. Lett. **104**, 136801 (2010).

Tunnelling AFM study of the local density of states in the self assembled In(Ga)As/GaAs(001) quantum dots and rings

D O Filatov[1,3], M A Lapshina[1], M A Isakov[1], P A Borodin[2] and A A Bukharaev[2]

[1] University of Nizhny Novgorod, 23 Gagarin Ave. Nizhny Novgorod 603950 Russia
[2] Zavoisky Physical-Technical Institute, Kazan' Scientific Center, Russian Academy of Sciences, 10/17 Sibirskii Trakt, Kazan' 420029 Russia

E-mail: dmitry_filatov@inbox.ru

Abstract. The local density of states (LDOS) in the self assembled In(Ga)As/GaAs(001) quantum dots (QDs) and quantum rings (QRs) was investigated by Tunnelling Atomic Force Microscopy (AFM) at 300K. The peaks in the differential tunnel spectra of the metal coated AFM probe contact to the QDs (QRs) have been observed. These peaks were attributed to the quantum confined states in the QDs (QRs). The patterns of the current images of the QDs (QRs) at the gap voltages corresponding to the maxima of the tunnel spectra were found to reflect the LDOS spatial distribution of the respective quantum confined states. Several electron and hole states were identified.

1. Introduction

Scanning Tunneling Microscopy (STM) is used for the investigation of the morphology and of the atomic structure of the self-assembled semiconductor nanostructures extensively. Recently, STM in Ultra High Vacuum (UHV) has been applied also to the investigation of the spatial distribution of the local density of states (LDOS) in the quantum semiconductor heterostructures. In particular, Cross-Sectional STM has been applied to the investigations of the GaSb/InAs(001) quantum wells (QWs) [1] and of the self-assembled InAs/GaAs(001) quantum dots (QDs) [2]. In [3] the surface InAs/GaAs(001) QDs grown by Molecular Beam Epitaxy (MBE) have been studied by UHV STM *in situ*. The peaks related to the quantum confined states in the QDs have been observed in the differential conductivity $\sigma_d = dI_t/dV_g$ spectra of the STM tip contact to the QDs (here I_t is the tip current and V_g is the gap voltage). The $\sigma_d(x, y)$ images of the QDs (where x and y are the tip coordinates on the sample surface) recorded at the values of V_g corresponding to the peaks in the $\sigma_d(V_g)$ spectra were found to correlate with the probability density patterns $|\chi(x, y)|^2$ where $\chi(x, y)$ are the envelope electron wavefunctions calculated in [4] for the pyramidal InAs/GaAs(001) QDs defined by the (101) facets.

Here we report on the first time application of Tunnelling Atomic Force Microscopy (AFM) [5] to the *ex situ* investigation of the LDOS in the self-assembled In(Ga)As/GaAs(001) QDs and quantum rings (QRs) grown by Atmospheric Pressure Metal Organic Vapor Phase Epitaxy (AP MOVPE). The samples were covered by the native oxide formed during the sample transfer from the growth chamber to the AFM one through the ambient air that makes the STM studies of these samples hardly possible.

[3] The corresponding author.

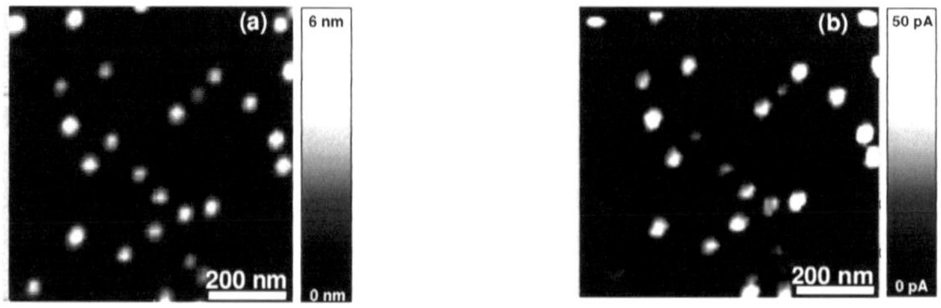

Figure 1. The AFM (a) and current (b) images of the InAs/GaAs QDs. $V_g = -3.7$ V.

2. Experimental

The samples were grown on the n^+- and p^+-GaAs(001) substrates (for the investigations of the conduction band states in the QDs (QRs) and of the valence band ones, respectively). The GaAs buffer layers ≈ 200 nm thick were doped by Si (for the n^+ substrates) or by Zn (for the p^+ ones) heavily (the impurity concentration ~ 10^{18} cm^{-3}). The 3 nm thick undoped GaAs spacer layers were grown between the buffers and the InAs QDs. The latter were grown at 530°C, the nominal thickness of InAs was ≈ 5 monolayers (ML). In order to grow the QRs, the QDs were capped by a 10 nm thick GaAs cladding layer at 600°C. In AP MOVPE the QRs grow just during capping [6] unlike MBE where the post-growth annealing is applied usually [7]. The In molar fraction in the QR material was estimated from the photoluminescence and photoelectric spectra to be 0.35–0.4.

The Tunneling AFM experiments were carried out in Omicron® MultiProbe P™ UHV system at 300K. The sample was scanned by a conductive Si probe coated by Pt or by W_2C in the contact mode. Simultaneously with the acquisition of the AFM images, the current ones $I_t(x, y)$ at the preset values of V_g were acquired. In another mode, the $I_t(V_g)$ curves were acquired in every point of the scans.

3. Results and discussion

The AFM and current images of an InAs/p^+-GaAs QD sample are presented in figure 1. An increasing of I_t when the AFM tip contacts the QD surface had been observed that has been attributed to the electron tunnelling between the AFM tip and the conductive buffer through the quantum confined states in the QDs. In figure 2 the $\sigma_d(V_g)$ spectra of the InAs QDs and of the InGaAs QRs are presented. These spectra have been obtained from the measured $I_t(V_g)$ curves of the probe-to-sample contact by numerical differentiation with nonlinear smoothing. The peaks in the $\sigma_d(V_g)$ spectra were related to the quantum confined states in the QDs (QRs) [3]. Comparing the $\sigma_d(V_g)$ spectra of the QDs on the n^+-GaAs and on p^+-GaAs presented in figures 2 (a) and (b), respectively one can note that the peaks related to the quantum confined states in the n-type QDs are well resolved while in the latter case a nearly exponential curves with the weak oscillations only have been observed. This could be explained noting that according to [4], the energy spacing between the electron levels in the QDs with the base size length b = 15–20 nm is ~100 meV while the one for the hole levels is 10-20 meV.

In order to associate the positions of the peaks in the $\sigma_d(V_g)$ spectra with the quantum confined level energies one must take into account the voltage drop on the space charge region between the QDs

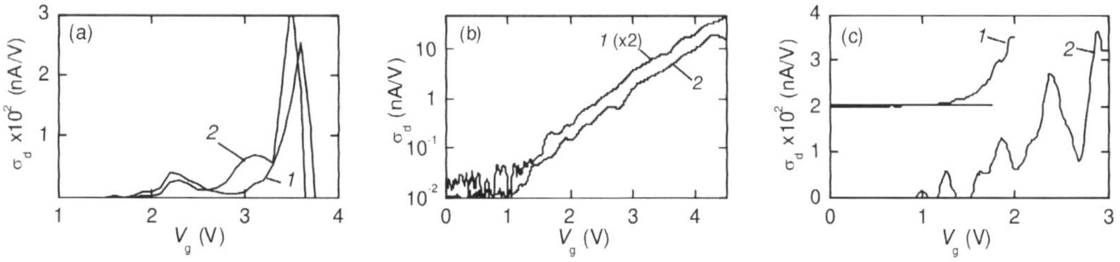

Figure 2. The $\sigma_d(V_g)$ spectra of the QDs on n^+-GaAs (a) and on p^+-GaAs (b); the same for the QRs (c). The curve numbers associate them to the points of measurement in figures 3 (a, e) and 4 (a).

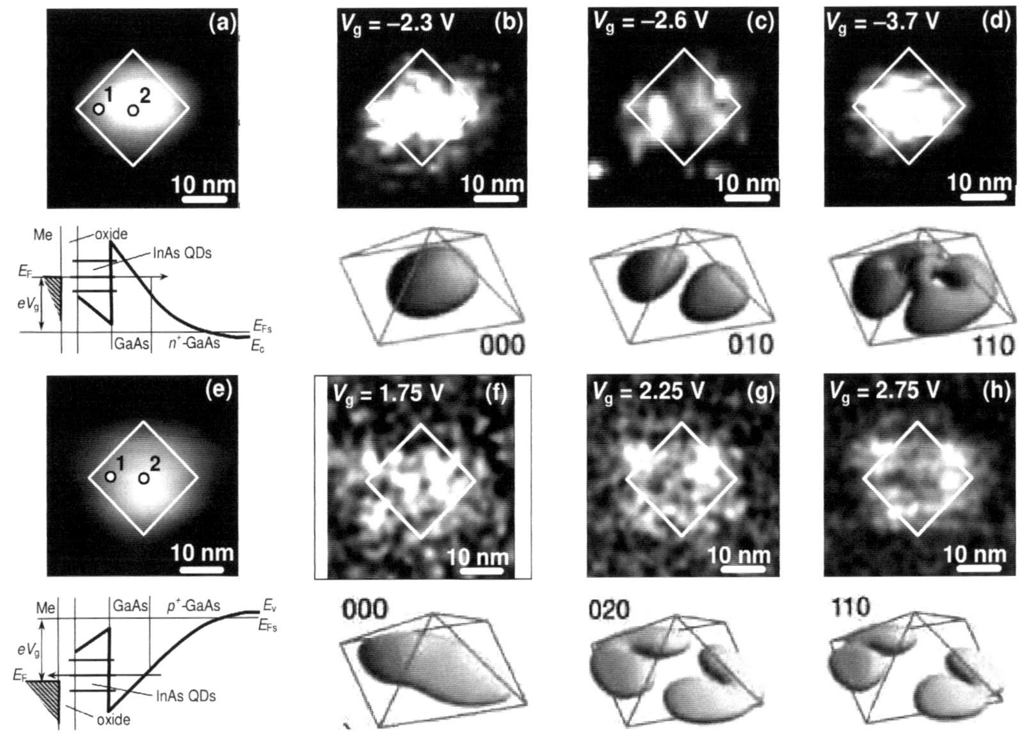

Figure 3. The AFM (a, e) and current (b—d, f—h) images of the individual InAs QDs on n^+-GaAs (a—d) and on p^+-GaAs (e—h). Below the $|\chi(x, y, z)|^2 = 0.65$ surfaces calculated in [4] for the respective electron and hole states $n_1n_2n_3$ in a pyramidal InAs/GaAs(001) QD with b = 16 nm are presented[4].

and the buffer layer [3] and on the surface states at the semiconductor/native oxide interface. Using Hasegawa's model [8], we have found that ≈1/2 of V_g drops on the surface states.

The AFM and current images of the individual QDs are presented in figure 3. The AFM images of the QDs are rounded and enlarged compared to the actual QD sizes (b = 14-16 nm) that was attributed to the convolution artifact [9]. The patterns of the $I_t(x, y)$ images at the V_gs corresponding to the maxima of the $\sigma_d(V_g)$ spectra reflect the spatial distribution of the LDOS in the (x, y) plane:

$$\rho_E(x, y) \propto \sum_{n_1n_2n_3=0}^{N_1N_2N_3} \left|\chi_{n_1n_2n_3}(x, y)\right|^2 . \tag{1}$$

Here the envelope wavefunctions were considered to be twofold spin-degenerated. The summation in (1) was taken over the states below the Fermi level in the probe coating material E_F, i. e. $E_{n_1n_2n_3} < E_F + eV_g$ for the QDs on n^+-GaAs, see the inset in figure 3 (a). The above condition defines the upper limits of the summation N_1, N_2, and N_3 in (1). The same holds for the QDs on p^+-GaAs with altering the energy sign for the holes, see the inset in figure 2 (e).

Following [3], the current images of the QDs were compared to the $|\chi|^2$ patterns of the quantum confined electron and hole states in the InAs/GaAs(001) pyramidal QDs calculated in [4]. The current images were more noisy than the STM ones in [3] that was attributed to the nonuniformity of the native oxide thickness. Nevertheless, several electron and hole states in the QDs were identified, the respective images of the $|\chi(x, y, z)|^2 = 0.65$ surfaces [4] are presented in figure 3.

---

[4] Reprinted figures with permission from [4]. Copyright (1999) by the American Physical Society.

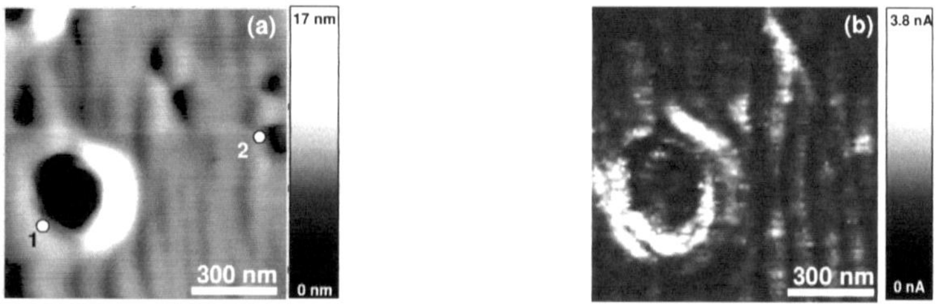

Figure 4. The AFM (a) and current (b) images of the InGaAs/GaAs QRs. $V_g = -1.37$ V.

The AFM and current images of the QRs are presented in figure 4. The QRs of various diameters D = 150-400 nm were observed. The smaller QRs arised from the transformation of the coherent InAs QDs during capping [7] while the larger ones originated from the larger relaxed InGaAs clusters formed by coalescence of the smaller InAs QDs during AP MOVPE growth [6]. Nevertheless, the larger rings could be classified as the QRs as well in spite of their relatively large sizes since the quantum size effect related steps had also been observed in their $\sigma_d(V_g)$ spectra (see figure 2 (c), curve 1).

The observation of the angular patterning in the current images of the QRs was attributed to the asymmetry of the QRs' shape. Otherwise, in the case of a circularly symmetric potential (Curie group of symmetry C_∞) no such patterning should be observed since in this case the angular part of the envelope wavefunction is expressed in the polar coordinates (ρ, φ) by the term $\exp(il\varphi)$ where i is the imaginary unity and $l = 0, \pm1,...$ is the angular quantum number, and $|\exp(il\varphi)|^2 \equiv 1$ for any φ. In addition, one should take into account the effect of the piezoelectric field in the strained InGaAs [4], which also reduces the potential symmetry downto C_{2v} even in a perfectly round QRs.

4. Conclusion

The results of the presented study demonstrate applicability of Tunnelling AFM to the *ex situ* investigations of the LDOS of the quantum confined states in the In(Ga)As/GaAs nanostructures covered by the native oxide. The observed patterns of the current images agree with the results of the QD eigenfunction calculations present in the literature. The current images of the QRs were related to the asymmetry of the confining potential.

Acknowledgments

The present work was supported in part by Russian Federal Agency for Education (NK-346P-25). The authors are grateful to B N Zvonkov (Research Institute for Physics and Technology, N I Lobachevskii University of Nizhny Novgorod) for growing the samples for investigations.

References

[1] Suzuki K, Kanisawa K, Janer C, Perraud S, Takashina K, Fujisawa T and Hirayama Y 2007 *Phys. Rev. Lett.* **98** 136802
[2] Grandidier B, Niquet Y M, Legrand B, Nys J P, Priester C, Stievenard D, Gerard J M and Thierry-Mieg V 2000 *Phys. Rev. Lett.* **85** 1068
[3] Maltezopoulos T, Bolz A, Meyer C, Heyn C, Hansen W, Morgenstern M and Wiesendanger R 2003 *Phys. Rev. Lett.* **91** 196804
[4] Stier O, Grundmann M and Bimberg D 1999 *Phys. Rev.* B **59** 5688
[5] Olbrich A, Ebersberger B and Boit C 1998 *Appl. Phys. Lett.* **73** 3114
[6] Baidus' N V, Zvonkov B N, Filatov D O, Guschina Yu Yu, Karpovich I A and Zdoroveischev A V 2000 *J. Surf. Inv.* 7(2004) 71
[7] Lorke A and Luyken R J 1998 *Phys.* B **256** 424
[8] Hasegawa H and Sawada T 1983 *Thin Solid Films* **103** 119
[9] Bukharaev A A, Berdunov N V, Ovchinnikov D V and Salikhov K M 1998 *Scanning Microsc.* **12** 225

Resonant tunnelling through individual Au nanoclusters embedded in ultrathin SiO_2 films studied by Tunnelling AFM

D O Filatov[1,3], M A Lapshina[1], D A Antonov[1], O N Gorshkov[1], A V Zenkevich[2] and Yu Yu Lebedinskii[2]

[1] University of Nizhny Novgorod, 23 Gagarin Ave. Nizhny Novgorod 603950 Russia
[2] Moscow Engineering Physics Institute, 31 Kashirskoe Ch., Moscow 115409 Russia

E-mail: dmitry_filatov@inbox.ru

Abstract. The observation of the resonant electron tunnelling through the individual Au nanoclusters (NCs) embedded in the ultrathin (4–5 nm thick) SiO_2 films on *n*-Si(001) substrates by Tunnelling Atomic Force Microscopy (TUNA) at 300K is reported. The nanocomposite SiO_2:NC-Au structures with Au NCs 3–7 nm in lateral size and 1–2 nm in thickness were formed by Pulsed Laser Deposition (PLD). The spots of increased probe current (or the current channels) on the sample surface 3–10 nm in diameter attributed to the electron tunnelling through the individual Au NCs have been observed. The *I*—*V* curves of the probe-to-sample contact measured in the current channels having larger (5–10 nm) size exhibited Coulomb staircase while those measured in the smaller channels exhibited the peaks attributed to the resonant tunnelling through the Au NCs less than 1 nm in thickness. In order to explain the experimental results we had employed a double barrier Pt/SiO_2/Au/SiO_2/Si structure model, which has provided a satisfactory description of the obtained experimental data.

1. Introduction

The metal nanoclusters (NCs) embedded into the thin dielectric films have attracted increased attention in recent years, in particular, as the promising materials for the nonvolatile memory applications [1] as well as for the applications in the novel nanoelectronic devices utilizing the single-electron and/or quantum size effects [2]. From this point of view, the investigations of the electron transport in the metal NC systems are important. Scanning Tunnelling Microscopy/Spectroscopy (STM/STS) has been widely used for the investigations of the electron transport through the metal NCs on the conductive substrates [3]. However, with respect to the metal NCs dispersed in the dielectric films the application of STM/STS is limited to the case when the NCs' concentration is high enough, i.e. when the films possess sufficient percolation conductivity [4, 5]. In this case, keeping STM feedback is provided by the continuous switching of the electrons tunnelling path through the film (hereinafter referred to as the current channel) from one chain of the NCs to another.

Since its development, Tunnelling Atomic Force Microscopy (TUNA) technique [6] has been mainly applied to study defects in thin dielectric films [4]. In [8] TUNA was applied to investigate the electron transport via Zr NCs produced by ion implantation in $Zr(Y)O_2$ films on p^+-Si substrates. The peaks observed in the *I*—*V* curves of the probe-to-sample contact were attributed to the resonant tun-

[3] The corresponding author.

nelling through the NCs of the highest resistance in the current channels (so called "bottleneck clusters" [4]).

In the present work, we used TUNA technique to investigate the electronic properties of individual Au NCs embedded in ultrathin (4-5 nm) SiO_2 films on the *n*-Si substrates.

2. Experimental

The ultrathin SiO_2:nc-Au films were formed on the *n*-Si(001) substrates by Pulsed Laser Deposition (PLD) in combination with plasma oxidation at 300K in a single vacuum cycle. Chemically cleaned *n*-Si substrates were pre-oxidized in the glow discharge oxygen plasma up to the underlying oxide thickness d_u = 1–2 nm and followed by the co-deposition of Si—Au mixture layer by PLD in UHV from the elemental Au and Si targets . The deposited Si—Au layer was further oxidized using the same glow discharge plasma procedure which leads to the segregation of Au from the oxidizing Si to form Au NCs. The size and shape of NCs was defined by the Si/Au concentration ratio and is limited by the lack of the long range thermal diffusion at 300K. Finally, the NCs were capped by the SiO_2 cap layer with the thickness d_c = 1–2 nm combining Si deposition in UHV and plasma oxidation. All the steps of the SiO_2:NC-Au structure formation (described in details in [9]) were monitored *in situ* by X-ray Photoelectron Spectroscopy (XPS) to control the chemical state and thickness of each layer. The technique described above allows forming the NC arrays arranged in a single sheet between the underlying and cladding layers, the thicknesses of which can be controlled precisely (unlike the NCs made by the ion implantation where some scatter in the NCs' positions with respect to the film boundaries takes place always). Using the same growth procedure and additional amorphous graphite supporting layer (also produced by PLD), similar samples for Transmission Electron Microscopy (TEM) investigations were prepared on the cleaved NaCl(001) facets.

The TUNA investigations were carried out in Omicron® MultiProbe™ S UHV system at 300K. The sample surface was scanned using Pt coated Si probe in the contact mode. The *I*—*V* curves of the probe-to-sample contact were acquired in every point of the scan.

3. Results and discussion

A plain view TEM image of SiO_2(1.5 nm)/nc-Au/SiO_2(1.5 nm) film grown on NaCl(001) is presented in figure 1 (a). The lateral size *D* of the Au NCs was estimated from this image to be 3.0–6.5 nm. The NCs' height *h* calculated from the overall Au amount in the film measured by Rutherford Backscattering Spectrometry (RBS) was 1–2 nm. These observations indicate that Au NCs formed in this process have a flattened shape.

A TUNA image of the SiO_2(1.5 nm)/NC-Au/SiO_2(1.5 nm)/*n*-Si film is presented in figure 1 (c). The inverted contrast in this image (the larger the probe current I_t the darker is the spot) corresponds to the negative polarity of the bias voltage V_g between the probe and the sample which, in turn, corresponds to the injection of the electrons from the Pt coated tip into the *n*-Si substrate through the nanocomposite film. It is worth noting that the TUNA image correlates with the topographic one presented in figure 1 (b) weakly (the Pirson's correlation coefficient R_P≈0.015 for this particular pair of images).

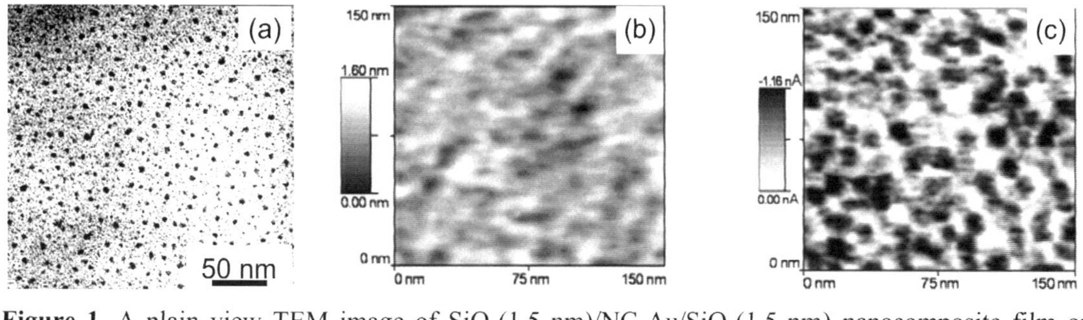

Figure 1. A plain view TEM image of SiO_2(1.5 nm)/NC-Au/SiO_2(1.5 nm) nanocomposite film on NaCl(001) (a); the AFM (b) and the TUNA (c) images of the similar film on *n*-Si(001); V_g = –2.5 V.

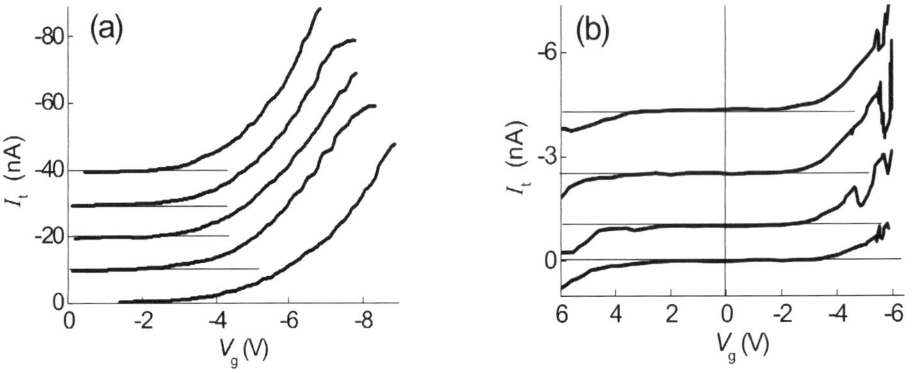

Figure 2. The *I—V* curves measured with a Pt-coated Si AFM tip on the SiO_2/NC-Au/SiO_2/*n*-Si structure in different current channels of larger(a) and smaller (b) sizes.

The spots of 3–10 nm in size in the TUNA image (the current channels) were attributed to the electron tunnelling through the individual Au NCs [10]. The *I—V* curves of the probe-to-sample contact measured in the channels of the larger size (5–10 nm) presented in figure 2 (a) exhibited weakly expressed Coulomb staircase. Earlier, the Coulomb blockade has been observed at 300 K in similar structures using TUNA [14] and was discussed in more details therein. In the present work, we used the model of an ellipsoidal NC (taking into account the flattened shape of the NCs as revealed above) between two flat electrodes to estimate the size of NCs from the Coulomb staircase period. Assuming $h = 1$ nm, we obtained $D = 3$–5 nm that is consistent with the TEM data.

The *I—V* curves measured in the channels of the smaller (3–5 nm) size are shown in figure 2 (b). The peaks at the negative V_g were attributed to the resonant tunnelling through the Au NCs. The interpretation of the tunnelling spectra was based on a double barrier Pt/SiO_2/Au/SiO_2/Si structure model (figure 3) similar to the one described in [11]. The model was based on the exact solution of one-dimensional Schrödinger's equation in the effective mass approximation. The applicability of the one-dimensional model is justified by the flattened shape of the NC, so that they can be treated as the thin disks and the size quantization in the film plane can be neglected. The band diagrams of the model structures were calculated by solving the one-dimensional Poisson's equation taking into account the band bending at the *n*-Si/SiO_2 interface (figure 3). The NCs were assumed to be completely charge depleted due to the quantum size effect that allowed us to neglect the screening of the external electric field by the electrons confined in the NC.

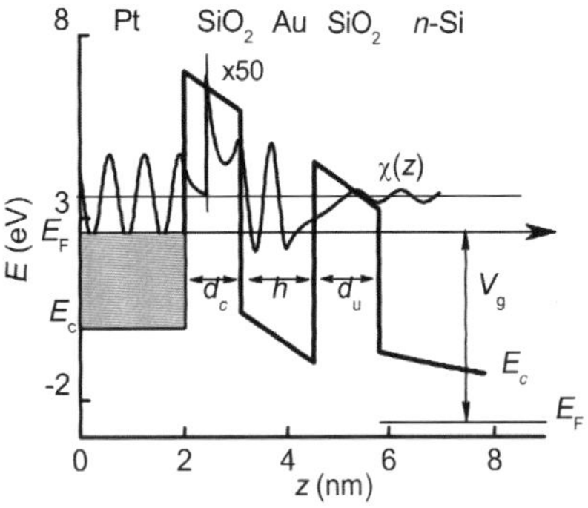

Figure 3. The calculated band diagram of a Pt/SiO_2/NC-Au/SiO_2/*n*-Si structure for $V_g = 5.7$ V; a calculated envelope wavefunction $\chi(z)$ for a resonant state.

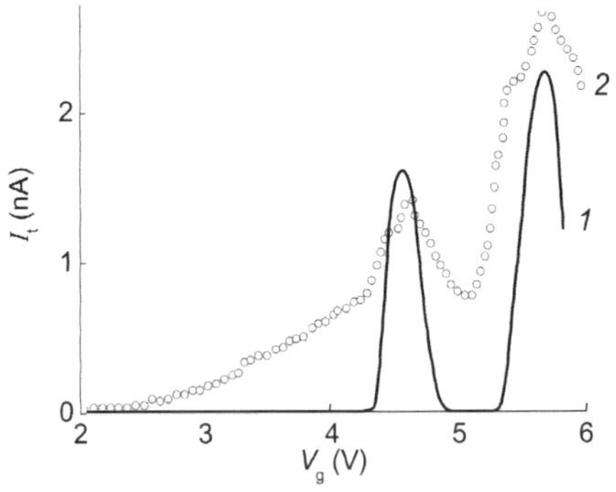

Figure 4. The calculated (1) and measured (2) *I—V* curves of the contact of an AFM probe to the SiO_2(1.5 nm)/SiO_2:NC-Au/SiO_2(1.8 nm)/*n*-Si structure.

Earlier, STM/STS and Auger Electron Spectroscopy (AES) investigations revealed that similar Au NCs formed on the highly oriented pyrolitic graphite can undergo metal-to-insulator transition at $h<1$ nm, and an energy gap was observed in their tunnel spectra [12]. In this case, the electron-electron scattering inside the NCs would not affect the coherent tunnelling through them. However, it is not the case for the NCs of larger size ($\geq$ 2 nm), which are filled with electrons.

The typical fitting of a measured *I—V* curve of the probe-to-sample contact with a model one is presented in figure 4, d_u, d_c, and h were the fitting parameters. The best fit was achieved at d_u = 1.4 nm, d_c = 1.1 nm, and h = 1.3 nm (the band diagram in figure 3 has been calculated just for these values) which is in satisfactory agreement with XPS [12] and TEM data despite the fact that the described model is rather crude. Particularly, the resonant peaks in the calculated *I—V* curve (figure 4) are narrower than those in the measured one which should be attributed to the inhomogeneous broadening due to the non-uniformity of the actual NC thickness. Also, one can notice the splitting of the resonant peaks in the measured *I—V* curve that can be ascribed to the stepwise fluctuations in the NC thickness. Nevertheless, from the comparison of the results of the modelling with the experimental data one can conclude that the used model provides satisfactory description of the electron tunnelling through the small NCs and that the peaks observed in the measured *I—V* curves originate from the resonant tunnelling effect.

4. Conclusion

In the present study, the TUNA technique was applied to observe the resonant electron tunnelling through the individual Au NCs $\leq$1 nm in thickness embedded in the ultrathin SiO_2 films on the Si substrates. The charge depletion of the Au NCs was found to be an essential condition for the manifestation of the resonant tunnelling. The observed effect can be exploited in a nanoelectronic devices based on the single metal NC embedded into the ultrathin dielectric films.

Acknowledgments

This work was supported in part by Russian Federal Agency for Education (NK-346P-25). The authors are grateful to A A Timofeev (Moscow Engineering Physics Institute) and to P S Chernykh (Research Institute for Nuclear Physics, Moscow State University) for the TEM and RBS measurements, respectively.

References

[1] Ren F, Heng X, Guang X, Cai X, Wang J B and Jiang C Z 2009 *Appl. Phys.* A **96** 317
[2] Cerofolini G F, Arena G, Camalleri M, Galati C, Reina S, Renna L, Mascolo D and Nosik V 2005 *Microel. Eng.* **81** 405
[3] Binns C 2001 *Surf. Sci. Rep.* **44** 1
[4] Bar-Sadeh E, Goldstein Y, Zhang C, Deng H, Abeles B and Millo O 1994 *Phys. Rev.* B **50** 8961.
[5] Imamura H, Chiba J, Mitani S and Takanashi K 2000 *Phys. Rev.* B **61** 46
[6] Olbrich A, Ebersberger B and Boit C 1998 *Appl. Phys. Lett.* **73** 3114
[7] Yanev V *et al.* 2008 *Appl. Phys. Lett.* **92** 252910.
[8] Antonov D A, Gorshkov O N, Kasatkin A P, Maximov G A, Saveliev D A and Filatov D O 2004 *PLDS* 1/2(2004) 139
[9] Zenkevich A V, Lebedinskii Yu Yu, Timofeev A A, Nevolin V N, Antonov D A, Filatov D O and Maximov G A 2008 *J. Adv. Mat.* 04(2008) 5
[10] Antonov D A, Filatov D O, Zenkevich A V and Lebedinskii Yu Yu 2007 *Bulletin of RAS: Physics* **71** 56
[11] Lapshina M A, Filatov D O, Antonov D A and Barantsev N S 2009 *J. Surf. Inv.* **3** 559
[12] Borman V D, Borisyuk P V, Lebid'ko V V, Pushkin M A, Tronin V N, Troyan V I, Antonov D A and Filatov D O 2006 *JETP* **129** 1

Charge dynamics in two-electron quantum dots

J. Särkkä and A. Harju
Helsinki Institute of Physics and Department of Applied Physics,
Aalto University, FI-02150 Espoo, Finland
E-mail: jani.sarkka(at)tkk.fi

Abstract. We investigate charge dynamics in a two-electron double quantum dot. The quantum dot is manipulated by using a time-dependent external voltage that induces charge oscillations between the dots. We study the dependence of the charge dynamics on the external magnetic field and on the periodicity of the external potential. We find that for suitable parameter values, it is possible to induce both one-electron and two-electron oscillations between the dots.

1. Introduction

The search for a working realization for a quantum bit has been going over a decade. The proposal of Loss and DiVincenzo of using spin of a single electron trapped in a quantum dot [1] has been one of the most popular candidates for a qubit. Also a setup of spins of two electrons in a quantum dot has been studied. The quantum dot confinement potential may be realized by surface acoustic waves, where a single or several electrons are trapped in a surface acoustic minimum. The electrons move along the surface acoustic wave, making possible a realization of a flying qubit [2, 3, 4]. Recently Kataoka et al. [5] observed coherent dynamics of a single-electron wave function in a surface acoustic wave quantum dot. When the electron in a moving quantum dot enters a region with different confinement potential, the sudden change in the potential induces excitations into the higher energy states. This causes tunneling of electrons into an adjacent empty channel. The charge in this channel causes a measurable current. The rapid control of external gate voltage in two-electron double quantum dots has been demonstrated by Petta et al. [6, 7].

Motivated by these experiments, we study numerically charge dynamics in two-electron quantum dots operated in a similar fashion. Our studies show that charge dynamics may be manipulated by choosing suitable values for parameters.

2. Model and Method

We model the system with two-electron Hamiltonian

$$H = \sum_{i=1}^{2} \left(\frac{\left(-i\hbar\nabla_i - \frac{e}{c}\mathbf{A} \right)^2}{2m^*} + V(\mathbf{r}_i) + g^*\mu_B \mathbf{B}\cdot\mathbf{s}_i \right) + \frac{e^2}{\epsilon r_{12}}, \tag{1}$$

where the confinement potential (illustrated in Fig. 1) is

$$V(\mathbf{r}) = \frac{\hbar}{2}\omega^2[x^2 - \lambda\tanh(x/\delta)x - dx] + \frac{\hbar}{2}\omega^2 y^2. \tag{2}$$

Here d is the detuning parameter, which describes the linear dependence of the potential on the x-coordinate and has length as its unit. The confinement strength is $\hbar\omega$= 3 meV, δ=10 nm, and the two potential minima are at points $x = \pm L$, $y = 0$ if we set

$$\lambda = L\delta/[L(1/\cosh^2(L/\delta)) + \delta \tanh(L/\delta)]. \tag{3}$$

We initialize the system so that both electrons are in the same potential minimum. This is done by setting the position of the potential minima to L=80 nm and the initial value for the detuning parameter d=4.6 nm (dashed line in Fig. 1), creating a ground state where both electrons are in the right dot (see Fig. 2). Then, we lower the detuning in 5 ps to d=1.1 nm and simultaneously move the positions of the potential minima to L=40 nm (solid line in Fig. 1). For these values, the ground state of a single electron in the right dot is degenerate with the first excited state of a single electron in the left dot (see Fig. 3). Hence, charge oscillations between the two dots are now possible.

If the detuning parameter is lowered to d=0 nm instead of 1.1 nm, the two-electron ground states become degenerate in both dots. This enables the oscillation of both electrons between the dots. We also simulate a scheme with oscillating detuning d=2.8 nm+1.8 nm $\times \cos(2\pi t/\tau)$, where the period τ has values between 28 ps and 66 ps.

For numerical analysis of the problem, we discretize the four-dimensional Schrödinger equation using finite difference method. We solve the ground state of the system using Lanczos diagonalization. The time evolution $\psi(t + \Delta t) = \exp(-iH\Delta t/\hbar)\psi(t)$ is calculated using Krylov subspace methods [8, 9].

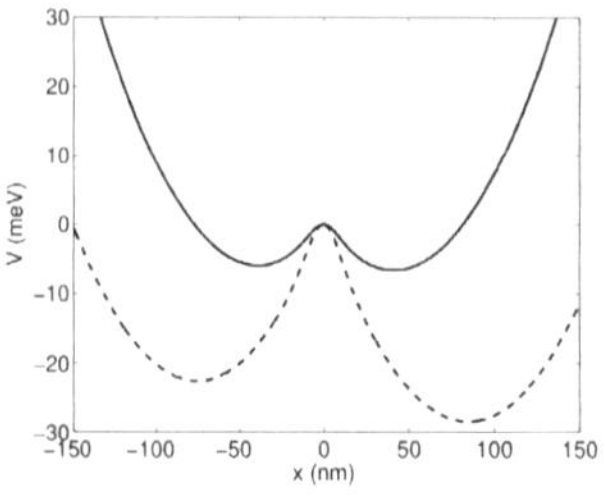

Figure 1. The confinement potential with L=80 nm, d=4.6 nm (dashed) and with L=40 nm, d=1.1 nm (solid).

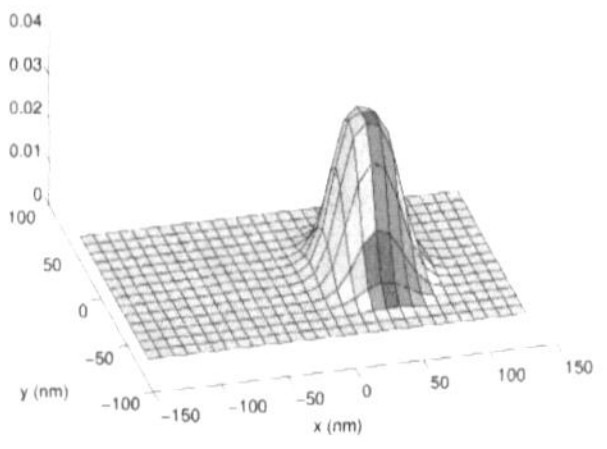

Figure 2. The initial charge density when the whole charge is in the ground state of the right dot.

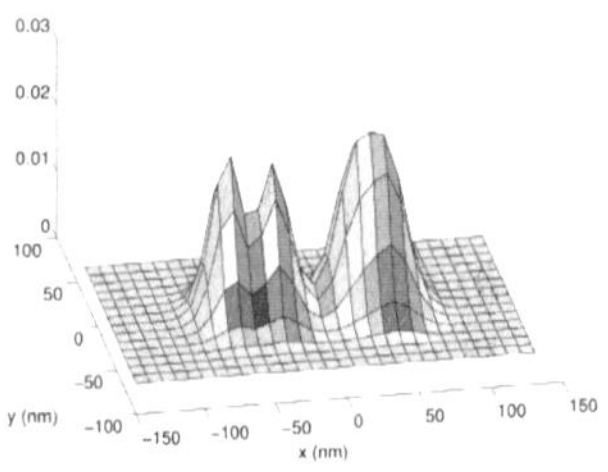

Figure 3. Charge density when half of the charge is in the excited state of the left dot.

3. Results and Discussion

We study the effect of detuning on the charge oscillations using several detuning values, illustrated in Fig. 4. We sweep the detuning from 4.6 nm to its final value in 5 ps, and after that keep the detuning constant in a zero magnetic field. For zero detuning, both electrons oscillate between the dots with period $\tau \approx 300$ ps, as the ground states containing two electrons in one dot are degenerate. When the final detuning is changed to 0.35 nm, the charge dynamics changes completely, only a fraction of the charge goes to the other dot. This is due to the fact that now the two-electron ground states in the dots are not degenerate. When the final detuning is raised to 0.7 nm, more charge is moving to the other dot. When the final detuning is 1.1 nm, we have a situation where the first excited state of one electron in the first dot is degenerate with the ground state of two electrons in the second dot. Now one electron oscillates between the dots with significantly faster oscillations than in the case of degenerate ground states in the dots, having a period τ=20 ps. Using this detuning, it is possible to induce faster charge dynamics.

The switching speed of the detuning (5 ps) is chosen so that the behavior of the charge dynamics is sufficiently smooth. If the switching is done in shorter time, the overall dynamics is similar, but shorter scale oscillations with a small amplitude emerge. This phenomenon is depicted in Fig. 5, where we have calculated the dynamics corresponding to the uppermost curve in Fig. 4 using switching times between 1 and 5 ps. The amplitude of the shorter scale oscillations increases slightly with decreasing switching speed, but the period and amplitude of the longer scale oscillations remain the same.

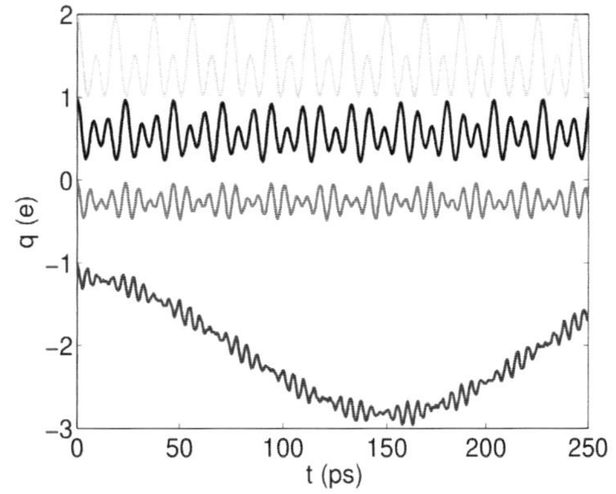

Figure 4. Charge dynamics for constant detuning parameters from down to up d=0 nm (blue), 0.35 nm (red), 0.7 nm (black), and 1.1 nm (green). The detuning was switched in 5 ps. Adjacent curves below the uppermost are offset by -1 for clarity.

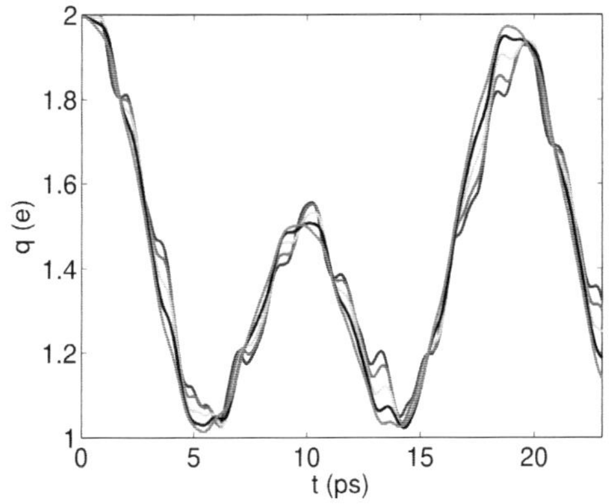

Figure 5. Charge dynamics for zero magnetic field and detuning d=1.1 nm for different detuning switching times: τ=1 ps (blue), 2 ps (red), 3 ps (green), 4 ps (black), and 5 ps (magenta).

Next, we apply an oscillating detuning potential and also a non-zero magnetic field. The Zeeman term in the Hamiltonian (Eq. (1)) is small compared to the exchange energy, and the electrons remain in the $S = 0$ state. We use two different periods for the detuning voltage, namely τ=33 ps (Fig. 6) and τ=66 ps (Fig. 7). In Fig. 6, one can see that for small magnetic fields there exists a slower oscillation in a 200 ps timescale, and a faster oscillation having the same period as the detuning voltage. When the magnetic field increases to 0.3 T, the one-electron states in the dots are not degenerate anymore. Consequently, the charge oscillations have smaller amplitude. For a stronger magnetic field of 1 T, we find that the shorter scale oscillations are damped, but the longer scale oscillations have now larger amplitude. In Fig. 7, same parameters are used as in Fig. 6, only with the exception that the detuning oscillation period is now two times as long. The resulting charge oscillations are nonetheless quite different. The shorter scale oscillations have now a smaller amplitude, and the charge dynamics is now determined by the longer scale oscillations. For zero magnetic field, the charge stays mainly in the other dot. Only momentarily there is a significant net charge in the other dot. For magnetic field value 0.03 T, the oscillations are completely different. Now the one-electron charge moves completely to the other dot and back, having a period around 800 ps. When the magnetic field values increase, the longer scale oscillations vanish as the system is not degenerate anymore. The shorter-scale oscillations dominate the dynamics, which becomes more chaotic with increasing magnetic field. We also studied other periods for the detuning, but for values that where not multiples of 33 ps in zero magnetic field, we did not observe any longer scale oscillation, and the charge dynamics was more chaotic. Obviously, for period 33 ps we have a kind of eigenfrequency for the system, although the period differs from the period 20 ps of the oscillations in the case of a constant detuning parameter in Fig. 5.

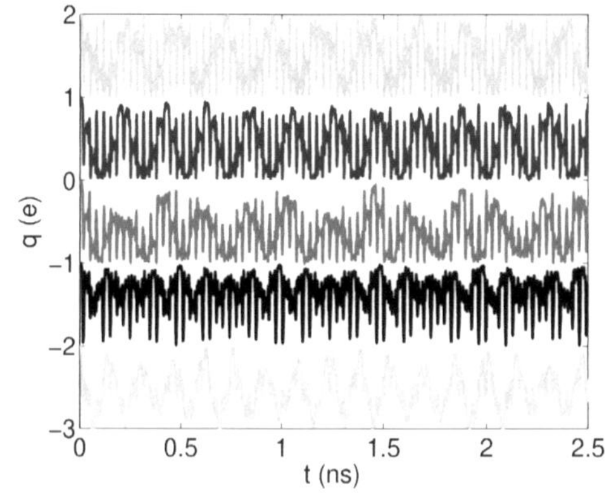

Figure 6. Charge dynamics for detuning oscillation period τ=33 ps, detuning 1.1 nm and different magnetic fields from up to down: B=0 T (green), 0.03 T (blue), 0.05 T (red), 0.3 T (black), and 1 T (cyan). Adjacent curves after the first are offset by -1 for clarity.

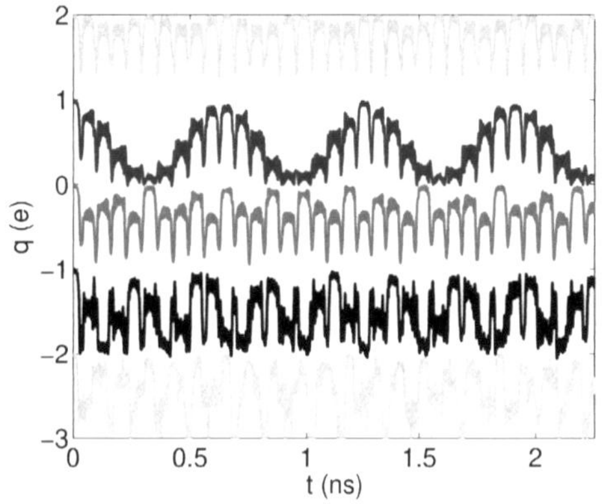

Figure 7. Charge dynamics for detuning oscillation period τ=66 ps, detuning 1.1 nm and different magnetic fields from up to down: B=0 T (green), 0.03 T (blue), 0.05 T (red), 0.3 T (black), and 1 T (cyan). Adjacent curves after the first are offset by -1 for clarity.

4. Summary

In this work, we studied charge dynamics in a two-electron double quantum dot. We found parameters for controlling the detuning voltage that enable charge oscillations between the dots. By calculating numerically the two-electron dynamics, we observed two different oscillations, a slow oscillation of two unit charges and a rapid oscillation of a single unit charge between the dots. The time used in switching the detuning to a constant value did not affect the charge dynamics. For oscillating detunings, regular and chaotic dynamics were found.

5. Acknowledgments

This work was supported by the Academy of Finland through its Centers of Excellence program (2006-2011).

References

[1] Loss D and DiVincenzo D P 1998 *Phys. Rev. A* **57** 120–126
[2] Shilton J M, Talyanskii V I, Pepper M, Ritchie D A, Frost J E F, Ford C J B, Smith C G and Jones G A C 1996 *J. Phys. Condens. Matter* **8** L531–L539
[3] Barnes C H W, Shilton J M and Robinson A M 2000 *Phys. Rev. B* **62** 8410–8419
[4] Astley M R, Kataoka M, Ford C J B, Barnes C H W, Anderson D, Jones G A C, Farrer I, Ritchie D A and Pepper M 2007 *Phys. Rev. Lett.* **99** 156802
[5] Kataoka M, Astley M R, Thorn A L, Oi D K L, Barnes C H W, Ford C J B, Anderson D, Jones G A C, Farrer I, Ritchie D A and Pepper M 2009 *Phys. Rev. Lett.* **102** 156801
[6] Petta J R, Johnson A C, Taylor J M, Laird E A, Yacoby A, Lukin M D, Marcus C M, Hanson M P and Gossard A C 2005 *Science* **309** 2180–2184
[7] Petta J R, Lu H and Gossard A C 2010 *Science* **327** 669–672
[8] Särkkä J and Harju A 2008 *Phys. Rev. B* **77** 245315
[9] Särkkä J and Harju A 2009 *Phys. Rev. B* **80** 045323

Quantum Dots 2010 IOP Publishing
Journal of Physics: Conference Series **245** (2010) 012020 doi:10.1088/1742-6596/245/1/012020

Vertically coupled non-uniform quantum rings with two separated electrons in threading magnetic field

J. H. Marín[1], F. A. Rodriguez-Prada[2], and I. D. Mikhailov[2]

[1]Escuela de Física, Universidad Nacional de Colombia, Medellín, Colombia A.A. 3840
[2]Escuela de Física, Universidad Industrial de Santander, Bucaramanga, Colombia A.A. 678

E-mail: jhmarca65@gmail.com ; jhmarin@unal.edu.co

Abstract. We propose a simple method for calculating the energy spectrum of two separated electrons in vertically coupled narrow quantum rings with non-uniform cross-sections heights. We present numerical results for energies of some low-lying levels as a function of the magnetic flux for a particular case, in which one of two rings is uniform and other has a locally distorted height. The effect of the distortion on the Aharonov-Bohm oscillations is studied.

1. Introduction

The self-assembled quantum rings (QRs) are the subject of extensive theoretical and experimental studies [1] due to their particular topology, which leads to the Aharonov–Bohm (AB) effect consisting in the appearance of the persistent currents and the energy oscillations in an increasing magnetic field, applied along the ring axis [2]. This effect is clearly manifested only in a narrow and homogeneous ring [3]. In 1D QRs the two-electron problem is separated exactly both for a single QR [4] and two vertically stacked QRs [5]. The QR's properties are changed drastically if the ring is non-uniform. It has been shown recently, that any smooth and tiny variations in the 2D ring width or curvature may produce quenching of the AB oscillations of the lower energy levels [6] due to the electron localization near the defect. The samples of 3D rings, used in the experiment for observing the AB effect, have finite but very small and variable thicknesses with drastically different crystalline shapes [7]. The properties of a thin 3D ring is similar to those of a 2D ring, only if the 3D ring thickness is uniform, otherwise the potential energy corresponding to the electron in-plane motion is non-homogeneous. For example, the increase of the ring height from 4 nm up to 4.5 nm lowers the energy of the electron about 20%. This effect may be more interesting for the model of two-electron vertically stacked QRs when the strongly correlated electrons should be mainly located on the opposite sides of the rings. Therefore the localization of one of them should lead to the localization of the other one and to the quenching of the AB oscillations. In the present work, we analyze the model of two separated electrons, in two coaxial and vertically stacked thin QRs, schematically presented in figure 1, where the lower and upper rings have the same width w and different centerline radii $R_1 = R$ and $R_2 = \alpha R$, respectively, being the separation between them $d = \beta R$. The external magnetic field $\vec{B}$ is applied along the z-axis. The lower ring is considered as homogeneous with fixed thickness h_1 ($h_1 = h$), while the thickness of the

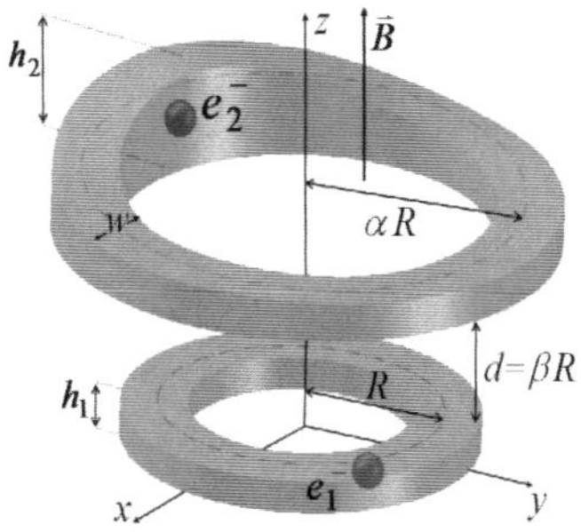

Figure 1. Scheme of vertically coupled uniform and non-uniform quantum rings

upper ring is supposed to be dependent on the polar angle as $h_2 = h/\sqrt{1+\sigma^2\vartheta_2^2}$; $(-\pi < \vartheta_2 < \pi)$. The smaller the parameter σ, the smoother is the surface of the upper ring.

2. Theory

In the effective mass approximation the dimensionless two-electron Hamiltonian can be written as:

$$H = \sum_{j=1,2} H_0(\mathbf{r}_j) + \frac{2}{|\mathbf{r}_1 - \mathbf{r}_2|}; \quad H_0(\mathbf{r}_j) = -\Delta_j + V_j(\mathbf{r}_j) + \frac{\gamma^2\rho_j^2}{4} - i\gamma\frac{\partial}{\partial\vartheta_j} \tag{1}$$

Here $V_1(\mathbf{r}_1)$ and $V_2(\mathbf{r}_2)$ are infinite-barrier confinement potential in the lower and in the upper rings, respectively, and the effective Bohr radius $a_0^* = \hbar^2\varepsilon/m^*e^2$, the effective Rydberg $R_y^* = e^2/2\varepsilon a_0^*$ and $\gamma = e\hbar B/2m^*cR_y^*$ are taken as units of length, energy and the conventional dimensionless magnetic field strength, respectively. For the sake of the mathematical convenience, we consider below only quasi-1D rings with large radius–width and width-thickness aspect ratios, i.e. $h_j << w << R_j$, for which the two-electron wave function can be separated by using adiabatic approximation (AA) [8] and represented in the form: $\Psi(\mathbf{r}_1,\mathbf{r}_2) = \psi(\vartheta_1,\vartheta_2) f^{(f)}(\rho_1,z_1,\rho_2,z_2,\vartheta_2)$. Here $\psi(\vartheta_1,\vartheta_2)$ is the variational function that describes the slow electrons rotation around z-axis and $f^{(f)}(\rho_1,z_1,\rho_2,z_2,\vartheta_2)$ the one-particle ground state wave function which describes the transversal fast motion of two non-interacting electrons in planes through the z axis with polar angles ϑ_1 and ϑ_2, respectively. The condition $h_j << w$ allows us to apply the AA in order to separate z_j and ρ_j coordinates and to find this function in an analytical form. Starting from the Schrödinger variational principle [9] we find the energies of the low-lying states by minimizing the functional:

$$E[\Psi] = \left\langle f^{(f)}\psi \middle| H \middle| f^{(f)}\psi \right\rangle \Big/ \left\langle f^{(f)}\psi \middle| f^{(f)}\psi \right\rangle \to \min \tag{2}$$

Taking the functional derivative with respect to ψ one can obtain the wave equation for $\psi(\vartheta_1,\vartheta_2)$ [8].

$$\hat{H}_{sl}\psi = \varepsilon\psi; \quad \hat{H}_{sl} = -\left(\frac{\partial}{\partial\vartheta_1} - i\Phi\right)^2 - \frac{1}{\alpha^2}\left(\frac{\partial}{\partial\vartheta_2} - i\alpha^2\Phi\right)^2 + \tilde{V}(\vartheta_1 - \vartheta_2) + \frac{\pi^2\sigma^2R^2}{h^2}\vartheta_2^2 + a\Phi^2$$
$$\tilde{V}(\vartheta) = 2R\left[\beta^2 + (1-\alpha)^2 + 4\alpha\sin^2\vartheta/2 + a\right]^{-\frac{1}{2}}; \quad \varepsilon = \left(E - \frac{2\pi^2}{h^2} - \frac{2\pi^2}{w^2}\right)R^2; \quad a = \frac{w^2}{12R^2} \tag{3}$$

Here ε is normalized energy, Φ is the magnetic flux through interior of the lower ring expressed in term of the magnetic flux unit $\Phi_0 = hc/e$, i.e. $\Phi = B\cdot\pi R^2/\Phi_0$ and the terms of order h^2/R^2 and superior are neglected. The equation (3) in the centre-of-mass $\Theta = (\vartheta_1 + \alpha^2\vartheta_2)/(1+\alpha^2)$ and relative $\vartheta = \vartheta_1 - \vartheta_2$ coordinates can be rewritten as:

$$\hat{H}_{sl}\psi(\vartheta,\Theta) = \varepsilon\psi(\vartheta,\Theta); \quad \hat{H}_{sl} = \hat{H}_\Theta + \hat{H}_\vartheta - \frac{2\pi^2\sigma^2R^2}{h^2p}\vartheta\Theta; \quad p = 1+\alpha^2; -\pi < \vartheta < \pi; \quad -2\pi < \Theta < 2\pi$$
$$\hat{H}_\Theta = -\frac{1}{p}\frac{\partial^2}{\partial\Theta^2} + 2i\Phi\frac{\partial}{\partial\Theta} + p\Phi^2 + \frac{\pi^2\sigma^2R^2}{h^2}\Theta^2; \hat{H}_\vartheta = -\frac{p}{\alpha^2}\frac{\partial^2}{\partial\vartheta^2} + \tilde{V}(\vartheta) + \frac{\pi^2\sigma^2R^2}{h^2p^2}\vartheta^2 \tag{4}$$

We solve the equation (4) by means of the exact diagonalization method, using as the base functions the products of $\mathrm{P}_M(\Theta) = \sum_{n=-\infty}^{\infty} c_{n,M}e^{in\Theta}$ and $\Upsilon_m(\vartheta) = \sum_{n=-\infty}^{\infty} b_{n,m}e^{in\vartheta/2}$, eigenfunctions of the centre-of-mass $\hat{H}_\Theta$ and relative $\hat{H}_\vartheta$ parts of the Hamiltonian, for different values of centre-of-mass, $M = 0,\pm1,\pm2,\ldots$ and relative angular, $m = 0,\pm1,\pm2,\ldots$ momenta. The corresponding matrix elements of Hamiltonian $\hat{H}_{sl}$ by using these basis functions can be calculated analytically. We checked the accuracy of our method by comparing our results with those of references [4, 8] for 1D uniform two-electron QRs with different radii. Discrepancies between our results and those from references [4,8] for the first 20 low-lying energy states is less than 5% when our Fourier series converge.

3. Results

We have performed numerical calculations of two-electron normalized energies $\varepsilon = E \cdot R^2$ as a function of the magnetic flux and for different geometric parameters of the ring in order to analyze the effect of the ring dimension and the smoothness of the ring surface. First, we calculate the energies one-electron single QR with smooth distortion of the surface thickness. Results are shown In Fig. 2.

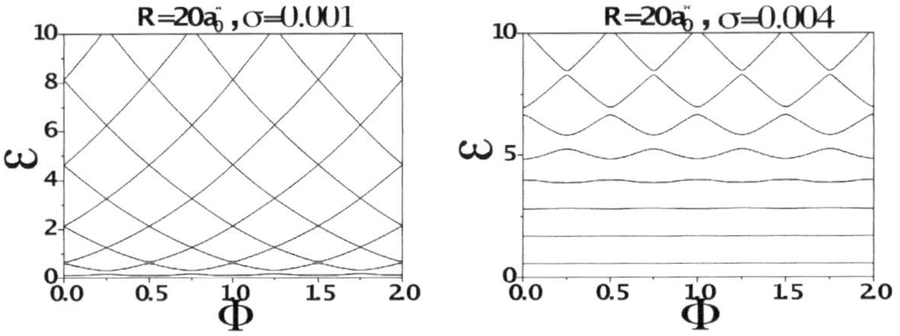

Figure 2. Renormalized energies of the one-electron low-lying states as functions of the magnetic flux for a single ring with two different radii and smoothness of the with parameters $h = 0.2$

It is seen that a very small increase of the surface non-homogeneity, given by the variation of the parameter σ from 0.001 to 0.004 has produced a quenching of the AB oscillations of the lower energy levels similarly to one, found previously in the references [6] for non-uniform 1D and 2D rings. The physical reason for the flattening of the energy levels is similar to one given in references [6] for 2D non-uniform rings. The higher regions of the circular ring are more favourable for the electrons, i.e., they correspond to a lower effective potential. Therefore, low energy electrons are confined near the point corresponding to $\vartheta_2 = 0$, wherein the height attains its maximum value. Since the ring is thin, small variations in the height produce large variations in the transversal energy, thus confining the electron.

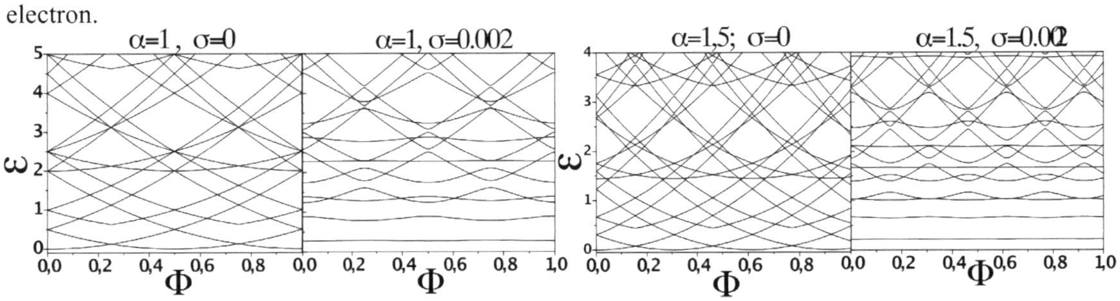

Figure 3. Effect of non-homogeneity on the two-electron energy levels for QRs with same ($\alpha = 1$), and different ($\alpha = 1.5$) radii. Other parameters of QRs are $R = 20a_0$*, $w = 1a_0$*, $h = 0.2a_0$*, $\beta = 0.1$

The energy levels of two electrons in vertically coupled QRs as a function of magnetic flux have been plotted in Fig. 3. We choose the radius of the lower ring $R = 20a_0$* and the separation between rings very small $(\beta = 0.1)$ in order to compare our results with those from reference [4], where it has been shown that in two-electron 1D QR such dependencies are given by a set of curves with oscillations and multiples crossovers similar to those of an image of the fingerprints. As magnetic fingerprints are changed with the increase of the radius, it has been concluded that the Coulomb interaction of two electrons in 1D QRs of quantum size effects and magnetic fingerprints. A similar analysis for 3D non-uniform two-electron structures is more complicated because it involves various additional factors, related to a wider variety of possible rings morphologies. The curves presented on the left side of Figure $3(\alpha = 1, \sigma = 0)$ for two identical, uniform, very thin and sufficiently narrow rings, located closely one to another, are practically coincide with those presented in Figure 6 of Reference [4] for two-electron 1D QR. As one of the two uniform rings has the radius 50% larger$(\alpha = 1.5, \sigma = 0)$, the picture of the magnetic fingerprints has more intersections of the energy level curves and it looks like a superposition of fingerprints of rings with radii $20a_0$ * and $30a_0$ *. One can see that the number of the energy levels, with quenched AB oscillations, increases with a rate, which is almost independent of the radii ratio, when the non-homogeneity parameter grows from 0 to 0.002

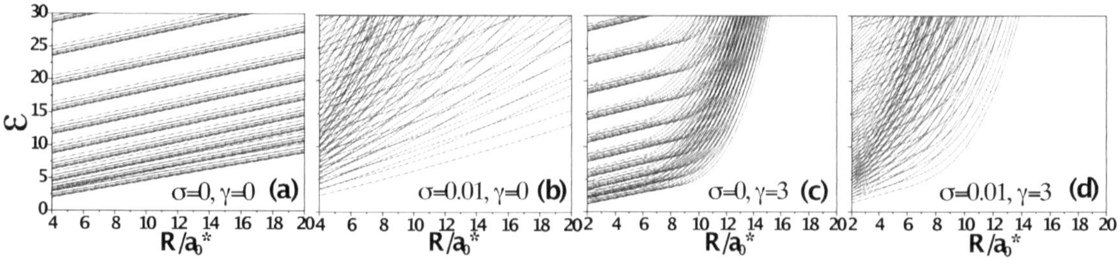

Figure 4. Renormalized energies as functions of radii vertically stacked rings in zero- (figures (a) and (b)) and in a strong magnetic field ($\gamma = 3$, $B \approx 20T$, figures (c) and (d)). Structure have the parameters $\alpha = 4$, $\beta = 0.1$, $h = 0.01$, $w = 0.1$

Finally, in Figure 4 we show the dependencies of normalized energies ($\varepsilon = E \cdot R^2$) on the lower (uniform) ring radius R, in zero and strong magnetic fields $(\gamma = 3)$for the case as the radius of the lower ring is much smaller ($\alpha = 4$) than of the upper. For radii given in Figure 4, the contribution of the normalized potential energy, which depends on the radius linearly, in the normalized Hamiltonian is larger than the contribution of the normalized kinetic energy, which is independent of radius. Therefore, the dependence of all energies on the radius is almost linear. It is seen from Figure 4 that in zero magnetic field case the levels are arranged into bands of sublevels with the same slopes, while when one of the rings is non-uniform the lines have two different slopes, providing many crossovers. External magnetic field adds diamagnetic term in normalized energy proportional to R^4. As the result the abrupt increase of the energies is observed for large radii as a strong magnetic field is applied.

4. Conclusion

We have calculated the energies of two-electron vertically stacked rings with different radii in a threading magnetic field, focusing our attention on the effects of the rings thickness non-uniformity. For the model with only one non-uniform ring, we have obtained new results. We show that in rings with different radii, the curves of the higher energies as functions of the magnetic flux, give a picture with higher number of crossovers than in the rings with equal radii, due to the presence of the Aharonov–Bohm oscillations with two different periods, while the oscillations of the lower energies are quenched similarly in both cases.

Acknowledgments

This work was financed by Universidad Nacional de Medellín (UN) through the DIME Cod. QUIPU 20101007748 and the Excellence Center of Novel Materials ECNM, under Contract No. 043-2005

References

[1] Lorke A, Luyken R J, Govorov A O, Kotthaus J P, Garcia J M and Petroff P M, 2000 *Phys. Rev. Lett.* **84** 2223; Reimann S M and Manninen M 2002 *Rev. Mod. Phys.* **74** 1283
[2] Römer R A and Raikh M E 2000 *Phys. Rev. B* **62** 7045; Govorov A O, Ulloa S E, Karrai K and Warburton R J, 2002 *Phys. Rev. B* **66** 081309(R)
[3] Wendler L, Fomin V M and Chaplik A V 1994 *Superlatt. and Microstructure.* **16** 311; Wendler L, Fomin V M 1995 *Phys. Rev.* **B 51** 17814; Wendler L, Fomin V M, Chaplik A V and Govorov A O 1996 *Phys. Rev. B* **54** 4794; Wendler L and Fomin V M 1995 *Phys. Rev.* B **51** 17814
[4] Zhu J-L, Dai Z and Hu X 2003 *Phys. Rev. B* **68** 045324
[5] Marín J H, García F and Mikhailov I D 2006 *Braz. Jour. of Phys.* **36**(3B) 940
[6] Pershin Y V and Piermarocchi C 2005 *Phys. Rev. B 72* 195340; Bruno-Alfonso A and Latgé A 2008 *Phys. Rev.* **B77** 205303
[7] Haft D, Schulhauser C, Govorov A O, Warburton R J, Karrai K, García J M; Schoenfeldd W, Petroff P M 2002 *Physica **E 13*** 165; Kuroda T, Mano T, Ochia T, Sanguinetti S, Sakoda K, Kido G and Koguchi N 2005 *Phys. Rev. B 72* 205301
[8] Marín J H, Gutiérrez W and Mikhailov I D 2010 *J. Phys.: Conference Series* **210** 012145

Time-dependent transport through a correlated quantum dot with magnetic impurity

D Becker[1], **S Weiss**[2], **J Eckel**[3], **M Thorwart**[1,3], **and D Pfannkuche**[1]

[1]I. Institut für Theoretische Physik, Universität Hamburg, D-20355 Hamburg, Germany
[2]Niels Bohr Instituttet, Nano-Science Centret, Universitetsparken 5, DK-2100 Copenhagen, Denmark
[3]FRIAS, Albert-Ludwigs-Universität Freiburg, Albertstr.19, D-79104 Freiburg, Germany

E-mail: daniel.becker@physnet.uni-hamburg.de

Abstract. We investigate electronic- and spin transport through a single level quantum dot with magnetic impurity in a symmetric forward bias setup. On the quantum dot, electrons either interact with each other due to Coulomb interaction or with the spin 1/2 magnetic impurity. For certain configurations, the tunnel coupling to the leads induces an exponential relaxation of the impurity spin, which has been prepared in a polarized state initially. Furthermore, we study the influence of the nonequilibrium transport current on the relaxation dynamics. We obtain the respective numerical result by means of the iterative summation of path integral (ISPI) scheme. Within this approach, observables of interest are calculated from a functional derivative with respect to appropriate source terms in the Keldysh partition function. The real-time path integral extends over all possible paths (i) of the impurity spin and (ii) of the Ising like fluctuating spin fields we have to introduce in order to decouple the quartic interaction term of the Anderson model. The ISPI scheme allows us to sum up all paths including the time non-local self energies of the leads.

1. Introduction

Understanding nonequilibrium quantum transport in correlated nanoscale systems is presently in the focus of condensed matter research. Nanoscale quantum systems with an internal magnetic or spin structure such as nanomagnets, spin-doped quantum dots or single-electron transistors are the prime model systems for the complex physics involved. Those nonequilibrium conditions are induced by the coupling to a macroscopic (quantum or classical) environment. Recent physical questions that arise connect different research fields such as fundamental quantum mechanics, possible realizations of quantum computers and the design of novel (spin-)electronic devices [1, 2, 3].

In this article, we study the real-time quantum transport behaviour of a single-level quantum dot (SLQD) with an additional fixed magnetic spin-1/2 impurity on the dot. Besides the magnetic electron-impurity interaction, Coulomb correlations between dot electrons with mutual spins are taken into account. The ISPI [4] scheme allows us to study numerically the system under nonequilibrium conditions, when all appearing energy scales as Coulomb coupling, temperature, and magnetic interaction are of the same order of magnitude. We present results for the relaxation of the impurity spin and find an exponential decay as a function of time.

2. Model System

The model system consists of a SLQD, which is coupled to two metallic leads L and R via identical tunnelling barriers and a gate electrode to adjust the dot's electrostatic potential $\Phi_{\rm dot}$. Inside the SLQD resides a spatially fixed spin 1/2 magnetic impurity $\boldsymbol{M}$, which interacts with the total spin of the dot electrons. Since the two-electron state $|{\rm d}\rangle$ has to be a singlet with total spin zero, this electron-impurity interaction is only present for single occupation. The model allows for Coulomb repulsion in electron state $|{\rm d}\rangle$ with scalar energy parameter $U \neq 0$ as well as a lifted degeneracy of electron- $|\sigma\rangle$ and impurity (spin) states $|\tau\rangle$ with $\sigma, \tau = \pm 1 (= \uparrow, \downarrow)$. We describe the system via the Anderson-type Hamiltonian $\hat{H} = \hat{H}_D + \hat{H}_{\rm leads} + \hat{H}_T + \hat{H}_{\rm INT}$, with the individual parts $\hat{H}_D$ for the SLQD (without impurity), $\hat{H}_{\rm leads} = \sum_p \hat{H}_p$ with $p = {\rm L}, {\rm R}\, (= \pm 1)$ representing the leads, the tunnelling/hybridization operator $\hat{H}_T$, and the electron-impurity interaction $\hat{H}_{\rm INT}$. The dot-part without impurity is given by $\hat{H}_D = \sum_\sigma \epsilon\, \hat{d}_\sigma^\dagger \hat{d}_\sigma + U \hat{d}_\uparrow^\dagger \hat{d}_\uparrow \hat{d}_\downarrow^\dagger \hat{d}_\downarrow$, where $\hat{d}_\sigma^\dagger$ is the creation operator for dot electrons with spin σ and ϵ is the level energy. The leads are described as free electron gases with $\hat{H}_p = \sum_{\boldsymbol{k}p\sigma} \epsilon_{\boldsymbol{k}p}\, \hat{c}_{\boldsymbol{k}p\sigma}^\dagger \hat{c}_{\boldsymbol{k}p\sigma}$ and tunnelling between leads and dot by $\hat{H}_T = \sum_{\boldsymbol{k}p\sigma} \gamma \hat{d}_\sigma^\dagger \hat{c}_{\boldsymbol{k}p\sigma} + H.c.$, where $\hat{c}_{\boldsymbol{k}p\sigma}^\dagger$ creates electrons in lead p with spin σ, momentum $\boldsymbol{k}$, and energy $\epsilon_{\boldsymbol{k}p} = \epsilon_{\boldsymbol{k}} - \mu_p$. The chemical potentials $\mu_p = pV_{\rm bias}/2$ are tuned by a symmetrically applied bias voltage $V_{\rm bias}$.

We model the spin-spin interaction between electron and impurity with Hamiltonian

$$\hat{H}_{\rm INT} = J \sum_{\sigma=\pm 1} (\hat{\tau}_z \sigma \hat{d}_\sigma^\dagger \hat{d}_\sigma + \hat{\tau}_\sigma \hat{d}_{-\sigma}^\dagger \hat{d}_\sigma / 2), \tag{1}$$

where J is the coupling strength. The longitudinal part (first term) of the interaction in (1) accounts for the shift of energy depending on the mutual spin alignment, whereas the transversal (second) term covers mutual spin flips of the impurity and electrons. The $\hat{\tau}_\varkappa$ with $\varkappa \in \{x, y, z, +, -\}$ denote the Pauli matrices for the impurity spin.

To calculate the non-equilibrium behaviour, we adapt the approach of the iterative summation of the path integral [4] to the present model. It is particularly suitable for systems with no small energy parameter and intermediate to long propagation times. We start by expressing the Keldysh partition function $\mathcal{Z}$ (see [5]) as path integral over spin fields and fermionic Grassmann variables. To this end, we discretise time propagation along the Keldysh contour $\mathcal{C}$ by introducing N equidistant vertices each on the forward and backward branch. This defines a discretisation step δ_t. Between two vertices on the *forward* branch, the system is propagated with time-evolution operator $\hat{U}_{\delta_t} = \exp\{-\boldsymbol{i}\hat{H}\delta_t\} = 1/2 \sum_{s=\pm 1} \exp\{-\boldsymbol{i}\hat{\mathcal{H}}\delta_t\}$ (operator $\hat{U}_{\delta_t}^\dagger$ on the *backward* branch). The Ising-like spin field s is introduced during a discrete Hubbard-Stratonovich transformation [4], by which we map the Hamiltonian $\hat{H}$ of Coulomb interacting electrons to $\hat{\mathcal{H}}$, which describes a system of free fermions coupled to s. $\hat{\mathcal{H}}$ is obtained by replacing the *quartic* $\hat{H}_{\rm dot}$ with the *quadratic* operator $\hat{\mathcal{H}}_{\rm dot} = \sum_\sigma (\epsilon_s^U + \boldsymbol{i}\lambda\sigma s) \hat{d}_\sigma^\dagger \hat{d}_\sigma$, where $\epsilon_\sigma^U = \epsilon_\sigma + U/2$ and $\lambda = \sinh^{-1} \sqrt{\sin(U\delta_t/2)} + \boldsymbol{i} \sin^{-1} \sqrt{\sin(U\delta_t/2)}$.

At every vertex, unity operator partitions of coherent states for dot (d) and lead (c) electrons as well as for the discrete impurity spin basis ($\mathbb{1} = \sum_\tau |\tau\rangle\langle\tau|$) are inserted. Hence, in addition to the integration over Grassmann fields c and d, the resulting path integral

$$\mathcal{Z} = \sum_{\tau_1,\ldots,\tau_{2N}} \sum_{s_1,\ldots,s_{2N}} \int \mathcal{D}[d\bar{d}c\bar{c}]\, (-1)^\ell \left(\frac{-\boldsymbol{i}J\delta_t}{2}\right)^m \boldsymbol{e}^{\boldsymbol{i}S[d\bar{d}c\bar{c}\tau s]} \prod_{j \in T_{\rm flip}^-} \bar{d}_{-\tau_j}^{j+1} d_{\tau_j}^j \prod_{k \in T_{\rm flip}^+} \bar{d}_{\tau_k}^k d_{-\tau_k}^{k-1} \tag{2}$$

has to be summed over all tupel (paths) of the spin fields τ and s. Equation (2) is exact up to a normalization constant, which does not contribute to observables and therefore is omitted. For each impurity path, m gives the number of flips (adjacent spins with opposite orientations), of which ℓ lie on the backwards contour. Since each term on the rhs of equation (2) is essentially an expectation value of a dot operator polynomial, only paths with even m (hence, an equal numbers of $\bar{d}_\sigma$ and d_σ fields) contribute. The $T^{\pm}_{\text{flip}}$ are sets of time indices (relating to real-time later vertex) of flips on the forward and backward branch in ascending order. The action reads

$$S = \int_C \mathbf{d}t \sum_{\chi \in \{\sigma\},\{\boldsymbol{k}p\sigma\}} \bar{f}_\chi(t) \boldsymbol{i}\partial_t f_\chi(t) - \mathcal{H}[d\bar{d}c\bar{c}\tau s] \qquad f = c \text{ if } \chi \in \{\boldsymbol{k}p\sigma\}, \text{ else } f = d, \tag{3}$$

where $\mathcal{H}$ is the result of replacing every operator in $\hat{\mathcal{H}}$ by a corresponding time-dependent field. To get the expectation value of an operator $\hat{O}$, we add the source term $-\eta\, O[d\bar{d}c\bar{c}\tau s]$ to S and arrive at the *generating function* $\mathcal{Z}[\eta]$. Again, the function O is obtained by substituting fields for operators. We can then write $\langle \hat{O} \rangle = -\boldsymbol{i}(\mathbf{d} \ln \mathcal{Z}[\eta]/\mathbf{d}\eta)|_{\eta=0}$. In the following, we are interested in the impurity orientation $\langle \hat{\tau}_z \rangle$ but also the charge current $\langle \hat{I} \rangle$ is implemented.

Since the integral over paths of Grassmann fields in $\mathcal{Z}[\eta]$ is Gaussian, they can all be integrated out, if we assume that the leads are in thermal equilibrium. All calculations are done in the wide-band limit for the leads' density of states $\rho(\epsilon_{\boldsymbol{k}p}) = \rho(\epsilon_{\text{F}}) = const.$ with Fermi energy ϵ_{F}, that is absorbed in the coupling $\Gamma := 2\pi\rho(\epsilon_{\text{F}})\,|\gamma|^2$. In the final expression for the generating function

$$\mathcal{Z}[\eta] = \sum_{\{\tau\},\{s\}} \prod_\sigma (-1)^\ell \left(\frac{J\delta_t}{2}\right)^m \boldsymbol{e}^{\boldsymbol{i}S_M} \det \boldsymbol{i}(G_\sigma)^{-1} \det \Xi_\sigma, \tag{4}$$

the determinants and the sum over paths $\{\tau\}$ and $\{s\}$ can be calculated iteratively, provided that not *both* V_{bias} *and* temperature T are zero or very small (see [4]). This is possible due to an exponential decay in real-time distance $|t - t'|$ of the free leads' Green's function, which sets the dominant time-scale for dynamical correlations in the central region. Depending on the choice of observable $\hat{O} = \hat{I}, \hat{\tau}_z$, the effective Green's function G_σ for spin σ dot electrons or the impurity action $S_{\boldsymbol{M}}$ carries the η-dependence. The spin flip matrices Ξ_σ have dimension $m/2$ and are constructed from G_σ as $\Xi^{kl}_\sigma = G_\sigma\{j_{2k}, j_{2l-1}\}$ for $\sigma = \tau_1$, $j_x \in T_{\text{flip}} = T^{+}_{\text{flip}} \cup T^{-}_{\text{flip}}$, and $G_\sigma\{j, j'\} = G_\sigma(t_j, t_{j'})$. For $\sigma = -\tau_1$, we have $\Xi^{kl}_\sigma = G_\sigma\{j_{2k-1}, j_{2l}\}$.

3. Results

We studied the dynamics of the impurity spin orientation $\langle \hat{\tau}_z \rangle$ in the presence of dot electrons, which are tunnelling on and off the dot and interact with the impurity. The initial state is set to be an empty dot with impurity wave function $|\Uparrow\rangle$. Figure 1(a) shows the real-time behaviour of $\langle \hat{\tau}_z \rangle$ for $\epsilon = 0\Gamma$, $U = 0.5\Gamma$, $T = \Gamma$, and $eV_{\text{bias}} = 0.6\Gamma$ and different values of the J. It can be seen, that the impurity spin decays exponentially towards an unpolarized state with $\langle \hat{\tau}_z \rangle(\infty) = 0$. The decay time decreases with larger interaction strength J. In figure 1(b), the influence of the bias voltage V_{bias} on the speed of this relaxation process is shown for $J = \Gamma$ (other parameters as in 1(a)). Although we find a weak increase with growing values of V_{bias} and, consequently, increasing current through the dot (see inset in 1(b)), it is not a non-zero current that causes relaxation. Rather, it already occurs for $V_{\text{bias}} = 0$ due to the lead-dot coupling.

The exponential decay of $\langle \hat{\tau}_z \rangle$ can be explained by a simple dynamical picture. As a result of the lead-dot hybridization, electrons tunnel stochastically on and off the dot. While a single electron is present, it interacts with the impurity via Hamiltonian (1), leading to mutual electron-impurity spin flips. The probability that a dot electron can induce flips during its stay on the dot increases with the rate J/Γ. The value of J is connected to the number of flips per time unit, while Γ is proportional to the time between two tunnelling events and hence to the number of transitions between flipping single-electron and non-flipping two- or zero-electron states.For the chosen parameters, the stationary state of dot and impurity spins is unpolarized. Hence, electrons of both spin directions are present on the dot with equal probability and dynamically

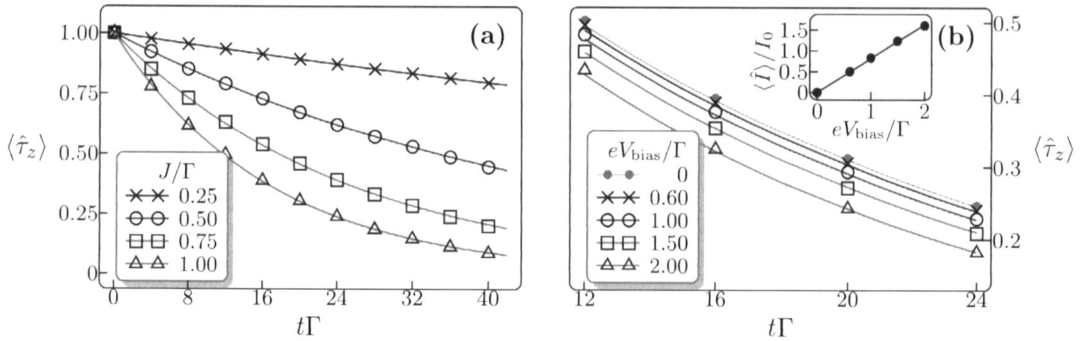

Figure 1. **(a)** Real-time relaxation of impurity spin orientation $\langle \hat{\tau}_z \rangle$ for different values of the electron-impurity interaction strength J ($\epsilon = 0\Gamma$, $U = 0.5\Gamma$, $T = \Gamma$, and $eV_{\rm bias} = 0.6\Gamma$). In this configuration, a small unpolarized current flows through the dot. Since the impurity spin $\boldsymbol{M}$ interacts with electrons of both spin directions with equal probability, the initial state with $\langle \hat{\tau}_z \rangle(0) = 1$ decays exponentially. All plot marks are numerical results, the solid lines are fitted curves with the decay time as single fit parameter. **(b)** Influence of the bias voltage $V_{\rm bias}$ on the relaxation of the spin polarization. For all plots J is equal to Γ (all other parameters as in (a)). With increasing values of $V_{\rm bias}$, the polarization decays faster. The inset shows the current $\langle \hat{I} \rangle$ for each value of $eV_{\rm bias}$ with $I_0 = e\Gamma/h$.

decrease the spin polarization to zero. Growing values of $eV_{\rm bias}$ between zero and 2Γ result in a rising current and a slight relative increase of the time ratio to find the dot in a spin flipping single-electron state, compared to the lower bias situation. This is the reason for the dependence of the spin relaxation time on $V_{\rm bias}$ that is shown in figure 1(b).

4. Conclusions

We investigated the real-time dynamics of a SLQD with spin-1/2 magnetic impurity. An initially polarized impurity spin shows long-time transient behaviour, when coupled to metallic leads. The relaxation time mainly depends on the strength of the spin-spin interaction J compared to the coupling Γ. Bias voltage and current have a considerably smaller impact on the relaxation time. In particular, even in the case of $V_{\rm bias} = 0$, the polarization will decay due to electrons that tunnel stochastically between dot and leads. We provided a fully quantum mechanical description of this model system, which is of particular interest, since it has a sufficiently rich internal structure to study the correlated real-time dynamics of electron current and magnetic structure. Our exact method allows to tackle the most difficult and so far unexplored regime, where the Coulomb and the magnetic interaction are of comparable size and no small parameter exists. These results open the way for a direct and efficient control of the magnetization of a quantum dot by a nonequilibrium current.

References

[1] Kouwenhoven L P, Markus C M, McEuen P L, Tarucha S, Westervelt R M and Wingreen N S 1997 *Mesoscopic Electron Transport, edited by L. L. Sohn, L. P. Kouwenhoven, and Gerd Schön* (Dordrecht, Boston, London: Kluwer Academic Publishers)
[2] Bird J P (ed) 2003 *Electron Transport in Quantum Dots* (Dordrecht, Boston, London: Kluwer Academic Publishers)
[3] Wolf S A, Awschalom D D, Buhrman R A, Daughton J M, von Molnár S, Roukes M L, Chtchelkanova A Y and Treger D M 2001 *Science* **294** 1488–1495
[4] Weiss S, Eckel J, Thorwart M and Egger R 2008 *Phys. Rev. B* **77** 195316
[5] Kamenev A and Levchenko A 2009 *Adv. Phys.* **58** 197

Theory of Piezoelectric Fields in InGaAs Site-Controlled Quantum Dots

Sorcha B. Healy[1] and Eoin P. O'Reilly[1,2]

[1]Tyndall National Institute, Lee Maltings, Cork, Ireland

[2]Department of Physics, University College Cork, Cork, Ireland

eoin.oreilly@tyndall.ie

Abstract. We present a surface integral method to calculate the piezoelectric polarisation potential due to a strained quantum dot (QD) grown on a (111)-oriented substrate. We show that the potential is reduced in a QD compared to a strained quantum well of the same height, but that there still remains a significant polarisation field across (111)-oriented InGaAs/GaAs QD structures. We use the 8-band **k.p** method to show that these piezoelectric fields must be included to explain the measured interband transition energies in a series of site-controlled QDs with increasing height and In composition.

1. Introduction

Most interest in InGaAs/GaAs quantum dots (QDs) has to date focused on dots grown on an (001) substrate. Such dots experience a small built-in piezoelectric potential, due to the shear components of strain, ε_{ij} $(i \neq j)$. It is well established that there can be a large piezoelectric polarisation, due to the shear strain, across a strained quantum well structure grown on a (111) substrate [1]. A large piezoelectric potential may also be expected in (111)-oriented quantum dots. Very high quality growth has recently been demonstrated of InGaAs/GaAs site-controlled QDs, grown by metal-organic vapour phase epitaxy on GaAs (111)B substrates, pre-patterned with inverted pyramids [2]. These dots offer both high uniformity and spectral purity. The inhomogeneous broadening of excitonic emission from an ensemble of QDs was found to be unusually narrow, with a standard deviation of 1.19 meV and the spectral purity of emission lines from individual dots was found to be very high (18–30 μeV). It is therefore timely to undertake a theoretical analysis of the electronic structure of such dots. We investigate here the role of the built-in piezoelectric potential in such structures. We present a surface integral method to determine the polarisation potential, which we then use to estimate the potential variation in (111)-oriented strained QDs. We then use the 8-band **k.p** method to calculate the effect of the piezoelectric potential in site-controlled QDs, showing that the built-in polarisation field has only a small effect in thin QDs with low In composition, but becomes of increasing importance, and must be taken into account, as the In composition and dot height are increased.

2. Derivation of Polarisation Potential in a (111)-oriented Quantum Dot

We derive here a semi-analytical expression to estimate the built-in polarisation potential in a (111)-oriented quantum dot. We follow the approach used by Davies [3]. This approach has been used previously to identify the key features of the piezoelectric potential in (001)-oriented QDs [3], providing results which reproduce well the trends found in more realistic atomistic calculations of both strain and piezoelectricity. The method omits any nonlinear dependence of the piezoelectric coefficients. Such terms are expected to be more important with increasing strain and In composition [4]. We expect, because we are considering dots with relatively low indium composition and strain, that the linear terms are still of greatest importance in the dots considered here. Davies showed that the polarisation potential $\varphi(\boldsymbol{r})$ at $\boldsymbol{r} = (x,y,z)$ can be given by an integral over the volume of the dot for a strained dot of constant composition buried in an infinite matrix:

$$\varphi(\boldsymbol{r}) = -C\int_{\text{Dot}} f(\boldsymbol{r},\boldsymbol{r}')\text{d}^3\boldsymbol{r}'; \quad C = \frac{9e_{14}\varepsilon_0}{4\pi\varepsilon_0\varepsilon_r}\frac{1+\nu}{1-\nu}; \quad f(\boldsymbol{r},\boldsymbol{r}') = \frac{(x-x')(y-y')(z-z')}{|\boldsymbol{r}-\boldsymbol{r}'|^5} \tag{1}$$

where e_{14} is the piezoelectric coefficient, ε_0 is the misfit strain of the dot to the matrix, ε_r is the dielectric constant, and ν is Poisson's ratio. The calculations assume that the elastic behaviour is linear and isotropic, with the same elastic constants everywhere, and also that e_{14} and ε_r have the same value in the dot and the matrix. Using the divergence theorem, the volume integral in Eq. (1) can be expressed as a surface integral by choosing a vector field $G(\boldsymbol{r},\boldsymbol{r}')$ for which $\nabla.G(\boldsymbol{r},\boldsymbol{r}') = f(\boldsymbol{r},\boldsymbol{r}')$. Davies introduced a number of functions G which satisfy this requirement, the simplest for a (001)-oriented dot being of the form $G(\boldsymbol{r},\boldsymbol{r}') = G_z(\boldsymbol{r},\boldsymbol{r}')\hat{z}$. This form of G is very useful to analyse dots grown on a (001) substrate, because it allows the piezoelectric potential to be evaluated by integrating over the (001)-oriented dot faces. We now derive a form of G suitable to analyse QDs grown on a (111) substrate.

We first rotate the basis vectors so that x_3 points along the (111) direction, with the new unit vectors defined in terms of the original Cartesian coordinates as

$$\hat{x}_1 = \frac{1}{\sqrt{6}}(1,1,-2); \qquad \hat{x}_2 = \frac{1}{\sqrt{2}}(-1,1,0); \qquad \hat{x}_3 = \frac{1}{\sqrt{3}}(1,1,1) \tag{2}$$

When we transform the integrand in Eq. (1) to be a function of the rotated coordinates, we obtain a large and unwieldy form for $f(\boldsymbol{r},\boldsymbol{r}')$. It is however possible to then find a function $G_3(\boldsymbol{r},\boldsymbol{r}')\hat{x}_3$ such that $\frac{\partial G_3}{\partial x_3'} = f$ with the potential given by $\varphi(\boldsymbol{r}) = C[I_1 + 5I_2 + I_3]$, where C is given in Eq. (1), and

$$I_1 = \int \frac{(x_3 - x_3')^2 \hat{n}_3.dS'}{|\boldsymbol{r}-\boldsymbol{r}'|^3}; \quad I_2 = \int \frac{\hat{n}_3.dS'}{|\boldsymbol{r}-\boldsymbol{r}'|}; \quad I_3 = \int \frac{\sqrt{2}(x_3 - x_3')[3|\rho-\rho'|^2 + 2(x_3 - x_3')^2]\cos(3\theta)\hat{n}_3.dS'}{|\rho-\rho'||r-r'|^3} \tag{3}$$

where $\rho = (x_1,x_2)$ and θ is the angle between $\rho - \rho'$ and the x_1 axis. At $x_1 = x_2 = 0$ (i.e. along the central x_3 axis of the dot) I_3 will integrate to zero if the dot is of sufficiently high symmetry. For a cylinder, truncated cone or indeed triangular pyramid, this is the case.

We have shown previously when studying the built-in potential in III-N wurtzite dots that the integrals I_1 and I_2 have analytical solutions along the central axis ($x_1 = x_2 = 0$) for a wide range of dot shapes [5]. We therefore use Eq. (3) to compare the piezoelectric potential in a (111)-oriented QD with that in a (111)-oriented quantum well (QW) of the same height. We consider the piezoelectric potential in a cylindrical QD of height h and radius R as a function of $f = h/R$, where $f = 0$ for a QW and $f \sim 0.04 - 0.25$ for the site-controlled QDs which we consider below. The potential drop $\varphi^0_{\max}$ across a (111)-oriented strained QW of height h is given from Eq.(3) as

$$\varphi^0_{\max} = \frac{2e_{14}h\varepsilon_0}{\sqrt{3}\varepsilon_0\varepsilon_r}\frac{1+\nu}{1-\nu} \tag{4}$$

We can solve Eq.(3) for finite R to show that the piezoelectric potential drop $\varphi_{\max}$ between the top and bottom surface of an isolated cylindrical dot will vary with f for small f as

$$\varphi_{\max} = \varphi_{\max}^{0}\left(1 - \frac{9}{8}f\right) \tag{5}$$

The solid lines with symbols in Figure 1 show the calculated variation of peak-to-peak potential as a function of dot radius, R, for a dot height h of 2 nm and of 8 nm, along with a fit to the calculated variation using Eq. (5). It can be seen that Eq. (5) is within 5% of the calculated value until $f = 0.36$ (R = 22 nm for h = 8 nm in Fig. 1). Two factors modify the strain-induced polarization potential in a QD: first, the effect of the finite dot size, which reduces the size of the region in which there is a piezoelectric-induced polarisation, and second the effects of the change in strain distribution in the QD compared to a QW of the same height. $\varphi_{\max}$ would vary only as $\varphi_{\max}^{0}$ $(1 - 0.5f)$ if the strain distribution was identical in the dot and in the well. This is a much weaker variation than that given in Eq. 5, showing that the strain redistribution in the dot compared to the well makes a significant contribution to the overall reduction in the piezoelectric potential in the dot.

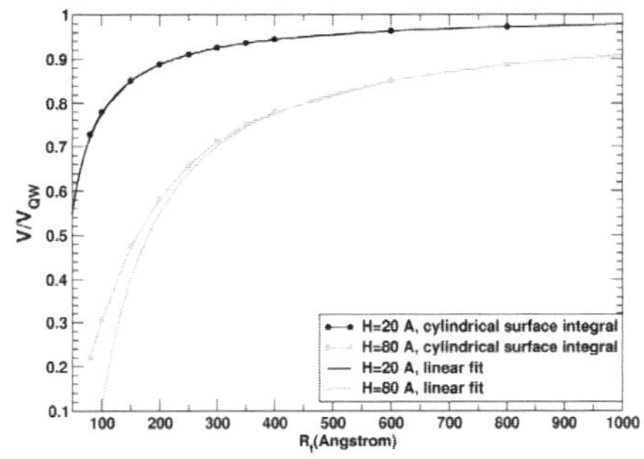

Figure 1: Calculated variation with radius of potential drop across a cylindrical dot of height 2 nm (black line) and 8 nm (green lines) compared with the analytical expression of Eq. (5).

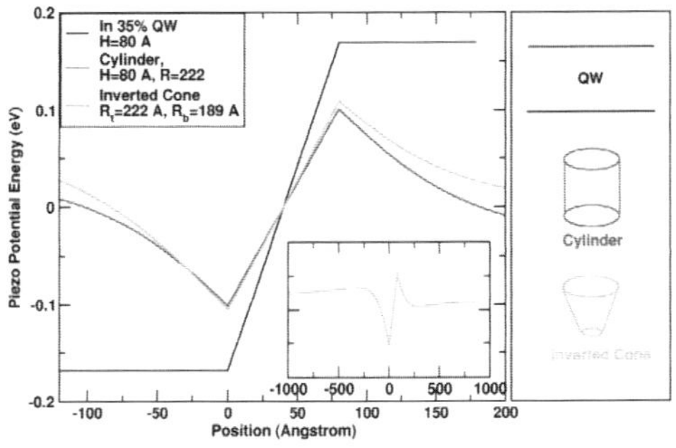

Figure 2: Line-scan of piezopotential across a 8nm $In_{.35}Ga_{.63}As$ QW compared to cylindrical and inverted cone QDs. Inset: Inverted cone over wider length range.

Fig. 2 shows the calculated variation of the polarization potential across an 8nm $In_{.35}Ga_{.65}As/GaAs$ QW, as well as the potential along the x_3 axis through the centre of a cylinder of the same height, and radius R = 22 nm, and for a 8 nm high truncated cone with radius R_t = 22 nm and R_b = 18.9 nm on the top and bottom surfaces respectively. The calculations in Fig. 2 assume a linear variation of piezoelectric coefficient e_{14} with In concentration. The reduction in potential difference across the dots compared to that across a QW of the same height and composition can be clearly observed. Nevertheless, we still calculate a potential drop of order 200 meV, equivalent to a built-in field of 2.5 $\times$ 10^5 V cm^{-1} across the QDs. We show below that polarisation fields of this scale can have a significant impact on the ground state transition energy in the QD.

The inset in Fig. 2 shows the potential variation over a larger distance along the x_3 axis of the conical QD. We see that the potential outside the QD returns towards zero, and can even change sign at larger distances from the dot. This is to be contrasted with the QW, where the potential remains constant outside the well. The change in sign of the potential at large x_3 values arises due to the strain distribution in the matrix material surrounding the dot. Because the dot expands along the x_3 axis, the surrounding matrix material must shrink along this direction, thus accounting for the observed variation in polarisation field and potential.

3. Transition Energies in Site-controlled InGaAs QDs

We have developed an 8-band **k.p** model to calculate the electronic structure in (111)-oriented QD structures. Details of the method can be found in [6]. The initial calculations which we presented using this method did not include the piezoelectric potential [7]. The results of these calculations were compared with measurements of the ground state transition energy in a series of site-controlled InGaAs/GaAs QDs with increasing In composition and nominal dot height. The nominal growth rate is lower than the actual growth rate for a site-controlled QD, with the dot growing at a faster rate than the surrounding side walls. We found good agreement in our initial calculations between the measured transition energies in a series of dots with nominal height of 0.5nm when we assumed the actual dot growth rate to be ~2.5 times faster than the nominal rate. However our calculations, without the piezoelectric potential, overestimated the transition energies for taller dots.

Figure 3 compares the calculated (open symbols) and measured transition energies (solid symbols) for dots of various In compositions and heights, with the piezoelectric potential included in the calculations. The previous calculations underestimated by about 50 meV the transition energy of 35% In dots with nominal height of 2 nm. It can be seen that there is now good agreement between theory and experiment across the full composition and height range considered, confirming the importance of including piezoelectric effects in (111)-oriented InGaAs QDs with higher In composition. The dots were prepared for measurement in two different ways: either by surface-etching or back-etching. A systematic reduction in transition energy was observed in the back-etched samples [2], so the back-etched samples in Fig. 3 have all been shifted upwards in energy to account for this processing-induced shift in emission energy.

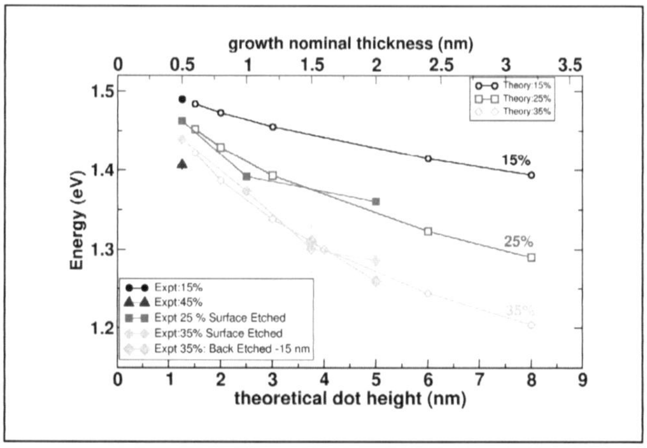

Figure 3: Comparison of theoretical (pen symbols) and experimental (filled symbols) interband transition energies for inverted pyramidal site-controlled dots of varying heights and In concentration.

4. Conclusions

In summary, we presented a simple analytical framework to estimate the size of the piezoelectric potential in (111)-oriented strained quantum dots. We showed that although the field is reduced compared to the value in a quantum well of the same composition it can still significantly influence the transition energy in dots with higher In composition, in agreement with the results of recent experimental measurements on a series of (111)-oriented site-controlled InGaAs/GaAs QD structures.

Acknowledgements This work was supported by Science Foundation Ireland. We thank Stefan Schulz, Emanuele Pelucchi and Paudie Scanlon for useful discussions.

References

[1] D. L. Smith and C. Mailhiot, Rev. Mod. Phys. **62**, 173 (1990)
[2] L. O. Mereni, V. Dimastrodonato, R. J. Young, E. Pelucchi, Appl.Phys.Lett. **94**, 223121 (2009)
[3] J.H. Davies, J. Appl. Phys. **84**, 1358 (1998)
[4] G. Bester, X. Wu, D. Vanderbilt and A. Zunger, Phys.Rev. Lett. **96**, 187602 (2006)
[5] D.P. Williams, A.D. Andreev, E.P. O'Reilly and D.A. Faux, Phys. Rev. B. **72**, 235318 (2005)
[6] S.B. Healy and E.P. O'Reilly, IEEE J. Quantum Electron. **46**, 742 (2010)
[7] S. B. Healy *et al.*, Physica E (to be published; doi:10.1016/j.physe.2009.11.099)

Journal of Physics: Conference Series **245** (2010) 012023 doi:10.1088/1742-6596/245/1/012023

Spin transitions in semiconductor quantum rings

Benjamin Baxevanis and Daniela Pfannkuche

I. Institute for Theoretical Physics, University of Hamburg, 20355 Hamburg, Germany

E-mail: `benjamin.baxevanis@physik.uni-hamburg.de`

Abstract. We adopt the path integral Monte Carlo method to accurately resolve the total spin of the ground state of electrons confined in a quantum ring with different geometries. Using this method, an evaluation of the ground state of three electrons in a ring shows a spin transition to the fully polarized state by increasing the radius and thereby enhancing the Coulomb interaction. The total spin of the ground state is determined by the mutual interplay of confinement and electron-electron interaction. An analysis of the four-electron ring demonstrates that in this case no spin transitions take place. Furthermore, the effect of geometric distortion of the ring on its ground state has been investigated. Elliptically deforming the ring breaks the symmetry of the system and leads to the removal of orbital degeneracy. For strong distortion the splitting between hybridized states is sufficient to overcome the exchange-energy saving associated with a higher spin state. We have found that this effect removes the polarization of three electrons. Even in a four-electron ring the ground state is forced by the distortion to be unpolarized and thus suppressing the Hund's rule ground state.

1. Introduction

The interest in spin effects in semiconductor nanostructures has grown during the last decades due to their possible application as spintronic devices. Most prominent representatives are quantum dots which can be manufactured to contain a defined small number of electrons. Particularly methods for manipulating and preparing the spins in nano devices are required to build spintronic devices. Among other applications, the use of quantum dots as spin filters, memory devices and qubits in quantum computers are proposed [1, 2].

Ring shaped quantum dots, referred to as quantum rings, are a promising class of nanostructures because the electrons are additionally confined on a circular orbit which will be shown to enhance spin effects. Different radii and ring widths can be experimentally realized depending on the process of manufacture [3, 4]. The ground-state transition solely induced by changing the geometry of the ring is a salient question. Recent theoretical investigations indicated that changes in ring size are accompanied by an alteration of the total spin of the ground state [5, 6]. We have performed a comprehensive numerical study of the dependence of the ground-state spin on two ring parameters which are diameter and elliptical distortion. The results showed in particular that the system can be driven to ferromagnetism by increasing the diameter.

2. Model and method

The Hamiltonian of N interacting electrons in a quantum ring is written as

$$\mathcal{H} = \sum_{i}^{N} \mathcal{H}_i^{(1)} + \sum_{i<j}^{N} \frac{e^2}{\kappa |\mathbf{r}_i - \mathbf{r}_j|}, \tag{1}$$

where κ denotes the dielectric constant of the bulk and the ring is assumed to be perfectly flat, i.e. the electron motion is strictly two-dimensional. The one-particle Hamiltonian $\mathcal{H}_i^{(1)}$ describes an electron confined in radial direction by a displaced harmonic confinement

$$\mathcal{H}_i^{(1)} = \frac{\mathbf{p}_i^2}{2m^*} + \frac{1}{2} m^* \omega_0^2 (r_i - r_0)^2, \tag{2}$$

where r_0 is the radius of the ring, r_i the radial coordinate of the electron and m^* its effective mass. The electrons move freely in angular direction and are bounded in radial direction. We may relate the confinement frequency ω_0 to the width of the ring defined as $W = 2l_0$, where $l_0 = \sqrt{\hbar/m^*\omega_0}$ is the characteristic length of the displaced harmonic oscillator. In the following, lengths and energies are given in effective Bohr radii $a_0^* = \frac{\kappa\hbar^2}{m^* e^2}$ and effective Hartree $E_H^* = \frac{\hbar^2}{m^* a_0^{*2}} = \frac{e^2}{\kappa a_0^*}$, respectively. For example, taking GaAs rings the units of length and energy are $a_0^* = 9.8$nm and $E_H^* = 12$meV. The ring radius r_0 of experimentally realized quantum rings is of the order of 10 to 100 nm while the confinement $\hbar\omega_0$ varies between 1 up to a few meV [3, 4].

The purpose of our calculations is to determine the ground state of the Hamiltonian (1) during the crossover between different ring geometries. We utilize the path integral Monte Carlo method which yields numerically exact results. The partition function for N electrons at an inverse temperature $\beta = 1/k_B T$ with k_B the Boltzmann constant reads [7, 8]

$$Z = \frac{1}{N!} \sum_{\mathcal{P}} (-1)^p n(\mathcal{P}) \int d\mathbf{R} \langle \mathbf{R} | e^{-\beta\mathcal{H}} | \mathcal{P}\mathbf{R} \rangle, \tag{3}$$

with

$$n(\mathcal{P}) = \sum_{\sigma_1 \dots \sigma_N = \pm\frac{1}{2}} \delta_{\sigma_1, \mathcal{P}(\sigma_1)} \dots \delta_{\sigma_N, \mathcal{P}(\sigma_N)}. \tag{4}$$

A set of coordinates $\mathbf{R} = (\mathbf{r}_1, \dots, \mathbf{r}_N)$ is introduced with $\mathbf{r}_i$ being the position of the i-th particle with spin projection σ_i. $\mathcal{P}$ denotes a permutation (with parity p) of the particle labels. In order to numerically evaluate the partition function, the diagonal elements of the density matrix $e^{-\beta\mathcal{H}}$ are expanded in a path integral

$$\langle \mathbf{R} | e^{-\beta\mathcal{H}} | \mathcal{P}\mathbf{R} \rangle = \int d\mathbf{R}_1 \int d\mathbf{R}_2 \dots \int d\mathbf{R}_{M-1} \times \langle \mathbf{R} | e^{-\frac{\beta}{M}\mathcal{H}} | \mathbf{R}_1 \rangle \dots \langle \mathbf{R}_{M-1} | e^{-\frac{\beta}{M}\mathcal{H}} | \mathcal{P}\mathbf{R} \rangle, \tag{5}$$

where M is referred to as the number of time slices. Using (5), the partition function can be evaluated by statistically sampling the discrete path $\{\mathbf{R}, \mathbf{R}_1, \dots, \mathbf{R}_{M-1}, \mathcal{P}\mathbf{R}\}$ [8]. Thermodynamic observables are computed using the partition function, specifically the free energy

$$E = -\frac{\partial \ln Z}{\partial \beta}. \tag{6}$$

In the following, the temperature is set to $k_B T = 0.1\hbar\omega_0$ which is low enough to ensure that the partition function receives contributions mainly from the lowest energy state but remains still computationally feasible. The objective is to discriminate the lowest energy states by their total spin which is preserved in the system. This can be achieved by diagonalizing the density matrix in the space of S^2 states by employing group theory which yields new coefficients $n(\mathcal{P})$ in equation 4 [9]. In the case of two electrons, for instance, the density matrix can be decomposed in two sectors corresponding to $S = 0$ and $S = 1$. By taking the partial trace relative to a basis with fixed total spin S, energies E_S can be obtained. The total spin of the ground state can be accurately specified monitoring the sign of the energy difference $\Delta_{S,S'} = E_{S'} - E_S$. The ground state exhibits a transition of its total spin from S to S' if the sign of $\Delta_{S,S'}$ becomes negative.

3. Results

3.1. Circular quantum ring

The dominant interaction in a quantum ring strongly depends on the radius. The kinetic energy of a narrow circular ring scales roughly with the inverse square of the radius, whereas the interaction energy drops inversely with increasing ring radius. The continuous crossover from a weakly to a strongly interacting system is explored by calculating the energy differences $\Delta_{S,S'}$ as can be seen in figure 1 (a). The confinement energy $\hbar\omega_0 = 2.0E_H^*$ is kept fixed while tuning the ring radius. A spin transition of the ground state is seen in the case of three electrons populating the ring while for four electrons no transition occurs. Regarding small radii, the ground state is $S = \frac{1}{2}$ which can be understood considering the energy levels of the system. Supposing that the radial motion of the electrons is much slower than the circular motion ($\omega_0 r_0^2 > 1$), the energy levels are given by a rigid rotator with a two-fold degeneracy (left/right motion) and a non-degenerate ground state. The weakly interacting electrons occupy the lowest single-particle levels giving rise to the $S = \frac{1}{2}$ ground state. Increasing the ring radius leads to a spin transition to $S = \frac{3}{2}$. The gap between the single-particle energy states is reduced leading to a fully polarized system which minimizes the exchange energy. Finally, the energy difference between all spin-states vanishes for large radii while approaching a classical limit. Such spin transition, however, is absent in a four-electron system (cp. figure 1 (a)). The ground state is given by Hund's rule with a total spin of $S = 1$ over the whole range of possible radii. The lowest single-particle levels are occupied by the electrons. In the degenerate level, however, the electrons maximize their spin reducing the exchange energy. Single-particle states with higher angular momentum are not occupied because their energy is too large. In the limit of large radii, again, all spin states coincide energetically, but without an alteration of the ground-state spin.

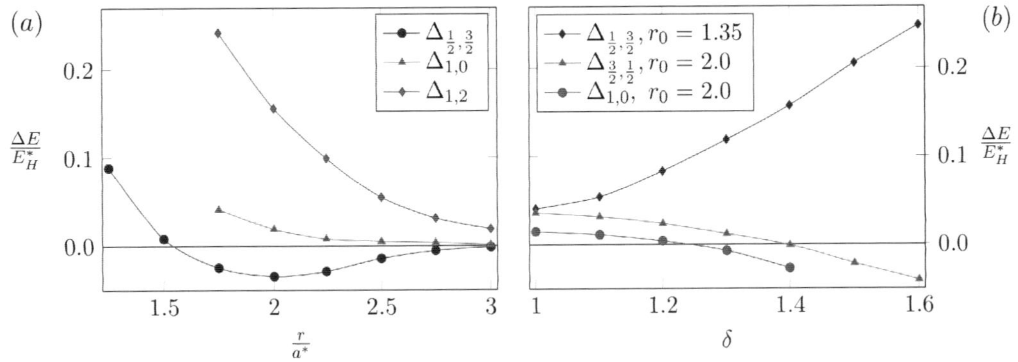

Figure 1. The energy difference $\Delta_{S,S'} = E_{S'} - E_S$ in a three and four electron ring depending on (a) the ring radius r_0 and (b) the elliptical deformation δ. In the former case, only the three electron system exhibits a spin transition $S = \frac{1}{2} \to S = \frac{3}{2}$ while the four electron ring retains the $S = 1$ ground state. In the latter case, the quantum ring can be completely depolarized for large radii ($r_0 = 2.0$).

3.2. Distorted quantum ring

In this subsection we investigate quantum rings with elliptical shape which form due to crystal strain. In our model, the coordinates are subject to the transformation

$$x^2 \rightarrow \frac{x^2}{\delta} \text{ and } y^2 \rightarrow \delta \cdot y^2, \tag{7}$$

i.e. one semi-axis is stretched whereas the other semi-axis is shortened. Two situations can be observed in the three electron system (cp. figure 1 (*b*)). In the first case, the ring radius is chosen to be $r_0 = 1.35a^*$ so that the ground-state spin is $S = \frac{1}{2}$. The distortion of the ring even increases the energy splitting between the $S = \frac{1}{2}$ ground state and the excited $S = \frac{3}{2}$ state. Deformation of the ring reduces the symmetry of the system and thus removes the rotational degeneracy of the single-particle energies. As a consequence, the $S = \frac{1}{2}$ state mainly occupies the two lowest single-particle states and reduces its energy. In contrast, in the $S = \frac{3}{2}$ configuration a third single-particle energy level is populated. The three electrons, however, can be initially prepared in a spin-polarized state choosing a radius of $r = 2.0a^*$. As in the previous case, the system can be forced to an unpolarized ground state as the ring gets distorted.

Regarding four electrons (figure 1 (*b*)), only the energy difference between $S = 1$ and $S = 0$ is given since the $S = 2$ energy increases monotonically during the distortion process. However, a spin transition of the ground state is taking place as the distortion proceeds. The Hund's rule ground state of the circular ring is replaced by an $S = 0$ ground state. In this state the lowest two single-particle energies are occupied. The energy lowering of single-particle states is sufficient to overcome the exchange-energy savings associated with the Hund's rule $S = 1$ state [10]. In both discussed cases (ring with three and four electrons) the deformation leads to a depolarized ground state. The described effects are related to the competing energy scales of the confinement and the Coulomb interaction. With an elliptical deformation an additional parameter emerges which tunes the gap between previously degenerate single-particle energy levels. Two regimes can be distinguished. In the first case, the system tries to reduce the Coulomb energy leading to a polarized spin configuration associated with exchange-energy saving. In the second case, the ground state of the system is determined by the large spacing between the single-particle energy levels. The system tends to occupy the lowest energy levels and thus minimizes the total spin. In this way the system can be tuned between a polarized and unpolarized ground state.

4. Discussion

In a quantum ring the spin of the ground state strongly depends on ring geometry. The quantum ring can be prepared in a polarized as well as a depolarized ground state for certain electron numbers. The findings are not only important for equilibrium properties but also for transport through rings. Especially, the occurring polarization of the ring can lead to negative differential conductances and even to the phenomenon of spin blockade in transport measurements [11].

References

[1] Loss D and DiVincenzo D P 1998 *Phys. Rev. A* **57** 120
[2] Recher P, Sukhorukov E V and Loss D 2000 *Phys. Rev. Lett.* **85**, 1962
[3] Lorke A, Luyken J R, Govorov A O, Kotthaus J P, Garcia J M and Petroff P M 2000 *Phys. Rev. Lett.* **84** 2223
[4] Fuhrer A, Luscher S, Ihn T, Heinzel T, Ensslin K, Wegscheider W and Bichler M 2001 *Nature* **413** 822
[5] Yang N, Zhu J L and Dai Z 2008 *J. Phys.: Condens. Matter* **20** 295202
[6] Saiga Y, Hirashima D S and Usukura J 2007 *Phys. Rev. B* **75** 045343
[7] Feynman R P and Hibbs A R 1965 *Quantum Mechanics and Path Integrals* (New York: McGraw Hill)
[8] Ceperley D M 1995 *Rev. Mod. Phys.* **67** 279
[9] Lyubartsev A P and Vorontsov-Velyaminov P N 1993 *Phys. Rev. A* **48** 4075
[10] Tews M, Pfannkuche D 2002 *Phys. Rev. B* **65** 073307
[11] Weinmann D, Häusler W and Kramer B 1994 *Phys. Rev. Lett.* **74** 984

A superconducting resonator designed for coupling to spin based qubits in quantum dots

G Allison[1,2], A Oiwa[1,3], S Kumar[4], D DiVincenzo[4], M Ketchen[4], K Hirakawa[3,5,6], H Takayanagi[3,7] and S Tarucha[1,6]

[1]Department of Applied Physics, The University of Tokyo, 7-3-1 Hongo, Bunkyo-ku, Tokyo 113-8656, Japan
[2]Department of Physics, Princeton University, Princeton, NJ 08544, USA
[3]JST CREST, 4-1-8 Hon-cho, Kawaguchi-shi, Saitama 332-0012, Japan
[4]IBM T. J. Watson Research Center, Yorktown Heights, NY 10598, USA
[5]IIS, The University of Tokyo,4-6-1 Komaba, Meguro-ku, Tokyo 153-8505, Japan
[6]INQIE, The University of Tokyo, 4-6-1 Komaba, Meguro-ku, Tokyo 153-8505, Japan
[7]Department of Physics, Tokyo University of Science, 1-3 Kagurazaka, Shinjuku-ku, Tokyo 162-8601, Japan
giles@meso.t.u-tokyo.ac.uk

Abstract. We present the implementation of a system for the quantum state transfer between electron spin and photons exploiting the coupling of an InAs self-assembled quantum dot to a superconducting resonator cavity.

1. Introduction

Manipulation of the electron spin in a quantum dot has long been proposed as a suitable candidate for a quantum bit (qubit) due to the long spin coherence times involved [1]. However, electron spin is an unsuitable property for long distance interactions between qubits, thus, for any large scale quantum computing network, it is necessary to transfer information between different quantum states. In previous studies, several types of two-level systems, such as superconducting plasmons and transmons [2-4], have been coupled to the discrete photon modes in high quality resonators. This has lead to the recent demonstration of a two qubit superconducting processor and implementation of quantum search algorithms [5]. Microcavity systems [6] and photonic crystals [7] coupled to the exciton states of a quantum dot have also been demonstrated in the strong coupling regime. In this work we propose a system for the quantum state transfer between electron spin and photons exploiting the coupling of an InAs self-assembled quantum dot (SAQD) to a resonator cavity [8]. The integral part of our device is the strong spin-orbit interaction (SOI) of InAs SAQDs, which realizes the coupling between the photon field and electron spin in the cavity [9]. The spin coherence times of InAs based SAQDs are longer than that of superconducting qubits [10-11].

In this work we present fabrication methods and characterization measurements of quarter-wavelength coplanar waveguide superconducting resonators coupled to SAQDs. Niobium is deposited using a sputtering method onto a GaAs substrate and high q-factor resonators are patterned using electron beam lithography (EBL) and reactive ion etching (RIE) techniques. A novel method has been devised to protect SAQDs on the surface of the GaAs during fabrication and for accurate positioning such that a single SAQD is sufficiently close to an antinode of the first harmonic standing wave of the cavity. The magnitude of the SOI in such InAs SAQDs is large (~100 μeV [12]) resulting in an electron Landé g-factor of 4-5, thus, a small magnetic field, B~0.1 T, brings into resonance with the bare frequency of the resonator (f_0=5-7 GHz) where vacuum Rabi splitting is observed and qubit-cavity coupling strength determined.

2. Background

Theoretical studies of electron spins in nanowire quantum dots placed within transmission line resonators have revealed possible coherent manipulation, storage and readout of quantum information due to the spin-orbit interaction [9]. This all electrical system has the possibility of coupling quantum dot spins in different quantum dots through the resonator modes. InAs self-assembled quantum dots should show a similar behaviour and have a great advantage due to their large spin-orbit interaction. In [9] it is shown that the Jaynes-Cummings [13] coupling of this system is given by

$$M = 0.25eEH\frac{E_z}{\Delta E_0}\frac{L}{\lambda_{SO}} \tag{1}$$

where e is the electron charge, H and L are the height and length of the SAQD respectively, E_z the Zeeman energy, ΔE_0 the energy level, and λ_{SO} the spin orbit coupling length. E is the effective electric field in the cavity. SAQDs with L~200nm have been shown to have Landé g-factor 3-8 [14], therefore a magnetic field of B~0.1T is required to bring the dot to the resonant frequency of 5-7 GHz. The strong spin orbit coupling Δ_{SO} ~150 μeV [12] results in a coupling of M=0.3-4 MHz. The coupling constant value, however, can be widely altered by the choices of the QD parameters.

3. Sample fabrication & measurement method

The uncapped SAQDs are grown on an insulating GaAs substrate and are randomly positioned with varying size up to a length of ~300nm and height ~40nm [15]. The first stage of the sample fabrication uses atomic force microscopy to find a SAQD of suitable size. Secondly, EBL techniques are used to deposit a thin (100nm) film of SiO_2 which covers just the SAQD of interest and will act as an insulating buffer layer in the latter processing stages. Next, all the unwanted SAQDs are removed using an H_2SO_4:H_20 (1:5) etch for a few seconds. A thin (200nm) layer of Niobium is then deposited on the sample surface using a sputtering technique before EBL and RIE techniques are used to form the transmission line and quarter wavelength resonators.

There are two factors crucial to the performance of the device; firstly the resonator must be accurately positioned so that the SAQD lies in the resonator cavity close to an antinode of the first harmonic, and secondly, the RIE rate and film thickness must be accurately determined to ensure that the SAQD is not damaged during this process.

Ten quarter wavelength resonators of width 6μm and different lengths, and thus different resonant frequencies, of the order 4mm are capacitively coupled to a superconducting feedline and in turn each resonator is coupled to a SAQD. A gate electrode is added to give independent control of the electron number and state. A schematic of the resonator design is shown in figure 1. The transmission through the feedline is measured as a function of frequency by a vector network analyser (Agilent model E5071C) and a dip in the transmission is observed at the resonant frequency of each resonator. Cryogenic amplifiers and attenuators are used to control the RF power as shown in figure 2 and the sample is cooled to T~50mK using a dilution refrigerator placed within a superconducting magnet.

In the work presented here we measure the properties of a superconducting Nb resonator fabricated using these techniques, however, this device currently contains no SAQDs. The purpose of these tests is to check the quality, power and magnetic field dependence of our resonators to ensure the feasibility of the proposal for quantum state transfer.

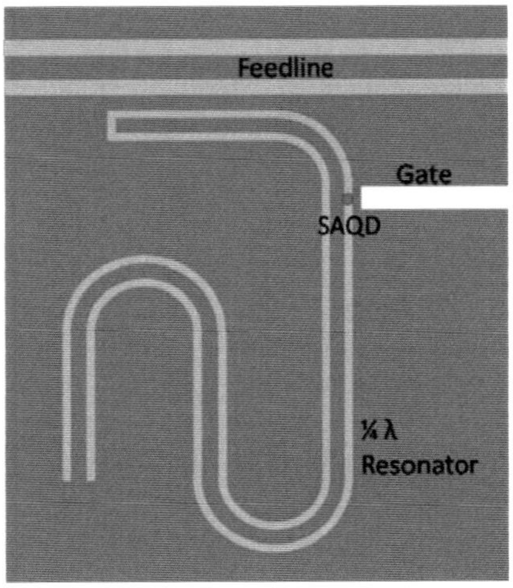

Figure 1. Schematic of the quarter wavelength design including SAQD, gate electrode and transmission feedline.

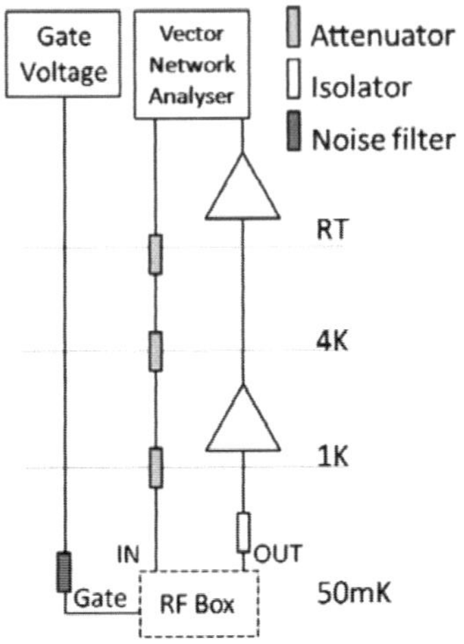

Figure 2. Schematic circuit diagram.

4. Results

High quality factor resonators are required in order to reach the strong coupling regime, thus it is extremely important that our fabrication techniques result in a good quality resonator. In this section we show characterisation measurements of our resonators at low power and low temperature. The resonances have quality factor of the order 3,000 (estimated from the full width at half height, a more accurate fit can be obtained using [16]). The resonances were observed at powers below -100dBm, the power at which the single photon excitation condition is reached (see figure 3).

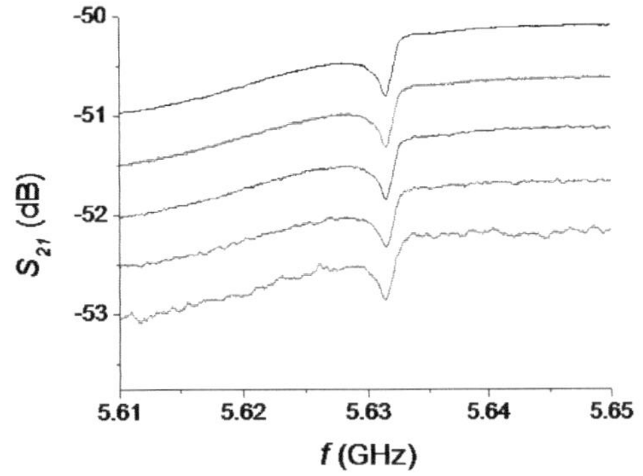

Figure 3. Example of the transmission of a resonance as a function of frequency at powers -60 to -100dBm in steps of 10dBm (top to bottom, offset for clarity).

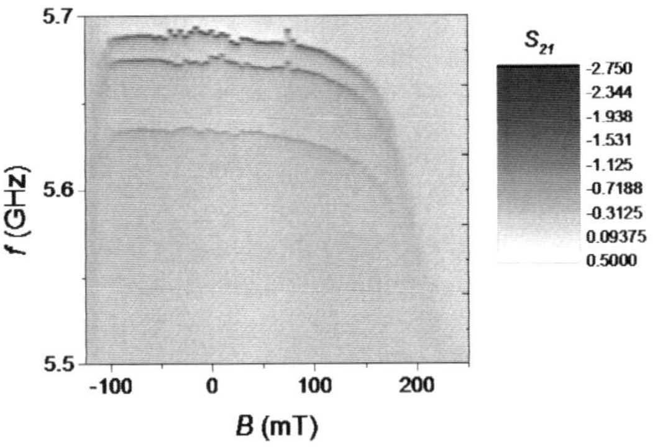

Figure 4. Magnetic field dependence of three resonances.

The resonances were measured as a function of magnetic field as shown in figure 4. The resonant frequency decreases weakly with magnetic field for B<0.15T, which is followed by a fairly rapid decrease before disappearing altogether at B~0.25T. The origin of the rapid disappearance of the resonance is likely to be due to kinetic inductance [17]. For a SAQD with a g-factor of 3-5 we expect coupling at B~0.1T, well below this critical field. It should be noted that the disappearance of the resonances is not due to the loss of superconductivity of the Nb; in separate measurements of the dc transport of a device with the same geometry as a typical resonator, the superconducting behaviour is observed in fields as large as B=3T. The data show a non-monotonic background due to reflections in the transmission line, however, in figure 4 this background has been removed to a good approximation by subtracting the high magnetic field data.

5. Conclusion

We have proposed a system for the quantum state transfer between photons and electron spins. In our preliminary measurements we have shown that our resonator devices work in the magnetic field range required for the Zeeman energy of a SAQD to equal the resonant energy of the device. We have also shown that the quality factor is sufficient to produce a measurable coupling between dot and resonator. The next stage is to implement the SAQD into the device and to measure the coupling strength between dot and resonator.

6. Acknowledgements

We acknowledge financial support from Grant-in-Aid for Scientific Research S (No. 19104007) (S.T.) and A (No. 21244046) (A.O.) and supported by DARPA QuEST grant HR0011-09-1-0007 (G.A. and S.T.).
This work was partially supported by Strategic Internation Cooperative Program (Joint Research Type), JST.

References

[1] Koppens F H L, Buizert C, Tielrooij K J, Vink I T, Nowack K C, Meunier T, Kouwenhoven L P and Vandersypen L M K 2006 *Nature* **442** 766
[2] Blais A, Huang R-S, Wallraff A, Girvin S M and Schoelkopf R J 2004 *Phys. Rev. A* **69** 062320
[3] Wallraff A, Schuster D I, Blais A, Frunzio L, Huang R-S, Majer J, Kumar S, Girvin S M and Schoelkopf R J 2004 *Nature* **431** 162
[4] Fink J M, Bianchetti R, Baur M, Göppl M, Steffen L, Filipp S, Leek P J, Blais A, and Wallraff A 2009 *Phys. Rev. Lett.* **103** 083601
[5] DiCarlo L, Chow J M, Gambetta J M, Bishop L S, Johnson B R, Schuster D I, Majer J, Blais A, Frunzio L, Girvin S M and Schoelkopf R J 2009 *Nature* **460** 240
[6] Reithmaier J P, Sęk G, Löffler A, Hofmann C, Kuhn S, Reitzenstein, S Keldysh L V, Kulakovskii V D, Reinecke T L and Forchel A 2004 *Nature* **432** 197
[7] Hennessy K, Badolato A, Winger M, Gerace D, Atatüre M, Gulde S, Fält S, Hu E L and Imamolu A 2007 *Nature* **445** 896
[8] Imamoglu A, Awschalom D D, Burkard G, DiVincenzo D P, Loss D, Sherwin M and Small A 1999 *Phys. Rev. Lett.* **83** 4204
[9] Trif M, Golovach V N and Loss D 2008 *Phys. Rev. B* **77** 045434
[10] Greilich A, Shabaev A, Yakovlev D R, Efros Al. L, Yugova A, Reuter D, Wieck A D and Bayer, M 2007 *Science* **317** 1896
[11] Greilich A, Yakovlev D R, Shabaev A, Efros Al L, Yugova I A, Oulton R, Stavarache V, Reuter D, Wieck A and Bayer M 2006 *Science* **313** 341
[12] Takahashi S, Igarashi Y, Deacon R S, Oiwa A, Shibata K, Hirakawa K and Tarucha S 2009 *J. Phys. Conf. Ser.* **150** 022084; Takahashi S, Deacon R S, Yoshida K, Oiwa A, Shibata K, Tokura Y and Tarucha S 2009 *cond.mat.* 0912.3088, in press *Phys. Rev. Lett.*
[13] Jaynes E and Cummings F 1963 *Proc. IEEE* **51** 89
[14] Igarashi Y, Jung M, Yamamoto M, Oiwa A, Machida T, Hirakawa K and Tarucha S 2007 *Phys. Rev. B* **76** 081303(R) and reference 8 therein.
[15] Shibata K, Jung M, Cha K M, Sotome M, and Hirakawa K 2009 *Appl. Phys. Lett.* **94** 162107
[16] Kumar S, Gao J, Zmuidzinas J, Mazin B A, LeDuc H G, Day P K 2008 *Appl. Phys. Lett.* **92** 123503
[17] Healey J E, Lindström T, Colclough M S, Muirhead C M, Tzalenchuk A Ya 2008 *Appl. Phys. Lett.* **93** 043513

A Theoretical Analysis of Instantaneous Coulomb Renormalizations in a Single Quantum Dot Pump-Probe Experiment

J Huneke[1,2], T Kuhn[1], I D'Amico[2] and V M Axt[3]

[1] Institut für Festkörpertheorie, Westfälische Wilhelms-Universität Münster, Wilhelm-Klemm-Straße 10, 48149 Münster, Germany
[2] Department of Physics, University of York, Heslington York YO10 5DD, United Kingdom
[3] Institut für Theoretische Physik III, Universität Bayreuth, 95440 Bayreuth, Germany

E-mail: jan.huneke@uni-muenster.de

Abstract. We present a theoretical analysis of Coulomb-correlation effects on the optical spectra of a single negatively charged quantum dot. The considered initial and final many particle states are adapted and motivated by a recent two-color pump-probe experiment. We demonstrate that the lowest energy transition is noticeably affected by exchange interactions if the initial state consists of a hot trion. It turns out that for initial states as obtained after spin-conserving carrier relaxation new absorption and emission lines emerge.

1. Introduction

In a recent publication femtosecond two-color pump-probe experiments on a strongly confined and singly charged quantum dot (QD) have been reported that allowed for an analysis of the optically induced dynamics on a picosecond timescale [1]. In these experiments, a pump pulse initialized the QD by a resonant excitation of a higher shell while the probe pulse was in resonance with the lowest energy transition. It was found that for short positive delay times between pump and probe pulse the fundamental trion line (FTL) vanished. To explain this result instantaneous Coulomb renormalizations of the respective transition energies have been suggested. For larger positive delay times the experiment reported optical gain on the FTL, which was traced back to spin-conserving intraband relaxations.

We study, motivated by the above mentioned experiment, the influence of Coulomb-correlation effects on the FTL. In particular we will show that this absorption line is noticeably affected when a pump pulse creates an exciton in a higher shell. Furthermore, we show that after spin-conserving intraband relaxations of the excited carriers new optical transition lines around the FTL emerge which can be assigned to transitions from the trion state either into a single electron state or to a charged biexciton state.

2. Theoretical Model

In order to study the influence of Coulomb interactions on the optical transitions in a strongly confined QD we model the QD in the envelope-function approximation by confinement potentials for electrons and holes given by an anisotropic harmonic oscillator potential. We assume that

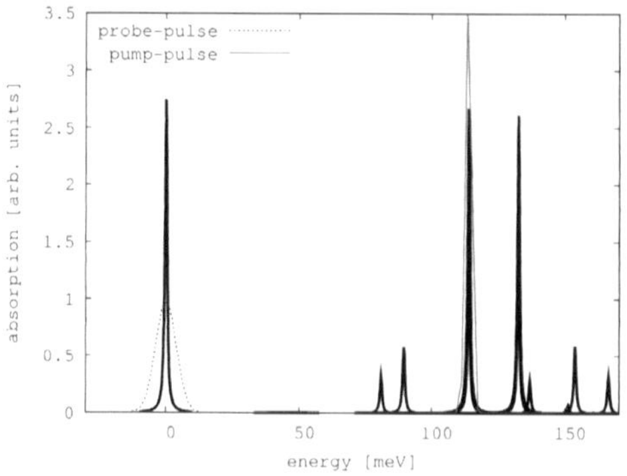

Figure 1. (color online) Absorption spectrum of a singly charged quantum dot. Energies are given with respect to the lowest energy transition. The spectral envelopes of the pump pulse (thin red line) and probe pulse (thin dashed blue line) used by the two color pump-probe experiment in Ref. [1] are also depicted.

the confinement in growth direction is much stronger than in the lateral direction. Therefore, only the lowest energy state in the growth (z-)direction is considered which is of heavy-hole type. The confinement lengths in the in-plane directions x and y are taken to be similar, such that the single particle states $|i\rangle$ are organized in shells referred to as $s, p, d, \ldots$. Furthermore, we assume identical envelopes for conduction-band electrons and valence-band holes, such that the Coulomb matrix elements for electron-electron-, hole-hole-, and electron-hole-interaction are the same for a given combination of oscillator quantum numbers i, j, k, l, i.e., $V^{ee}_{ijkl} = V^{eh}_{ijkl} = V^{hh}_{ijkl}$ [2, 3]. The Hamiltonian conserves the total number of electrons N_e and holes N_h as well as the spin of the electrons (S^e, S^e_z) and holes (S^h, S^h_z). As a consequence, the Hamiltonian in a basis of uncorrelated many particle states consists of subblocks for fixed numbers of electrons and holes and constant spin configurations. We have calculated the Coulomb-correlated multi-exciton states by numerically diagonalizing these subblocks taking into account the twelve lowest single particle states. Finally, the coupling to an external light field is described within the usual dipole and rotating wave approximation giving rise to the standard selection rules for excitonic transitions. The optical absorption or stimulated emission spectra resulting from this coupling are then evaluated using Fermi's golden rule. In order to account for various line broadening mechanisms we have replaced the δ-function in Fermi's golden rule by a Lorentzian. We consider for the following results a single CdSe QD of ellipsoidal shape with the lengths $l_x = 2.6$ nm, $l_y = 2.8$ nm and $l_z = 1.0$ nm. The conduction band mass $m_e = 0.13\ m_0$, the valence band mass $m_h = 0.45\ m_0$, as well the dielectric constant $\epsilon_r = 9.57$ are taken from Ref. [4].

3. Results

Let us first discuss optical transitions of a negatively charged QD from the ground state to a charged exciton (trion) state with two electrons and one hole. In Fig. 1 the calculated absorption spectrum including the 20 lowest trion states is plotted. The energies are given relative to the FTL. The absorption spectrum shows three main absorption lines. Our numerical results indicate that these states are dominated by a single uncorrelated electron-hole state with weak admixture of other states. The first line corresponds to the optically allowed FTL where all particles in the final state are in the s-shell, the two electrons having opposite spin projections forming a singlet state, i.e., $S^e = 0$. The second and third line are triplet states displaying excited trion transitions to the p-shell which are split due to the anisotropic in-plane confinement potential. Around these main excited transition lines a bunch of weaker absorption lines appears which correspond to strongly mixed singlet states. The comparison of our calculated spectra with the photoluminescence (PL) and the photoluminescence excitation spectra (PLE) measured in Ref. [1] reveals a good quantitative and qualitative agreement. This allows us to identify the electronic states which were resonantly excited in the pump-probe experiment. For this purpose, we have also plotted the spectral envelopes of the pump (red) and probe (thin dashed blue line)

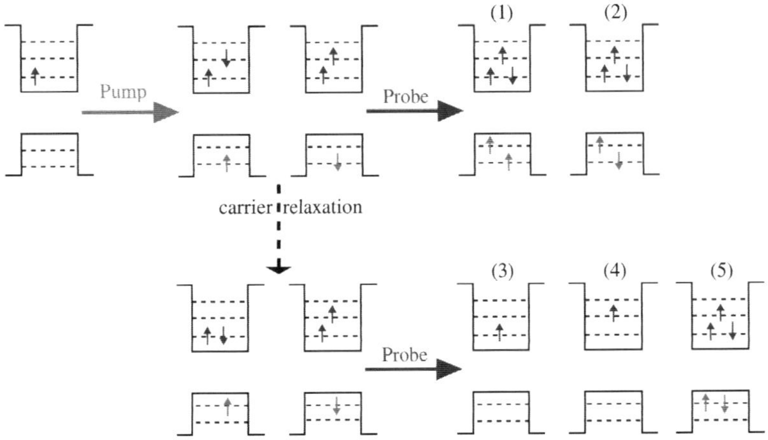

Figure 2. (color online) Schematic illustration of the analyzed pump-probe experiment. The pump pulse excites a negatively charged QD and creates a hot trion. The following probe pulse is in the spectral range of the FTL and measures optical transitions to a hot charged biexciton (states (1) and (2)). After spin-conserving carrier relaxation new optical transitions emerge resulting in stimulated emission to a single electron state (states (3) and (4)) and absorption to a charged biexciton in the ground state (state (5)).

pulses as used in Ref. [1].

When applying a π-pulse in resonance to a higher shell a hot electron-hole pair is created transferring the initial single electron state to a hot trion state. The electrons in this optically initialized QD may have parallel or antiparallel spins (see Fig. 2). The probe-pulse in the spectral region of the FTL now gives rise to transitions to a charged biexciton state with three electrons and two holes, where the holes may have spins with parallel or antiparallel orientation (states labeled (1) and (2) in Fig. 2).

Figure 3 shows in part (a) again the absorption spectrum of the unexcited QD, i.e., the FTL. In part (b) the spectrum measured by a probe pulse in the presence of the hot trion is plotted. We clearly see that the FTL is now split into two lines with approximately half of the original intensity. The lines correspond to the transitions to the charged biexciton states labeled (1) and (2) in Fig. 2. The energetically lower line results from the transition to the (hole-)triplet states, $S_h = 1$, whereas the higher line is related to the transition to the (hole-)singlet state, $S_h = 0$. Indeed, it turns out that the energy separation (~ 14 meV) is close to the hole exchange matrix element V^{hh}_{spsp} between holes in s- and p-states. Hence, the reported vanishing FTL may be traced back to strong exchange interactions between holes in different shells.

The exchange interaction is effective only for carriers in different orbital shells and, therefore, vanishes when the hot trion relaxes to its lowest energy level. The times for spin-conserving energy relaxation processes are typically of the order of several picoseconds [5, 6]. As is shown in Fig. 2, we now have to distinguish two cases. If the electrons in s- and p-shell have opposite spin orientations, both electron and hole will relax to their respective ground states. If they have the same spin orientation, electron relaxation is suppressed due to Pauli blocking and only the hole will relax. This has important consequences for the absorption spectrum of a subsequent probe pulse in the spectral range of the FTL. In the former case the electronic s-shell is completely filled and the pump pulse will give rise to a stimulated emission leading to the final single electron state (3). This is exactly the inverse process of the fundamental trion absorption line. Thus the corresponding spectrum shown in Fig. 3(c) shows a gain line at zero energy (red solid line). In the latter case both an emission and an absorption process are possible. The final state in the emission process is an electron in the p-shell (4) while the absorption process leads to a

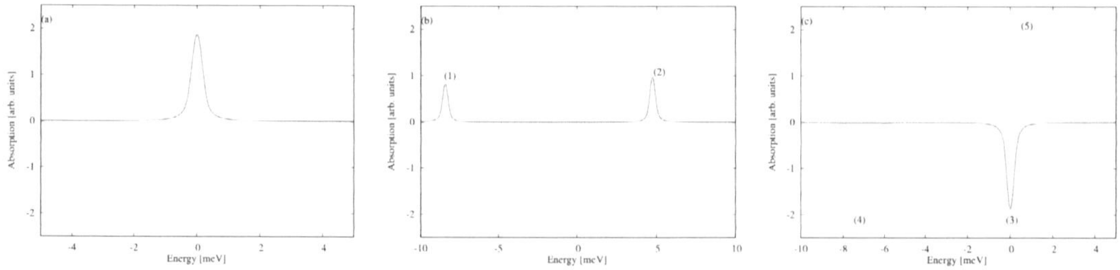

Figure 3. (Color online) Optical spectra in the region of the FTL resulting from a QD initially (a) in the negatively charged ground state, (b) in a hot trion state, and (c) after spin-conserving relaxation (see also Fig. 2). The red solid curve in (c) refers to the initial state with both electrons in the s-shell, the green dashed line to the state with one electron in the s- and one in the p-shell.

charged biexciton in the ground state (5). As can be seen in Fig. 3(c) (green dashed line) the emission gives rise to a gain line (4) which is strongly blue-shifted with respect to the FTL. Here the electron-electron exchange interaction considerably lowers the energy. The absorption line corresponding to the transition to the charged biexciton is ~ 1 meV red-shifted with respect to the FTL showing that here exchange interactions do not strongly affect the electronic states and Coulomb-correlations are rather small. It should be mentioned that this transition has not been observed in the discussed pump-probe experiment Ref. [1], although it is very close to the FTL. However, it turns out that the exact spectral position of this line strongly depends on geometry parameters, in particular the relative confinements length of electron and hole and it may easily be shifted outside of the spectral window detected in the experiment.

4. Conclusions

We have analyzed the influence of Coulomb-renormalizations on the optical spectra of a single negatively charged QD. The considered initial and final Coulomb-correlated multi-exciton states have been chosen according to states excited in a two-color pump-probe experiment. We have shown that the FTL is noticeably affected by exchange interactions as long as carrier relaxation has not started. This result may explain a reported vanishing of this lowest energy transition in a recent experiment. For initial states resulting from a spin-conserving energy relaxation new optical transition lines emerge exhibiting both absorption and emission.

J. H. thanks for financial support from Deutscher Akademischer Austauschdienst (DAAD).

References

[1] Sotier F, Thomay T, Hanke T, Korger J, Mahapatra S, Frey A, Brunner K, Bratschitsch R and Leitenstorfer A 2009 *Nature Physics* **5** 352–356

[2] Hawrylak P 2003 *Solid State Communications* **127** 793–798

[3] Baer N, Gartner P and Jahnke F 2004 *The European Physical Journal B-Condensed Matter and Complex Systems* **42** 231–237

[4] Madelung O 2004 *Semiconductors: data handbook* (Springer Verlag)

[5] Htoon H, Kulik D, Baklenov O, Holmes Jr A, Takagahara T and Shih C 2001 *Physical Review B* **63** 241303

[6] Alèn B, Karrai K *et al.* 2004 *Physica E: Low-dimensional Systems and Nanostructures* **21** 395–399

Journal of Physics: Conference Series **245** (2010) 012026 doi:10.1088/1742-6596/245/1/012026

Identifying spins states on self assembled Si/SiGe quantum dots by means of ESR

F. Lipps, F. Pezzoli, M. Stoffel, Ch. Deneke, J. Thomas, A. Rastelli, V. Kataev, O. G. Schmidt, B. Büchner

Institute for Solid State and Materials Research, IFW Dresden, P. O. Box 270116, D-01171 Dresden, Germany

E-mail: `f.lipps@ifw-dresden.de`

Abstract. Self-assembled Si/SiGe quantum dots were systematically studied with ESR at 9.56 GHz. A set of multilayer structures were grown with molecular beam epitaxy with different vertical spacing between quantum dot layers and different dot sizes. An extensive characterization with transmission electron microscopy and atomic force microscopy provides good structural information of the ensembles of quantum dots. Furthermore the samples were characterized by photoluminescence. Two ESR signals were identified with g-factors around 1.9992 and 1.9994. Both signals show a small anisotropy in g-factors and linewidth with respect to the magnetic field along growth and in-plane directions. Spin-relaxation times T_1 are determined with continuous wave ESR saturation measurements to be of the order of 10 μs. The spin dephasing times yield values up to 500 ns. Illumination with sub-bandgap light changes the relative intensity of the two signals. Based on calculations of the electronic structure of the heterostructures a qualitative model allows us to relate the two ESR signals to s- and p-like-states on the quantum dots.

1. Introduction

The spin-orbit coupling in IV-IV semiconductors is fairly weak. Therefore long spin lifetimes and coherence times are expected, opening opportunities for spintronic applications in Si-based systems. Because of the indirect bandgap of Si and Ge, the use of optical methods, which have proved powerful to study spin states in the case of III-V materials, is limited. Electron spin resonance (ESR) is a technique, which allows to determine static and dynamics of spin ensembles also for Si-based heterostructures [1]. Specifically, continuous wave ESR can provide information about the g factor as well as relaxation and dephasing times of a spin system.

2. Results

Two sets of samples were grown by means of molecular beam epitaxy (MBE) at temperatures of 600 °C and 700 °C respectively. The samples consist of fourfold-stack of SiGe quantum dots embedded in a Si matrix. The spacing between the quantum dots is varied from 20 to 100 nm. More details are given elsewhere [2].

ESR experiments were performed with a Bruker EMX spectrometer around a frequency of 9.56GHz in a resonator with optical access at temperatures around 4 K.
Because of the small spin-orbit coupling in Si-based systems the g factor of electrons is expected

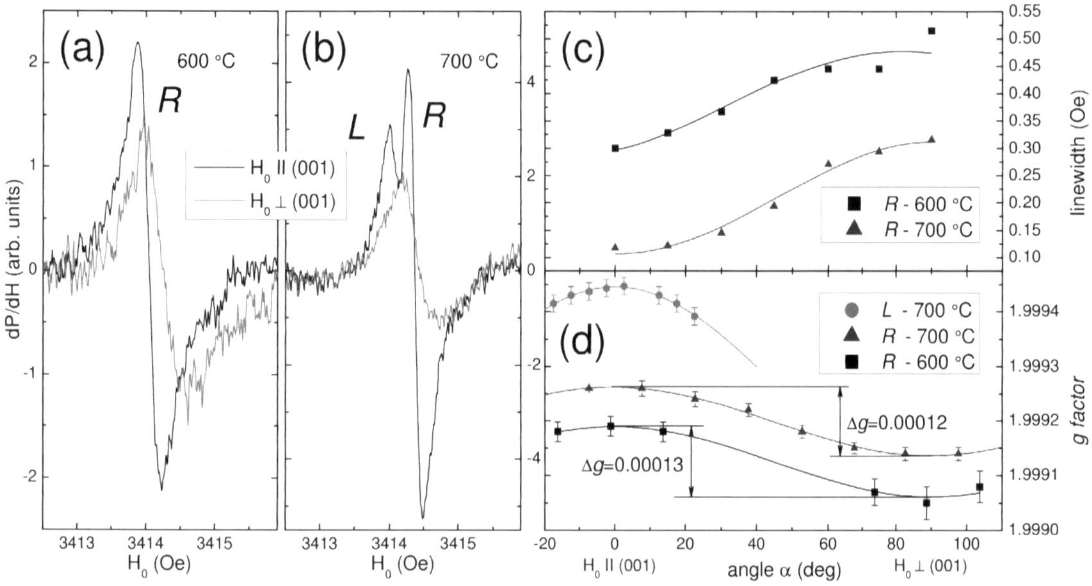

Figure 1. (a) ESR spectra for samples grown at 600 °C and (b) 700 °C with the magnetic field in (001) growth direction and perpendicular to it. Only a broad single peak is observed for 600 °C, while there are two peaks for 700 °C-sample. (c) linewidth and (d) g factor as a function of magnetic field direction. The g factor can only be determined in the range where the lines are resolved

to be close to the g factor of the free electron $g \approx 2$. ESR signals could be identified in all quantum dot samples with a g factor of about 1.999. Careful measurements of reference samples without quantum dots confirmed that these signals are indeed associated with the quantum dots (for details see [2]). In this paper we will focus on samples with 70 nm vertical spacing. While there is only a single peak visible for the samples grown at 600 °C around a g factor $g_{\|} = 1.9992$ [Fig.1(a)]two peaks are identified in the structures grown at 700 °C with g factors $g_{\|,R} = 1.9992$ and $g_{\|,L} = 1.9994$ - for a magnetic field applied in the (001) growth direction [Fig.1(b)]. From the linewidth the dephasing time T_2^* can be calculated [3] and amounts, for the 700 °C sample, to up to 500 ns. By measuring the ESR-intensity as a function of applied microwave power the relaxation time T_1 can also be calculated [3]. T_1 is estimated to be slightly shorter than 10 μs in all structures investigated.
As the external magnetic field is turned in-plane the lines broaden [Fig.1(c)] and in case of the former two peak spectra the peaks cannot be distinguished anymore [Fig.1(a),(b)]. However the different relative intensities make a characterization of the linewidth possible. Also the g factors show an anisotropic dependence regarding growth direction and quantum dot plane. However, this anisotropy is fairly small [Fig.1(d)] and is within the error of the measurements. The same holds for peak R in both structures.

The Si/SiGe quantum dots were modelled with the program NEXTNANO^3 [4]. Structural information extracted from transmission electron microscopy (TEM) and atomic force microscopy (AFM) were used to implement shape and intermixing profile of the quantum dots in the simulations. Elastic strain and single band effective mass calculations were carried out. In agreement with previous publications [5, 6] the band structure shows a type II band alignment. As a consequence electrons are confined at the apex of the dot in the lowest lying conduction band of

the Si matrix, the Δ_2 band, while holes are confined inside the SiGe dot.
The occupation of the quantum dots depends on the number of donors per dot. In addition it also depends on the ionization energies of those dopands, which determine the Fermi energy at a given temperature, with respect to the quantum dot eigenstates. The intrinsic (n type) background doping gives an average occupation on the order of 10 donors per dot. However, since the donor ionization energy (45 meV for phosphorus) is close to the lowest quantum dot eigenstate not all electrons are trapped on the dots.

For the two sample sets the Ge concentration in the dots and their size are different and

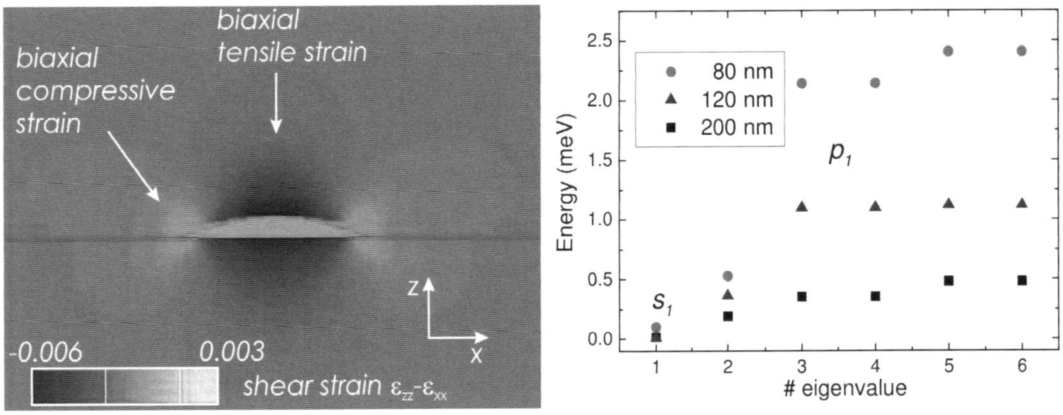

Figure 2. (a) strain profile of a modelled semi-parabolic quantum dot. The z-axis is along the (001) growth direction. (b) eigenvalues for different quantum dots, 80, 120 and 200 nm refer to their basediameters

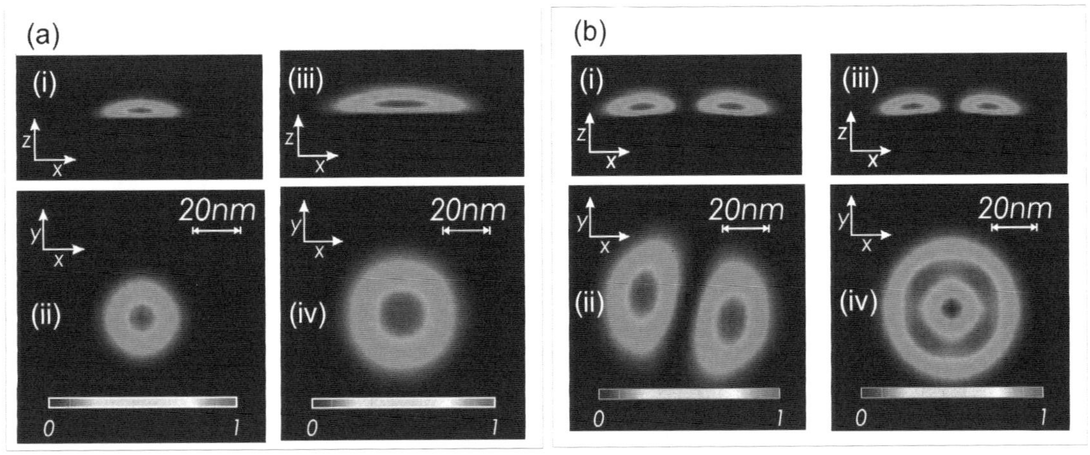

Figure 3. (a) probability wavefunctions of the s-state for 80 nm [(i),(ii)] and 120 nm dot basdiameter [(iii), (iv)]; probability wavefunction of p-state for magnetic field in xy-plane [(i),(ii)] and z-direction [(iii), (iv)] direction for dot basediameter of 120 nm

with that eigenstates and eigenfunctions differ. Samples grown at 600 °C show a base diameter of about 80 nm while samples grown at 700 °C are significantly larger having base diameters

of about 200 nm. Because of that the energy spacing between s- and p-states is significantly decreased from > 2 meV for the small dots to about 0.4 meV for the large ones (Fig. 2(b)). The structures consist of a large ensemble of dots where the individual energy levels vary slightly throughout the sample. At temperatures around 4 K, as used in the experiments, a thermal population of p-states, in addition to population of s-states, is therefore possible in the bigger dots. With an increase in size of the confining potential also the size of the electron wavefunction is increasing. However, the pancakelike shape (s-state) is preserved. This is nicely visible in Fig. 3(a), which shows the s-state probability wavefunctions of quantum dots with 80 nm [(i),(iii)] and 120 nm base diameter [(ii),(iv)].
While there is no influence of a magnetic field on the s-states the p-state is distorted when the magnetic field is turned from the plane [Fig. 3(b)[(i),(ii)]]to the growth direction [Fig. 3(b)[(iii),(iv)]].

The illumination with light below the bandgap of Si increases only the right peak R (not shown - see Ref. [2]). This can be understood as an ionization of donors or a direct generation of carriers from the quantum dot. With that an increase in the quantum dot occupation is reached. Therefore, the right peak R is identified as the quantum dot p-state.

3. Summary & Conclusion

A systematic ESR study on Si/SiGe heterostructures was performed. The ESR signals could be identified as s- and p-like-states at the quantum dots. g factor and relaxation times were characterized. The timescales for $T_1 \approx 10\mu s$ and $T_2 > 500ns$ are shorter than those of electrons on donors in Si [7], but sufficiently long to allow spin manipulation with established ESR techniques.

Acknowledgements

Support by DFG through FG-912 is gratefully acknowledged.

References

[1] Malissa H, Jantsch W, Chen G, Gruber D, Lichtenberger H, Schäffler F, Wilamowski Z, Tyryshkin A M and Lyon S 2006 *Materials Science and Engineering: B* **126** 172–175
[2] Lipps F, Pezzoli F, Stoffel M, Deneke C, Thomas J, Rastelli A, Kataev V, Schmidt O G and Büchner B 2010 *Phys. Rev. B* **81** 125312
[3] Poole Jr C P 1996 *Electron Spin Resonance: A Comprehensive Treatise on Experimental Techniques* 2nd ed (Dover Publications, Inc.)
[4] Birner S, Zibold T, Andlauer T, Kubis T, Sabathil M, Trellakis A and Vogl P 2007 *IEEE Trans. Electron Devices* **54** 2137–2142
[5] Schmidt O G, Eberl K and Rau Y 2000 *Phys. Rev. B* **62** 16715
[6] Grützmacher D, Fromherz T, Dais C, Stangl J, Müller E, Ekinci Y, Solak H H, Sigg H, Lechner R T, Wintersberger E, Birner S, Holy V and Bauer G 2007 *Nano Letters* **7** 3150–3156
[7] Feher G and Gere E A 1959 *Phys. Rev.* **114** 1245

Quantum entanglement in lateral GaAs/AlGaAs quantum dot molecules

V Křápek[1,2], P Klenovský[1], A Rastelli[2], O G Schmidt[2] and D Munzar[1]

[1]Department of Condensed Matter Physics, Masaryk University, Kotlářská 2, 61137 Brno, Czech Republic

[2]Institute for Integrative Nanosciences, IFW-Dresden, Helmholtzstr. 20, D01187 Dresden, Germany

E-mail: krapek@physics.muni.cz

Abstract. We calculate the excitonic structure of pairs of GaAs/AlGaAs quantum dots forming lateral molecules and obtain the entanglement of exciton states. The following advantages of the lateral geometry over the vertical one are found: (1) The energy structures of the dots forming a molecule can be in principle identical. (2) Comparable tunneling of electrons and holes ensures a high entanglement of antisymmetric excitons. A drawback of existing structures are very low tunneling energies, which make the entanglement vulnerable against differences in the sizes and shapes of both dots.

1. Introduction

Peculiar properties of the quantum computing allow to design algorithms superior to classic ones. A prominent example is Shor's algorithm factoring large numbers with a polynomial time complexity. Quantum algorithms rely on two-state quantum systems – qubits, and devices coherently manipulating qubits – quantum gates. They might be realized in various physical systems, among which semiconductor quantum dots (QDs) offer several advantages such as easy integration with existing electronic devices. The representation of the qubit by a charge confined in a vertical quantum dot molecule (QDM), with the position in the lower/upper dot representing the 0/1 state, has been proposed [1]. The proposal has been later criticized as based on non-realistic assumptions [2]. In particular, inevitable differences between the electronic structures of the individual dots lower the entanglement significantly. Another drawback is the difference between the tunneling energies for electrons and holes, which further reduces the entanglement of optically dark excitons.

Recent progress in the growth technology makes it possible to prepare *lateral* QDMs. The interaction between the laterally closely spaced dots has been reported for the InAs/GaAs system [3]. There are also GaAs/AlGaAs quantum dot pairs in which a mutual interaction is expected [4]. The main advantage of the lateral topology is that there is no *a-priori* obstruction in making both dots identical from the point of view of their electronic structure. In contrast, in the vertical arrangement the strain field in the lower dot and in the upper dot differs. By manipulating the dot sizes, we can put into resonance either the electron levels or the hole levels, but not both at the same time. No such restriction applies to the lateral topology. Lateral QDMs also have the advantage of comparable tunneling energies of electrons and holes [5].

In this paper we present calculations of the excitonic structure of lateral QDMs consisting of two overlapping cones and of the related entanglement. We briefly discuss the characteristic energies of QDMs and their impact on the entanglement: Coulomb attraction, tunneling, and detuning. Realistic calculations follow, which combine the eight-band $\vec{k} \cdot \vec{p}$ method for single particle states with the configuration interaction method; the typical shape and dimensions of the dots are taken from Refs. [4, 6].

2. Entanglement: Effect of characteristic energies

A quantum system consisting of two particles A and B in states $|\phi_A\rangle$ and $|\phi_B\rangle$ is described by the product state $|\phi_{AB}\rangle = |\phi_A\rangle|\phi_B\rangle$, e.g., $|0_A 0_B\rangle$. Applying the principle of superposition to the product states we can construct also states which cannot be expressed as products of individual particle states, a prominent example being $|0_A 0_B\rangle + |1_A 1_B\rangle$. These are so called *entangled* states. If the system is in an entangled state, the properties of the components are correlated. Here the composed system is an exciton and the components are an electron and a hole.

The entanglement of an exciton in a QDM can be explained in terms of an interplay of three characteristic mechanisms: Coulomb attraction, tunneling, and detuning. To study this effect we employ a simple tight binding (TB) model of a QDM [1, 2]. Let $|e_0\rangle$, $|e_1\rangle$ denote the states of electrons in individual dots labelled as 0 and 1 (representing the 0 and 1 of a qubit), $|h_0\rangle$, $|h_1\rangle$ denote the corresponding hole states. In the basis $|e_0 h_0\rangle$, $|e_0 h_1\rangle$, $|e_1 h_0\rangle$, $|e_1 h_1\rangle$, the Hamiltonian reads

$$\begin{bmatrix} E_0 - H_0 - U_{00} & -t_H & -t_E & 0 \\ -t_H & E_0 - H_1 - U_{01} & 0 & -t_E \\ -t_E & 0 & E_1 - H_0 - U_{10} & -t_H \\ 0 & -t_E & -t_H & E_1 - H_1 - U_{11} \end{bmatrix}$$

Here E_i and H_i are energies of electrons and holes in dot i, t_E and t_H are the tunneling matrix elements and $-U_{ij}$ are the Coulomb energies.

The number of parameters can be further reduced by the following assumptions. We introduce the *detuning* energy Δ due to the difference in the dot dimensions: $E_0 = E_1 + \Delta/2$, $H_0 = H_1 - \Delta/2$. We assume that $U_{00} = U_{11} = U_D$, $U_{01} = U_{10} = U_I$; $U = U_D - U_I$ is the *Coulomb* energy favoring the direct configurations over the indirect. Denoting $E_g = E_1 - H_1 - U_I$ we rewrite the Hamiltonian as

$$\begin{bmatrix} E_g + \Delta - U & -t_H & -t_E & 0 \\ -t_H & E_g + \Delta/2 & 0 & -t_E \\ -t_E & 0 & E_g + \Delta/2 & -t_H \\ 0 & -t_E & -t_H & E_g - U \end{bmatrix}$$

Following [1, 2], we classify the eigenstates according to the following fully entangled states: $|a\rangle = (|e_0 h_0\rangle + |e_1 h_1\rangle)/\sqrt{2}$ (direct, symmetric), $|b\rangle = (|e_0 h_1\rangle + |e_1 h_0\rangle)/\sqrt{2}$ (indirect, symmetric), $|c\rangle = (|e_0 h_1\rangle - |e_1 h_0\rangle)/\sqrt{2}$ (indirect, antisymmetric), $|d\rangle = (|e_0 h_0\rangle - |e_1 h_1\rangle)/\sqrt{2}$ (direct, antisymmetric). The degree of entanglement can be quantified via the entropy of entanglement $S = -\mathrm{Tr}\rho_A \log_2 \rho_A$ with $\rho_A = \mathrm{Tr}_B \rho$, where ρ is the density matrix of the exciton, ρ_A is the reduced density matrix of one component (e.g., electron), Tr_B is the partial trace over the other component (e.g., hole). The maximal value of the entropy of entanglement for the considered system is 1.

Figure 1 shows the entanglement of the two lowest excitonic states $|\alpha\rangle$ (with the dominant component $|a\rangle$) and $|\delta\rangle$ (with the dominant component $|d\rangle$) calculated for $U = 10$ meV, $\Delta = 1$ meV and variable t_E and t_H. We consider either $t_E = t_H$ (the ideal case) or $t_E = 3t_H$ (the largest difference found in our realistic model of lateral QDMs, see Fig. 2). For $|\alpha\rangle$ there exists an optimal tunneling for which the entanglement is the largest. For lower tunneling, the detuning prevents the formation of symmetric and antisymmetric orbitals and only one direct configuration dominates the state. For higher tunneling, the Coulomb term looses its importance and the contribution of indirect terms increases. The state $|\delta\rangle$ exhibits similar behavior with a slightly enhanced entanglement if t_E and t_H differ. However, if $t_E = t_H$, then entanglement of this state increases as tunneling increases and is fairly large.

The strategies to obtain a large degree of entanglement have to aim at $\Delta \ll t \ll U$. Exploiting of $|\delta\rangle$ is recommendable because of the larger entanglement as compared to $|\alpha\rangle$.

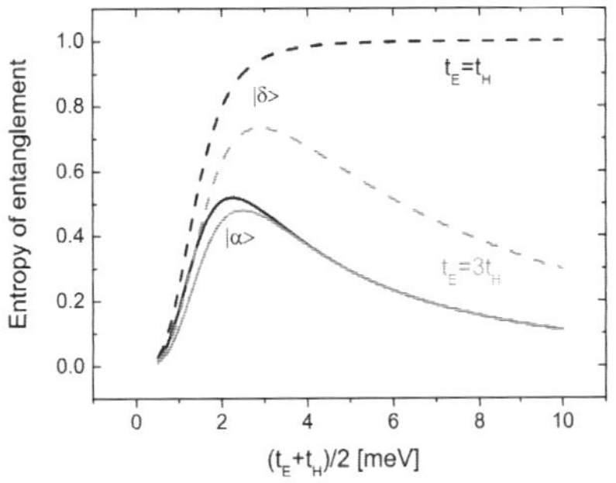

Figure 1. The entropy of entanglement of two lowest excitonic states: $|\alpha\rangle$ (full lines) and $|\delta\rangle$ (dashed lines) from a TB model. The values of the input parameters are $U = 10$ meV, $\Delta = 1$ meV. The values of t_E and t_H are either the same (blue lines) or $t_E = 3t_H$ (orange lines).

3. Realistic lateral molecules

The realistic calculations consist of two steps. First, we employ the eight band $\vec{k} \cdot \vec{p}$ method to obtain the single particle states. Second, these states are used as basis states for the configuration interaction method yielding the excitons. The values of the material parameters are taken from [7].

The model dots studied here correspond to two realistic systems which use GaAs as the dot material and AlGaAs as the host material: "molecules" [4] and "single dots" [6]. As described in Refs. [4, 6], the dots consist of pure GaAs, the bottom barrier consists of $Al_{0.45}Ga_{0.55}As$ and the top barrier of $Al_{0.35}Ga_{0.65}As$. The shape and dimensions of the structures are obtained with a rather good accuracy from atomic force micrographs. The dots forming the molecule are not completely separate (see Fig. 18 of Ref. [4] and inset of Fig. 2). "Molecules" are modeled as two reverted cones with the characteristic height h of 4 nm, the characteristic radius r of 30 nm, and the distance between the centers of the cones d of 32 nm. For "single dots" structure the values of the parameters are $h = 4$ nm, $r = 14$ nm, $d = 20$ nm; further, the tips of both cones are connected with a mound, which turns the hypothetic molecule into a single dot. However, there is still a lateral narrowing between the tips. In our study we adopted the shape of two reverted cones (we do not consider the connecting mound in "single dots"). We also allow d to vary in order to explore possibilities of improving the existing structures.

For very low values of d the QDM behaves as a single QD. Since the proposal of the qubit under consideration requires that the position of individual particles in one of the dots is well defined, we set the following criterion: The system of two cones forms a molecule if the probability density of the ground state in the area separating the cones is below 25 % of its maximum value. This criterion is fulfilled if $d > 14$ nm for $r = 14$ nm and $d > 21$ nm for $r = 30$ nm. To validate that such structures are molecules, we have calculated the effect of an electric field on them (not shown here). In a molecule it is possible to localize the charge carrier in one of the dots by adding a proper detuning (represented by the electric field), different from single dots. The molecular behavior of our structures was confirmed. Modifying the above mentioned criterion to 1 % leads to $d > 20$ nm for $r = 14$ nm and $d > 29$ nm for $r = 30$ nm, resulting in an extremely weak coupling between the dots. Therefore we have to accept a large probability density in the separation area as a drawback of considered lateral QDMs. The reason for this undesirable behavior is the shallow barrier between the dots, realized as the narrowing of the dot material only. This point should be considered in improving the growth technology.

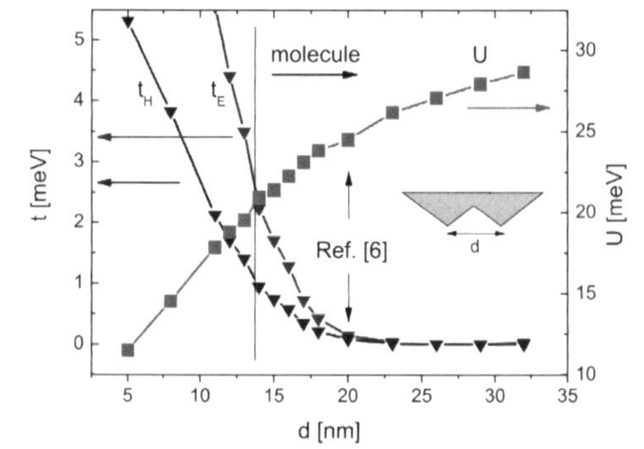

Figure 2. Tunneling energies t_E (brown) and t_H (blue) and the Coulomb energy U (green) as functions of d for a model QDM with $h = 4$ nm and $r = 14$ nm. The molecular limit (probability density in the separation area equal to 25 % of its maximum) is displayed as the vertical black line, the realistic d is denoted with the black arrows. The inset shows a schema of the model structure.

Figure 2 shows t_E, t_H and U as functions of d for $h = 4$ nm and $r = 14$ nm (the results for $r = 30$ nm are not shown). The electron tunneling for d corresponding to existing structures (32 and 20 nm for r of 30 and 14 nm, respectively) is below 0.15 meV, yielding a negligible entanglement for a realistic detuning. At the molecular limit, however, it increases towards 2.2 meV (r = 14 nm) and 1.5 meV (r = 30 nm). The t_E/t_H ratio varies between 2 and 3. This is a considerable improvement compared to vertical QDMs, where values above 10 are predicted [2]. The Coulomb energy U is around 20 meV for the distances d yielding a reasonable tunneling. Using the TB model and assuming a detuning $\Delta = 1$ meV, we obtain the maximum entanglement for t_E around 3 meV, a slightly larger value than that we obtained for simulated QDMs. We also calculated the expected detuning by varying the radius of a model dot. The distribution of r with a standard deviation of 0.25 nm (1 nm) yielded a detuning up to 1 meV (5 meV). For the molecule with $h = 4$ nm, $r_1 = 13.9$ nm, $r_2 = 14.1$ nm, $d = 14$ nm we obtained the entropy of entanglement $S = 0.17$ for $|\alpha\rangle$ and $S = 0.20$ for $|\delta\rangle$, while for zero detuning it was $S = 0.94$ for $|\alpha\rangle$ and $S = 0.99$ for $|\delta\rangle$. The minimization of the detuning is therefore essential.

4. Conclusion

In conclusion, existing lateral QDMs presented in Ref. [4] suffer from low tunneling energies (below 0.15 meV). Nevertheless, sufficiently large values (up to 2 meV) can be obtained for the dots brought closer to each other. The ratio t_E/t_H is considerably smaller than in vertical QDMs. The strategy to enhance the entanglement should consist of the following steps: (1) Reducing the distance between QDs in a QDM. (2) Reducing the lateral dimension of QDs and improving the barrier properties, resulting in a full separation of the QDs. (3) Minimization of the detuning by improving the homogeneity of QDs.

Acknowledgments

The work was supported by the Institutional research program MSM 0021622410, the GACR grant GA202/09/0676, and the DFG (FOR730).

References

[1] Bayer M, Hawrylak P, Hinzer K, Fafard S, Korkusinski M, Wasilewski Z, Stern O and Forchel A 2001 *Science* **291** 451
[2] Bester G, Zunger A and Shumway J 2005 *Phys. Rev. B* **71** 075325
[3] Beirne G, Hermannstadter C, Wang L, Rastelli A, Schmidt O and Michler P 2006 *Phys. Rev. Lett.* **96** 137401
[4] Wang L, Rastelli A, Kiravittaya S, Benyoucef M and Schmidt O G 2009 *Adv. Mater.* **21** 2601
[5] Wang L, Křápek V, Ding F, Horton F, Schliwa A, Bimberg D, Rastelli A and Schmidt O G 2009 *Phys. Rev. B* **80** 085309
[6] Rastelli A, Stufler S, Schliwa A, Songmuang R, Manzano C, Costantini G, Kern K, Zrenner A, Bimberg D and Schmidt O 2004 *Phys. Rev. Lett.* **92** 166104
[7] Vurgaftman I, Meyer J and Ram-Mohan L 2001 *J. Appl. Phys.* **89** 5815

Effect of magnetic field on the electron-nuclear spin dynamics in quantum dots

M. Yu. Petrov,[1] S. V. Yakovlev,[1] I. V. Ignatiev,[1] D. R. Yakovlev,[2,3] and M. Bayer[2]

[2] Physics Department, St.-Petersburg State University, 198504 St.-Petersburg, Russia
[1] Experimentelle Physik II, Technische Universität Dortmund, 44221 Dortmund, Germany
[3] A. F. Ioffe Physical-Technical Institute, Russian Academy of Sciences, 194021 St.-Petersburg, Russia

E-mail: m.yu.petrov@gmail.com

Abstract. We report on the theoretical study of the joint effect of the hyperfine interaction and an external magnetic field on the electron-nuclear spin dynamics in a quantum dot. The study is based on the *quantum-mechanical* description of the hyperfine interaction of an electron spin with a finite number of nuclear spins in the frame of the graded box model. This approach clearly demonstrates electron-spin depolarization in the nuclear spin fluctuations and its suppression by the external magnetic field and dynamic nuclear spin polarization.

1. Introduction

The dynamics of nuclear spin polarization in quantum dots (QDs) has been intensively investigated in recent years [1]. In the typical scenario of experiments, a series of pumping pulses of one helicity creates spin polarization of localized electrons. This polarization is a source of angular momentum, which can be transferred to the nuclear spin bath via electron-nuclear hyperfine interaction. Usually, the dynamic nuclear-spin polarization (DNP) can be observed in the Zeeman splitting of ground state optical transition in single-QD spectroscopy [2, 3, 4, 5]. In these experiments, one can measure the steady-state value of DNP. However, this information is not complete because there always appear also the nuclear spin fluctuations (NSF), the interaction with which leads to dephasing and decoherence of the electron spin [6, 7]. The dynamics of DNP also strongly depends on the NSF because the NSF determine what fraction of electron spin polarization can be stored. Therefore, a key point to understand nuclear spin dynamics is to understand the NSF.

Here, we theoretically study the effect of magnetic field on the electron-nuclear spin system of a QD. For this purpose, we extend the graded box model approach [8] such that it allows us to simulate jointly the hyperfine interaction of the electron spin with a nuclear spin bath as well as the effect of a magnetic field on this coupled system. The model allows us to qualitatively compare theoretical results with experimental measurements of electron spin polarization in (In,Ga)As/GaAs QDs.

2. Model

We consider a system comprising N nuclear spins $\hat{\boldsymbol{I}}_j$ coupled to one localized-electron spin $\hat{\boldsymbol{S}}$. In the presence of an external magnetic field $\boldsymbol{B}$ oriented along z axis of the Cartesian coordinate system, the

Hamiltonian of electron-nuclear interaction can be written as

$$\hat{\mathcal{H}} = \mu_B g_e B \hat{S}_z - \hbar \sum_{j=1}^{N} \gamma_j B \hat{I}_z^j + \sum_{j=1}^{N} A_j \left(\hat{\boldsymbol{S}} \cdot \hat{\boldsymbol{I}}_j \right) \tag{1}$$

where μ_B and g_e are the Bohr magneton and electron Landé g factor, γ_j and A_j are the nuclear gyromagnetic ratio and the hyperfine coupling of electron spin to jth nucleus. $A_j = v_0 A |\psi(\boldsymbol{R}_j)|^2$ where v_0, $|\psi(\boldsymbol{R}_j)|^2$, and A are the crystal unit-cell volume, the amplitude of electron density on the nuclear site $\boldsymbol{R}_j$, and A determines the hyperfine interaction strength, respectively. We consider a special regime of excitation, in which at the time-moments T_m, the electron spin is initialized in the pure quantum mechanical state characterized by the electron-spin density matrix $\rho_e = |\uparrow_z\rangle\langle\uparrow_z|$. We assume $T_{m+1} - T_m$ to be much larger than the period of electron spin precession about the effective nuclear field. In this regime, the electron and nuclear spins coherently interact during their relatively long correlation time, τ_c. One can describe the electron-nuclear spin motion using the density matrix formalism. In this case, the electron-nuclear density matrix is

$$\rho(T_m + t) = \exp[-\tfrac{i}{\hbar}\mathcal{H}t]\rho(T_m + 0)\exp[+\tfrac{i}{\hbar}\mathcal{H}t]. \tag{2}$$

The coherent evolution of the electron-nuclear spin system initiated by the preceding pulse is interrupted by the next electron spin initialization pulse at time T_{m+1}. This external action affects the electron spin but does not affect the nuclei, so that the nuclear density matrix, ρ_n, is continuous and can be extracted from the total density matrix by partially tracing over the electron-spin states: $\rho_n(T_{m+1}) = \mathrm{Tr}_e \rho(T_{m+1} - 0)$. Immediately after the pulse, the total density matrix is $\rho(T_{m+1} + 0) = \rho_n(T_{m+1}) \otimes \rho_e$.

We further assume that the nuclear spin state can only slightly vary during τ_c, which is small in comparison with the precession period of nuclear spins about external magnetic field. Hereafter, we ignore in Eq. (1) this slow precession.

To calculate Eq. (2) we use the graded box model approach [8], in which the nuclei can be divided in different groups, in each of which they are considered to be interacting with the same stength with the electron spin, i. e., we can introduce averaged hyperfine constants, $\mathcal{A}^{(k)}$, for each group. In this

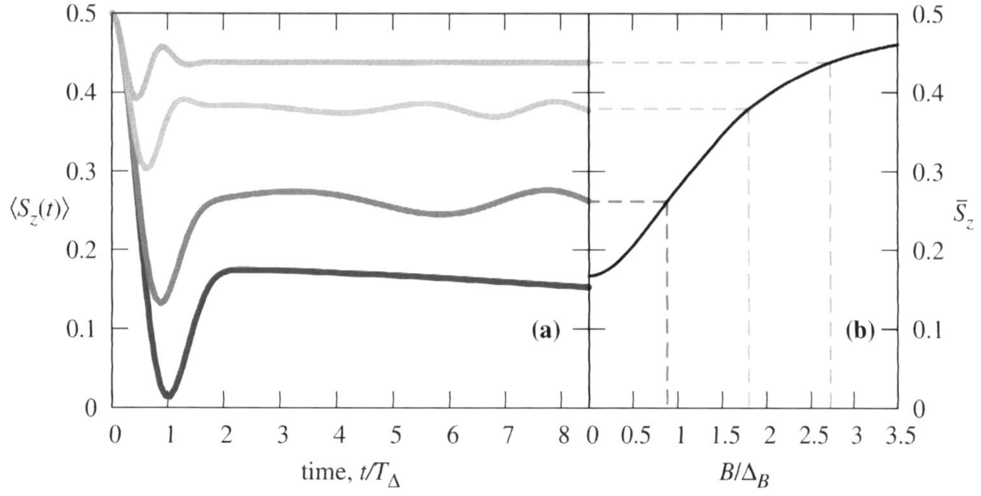

Figure 1. (a) Time-dependencies of electron spin polarization after first initialization for different external magnetic fields: dephasing in frozen NSF. (b) Magnetic field dependence of mean electron spin polarization.

approximation, Eq. (1) can be rewritten in the basis of collective spin momenta $\hat{\mathcal{I}}^k = \sum_{j\in(k)} \hat{\boldsymbol{I}}_j$ of the kth group of nuclei. Then for a number $n \ll N$ of nuclear groups, the Hamiltonian matrix is block-diagonal with the size of each block drastically smaller than the size of the total Hamiltonian matrix.

3. Results

The model above allows us to solve numerically Eq. (2) and to obtain the time dependence of density matrix $\rho(t)$. The time dependencies of electron-spin polarization, $\langle \hat{S}_z(t)\rangle = \mathrm{Tr}[\rho(t)\cdot \hat{S}_z]$, calculated for a model with $N = 12$ nuclear spins with $I = 3/2$ arrangeded in two nuclear groups are shown in Fig. 1(a) for different magnetic fields aligned along z axis. The panel gives the dynamics of electron spin initially polarized in the state $|\uparrow_z\rangle$ and interacting with unpolarized nuclear spins, whose initial state is characterized by a high temperature density matrix ($\rho_n(0) \propto \mathbf{1}$).

The time evolution obtained in these calculations agrees well with the phenomenological approach proposed by Merkulov *et al* [6]. In particular, the electron spin polarization rapidly relaxes to its minimum and then saturates at a steady-state value, which depends on the magnetic field magnitude. At zero magnetic field, the dephasing time, $T_\Delta = \hbar/(\mu_B g_e \Delta_B)$, depends on the statistics of frozen NSF, $w(\boldsymbol{B}_N)$, which can be described by a Gaussian with variance $\Delta_B = \sqrt{\left[\frac{2}{3}\sum_j I_j(I_j+1)A_j^2/(\mu_B g_e)^2\right]}$ for a large number of unpolarized nuclei [6]. The magnetic field dependence of steady-state electron spin polarization, $\bar{S}_z = 1/\tau \int_0^\tau \langle \hat{S}_z(t)\rangle dt$, is shown in Fig. 1(b). As one can see, the longitudinal magnetic field stabilizes the electron spin along the axis of initialization. It is interesting, that the half width of the dip in $\bar{S}_z$ near zero magnetic field is of the order of the NSF variance Δ_B. This result has been obtained earlier also in phenomenological calculations [6, 9].

To understand how the DNP affects the electron spin polarization, we have calculated similar magnetic field dependencies of $\bar{S}_z$ when the nuclear spins have been polarized by a sequence of electron-spin initializations with different orientation along z or $-z$ axis. To suppress some artefacts in the calculations of this nuclear spin training [8], a Poisson statistics has been used for each next excitation

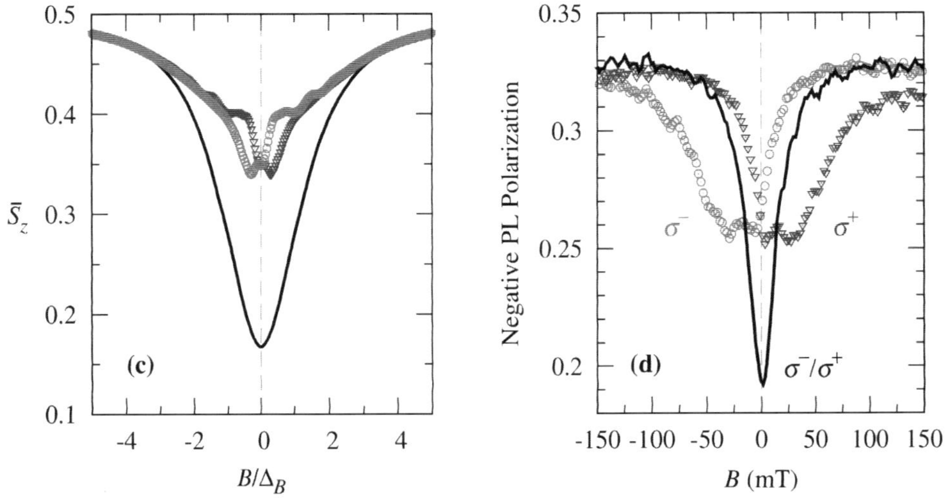

Figure 2. (a) Magnetic field dependencies of electron spin polarization for the cases of unpolarised nuclear spins (solid line) and polarized nuclear spin (symbols) for different signs of electron spin initialization. (b) Magnetic field dependencies of negative PL polarization measured for (In,Ga)As/GaAs QD ensembles. The curves are measured for alternatingly modulated excitation polarization (σ^-/σ^+) when no DNP has developed (solid line) as well as for constant circular polarization (σ^- or σ^+) creating DNP (circles and triangles, respectively).

pulse with average pump repetition period $\langle \Delta T \rangle = 8T_\Delta$. The results of these calculations are shown in Fig. 2(a). The appearance of DNP leads to the decrease of dip depth which illustrates the effect of electron spin stabilization. Besides, some small shift of the dip from the zero value of external magnetic field can be seen. This effect is due to the appearance of a regular DNP field which is subtracted from or added to the external magnetic field. One can also see an asymmetry of the dips in Fig. 2(a) calculated for trained nuclear spins. The reason for this asymmetry is that the nuclear polarization rate strongly depends on the applied magnetic field as confimred by additional numerical studies. There are also small peculiarities of these curves, which appear, most probably, because of the small number of nuclei in our model.

We compare these calculated curves with data measured experimentally [see Fig. 2(b)]. These experimental dependencies have been obtained by studying the photoluminescence (PL) of singly negatively charged (In,Ga)As/GaAs QDs annealed at 900 °C. The PL polarization is negative and reflects the spin polarization of resident electrons in these QDs. Further details can be found in Ref. [10]. As one can see, the theoretical modeling qualitatively reproduces the observed experimental behavior of electron-nuclear spin system. In particular, when the nuclei are unpolarized, the dip has the largest depth and is of symmetrical shape. When the nuclei become polarized, the experimental curves reveal a decrease of the dip depth and its shift from zero magnetic field, i. e., both effects obtained theoretically. However, there is also a quantitative difference between theory and experiment. In particular, the width of dip observed experimentally for unpolarized nuclei is much smaller than that obtained theoretically (in units of the dip shift when the nuclei are polarized). The reason for that is clear, as in the calculations only a small number of nuclear spins is considered. Therefore, the NSF, which are responsible for the dip, are drastically larger than those in real QDs and, correspondingly, give rise to the large dip width. Besides, the calculations show decreasing the dip width with pumping [see Fig. 2(a)], which is caused by NSF narrowing as it is already discussed in Ref. [8]. However, in contrast to the calculations, the experiment shows large broadening the dip at the permanent helicity of excitation polarization [see Fig. 2(b)]. Possible explanation for this difference is the large spread of DNP build-up from dot to dot in a QD ensemble the DNP as it is already discussed in Ref. [10]. This ensemble-related spread exceeds the theoretically predicted narrowing the NSF variance.

In conclusion, we have modeled the dynamics of electron-nuclear spin system in longitudinal magnetic field. A dip is observed in magnetic field dependence of electron spin polarization near zero field. This dip is shifted from zero field when nuclear spin system is polarized by the optical pumping. The predicted behavior of electron-nuclear spin system is supported by experiment. The narrowing of dip with pumping, however, is not observed experimentally, probably due to large spread of nuclear fields in the QD ensemble studied.

This work has been supported by the Deutsche Forschungsgemeinschaft, by the Russian Foundation for Basic Research, by the Ministry of Education and Science of the Russian Federation, and by the "Dynasty" foundation (MYP).

[1] *Spin Physics in Semiconductors* 2008 ed. Dyakonov M I, (Springer, Berlin)
[2] Gammon D, Efros Al L, Kennedy T A, Rosen M, Katzer D S, Park D, Brown S W, Korenev V L and Merkulov I A 2001 *Phys. Rev. Lett.* **86** 5176
[3] Lai C W, Malitinsky P, Badolato A and Imamoglu A 2006 *Phys. Rev. Lett.* **96** 167403
[4] Eble B, Krebs O, Lemaître A, Kowalik K, Kudelski A, Voisin P, Urbaszek B, Marie X and Amand T 2006 *Phys. Rev. B* **74** 081306
[5] Tartakovskii A I, Wright T, Russell A, Falko V I, Vankov A B, Skiba-Szymanska J, Drouzas I, Kolodka R S, Skolnick M S, Fry P W, Tahraoui A, Liu H.-Y and Hopkinson M 2007 *Phys. Rev. Lett.* **98** 026806
[6] Merkulov I A, Efros Al L and Rosen M 2002 *Phys. Rev. B* **65** 205309
[7] Khaetskii A, Loss D and Glazman L 2002 *Phys. Rev. Lett.* **88** 186802
[8] Petrov M Yu, Kozlov G G, Ignatiev I V, Cherbunin R V, Yakovlev D R and Bayer M 2009 *Phys. Rev. B* **80** 125318
[9] Petrov M Yu, Ignatiev I V, Poltavtsev S V, Greilich A, Bauschulte A, Yakovlev D R and Bayer M 2008 *Phys. Rev. B* **78** 045315
[10] Cherbunin R V, Verbin S Yu, Auer T, Yakovlev D R, Reuter D, Wieck A D, Gerlovin I Ya, Ignatiev I V, Vishnevsky D V, and Bayer M 2009 *Phys. Rev. B* **80** 035326

Quantum fluctuations in the low temperature magneto-resistance of bi-layer graphene

Y Ochiai[1], S Motooka[1], Y Ujiie[1], N Aoki[1], J P Bird[2] and D K Ferry[3]

[1]Graduate School of Advanced Integration Science, Chiba University, Chiba, Japan
[2]Department of Electrical Engineering, University at Buffalo, SUNY, Buffalo, USA
[3]Department of Electrical Engineering, Arizona State University, Tempe, USA

E-mail: ochiai@faculty.chiba-u.jp

Abstract. We report here the observation of quasi-periodic conductance fluctuations in the low temperature magneto-resistance of gated bi-layer (BL) field-effect transistors (FETs). Both correlation field analysis and conductance fluctuation behaviour suggests that the quantum transport in the BL-FETs closely resembles that in conventional semiconductor open quantum dots.

1. Introduction

There is considerable interest in the details of electrical transport in graphene because of its linear energy dispersion, and the chirality of its charge carriers [1, 2]. Indeed, there is exceptional promise of epitaxial graphene (on SiC) for integrated circuit applications. There have been many recent quantum transport studies of one or two-layer graphene structures. It has been reported that a considerably longer coherent length exists in the graphene samples [3] and the coherent nature can be discussed in comparison to the case of semiconductor quantum dots. The quantum transport in such systems has been tentatively analyzed in terms of Fabry-Perot interference and conductance fluctuations in quantum billiards, Aharonov Bohm (AB) interference in electron AB rings, and Coulomb oscillations and Fractal behavior in quasi-ballistic quantum dots [4-8]. We report here the observation of conductance fluctuations in the low temperature magneto-resistance (MR) of gated bi-layer (BL) graphene field effect transistors (FETs). We see no fractal behavior as discussed in our previous study of a sub-micron size quantum dot of BL graphene [8]. However, the correlation field analysis and conductance fluctuation behavior can be understood based on the corresponding behavior found for conventional open quantum dots [9].

2. Experiments

The BL-FET was fabricated by micromechanical cleavage of HOPG (SPI Grade-2) on a heavily doped Si/SiO_2 wafer. The graphene thickness was determined by AFM and Raman spectroscopy. Electrical contact was realized by forming 3 nm/30 nm Pt/Au electrodes whose pattern was defined by electron beam lithography. The dot size was approximately 5 μm × 5 μm, two orders of magnitude larger than our previous sub-micron size BL-FET [9]. We observe ripples and defects on the BL device, which is believed to decrease the carrier mobility [10]. Our previous estimates for BL graphene film [9] suggest a carrier concentration of 1.4×10^{12} cm^{-2} and a mobility of 2.0×10^{3} $cm^2V^{-1}s^{-1}$. We did not perform in-situ heating and thermal annealing, which would be expected to improve the homogeneity and the carrier mobility in the FETs. The MR of the devices was measured using lock-in detection with small excitation voltage (< 100 μV @ 17 Hz) in a He-3 cryostat refrigerator in applied normal magnetic fields up to 8 T.

3. Results

Typical MR of the graphene is shown in Fig. 1. The large variation of the MR might originate from the σ_{xy}-component of the quantum hall effect, as similar dependences were observed in the previous study of BL graphene samples [9]. The gate voltage dependence of the MR arises from the quantum coherent nature of the BL graphene. Therefore, subtraction of the rapid varying MR component has been performed and the quantum fluctuating part is shown in Fig. 2. It is clear that the quantum nature has been kept because of the existence of the magneto-fingerprint of the MR. The correlation field B_c of the fluctuations has been determined by the correlation function method based on the fluctuating MR as shown in Fig. 3, where the gate voltage dependence of B_c can be plotted as in Fig.4. We determine the mean value of the MR height in the conductance fluctuations (CF) from the original MR data in Fig. 2, which is discussed below to show the quantum coherent nature. From the gate voltage dependence of the MR at 4.2 K, we summarize the dependences of the B_c and mean-fluctuations together with the resistance value at zero magnetic field in Fig. 4. The maximum of the resistance probably corresponds to the Dirac point and the other two parameters shows a clear change near zero gate voltage, but the fact that this is located at a slightly negative bias may suggest a hole nature.

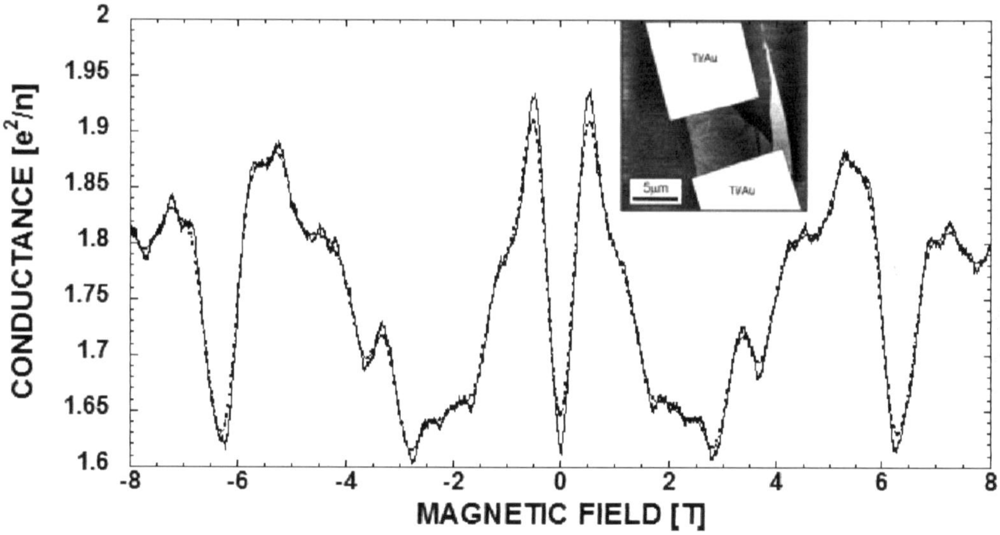

Figure 1. The magneto-resistance (MR) of the BL graphene at $V_g = 0$ V at 4.2 K. The solid line shows the original MR data while the dotted line shows the averaged MR obtained by smoothing. The inset shows an AFM image and of the BL graphene with the two-terminal Ti/Au electrode schematically illustrated.

4. Discussions

In our larger size BL-FET, we have not clearly observed fractal behavior in the transport. The coherence length is one third shorter than that in the smaller dot [9], so that the transport seems to belong in quasi-ballistic region. Scanning tunnelling spectroscopy indicated that the transport properties in an epitaxial graphene should be significantly influenced by atomic defects within the graphene lattice [11]. Also, a pronounced dielectric-screening effect has been observed in mono-layer graphene [12]. Here, comparing with previous study, we determine a slightly larger B_c and observe a smaller magneto-finger print of the MR. Discussing the above differences in terms of the quantum nature of the low temperature MR, they indicate that the fluctuations must originate from changes in many regular trapped orbits arising from the pointer states [13].

We observe a very interesting behavior of B_c, and the mean-fluctuations in the CF, as the back-gate voltage is varied. As shown in Fig. 4, near the Dirac point, the strength of the mean-fluctuations indicates a decrease of quantum coherency of the transport in the graphene because the decrease usually shows a slowdown of the degree of quantum nature and reduction of the coherent length in the graphene. However, also near the Dirac point, B_c decreases. As a simple and naive explanation, there seems to exist a contradiction in the relation between B_c and the mean-fluctuations in our present results. We need other evidence in order to fully explain these results.

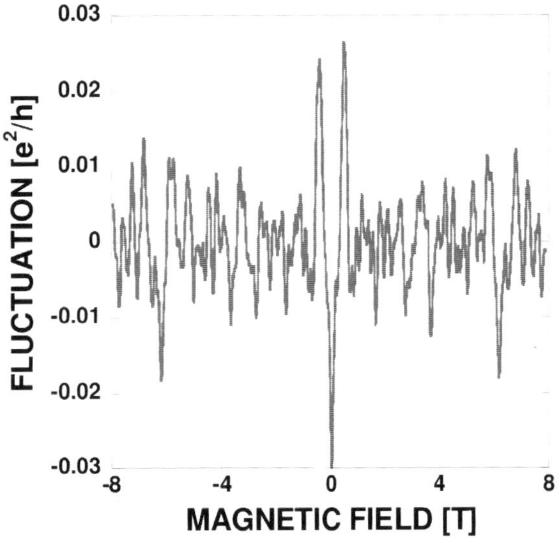

Figure 2. After subtraction from the MR data in the Fig. 1 are shown. In case of the CF, the magneto-finger print can clearly be observed.

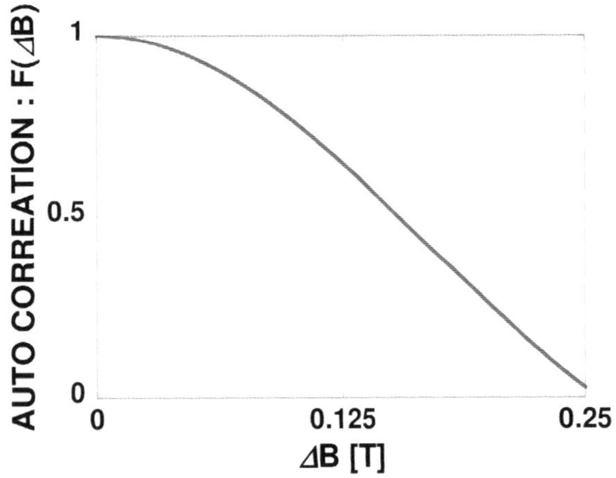

Figure 3. Auto correlation function calculated from the data in Fig. 2.

5. Summery

We have observed quasi-periodic conductance fluctuations in the low temperature MR of a BL graphene FET. Both of correlation field analysis and CF behavior suggests that the quantum transport in the BL-FET closely resembles that in conventional semiconductor open quantum dots. However, since there is a contradiction between B_c and CF, new explanation or evidence of the nano-scale transport in the graphene is needed.

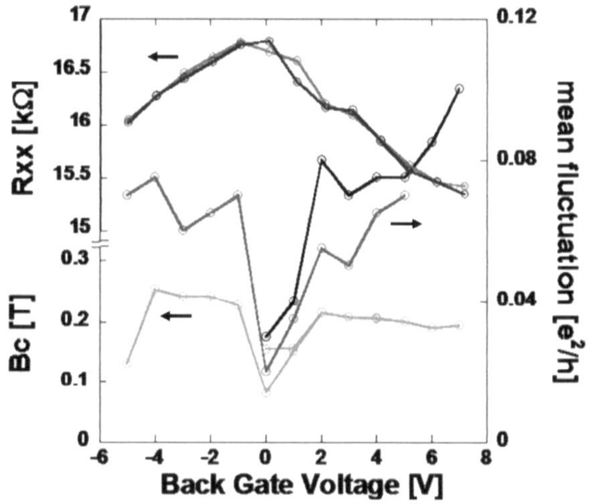

Figure 4. Correlation field B_c, mean-fluctuation of the MR and the zero-field resistance at 4.2 K are plotted in the same figure.

Acknowledgements
This work was supported in part by Grants-in-Aid for Scientific Research from the Japan Society for the Promotion of Science (JSPS) Nos. 19054016, 19204030 and 21656002.

References

[1] Novoselov K S, Geim A K, Morozov S V, Jiang D, Katsnelson M I, Grigorieva I V, Dubonos S V and Firosov A A 2005 *Nature* **438** 197
[2] Zhang Y, Tan Y W, Stormer H L, and Kim P 2005 *Nature* **438** 201
[3] Morozov S V, Novoselov K S, Katsnelson M I, Schedin F, Ponomarenko L A, Jiang D and Geim A K 2006 *Phys. Rev. Lett.* **97** 016801
[4] Miao F, Wijeratne S, Zhang Y, Coskun U C, Bao W and Lau C N 2008 *Science* **317** 1530
[5] Staley N E, Puls C P and Liu Y 2008 *Phys. Rev. B* **77** 155429
[6] Russo S, Oostinga J B, Wehenkel D, Heersche H B, Sobhani S S, Vandersypen L M K and Morpurgo A F 2008 *Phys. Rev. B* **77** 085413
[7] Ponomarenko L A, Schedin F, Katsnelson M I, Yang R, Hill E W, Novoselov K S, Geim A K 2008 *Science* **320** 356
[8] Ujiie Y, Motooka S, Morimoto T, Aoki N, Ferry D K, Bird J P and Ochiai Y 2008 *J. Phys.: Conf. Ser.* **109** 102035
[9] Ujiie Y, Motooka S, Morimoto T, Aoki N, Ferry D K, Bird J P and Ochiai Y 2009 *J. Phys.: Cond. Matt.* **21** 382202
[10] Krstić V, Obergfell D, Hansel S, Rikken G L J A, Blokland J H, Ferreira M S and Roth S 2008 *Nano Lett.* **8** 1700
[11] Rutter G M, Crain J N, Guisinger N P, Li T, First P N and Stroscio J A 2007 *Science* **317** 219
[12] Chen F, Xia J, Ferry D K and Tao N 2009 *Nano Lett.* **8** 2571
[13] Ferry D K, Akis R and Bird J P 2004 *Phys. Rev. Lett.* **93** 026803, and references contained therein

Proposal for a magnetic field induced graphene dot

P. A. Maksym[1], M. Roy[1], M. F. Craciun[2 4], S. Russo[2 4], M. Yamamoto[2], S. Tarucha[2] and H. Aoki[3]

[1]Department of Physics and Astronomy, University of Leicester, Leicester LE1 7RH, UK
[2]Department of Applied Physics, University of Tokyo, Hongo, Bunkyo-ku, Tokyo 113-8656, Japan
[3]Department of Physics, University of Tokyo, Hongo, Bunkyo-ku, Tokyo 113-0033, Japan

Abstract. Quantum dots induced by a strong magnetic field applied to a single layer of graphene in the perpendicular direction are investigated. The dot is defined by a model potential which consists of a well of depth ΔV relative to a flat asymptotic part and quantum states formed from the zeroth Landau level are considered. The energy of the dot states cannot be lower than $-\Delta V$ relative to the asymptotic potential. Consequently, when ΔV is chosen to be about half of the gap between the zeroth and first Landau levels, the dot states are isolated energetically in the gap between Landau level 0 and Landau level -1. This is confirmed with numerical calculations of the magnetic field dependent energy spectrum and the quantum states. Remarkably, an antidot formed by reversing the sign of ΔV also confines electrons but in the energy region between Landau level 0 and Landau level +1. This unusual behaviour gives an unambiguous signal of the novel physics of graphene quantum dots.

1. Introduction

Graphene is an astonishing material with a zero energy gap and massless charge carriers. This allows relativistic physics to be studied in the solid state and leads to many remarkable phenomena [1]. One of the most relevant to graphene dots is the quantum Klein paradox. This makes all potential barriers perfectly transparent to graphene charge carriers moving in one and two dimensions [2] so it is impossible to confine graphene carriers in an external potential well. However it is possible to confine them with a combination of electric and magnetic fields. Essentially, the magnetic field deflects the carriers so that they move in closed orbits and remain localised but this kind of confinement cannot be achieved with an arbitrary combination of electric and magnetic fields. The conditions needed have been analysed by Giavaras, Maksym and Roy [3] who have suggested a device structure that can be used to probe both confined and deconfined states in graphene dots. The present work is about a simpler structure that can be used to focus on confined states.

Conditional confinement is the one of the most important differences between graphene and conventional semiconductors. It is a direct consquence of the lack of mass. Roughly speaking, confined quantum states occur only when all the corresponding classical orbits are bounded but when massless particles move in a weak magnetic field *both* bounded *and* unbounded orbits occur at the same energy. Raising the magnetic field removes the unbounded orbits and enables confined states to occur. Details are given in ref. [3] for circularly symmetric scalar potentials,

[4] Present Address: School of Physics, University of Exeter, Exeter EX4 4QL, UK

$V = V_0 r^s$ and azimuthal vector potentials, $A_\theta = A_0 r^t$, where r is the radial distance and $s, t > 0$. In this case all the states are confined when $t > s$ but when $t = s$ the states are confined only when $c^2e^2A_0^2 > V_0^2$, where c is the Fermi velocity of the graphene sheet. Since V_0 and A_0 are experimentally tunable parameters this allows both confined and deconfined states to be probed in graphene dots and the interesting transition between the two regimes that occurs when $c^2e^2A_0^2 = V_0^2$ could also be probed experimentally.

Another important difference between graphene and conventional semiconductors is that graphene is gapless. Dots in conventional semiconductors are engineered so that the dot states are both confined and have energies in the band gap. Thus the dot states are spatially and energetically isolated from the rest of the system. The dot is an artificial atom and any probe of energy levels in the gap, for example charging experiments, is only sensitive to dot states and does not respond to bulk states. It is more difficult to achieve this in graphene because there is no natural gap. One option is to engineer a device structure that puts confined dot states in the gap between bulk Landau levels. However there is a problem. The potentials in a real graphene dot cannot rise indefinitely as a power law and must become flat asymptotically. So there are bulk Landau states outside the dot and it is necessary to ensure that the confined dot states occur in a region of low density of states away from the broadened bulk Landau levels. Giavaras, Maksym and Roy [3] showed that it is possible to achieve this while allowing the confinement-deconfinement transition to be observed provided that the scalar potential consists of a well which is separated from the asymptotic region by a barrier. However the barrier is not needed if one only wants to probe confined states. Then one can use a simple potential well as shown in figure 1.

A well in a uniform, perpendicular magnetic field clearly satisfies the confinement condition when the field is strong enough. The reason why it also satisfies the condition for energetic isolation is related to the unusual physics of the zeroth Landau level in graphene. In brief, the zeroth Landau level has zero energy and when a potential well of depth ΔV is applied to graphene the energy of states formed from the zeroth Landau level does not shift by more than about $|\Delta V|$. So dot states can be isolated energetically by choosing ΔV to put them into the middle of the Landau gap. The same result applies to antidot states, which are also confined because of the symmetry of the Dirac cone. So observations of the symmetry between dot and antidot confined states would give a direct experimental confirmation of the influence of the Dirac cone on the graphene dot energy spectrum. This physics is detailed in section 2 and numerical studies of the dot energy spectrum and quantum states are reported in section 3 where the properties of the graphene dot are compared with those of a typical electrostatic dot formed from GaAs.

2. Physics of Electron Confinement in the Zeroth Landau Level

The graphene sheet is taken to be in a uniform, perpendicular magnetic field, B and the azimuthal component of the magnetic vector potential at position $\mathbf{r}$ is $A_\theta = Br/2$, where θ is the azimuthal angle. The scalar potential is taken to be $V(r) = \pm(V_0 \cos(\pi r/2R) + V_1), r < 2R$, $V(r) = \pm(V_1 - V_0), r \geq 2R$, $V_0, V_1 > 0$, where the + sign corresponds to an antidot, the − sign corresponds to a dot and R is a size parameter. In terms of these parameters the depth of the dot and height of the antidot are given by $|\Delta V| = 2V_0$. Because the solution to the Laplace equation can be written as a Fourier series, this simple form is a crude model of the potential in a real dot.

The potentials have circular symmetry so the quantum states are eigenstates of angular momentum, $\hbar m$, where the angular momentum quantum number is the integer m. The graphene states are two-component states where each component gives the probability amplitude for an electron being on one of the two sub-lattices of the graphene sheet. The two-component envelope

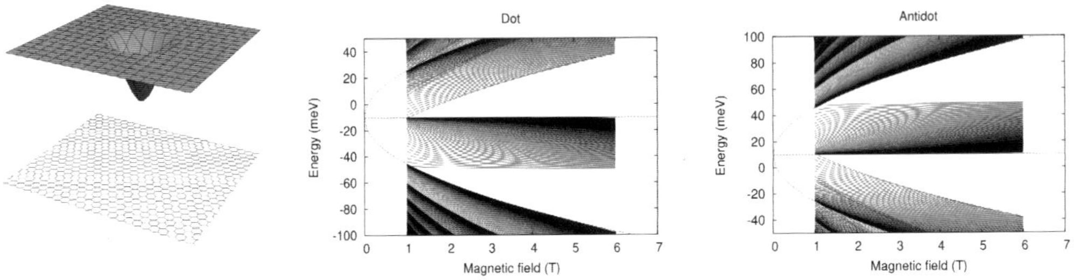

Figure 1. Physical model (left) and energy spectra of a dot (centre) and antidot (right).

function is $(\chi_1(r)\exp(i(m-1)\theta), \chi_2(r)\exp(im\theta))$ and the radial functions, χ_i satisfy [3]

$$\frac{V}{\gamma}\chi_1 - i\frac{d\chi_2}{dr} - i\frac{m}{r}\chi_2 - i\frac{e}{\hbar}A_\theta\chi_2 = \frac{E}{\gamma}\chi_1,$$
$$-i\frac{d\chi_1}{dr} + i\frac{(m-1)}{r}\chi_1 + i\frac{e}{\hbar}A_\theta\chi_1 + \frac{V}{\gamma}\chi_2 = \frac{E}{\gamma}\chi_2,$$

where $\gamma = 646$ meV nm and E is the energy. The Fermi velocity is given by $c = \gamma/\hbar$.

When $V = 0$ these equations give the graphene Landau levels of energy $E = \pm\gamma\sqrt{2N}/l$ where $l^2 = \hbar/eB$, l is the magnetic length, $N = n + (|m| + m)/2$ is the Landau quantum number and n is a radial quantum number. The zeroth Landau level states correspond to $n = 0$ and $m \leq 0$. Unlike the situation of a particle with mass, the energy of the zeroth Landau level is exactly zero and there is no zero point energy. In addition χ_1 vanishes and $\chi_2 \propto r^{|m|}\exp(-r^2/4l^2)$ so that the quantum states are localised on rings whose radius increases like $\sqrt{|m|}$.

Now consider what happens to the zeroth Landau level states in the presence of a non-zero potential. In this case χ_1 no longer vanishes but it can be shown that the zero point energy remains small [4]. Consequently the energy of the states is determined mainly by the local potential. In the case of a dot the lowest energy states are those of smallest spatial extent so if the dot is large enough that the $m = 0$ state fits into it, the energy of the lowest state in the dot is no more than ΔV below the asymptotic value of the potential. When m increases the states are localised in regions of weaker potential so their energy tends to increase with m. In the limit of very large m, the states merge into the bulk zeroth Landau level which is shifted from zero by the asymptotic value of the potential. These statements can be proved rigorously but the details are quite complicated [4]. This physics is the key to isolating a confined dot state energetically. If ΔV is chosen to be about half of the gap between Landau level 0 and Landau level -1, that is $\Delta V \sim \gamma/l\sqrt{2}$, the lowest dot state will be in the desired region of very low density of states.

To analyse the case of an antidot consider what happens when V is replaced by $-V$. It can be shown by direct substitution into the radial equations that if $(\chi_1(r)\exp(i(m-1)\theta), \chi_2(r)\exp(im\theta))$ is an eigenstate of energy E corresponding to potential V, then $(\chi_1(r)\exp(i(m-1)\theta), -\chi_2(r)\exp(im\theta))$ is an eigenstate of energy $-E$ corresponding to potential $-V$. Hence the antidot states are confined whenever the dot states are confined and the energy of the *highest* antidot state is no more than ΔV about the asymptotic value of the antidot potential. The fact that both dots and antidots confine electrons is unique to graphene and is an observable consequence of its Dirac cone band structure.

3. Numerical Studies of Dots and Antidots

To compute energies and states, the radial equations are made real by putting $f_1 = \chi_1$, $if_2 = \chi_2$, then f_1 and f_2 are found with the method outlined in ref. [3]. The model parameters are $V_0 = 20$ meV, $V_1 = 30$ meV and $R = 100$ nm. Hence $\Delta V = 40$ meV, about half the Landau gap at

5 T and $V(r)$ becomes ±10 meV outside the dot or antidot. Magnetic field dependent energy spectra are shown in figure 1. Each spectrum consists of a superposition of energy levels and the m range, $-150 \leq m \leq 10$, is chosen so that all states localised in the dot or antidot are included in the field range between 1 and 5.5 T. For comparison, the dashed lines show the Landau levels of an ideal graphene sheet. These levels are shifted by the asymptotic value of the potential, -10 meV in the case of the dot and +10 meV in the case of the antidot. The effect of the external potential is clearly to broaden the Landau levels. In the case of the dot, the zeroth Landau level at -10 meV is broadened into a band that occupies the region between -10 and -50 meV and the width of the band is very insensitive to the magnetic field. As expected the width of the band simply reflects the range of potential values: no zeroth Landau level state rises in energy above -10 meV and no zeroth Landau level state falls in energy below about -50 meV. In contrast, the upper edge of the broadened $N = -1$ Landau level falls in energy like $-\sqrt{B}$ and the lower edge of the broadened $N = +1$ Landau level rises in energy like $+\sqrt{B}$. Therefore the band of $N = 0$ dot states becomes energetically isolated as the field increases. With the present model parameters, the lowest dot state appears about in the middle of the gap when the field is about 5 T. In a real graphene sheet the Landau levels are broadened because of imperfections but the same principle applies and the magnetic field and potential can be chosen so that the lowest dot state lies in a region of extremely low density of states between broadened Landau levels. The right hand side of figure 1 shows the antidot spectrum. It has the same features as the dot spectrum except that the band of confined antidot states appears in the gap between the $N = 0$ and $N = 1$ Landau levels. The symmetry between the dot and antidot spectra is clearly visible. Figure 2 shows the lowest ($m = 0$) state in a dot and demonstrates that it is confined within the dot.

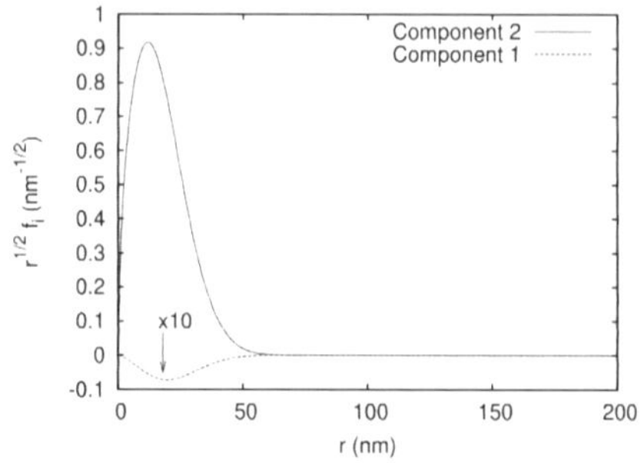

Figure 2. Two component radial envelope function for the $m = 0$ state in a dot.

In summary, the proposed graphene dot is able to confine electrons in a region that is isolated both spatially and energetically. The typical spacing between the dot levels at $B = 5$ T ranges from ~ 0.6 meV for $R = 100$ nm to ~ 2 meV for $R = 50$ nm. These sizes and level spacings are comparable to those for a GaAs electrostatic dot with confinement energy 4 meV in a magnetic field range of 3-10 T. The energy levels in the GaAs dot can be resolved by charging experiments, for example, so similar experiments on the graphene dot should be feasible.

Acknowledgments

This work was supported by the UK Royal Society. PAM is grateful for study leave from the University of Leicester and hospitality at the Department of Physics, University of Tokyo where part of this work was done. ST and MY acknowledge support from the Strategic International Cooperative Program (Joint Research Type), Japan Science and Technology Agency.

References

[1] A. K. Geim and K. S. Novoselov 2007 *Nature Materials* **6**, 183
[2] M. I. Katsnelson, K. S. Novoselov and A. K. Geim 2006 *Nature Phys.* **2**, 620
[3] G. Giavaras, P. A. Maksym and M. Roy, *J. Phys. Condens. Mat.* 2009 **21**, 102209
[4] G. Giavaras, P. A. Maksym and M. Roy, in preparation.

Journal of Physics: Conference Series **245** (2010) 012031 doi:10.1088/1742-6596/245/1/012031

The breakdown of Kohn's theorem in few-electron parabolic quantum dots doped with a single magnetic impurity Mn^{2+}

Nga T. T. Nguyen and F. M. Peeters

Departement Fysica, Universiteit Antwerpen, Groenenborgerlaan 171, B-2020 Antwerpen, Belgium

E-mail: nga.nguyen@ua.ac.be, francois.peeters@ua.ac.be

Abstract. The cyclotron resonance (CR) absorption spectrum is calculated for a II-VI parabolic quantum dot (QD) containing few electrons and a single magnetic dopant (Mn^{2+}). We find that Kohn's theorem no longer holds for this system and that the CR spectrum depends on the number of electrons inside the QD. The electron-Mn-ion interaction strength can be tuned for example by the magnetic field and by moving the Mn-ion to different positions inside the QD. We demonstrate that due to the presence of the Mn-ion the relative motion of the electrons couple with their center-of-mass motion through the electron-Mn-ion spin-spin exchange term resulting in an electron-electron interaction dependence of the magneto-optical absorption spectrum. At the ferromagnetic-antiferromagnetic transition we observe significant discontinuities in the CR lines.

1. Introduction

Quantum dots (QDs) doped with a single manganese ion (Mn^{2+} - spin 5/2) have been fabricated recently [1, 2] in CdTe based semiconductor. The single Mn-ion spin states were probed through multi-peak PL spectroscopy, and signatures of the exciton-Mn-ion spin exchange interaction were observed. Such QD systems are promising for future quantum computing application. For example the Mn-ion magnetic moments can be used to store information facilitating the readout functionalities in a quantum computing system.

Here we investigate a two-dimensional (2D) few-particle parabolic QD model system [3, 4]. We will examine the effect of the electron-electron (e-e) interaction on the magneto-optical absorption spectrum of such a few-electron QD in the presence of a single Mn-ion. Different absorption lines of allowed cyclotron resonance (CR) transitions were predicted as function of an external magnetic field.

We investigate the allowed CR transitions and their corresponding oscillator strength (OS). Our numerical results are for the experimental realized system Cd(Mn)Te where the lateral size is much larger than the height of the QD and consequently the system behaves like a quasi-2D QD. We find crossing and anti-crossing features that depend on the position of the Mn-ion in the QD and on the strength of the confinement potential. We will focus on the allowed transitions and the different CR lines that become possible due to the presence of the Mn-ion. The CR transitions now contain information not only of the electron state but also of the electron-Mn-ion (e-Mn) interaction and the magnetic state of the Mn-ion.

2. The model

The Hamiltonian of N_e interacting electrons $(\vec{r_i}, \vec{s_i})$ confined by a parabolic potential with confinement strength $\hbar\omega_0$ (tens of meV) embedded in a quasi-2D CdTe host QD doped with a single Mn-ion $(\vec{R}, \vec{M})$ in a perpendicular magnetic field $\vec{B}$ reads:

$$\hat{H} = \sum_{i=1}^{N_e} \left[\frac{1}{2m_e^*} \left(-i\hbar \vec{\nabla}_{\vec{r}_i} + e\vec{A}(\vec{r_i}) \right)^2 + \frac{1}{2} m_e^* \omega_0^2 \vec{r_i}^2 \right] + \sum_{i<j=1}^{N_e} \frac{e^2}{4\pi\epsilon_0\epsilon|\vec{r_i} - \vec{r_j}|}$$
$$+ \frac{1}{2}\hbar\omega_c (g_e m^* S_z + g_{Mn} m^* M_z) - J_c \sum_{i=1}^{N_e} \vec{s_i} \cdot \vec{M} \delta(\vec{r_i} - \vec{R}). \quad (1)$$

The vector potential $\vec{A} = B/2(-y, x, 0)$ is taken in the symmetric gauge. The confinement frequency ω_0 is related to the confinement length by: $l_0 = \sqrt{\hbar/m_e^*\omega_0}$ with $m^* = m_e^*/m_0 = 0.106$ the effective electron mass. $g_e = -1.67$ and $g_{Mn} = 2.02$ are the Landé g-factor of the host semiconductor and the Mn-ion, respectively. Electron cyclotron frequency is $\omega_c = eB/m_e^*$. The strength [5] of the e-Mn spin-spin exchange interaction is typically about $J_c = 1.5\,\mathrm{eV}\mathring{A}^2$. We use the set of parameters [1, 6] that is applicable for relatively small Cd(Mn)Te QDs. The dielectric constant $\epsilon = 10.6$, the effective Bohr radius $a_B^* = 52.9$ Å, and l_0 are of the order of tens of nm. For example, with $\hbar\omega_0 = 51.32\ meV$ we have $l_0 = 26.45$ Å.

We have adopted configuration interaction method to build up the Hamiltonian [6, 7] matrix and employed exact diagonalization [7] to obtain numerically the many-body eigenstates. Note that the total angular momentum L and spin S_z do not commute with $\hat{H}$ therefore they are no longer good quantum numbers and we have to consider the entire Hilbert space. This rapidly increases the size of the Hamiltonian.

The OS of the intra-band electron transitions (from ground state (GS) i to excited state j) $f_{ij} = 2\Delta E_{ij}/\hbar\omega_H \cdot |A_{ij}|^2/l_H^2$ are calculated using the transition amplitude [8]: $A_{ij} = \sum_{p=1}^{N_e} < \Psi_i(\vec{r_1}, ..., \vec{r_{N_e}})|r_p e^{\pm i\theta_p}|\Psi_j(\vec{r_1}, ..., \vec{r_{N_e}}) >$ with its transition energy $E_{ij} = E_j - E_i$. We define $\omega_H = \omega_0(1 + \Omega_c^2/4)^{1/2}$ where $\Omega_c = \omega_c/\omega_0$ and $l_H = [\hbar/(m_e^*\omega_H)]^{1/2}$ defines a new length. $\Psi_{i(j)}$ is the many-body wave function [7] of state $i\,(j)$ which is a linear combination of all possible N_c quantum state wave function $\psi_{i(j)}$, where $\psi_{i(j)}$ is a product of Slater determinants and spin component of the Mn-ion for configuration $i\,(j)$. The single-electron basis consists of the Fock-Darwin (FD) states in a magnetic field [7]. A sufficiently large number of FD levels are included to guarantee numerical convergency. All allowed transitions with OS exceeding 1% are retained. The following Lorentzian broadened formula: $\sigma_i(E) = \sum_j \Gamma_{ij}/\pi \cdot f_{ij}/[(E - E_{ij})^2 + \Gamma_{ij}^2]$ is used. Γ_{ij} is the broadening parameter that is taken to be about $0.01 \div 0.1$ meV in our numerical simulation of the absorption spectrum.

The presence of the Mn-ion leads to different arrangements ferromagnetic (FM) and antiferromagnetic (AFM) of the spins of the electron and the Mn-ion which are tunable by an external magnetic field B. This is due to the interplay between the Zeeman energies and the exchange interaction, which both depend on B. The FM phase is found for very small B when the exchange energy dominates over the Zeeman energies. In the other case the AFM phase is found. The presence of the Mn-ion leads to a mixing of the FD orbitals for the different eigenstates in contrast to the QD without a Mn-ion where those FD orbitals are the quantum eigenstates.

We show in Fig. 1 the OS and CR energies for $N_e = 1$ electron QD. The main transitions come from transitions to the two p-orbitals as typically oberved in QDs without Mn-doping. But in the presence of a Mn-ion we notice that the OS exhibits a discontinuity at $\Omega_c = 0.32$ where the FM-AFM transition takes place. I and II, respectively, corresponds to the transitions from the GS with $(n_r, l, s_z, M_z) = (0, 0, s_z, -5/2)$ to $(0, \mp 1, s_z, -5/2)$ as the major quantum state. The wave function of the GS changes at the FM-AFM from favoring a configuration where the

Journal of Physics: Conference Series **245** (2010) 012031 doi:10.1088/1742-6596/245/1/012031

electron spin has a large probability to be down to configuration with electron spin up. The other two CR-lines (III and IV) come from spin-spin exchange interaction corresponding to final states that have wave functions containing, respectively, $(0, -1, -1/2, -3/2)$ and $(0, 1, -1/2, -3/2)$ as their major FD states.

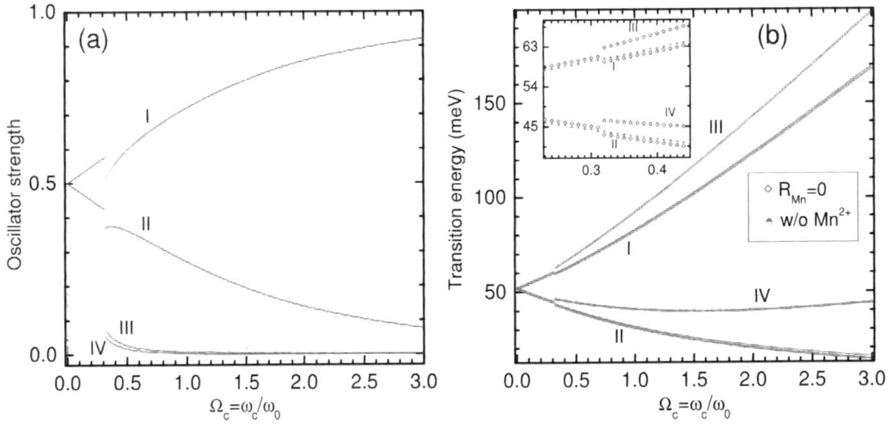

Figure 1. OS (a) and CR energies (b) of single-electron QD in case the Mn-ion is located at $R = 0$ as a function of magnetic field. The results without Mn-ion are added in the inset for reference.

Many-body effects in the FIR spectroscopy are examined in Fig. 2 where the CR energies for different N_e QDs are shown when the Mn-ion is located at $R = 0$ for two different $\Omega_c = 0.02$ (≈ 0.15 T) and 0.1. The number of allowed electron transitions depends on N_e. We refer to Ref. [9] for a detailed study on how the e-e interaction weakens the e-Mn exchange interaction and how it affects the FIR spectrum.

The N_e-dependence of the CR spectrum is due to a breaking of Kohn's theorem. The reason is that the center-of-mass motion couples with the relative motions and also with the Mn-ion through the e-Mn spin exchange. Note that the relative motions contain information about the e-e interaction. This fact leads to the existence of various IR absorption spectra for different N_e in which many transitions have different enegies.

For $N_e = 3$, the number of major CR lines is four in the FM phase. The contribution $(0, 1, -1/2, -3/2)$, which becomes dominant in the excited-state wave function, is ascribed to the presence of the double (circularly) right-polarized modes (of the spin-down s-electron and spin-up $p^+_{(n_r,l)=(0,1)}$−electron) as the first (sorted ascending in energy) two large-OS peaks and double left-polarized modes (the other two large-OS peaks) for $\Omega_c = 0.02$, and triple left-polarized modes for $\Omega_c = 0.1$ (the three largest-energy peaks). The very-low-OS (4% of the total OS) black solid peak is the left-polarized mode combining the $\sigma^+(\mathrm{s})$ of the spin-down s-electron and the p^+-electron where the configuration $(0, 1, -1/2, -3/2)$ strongly dominates the final state.

We note that for $N_e = 2, 4, ...$ electrons the FM phase is the GS up to lower B-fields as compare to the case $N_e = 1, 3, 5, ...$ electrons (see Ref. [7]). This fact leads to the appearance of only two major peaks for $N_e = 2$ and 4 electrons when $\Omega_c = 0.1$ which are the major transitions ($S_z = 0$) to the p-orbitals without exchanging spins. However, for N_e=4 the total spin can be such that we have triplet state for $\Omega_c = 0.02$ which results in a larger number of CR-lines in this case. Apparently, Fermi-Dirac statistics plays an essential role because the filling effect, as seen in the case $N_e = 4$ for the intermediate states $S_z = \pm 1$ and 0, influences directly the spin exchange interaction. The FIR absorption energies for $N_e = 5$ are different from those of $N_e = 3$ due to p-p (orbital) e-e interaction effect. For $\Omega_c = 0.02$ a left-polarized mode can be at a smaller energy as compared to a right-polarized mode where the left-polarized combined modes of the spin-down s-, spin-up p^+-, and spin-up $p^-_{(n_r,l)=(0,-1)}$-electron and of the two spin-up p-electrons are, respectively, the first and the fourth peak (sorted in energy). The existence of the very-low-OS peaks ($\approx 2 \div 5\%$ of the total OS) in all cases of N_e are evidence of the influence of the e-Mn interactions when the small component to the GS, e.g. (0,0,-1/2,-3/2) for $N_e = 1$ electron QD, becomes the dominant one in the excited states.

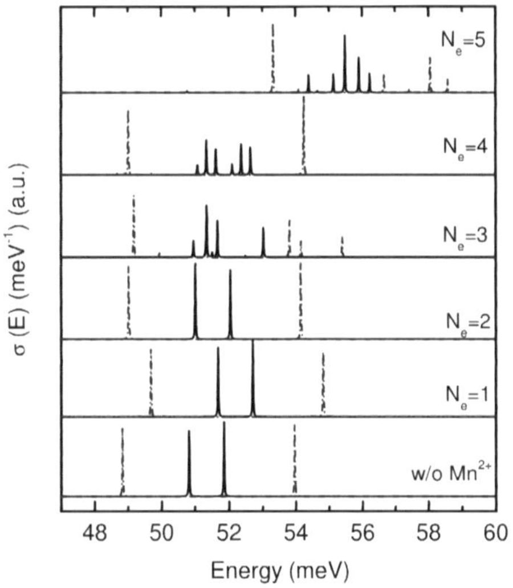

Figure 2. CR energies of a N_e-electron QD in case the Mn-ion is located at R=0 for two different fields $\Omega_c = 0.02$ (black solid) and 0.1 (blue dash-dotted). Absorption energies in case without the Mn-ion are added for reference.

Acknowledgments
This work was supported by FWO-Vl (Flemish Science Foundation), the EU Network of Excellence: SANDiE, and the Belgian Science Policy (IAP).

References
[1] Besombes L, Léger Y, Maingault L, Ferrand D, Mariette H, and Cibert J 2004 *Phys. Rev. Lett.* **93** 207403
[2] Léger Y, Besombes L, Fernández-Rossier J, Maingault L, and Mariette H 2006 *Phys. Rev. Lett.* **97** 107401
[3] Maksym P A, and Chakraborty T 1990 *Phys. Rev. Lett.* **64** 108
[4] Wojs A and Hawrylak P 1996 *Phys. Rev.* B **53** 10841
[5] Fernández-Rossier J and Brey L 2004 *Phys. Rev. Lett.* **93** 117201
[6] Qu F and Hawrylak P 2005, *Phys. Rev. Lett.* **95** 217206
[7] Nguyen N T T and Peeters F M 2008, *Phys. Rev.* B **78** 045321; *Phys. Rev.* B **78** 245311
[8] Geerinckx F, Peeters F M, and Devreese J T 1990, *J. Appl. Phys.* **68** 3435
[9] Nguyen N T T and Peeters F M 2009, *Phys. Rev.* B **80** 115335

Phase-diagram for the magnetic states of the Mn-ion subsystem in a magnetic quantum dot

Nga T. T. Nguyen and F. M. Peeters

Departement Fysica, Universiteit Antwerpen, Groenenborgerlaan 171, B-2020 Antwerpen, Belgium

E-mail: nga.nguyen@ua.ac.be, francois.peeters@ua.ac.be

Abstract. The interplay between two types of spin-spin exchange interaction (namely of the electron with the Mn-ions and the Mn-ions with each other) that are governed by the positions of the Mn-ions and the magnetic field is studied in the case of a Mn-ion doped CdTe quantum dot. We investigate the formation of different magnetic phases and the existence of frustrated magnetic states due to the dominant contribution of the Mn-Mn energy.

1. Introduction

Quantum dots (QDs) containing a single, two, three, etc magnetic ions (Mn^{2+}) have been realized and the interaction between the localized magnetic moments and the electron (or hole) in a single dot was investigated through e.g. local probe photoluminescence [1, 2]. In II-VI semiconductor based QDs such as CdTe or CdSe, the dots exhibit ferromagnetism where the number of carriers can be controlled separately from the number of Mn-ions in the QD.

Previously theoretical studies on Cd(Mn)Te QDs containing few Mn-ions include electronic properties of quasi-two-dimensional (2D) few-electron system at zero field [3], ferromagnetism in hard-wall-cubic noninteracting-electron structure [4], intra-band optical absorption spectrum for single-electron nanostructures [5], and magnetism [6, 7] in low-density Mn-ion system. Magnetism of different II-VI materials e.g. Cd(Mn)Se QDs consisting of a very low concentration of Mn-ions was recently realized [8, 9]. When injecting charged particles into these diluted magnetic semiconductor QDs, a long lifetime of excitons was measured which was a consequence of the exchange electron-Mn-ion (e-Mn) interaction. Note that in Cd(Mn)Se the Landé g-factor of electrons and the Mn-ions have opposite sign. This leads to strong magnetic anisotropy which is one of the important requirements of a memory element and was theoretically examined in e.g. Ref. [10] for symmetric QDs.

Here, we study a 2D Cd(Mn)Te single-electron parabolic QD subjected to an external magnetic field $\overrightarrow{B}$ using exact diagonalization. Such system has a relatively complex behavior due to the opposite signs of the g-factors $g_e \cdot g_{Mn} < 0$. Consequently, different local/system magnetic phases can be formed with respect to the spin orientation of the particles in the two exchange interaction extremes. Magnetic polaron states are obtained depending on the Mn-Mn interaction strength and the magnitude of the magnetic field (cyclotron frequency $\omega_c = eB/m_e^*$). Such a QD system can be described by the following Hamiltonian:

$$\hat{H} = \frac{1}{2m_e^*}\left(-i\hbar\overrightarrow{\nabla}_{\overrightarrow{r}} + e\overrightarrow{A}(\overrightarrow{r})\right)^2 + \frac{1}{2}m_e^*\omega_0^2\overrightarrow{r}^2$$

$$+\frac{1}{2}\hbar\omega_c\left[g_e m^* s_z + g_{Mn} m^*(M_{1z}+M_{2z})\right]$$
$$-J_c\overrightarrow{s}\cdot[\overrightarrow{M_1}\delta(\overrightarrow{r}-\overrightarrow{R_1})+\overrightarrow{M_2}\delta(\overrightarrow{r}-\overrightarrow{R_2}]$$
$$+J_{12}^{Mn}(\overrightarrow{R_{12}})\overrightarrow{M_1}\cdot\overrightarrow{M_2}. \quad (1)$$

The e-Mn and Mn-Mn exchange interactions are, respectively, the fourth and fifth term in Eq. (1). Their strength as a function of the relative position between two Mn-ions can be estimated from Fig. 1. The electrons interact with the Mn-ions by the contact spin-exchange interaction whose form can be expressed in term of second-quantization $J_{ij}(\overrightarrow{R}) = J_c\varphi_i^*(\overrightarrow{R})\varphi_j(\overrightarrow{R})$ with strength [11] $J_c = 1.5\,\mathrm{eV}\mathring{A}^2$. $\{\varphi_i\}$ are the Fock-Darwin single-electron states. The two Mn-ions interact with each other through the anti-ferromagnetic (AFM) spin exchange interaction [12, 13] whose strength $J_{12}^{Mn}(\overrightarrow{R_{12}}) = J_{12}^0\exp\{-\lambda(|\overrightarrow{R_{12}}|/a_0-1)\}$ decreases exponentially with their separation $|\overrightarrow{R_{12}}| = |\overrightarrow{R_1}-\overrightarrow{R_2}|$. $J_0^{12} = 0.5$ meV and $\lambda = 5.1$ are chosen according to experiment work [13] where the two Mn-ions are assumed to interact most strongly when they are nearest neighbors in the Mn-lattice with lattice spacing $a_0 = 5.4\mathring{A}$. Parameters applicable to Cd(Mn)Te QDs are used: the electron effective mass $m^* = m_e^*/m_0 = 0.106$, dielectric constant $\epsilon = 10.6$, effective Bohr radius $a_B^* = 52.9$ $\mathring{A}$ and $l_0 = \sqrt{\hbar/m_e^*\omega_0} \sim$ tens of nanometers, g_e=-1.67, and $g_{Mn} = 2.02$.

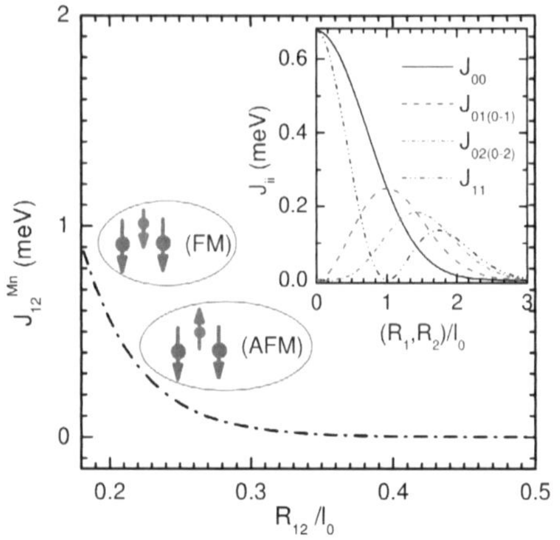

Figure 1. Electron-Mn (the inset) and Mn-Mn (big arrows) exchange interaction as a function of the separation between the Mn-ions. The strength of the e-Mn exchange interaction is evaluated as the diagonal elements in the exchange matrix for the *s*-, *p*-, and *d*-shell. The lower left insets show schematic plots of the FM and AFM phases in case of a single electron (small arrow). Subscripts of J in the inset refer to (n_r, l) as the radial and azimuthal quantum numbers.

We define ferromagnetic (FM) and AFM phases in accordance with the spin couplings of the two subsystems (of the electrons and the Mn-ions) with each other. In fact, at high magnetic fields the electrons and the Mn-ions have anti-parallel spins, which can be considered as the AFM phase. For small magnetic fields the FM phase is found mostly, at least in the case of $N_e = 1$ electron. The Mn-ions in their subsystem also exhibit different orderings (*AFM* and *FM*) in accordance with their spin orientations. We distinguish these local states with the FM and AFM phases of the system by writing them in *italic*.

2. Phase-diagram

The FM and AFM phase diagram for $N_e = 1$ electron [14] and $N_{Mn} = 1$ is summarized in Fig. 2(a) as a function of the Mn-ion position relative to the center of the QD and magnetic field. This result shows that the single-electron case samples the dominant FM region around the center of the dot. By moving the Mn-ion out of the center, the e-Mn interaction decreases and a slightly larger than zero field is sufficient to transit the system to the AFM phase.

Now we introduce a second Mn-ion. We found that the Mn-ion now becomes less sensitive to the presence of the field as compared to the one Mn-ion situation with the stable minimum value $M_z = -5/2$ [14]. Here, the Mn-ion(s) transit to spin $-5/2$ after the intermediate phases where M_z of the Mn-ions can be $-1/2$ or $-3/2$. The e-Mn and Mn-Mn exchange interaction compete most strongly in the very-small-field limit (see Fig. 1). Note that the e-Mn interaction strength is larger with increasing magnetic field. It is worth noting that the Mn-Mn exchange energy is only comparable with the e-Mn energy when they are positioned close to each other, say $R_{12} < 0.4l_0$. Plus, the Mn-ions can repel/attract the electrons differently if they are located asymmetrically with respect to the center of the QD. In case they are located symmetrically and for small Mn-Mn separation their spin exchange strength is enhanced and the Mn-ions stay in the *AFM* state. On the other hand when they equally attract the electron and the system they stay in the FM phase. Such a state is named *frustration*. To understand the physics that takes place with changing the magnetic field and the Mn-ion positions (symmetry and asymmetry) it is important to have information on the Mn-ion position dependence of the local *AFM* and *FM* states. Such information is made visible in Fig. 2(b). The phase diagram has a step-like form because the Mn-Mn interaction strength depends on the $M_{(1,2)z}$ which changes by steps of unity when changing their relative separation or their individual positions. In case the Mn-ions are very close to each other, say $R_{12} < 0.28l_0$, the subsystem stays in the *AFM* phase regardless they are symmetrically or asymmetrically positioned.

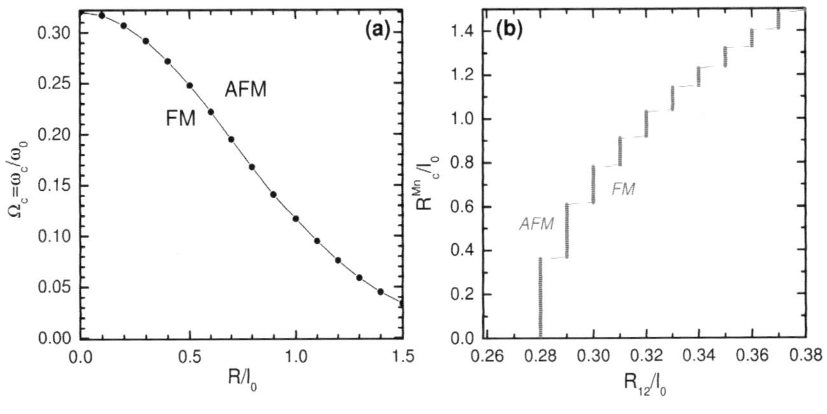

Figure 2. Magnetic phase diagram of a single-electron QD doped with one (a) and two (b) Mn-ions for zero field as function of the Mn-ion positions. $\overrightarrow{R_c^{Mn}} = (\overrightarrow{R_1} + \overrightarrow{R_2})/2$ is the Mn-ion center of mass coordinate.

Magnetic polaron state occurs when a single electron FM or AFM couples with a collection of *FM*-coupled Mn-ions. The FM polaron is found when the Mn-ion separation is engineered such that their magnetic coupling is *FM* and the e-Mn interaction is found FM. Therefore, such FM polaron can be obtained in the small-field limit. Because in the high-polarized situation, the electron will couple AFM with the Mn-ions and the system forms an AFM polaron.

We also calculate the ground-state (GS) magnetization $< M >_{GS} = -\partial < E_{GS} > /\partial B$ and the thermodynamic magnetization $< M(T) >= -\partial < E(T) > /\partial B$ for $T = 5.8$ K and show the results in Fig. 3 for a typical symmetric positioning of the Mn-ions $\overrightarrow{R_{1,2}} = (\pm 0.15 l_0, 0)$ as a function of magnetic field. The magnetic-field dependence of the magnetic polaron states (circles containing electron and two Mn-ions) is observed in different manners with and without T. The step appears each time the spins of either the electron or the Mn-ions transit to a smaller/larger intermediate state. In this case the magnetic polaron state starts at very small field, $\Omega_c = \omega_c/\omega_0 \approx 0$, the system exhibits a unique step that is the FM-AFM transition.

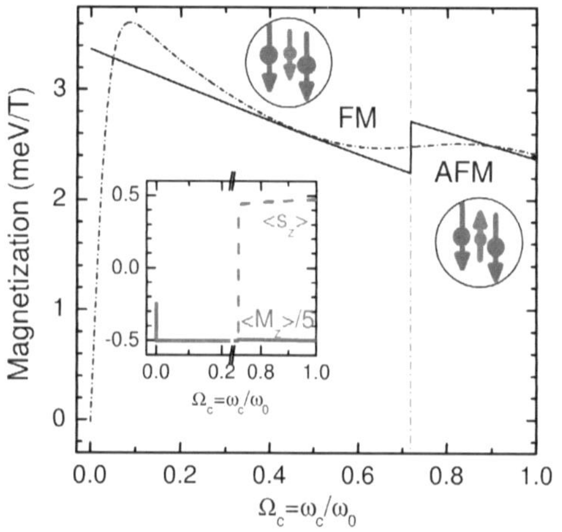

Figure 3. Magnetization as a function of magnetic field calculated for the GS energy (black solid) and $T = 5.8$ K (blue dash-dotted) in case $\overrightarrow{R_{1,2}} = (\pm 0.15 l_0, 0)$. Cyan dash-dotted separates the FM and AFM regions. Inset is the average of s_z and M_z for reference. Small and big arrows refer to the electron and Mn-ions, respectively.

Acknowledgments
This work was supported by FWO-Vl (Flemish Science Foundation) and the Belgian Science Policy (IAP).

References
[1] Besombes L, Léger Y, Maingault L, Ferrand D, Mariette H, and Cibert J 2004 *Phys. Rev. Lett.* **93** 207403
[2] Léger Y, Besombes L, Fernández-Rossier J, Maingault L, and Mariette H 2006 *Phys. Rev. Lett.* **97** 107401
[3] Qu F and Hawrylak P 2006 *Phys. Rev. Lett.* **96** 157201
[4] J. Fernández-Rossier and L. Brey 2004 *Phys. Rev. Lett.* **93** 117201
[5] Savić I and Vukmirović N 2007 *Phys. Rev.* B **76** 245307
[6] Abolfath R M, Hawrylak P, and Zutić I 2007, *Phys. Rev. Lett.* **98** 207203; *New J. Phys.* **9** 353
[7] Abolfath R M, Hawrylak P, and Zutić I 2008, *Phys. Rev. Lett.* **101** 207203
[8] Beaulac R, Archer P I, Liu X, Lee S, Salley G M, Dobrowolska M, Furdyna J K, and Gamelin D R 2008, *Nano Lett.* **8** 1197
[9] Beaulac R, Schneider L, Archer P I, Bacher G, and Gamelin D R 2009, *Science* **325** 973
[10] Cheng S J 2009, Phys. Rev. B **79** 245301
[11] Fernández-Rossier J and Brey L 2004 *Phys. Rev. Lett.* **93** 117201
[12] Furdyna J K 1988 *J. Appl. Phys.* **65** 29
[13] Shapira Y, Olivera Jr. N F 1987 *Phys. Rev.* B **35** 6888
[14] Nguyen N T T and Peeters F M 2008, *Phys. Rev.* B **78** 045321; *Phys. Rev.* B **78** 245311

Fast preparation and detection of Mn spin states in a magnetically doped quantum dot

Doris E. Reiter[1], **Vollrath Martin Axt**[2], **Tilmann Kuhn**[1]

[1]Institut für Festkörpertheorie, Westfälische Wilhelms-Universität, 48149 Münster, Germany
[2]Theoretische Physik III, Universität Bayreuth, 95440 Bayreuth

E-mail: Doris.Reiter@uni-muenster.de

Abstract. The spin dynamics of a single Mn spin in a single quantum dot under the excitation with ultra fast laser pulses is studied. We develop a switching protocol that is able to prepare stable superpositions of states with different Mn spin quantum numbers. Furthermore the optical signal in a pump-probe type setup is calculated which can be used to read out the Mn spin state.

1. Introduction

For applications in quantum information it is of crucial importance to be able to prepare both quantum states with well defined quantum numbers as well as stable coherent superposition states. An attractive system for corresponding applications is the spin of a single Mn atom in a II-VI quantum dot (QD) as has been fabricated recently [1], since the Mn spin showed very long decoherence times in optical experiments [2]. The optical manipulation of the Mn spin so far was based on incoherent relaxation or transfer mechanisms on the timescale of nanoseconds [3, 4]. We have recently proposed a protocol based on coherent manipulations by which spin eigenstates of the the Mn atom could be prepared on a picosecond time scale. In this paper we present an extension of this protocol to create coherent superpositions of these eigenstates.

2. Theoretical background

We model a single CdTe QD doped with a single Mn atom following the approach in Ref. [5]. We take into account one conduction band (CB) and two valence bands (VBs). Electrons in the CB have a total angular momentum of $S^e = \frac{1}{2}$ with $S^e_z = \pm\frac{1}{2}$, while the total angular momentum of the holes is $S^h = \frac{3}{2}$ splitting into the heavy hole (HH) states with $S^h_z = \pm\frac{3}{2}$ and the light hole (LH) states with $S^h_z = \pm\frac{1}{2}$. The HH-LH splitting is taken as 40 meV. Together the excitonic system comprises of 15 states including the ground state with no exciton $|0\rangle$, the single exciton states consisting of one HH and one electron $|H\pm 2\rangle$ and $|H\pm 1\rangle$, the single exciton states with one LH and one electron $|L\pm 1\rangle$ and $|L\pm 0\rangle$ (the sign in the latter refers to the LH spin), the biexciton states formed by either two HH or LH excitons $|HH0\rangle$ and $|LL0\rangle$, as well as four combined biexciton states composed of one HH exciton and one LH exciton $|HL\pm 2\rangle$ and $|HL\pm 1\rangle$. All states are labeled by their hole state and their total angular momentum. The Mn has six spin states with $M_z = \pm\frac{5}{2}, \pm\frac{3}{2}, \pm\frac{1}{2}$. The 90 states of the full system are then labeled $|X; M_z\rangle$ with the exciton state $|X\rangle$ and the Mn spin M_z. The electron and hole spin couple

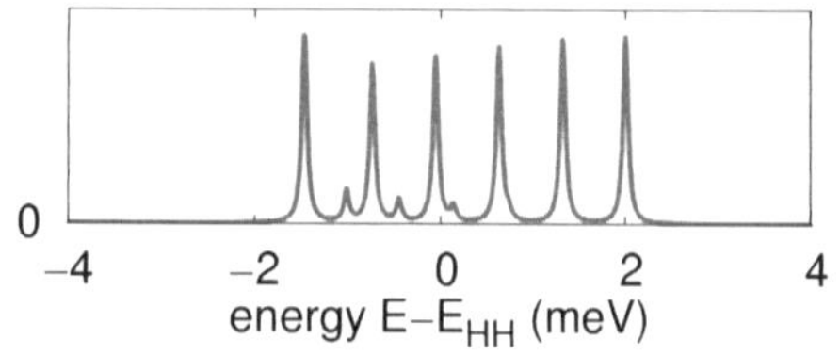

Figure 1. Calculated absorption spectrum at a magnetic field of $B_z = 6$ T for σ^- polarized light. In the initial state the six Mn spin states are taken to be equally occupied.

to the Mn spin via the exchange interaction $H_{exc} = j_e \vec{M} \cdot \vec{S^e} + j_h \vec{M} \cdot \vec{S^h}$. For the calculations $j_h = 0.423$ meV and $j_e = -j_h/4$ are taken. The exchange interaction can be separated into two types of terms: Ising type terms $\sim j_{e/h} M_z S_z^{e/h}$, which lead to an energy shift of the states, and spin flip terms $\sim j_{e/h}\frac{1}{2}(M_+ S_-^{e/h} + M_- S_+^{e/h})$. Due to the latter term simultaneous flips of the Mn spin and the electron/LH spin can occur. Note that a spin flip involving HHs is not possible by this term. Light coupling to the exciton system is treated in the standard dipole and rotating wave approximation. We consider Gaussian pulses with a full width at half maximum (FWHM) of 100 fs and a π pulse is set to create one exciton. The single exciton states can be classified into bright (optically active) with angular momentum ± 1 and dark (optically not active) states. Additionally the $e - h$ exchange interaction as well as Zeeman coupling to a magnetic field are taken into account. Optical signals to simulate a pump-probe setup, where a sequence of pump pulses excites the system, which is then probed by a weaker probe pulse with a FWHM of 10 ps, are calculated in the standard way. The spectrum is obtained by a Fourier transform with a damping of 20 ps.

3. Results

The absorption spectrum of a quantum dot containing a single Mn atom has a very characteristic feature: The exciton line of the lowest excitation splits into a set of six lines even at zero magnetic field according to the six orientations of the Mn spin [1]. A calculated absorption spectrum is shown in Fig. 1 for a magnetic field in Faraday configuration of $B_z = 6$ T and σ^- polarized light resonant of the HH exciton transition. The six peak structure is clearly observable. Each of the six peaks can be identified with the Mn spin state $M_z = +\frac{5}{2}$ to $M_z = -\frac{5}{2}$ from left to right [5]. Some small side peaks appear due to the coupling of electron and Mn spin.

In the time domain the spin flip terms manifest themselves as exchange induced Rabi oscillations between the initially excited state and the state, where electron and Mn spin have changed by one. Usually the two states involved are not in resonance, such that this oscillation has an amplitude smaller than one. However, by applying a series of 2π pulses a complete flip can be achieved [6]. Using this feature a switching scheme was developed to address selectively all states of the Mn spin [7]. In this paper a modified switching scheme to create superpositions between states with different Mn spin is proposed.

Starting from the initial state $|0; -\frac{5}{2}\rangle$ a σ^- polarized π pulse on the HH exciton transition excites the system to $|H-1; -\frac{5}{2}\rangle$. Note that the Mn spin is not affected by the laser excitation. In this state the electron can perform a spin flip while increasing the Mn spin by one, ending up in the state $|H-2; -\frac{3}{2}\rangle$. If the spin is only halfway flipped, the system is in a superposition between the states $|H-1; -\frac{5}{2}\rangle$ and $|H-2; -\frac{3}{2}\rangle$. Now a σ^+ polarized pulse on the LH exciton transition is applied. This does not act on state $|H-1; -\frac{5}{2}\rangle$, but excites the dark state $|H-2; -\frac{3}{2}\rangle$ into the combined biexciton state $|HL-1; -\frac{3}{2}\rangle$. In the latter state the CB state is filled, thus electrons and HH cannot change the Mn spin. Only the LH can flip and increase the Mn spin state, changing the state to $|HL-2; -\frac{1}{2}\rangle$. This flip is much faster than the electron flip, such that $|H-1; -\frac{5}{2}\rangle$ does not change. The combined biexciton $|HL-2\rangle$ is optically active and by two π pulses with σ^- polarization, one resonant on the HH transition and one resonant on the

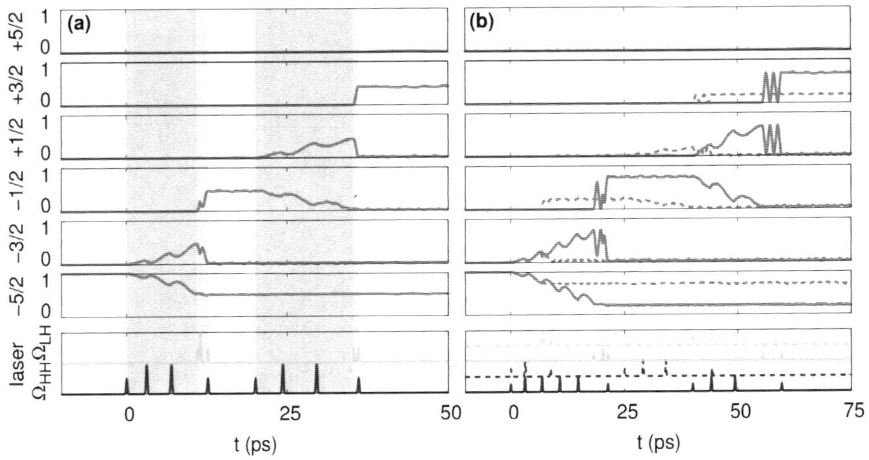

Figure 2. Occupations of the Mn spin eigenstates M_z (upper panels) under the influence of a laser pulse sequence resonant on the HH (Ω_{HH}) and the LH (Ω_{LH}) transition (lower panel) for a magnetic field of $B_z = 6$ T; (a) leading to an equal superposition between $M_z = -\frac{5}{2}$ and $M_z = +\frac{3}{2}$; (b) leading to a superposition of these states with weights 25% and 75%, respectively, (solid lines) or vice versa (dashed lines).

LH transition, it is de-excited to $|0; -\frac{1}{2}\rangle$. The π pulses likewise act on the other state $|H-1; -\frac{5}{2}\rangle$, turning the HH into a LH exciton via the ground state. Thus after the two π pulses the system is in the superposition of the states $|L-1; -\frac{5}{2}\rangle$ and $|0; -\frac{1}{2}\rangle$. Both states are stable, because in the first state the spins of all particles have their minimal value and thus cannot flip, while in the latter state no exciton is present. To switch the Mn spin further, the same combination of pulses can be used. While these pulses do not change the Mn spin state with $M_z = -\frac{5}{2}$, it increases the Mn spin $M_z = -\frac{1}{2}$ to $M_z = +\frac{3}{2}$. After the switching the exciton system goes back to the ground state for both states. Thus a coherent superposition between the states $|0; -\frac{5}{2}\rangle$ and $|0; +\frac{3}{2}\rangle$ has been established.

The resulting switching dynamics is shown in Fig. 2(a), where the occupations of the six Mn spin states and the applied laser pulse sequence resonant either on the HH transition (Ω_{HH}) or the LH transition (Ω_{LH}) are shown. For $t < 0$ $M_z = -\frac{5}{2}$ is fully occupied. Then the pulse sequence is started with a π pulse at $t = 0$. After two 2π pulses both $M_z = -\frac{5}{2}$ and $M_z = -\frac{3}{2}$ are half occupied and a π pulse resonant on the LH transition is applied. The occupation of $M_z = -\frac{3}{2}$ goes down to zero and $M_z = -\frac{1}{2}$ becomes occupied. When $M_z = -\frac{1}{2}$ is half occupied two π pulses act on the system. Then the occupations of both $M_z = -\frac{5}{2}$ and $M_z = -\frac{1}{2}$ are 0.5 and remain stable. With a π pulse at $t = 20$ ps the next switching step is initiated, in which the occupation of $M_z = -\frac{1}{2}$ goes to zero and via $M_z = +\frac{1}{2}$ finally $M_z = +\frac{3}{2}$ becomes occupied. Thus a superposition with equal weights of 50% between Mn spin states differing by four has been created. Figure 2(b) shows modifications of the pulse sequence resulting in a superposition of the same states with different weights (25% / 75% and vice versa).

The creation of a superposition state can also be followed in a pump-probe setup. Figure 3 shows probe spectra for σ^- polarized light on the HH exciton transition taken at different time delays τ during the switching sequence presented in Fig. 2(a). The time delay refers to the starting point of the pump pulse sequence. Thus negative time delays refer to the case when the probe pulse precedes the first pump pulse. Here no exciton is in the system and $M_z = -\frac{5}{2}$. Accordingly an absorption peak is observed in Fig. 2(a). The positions of the six peaks of the absorption spectrum in Fig. 1 are marked by thin vertical lines and indeed the single peak in the

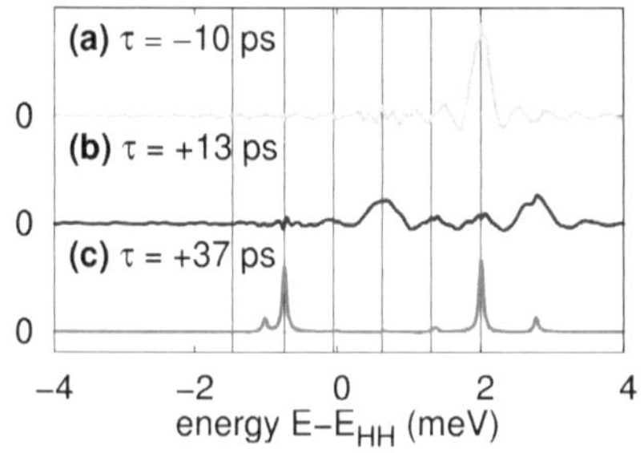

Figure 3. Probe spectra for different time delays τ as indicated for the switching sequence shown in Fig. 2(a).

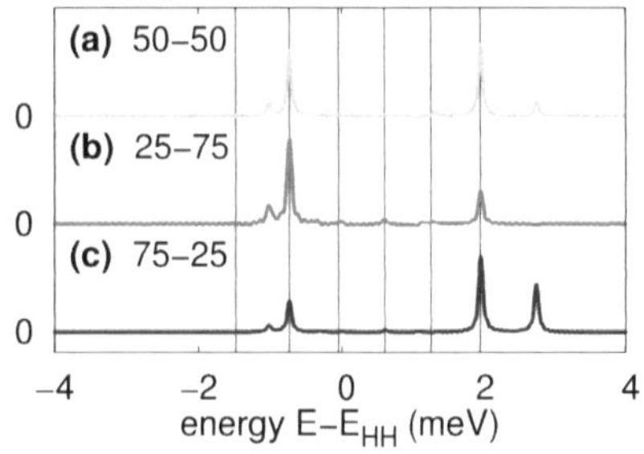

Figure 4. Probe spectra after the switching sequence for differently weighted superpositions as shown in Fig. 2.

probe spectrum matches the rightmost line. In addition, well known spectral oscillations caused by the following pulses are observed. The second probe spectrum (b) at $\tau = 13$ ps is taken, when the first stable superposition between $|L-1;-\frac{5}{2}\rangle$ and $|0;-\frac{1}{2}\rangle$ has been created. In the spectrum two peaks occur, one at the third line from the right, which corresponds to $M_z = -\frac{1}{2}$, while the other does not match one of the six lines. Because a LH was added to the system, the probe pulse does not act on the ground state to single exciton transition, but on the transition between LH exciton and combined biexciton. Thus the line is shifted by exchange interactions. Again spectral oscillations are superimposed. The spectrum (c) is taken at $\tau = 37$ ps, when a superposition between $M_z = -\frac{5}{2}$ and $M_z = -\frac{1}{2}$ is created. Two absorption peaks are seen at the second line from the left and the rightmost line in agreement with the Mn spin states.

Figure 4 finally shows probe spectra at a delay after the switching has taken place for the three cases with different weights of the Mn spin states. In each spectrum two peaks are seen corresponding to the respective Mn spin states. The strength of the peaks reflects the occupation of states: in (a) both peaks are equally high, in (b) the right peak, corresponding to $M_z = -\frac{5}{2}$ is much smaller than the left peak corresponding to $M_z = +\frac{3}{2}$, and in (c) it is vice versa.

4. Conclusions

We have proposed a switching protocol based on coherent manipulations of a single Mn spin in a QD to create coherent superpositions of Mn spin states. The creation of a superposition could be monitored in probe spectra, where two peaks appear, their strength reflecting the weights of the states of the superposition.

References

[1] Besombes L *et al.* 2004 *Phys. Rev. Lett.* **93** 207403
[2] Besombes L *et al.* 2008 *Phys. Rev.* B **78** 125324
[3] Le Gall C *et al.* 2009 *Phys. Rev. Lett.* **102** 127402
[4] Goryca M *et al.* 2009 *Phys. Rev. Lett.* **103** 087401
[5] Fernández-Rossier J 2006 *Phys. Rev.* B **73** 045301
[6] Reiter D E, Kuhn T and Axt V M 2009 *Phys. Status Solidi* B **246** 779
[7] Reiter D E, Kuhn T and Axt V M 2009 *Phys. Rev. Lett.* **102** 177403

Photo-induced conductance fluctuations in mesoscopic Ge/Si systems with quantum dots

N. P. Stepina[1], E.S. Koptev[1], A.V. Dvurechenskii[1], I.V. Osinnyukh[1], A.I. Nikiforov[1], D. Gruetzmacher[2], J. Moers[2], J. Gerharz[2]

[1] Institute of Semiconductor Physics, 630090 Novosibirsk, Russia
[2] Institute of Bio- and Nanosystems, Forschungszentrum Julich GmbH 52425 Julich, Germany

E-mail: stepina@isp.nsc.ru

Abstract. Nanometer sized SiGe quantum dot structures were used for the realization of infrared photodetectors working at extremely small photon fluxes. The sensitivity gain of the system to illumination is due to the fact that the hopping conductance via two-dimensional QD array strongly depends on the photoexcited charge trapped in the dots. Photo-induced conductance fluctuations in hopping regime were observed in both short/wide and long/narrow conductive channels. An optimized structure was determined by analyzing the time evolution of conductance in a frame of fluctuations count rate.

1. Introduction

A self-organizing growth technique, which includes a carefully controlled strained layer epitaxy at low (300°C) temperature, allows the formation of high density array of QDs with nanometer dimensions and homogeneous size distribution. The areal density of nanoclusters (4×10^{11} cm^{-2}) is large enough for the charge transport through such a system at low ($\leq$20K) temperatures being dominated by hole hopping between dots [1]. It was shown that the hopping conductance in a densely packed array of Ge islands crucially depends on the occupation of the QDs by holes [2]. Thus, changing the QD charge state by illumination is expected to cause a change in the conductance of the array. In macroscopic samples with Ge/Si QDs both negative and positive photoresponse depending on dot filling factor were observed [3] under illumination with wavelength between 0.9-1.55 μm. Long-time conductance dynamics (typically, $10^2 - 10^4$ seconds at T=4.2 K) has been revealed as well as after switching on and switching off the illumination, displaying a sluggish temporal dependence. It was found that the residual photoconductance could persist for several hours after the light switching off. Using these phenomena for mesoscopic samples, when the sample size is too small for self-averaging taking place, one can observe the physical processes corresponding to the unit events of network transformation. Thus, under weak illumination, a step-like change in the hopping conductance should be expected due to the charge changing by one carrier in one of the dots.

We present the experimental results of photo-induced conductance fluctuations in decananometer size QDs structures with different width and length of conductive channels.

Journal of Physics: Conference Series **245** (2010) 012034 doi:10.1088/1742-6596/245/1/012034

2. Samples and experiment

The samples were grown on a (001) p-Si substrate by molecular-beam epitaxy of Ge in the Stranskii-Krastanov growth mode. The growth temperature for 10 monolayers (ML) Ge layer was 300°C and the growth rate was 0.2 ML/s. To supply holes to the dots, a boron δ-doped Si layer was inserted 5 nm below the Ge QDs layer. Electron-beam lithography was used to create the conductance channel, with their width and length being changed in the range of 70-500 nm. Quasi-one-dimensional, square and short samples were prepared and measured. Al metal source and drain electrodes were deposited on the top of structure and heated at 480°C to form reproducible Ohmic contacts. The system was excited by fiber coupled laser with 1.55 μm wavelength and 1 mW power value. To get extremely small light intensity, initial laser power was attenuated up to 60 dB. Time-resolved measurements were carried out at temperature 4.2 K and small currents ($10^{-12} \div 10^{-14}$ A) allowing to be in Ohmic regime. Any possible contact influence was prevented by using a four-point scheme for measuring the resistance traces. Photo-emf of contacts under illumination did not exceed 1% from the QD responce. The temperature stability was controlled by Ge thermometer, the conductance change due to the temperature variation lay below the dark noise level.

3. Results and discussion

The scanning electron microscopy image of the sample with channel width w smaller than their length l is shown in a Fig1. The estimations of the correlation radius L_c of percolating network give their value of about micrometers. It means that all our samples has sizes smaller than L_c and conductance should reproduce the mesoscopic properties.

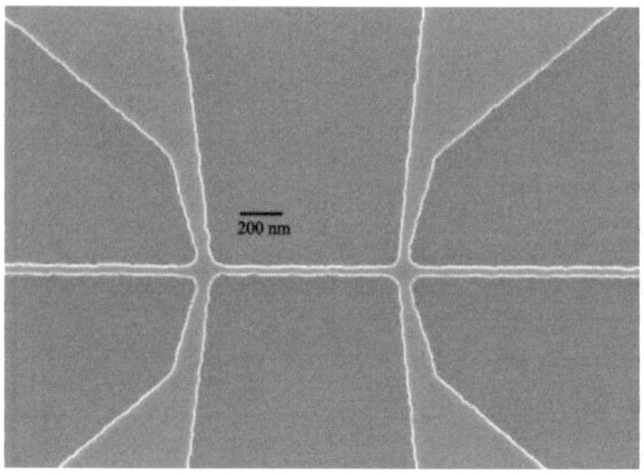

Figure 1. Scanned electron microscopy image of device with long and narrow conductive channel.

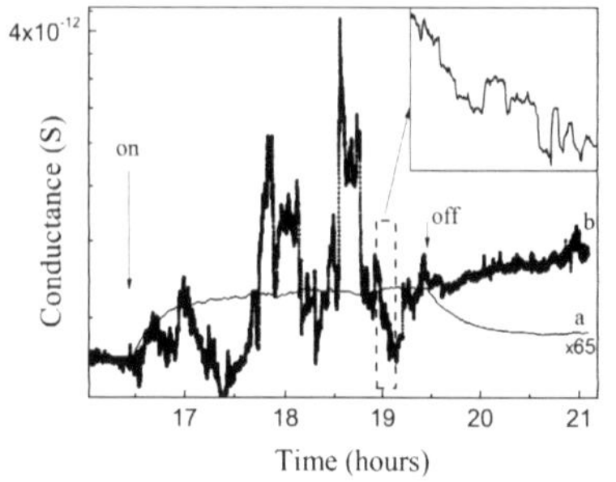

Figure 2. Conductance evolution of the mesoscopic sample with l=100 nm and w=70 nm. Inset is enhanced part of conductance curve.

Fig.2 shows the typical conductance traces for mesoscopic sample with w= 70 nm width and l=100 nm length of the conductive channel under 30 dB attenuation of laser light. In the large-area sample measured in parallel (curve a), illumination causes a monotonic change of the conductance with time. Unlike macroscopic structures, in small-area samples, where conduction takes place through a mesoscopic size hopping network, the conductance evolves with time (curve b) in discrete steps (inset to Fig.2). We suggest that these steps correspond to changing of the individual QDs filling factor due to absorption of a single photon. Under 1.55 μm wavelength illumination, every photon absorbed in QD excites one electron from the dot, caused the addition of a single positive charge into dot. Generated electrons can be captured by another dot; as the result, the redistribution of carriers among different QDs occurs. The resulting new potential landscape leads to a new conductive path providing change of the conductance with time.

The behavior of mesoscopic samples is determined not only by their size but also by the channel geometry. In the presence of large fluctuations of the random potential, the sample resistance will be determined by exponentially rare regions of space containing no localized states within a few $k_B T$ from Fermi level. In quasi-one-dimensional structures, energetically unfavorable hops cannot be avoided, as in opposite limit of short and wide samples, the percolating cluster will be shunted by exponentially rare isolated chains of hopes with conductances that are exponentially larger than the conductance of the infinitive cluster. These chains are formed by localized states located at anomalously close distances from one another. Then, the sample conductance is determined by the highest resistance link of the best-conducting chain (the "critical" hop). Fig3 shows the evolution of conductance with time for three samples with different sizes and geometry of conductive channels under 30 dB attenuation of light. The sample a) corresponds to the quasi-one-dimensional structure (l=500 nm, w=100 nm), sample b) is a square one (l=100 nm, w=100 nm) and sample c) has the smallest channel width (l=100 nm, w=70 nm). One can see that the square sample is characterized by the largest average sheet conductance (Fig3). This result is the evidence that localized states which shunt the infinite cluster appear at small channel length. Since the probability for the formation of such a chain is an exponential function of l, a small change in the channel length should lead to a substantial change in conductance. The width of the channel determines the probability of appearance of the best-conducting chain. The more the width the larger the number of such chains should be.

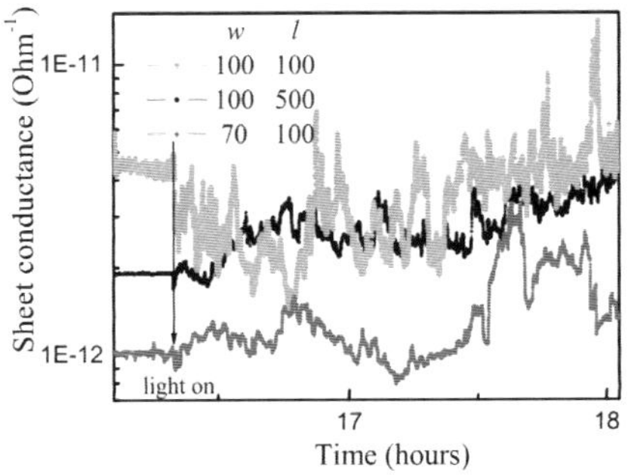

Figure 3. Conductance tracks for three samples with different size of conductive channels.

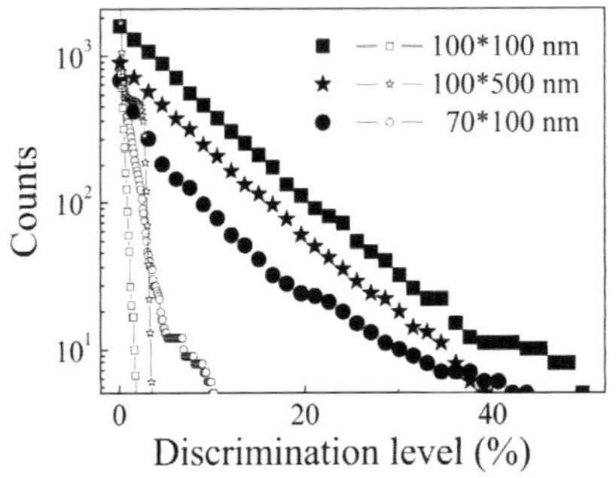

Figure 4. Number of counts at different discrimination levels for the samples in dark (open symbols) and under illumination (full symbols).

Experimental kinetics data for dark and photo-excited parts of time-resolved conductance were analyzed in the frame of count rates at different threshold (discrimination) levels of fluctuation amplitude. The rate of steps was estimated by the following simple procedure. We found the zeros of the current derivative (with respect to time) responsible for the fluctuations and then determined the value of $\Delta G = G_2 - G_1$ between every two extremums. The fluctuation amplitude was estimated as a percentage of ΔG to the G_1 value. Then, we count the number of fluctuations with amplitude larger that some discrimination level. The results of such procedure for conductance tracks depicted in Fig3 are shown in a Fig4. The range of time was chosen to be 10000 seconds. One can see that the dark noise does not exceed 4-10% (open symbols in Fig4) whereas the amplitude of conductance fluctuations under illumination reachs more than 50% of magnitude. The largest number of fluctuations is observed not in the sample with largest area (long sample), but in square sample. It means that this kind of samples is more sensitive to the photon flux and can be used for creating of a single-photon detector.

To understand whether the photodetector is really working in the photon counting mode, we analyzed the dependence of the number of counts on the discrimination level at different light intensities, the result of its analysis is shown in Fig. 6. The inset in Fig.6 depicts the number of fluctuations counted for all used light intensities at discrimination level (10%) restricting dark and light noise. The data shows that the number of steps is linearly dependent on the light intensity, as expected for a single-photon process. It is worth to be mentioned that the persistant photoconductance effect, which accounts for $\sim 10-20\%$ of the conductance value in macroscopic samples, keeps the conductance constant after switching off the light in mesoscopic samples. This is advantageous, because in contrast to conventional avalanch photodiodes here after detection of one photon no reset of the system to its initial state is needed for the detection of the next photon.

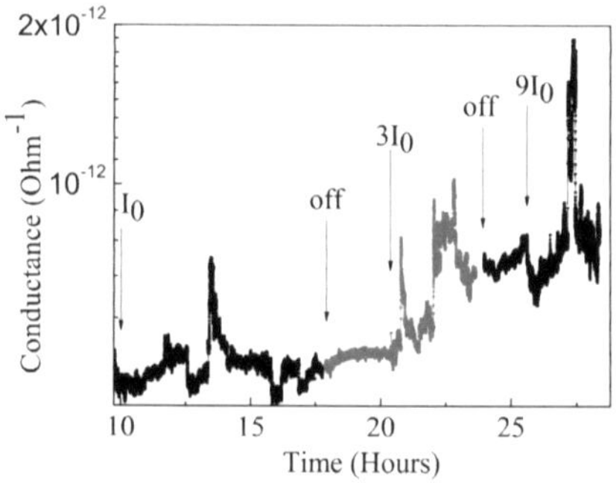

Figure 5. Conductance evolutions for sample c) at different light intensity I

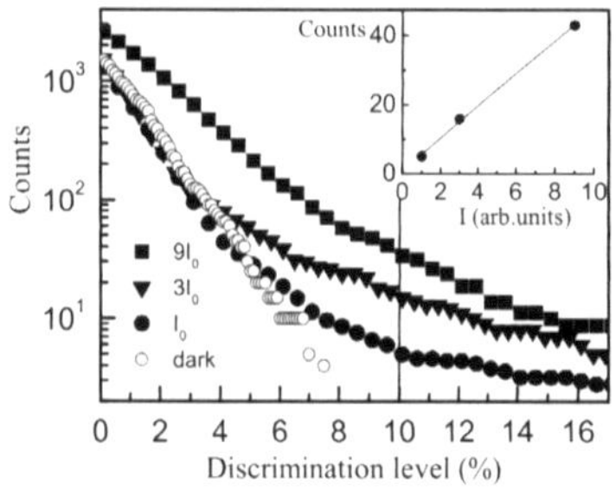

Figure 6. Number of counts for sample c) at different discrimination level. Inset-dependence of counts on I at level 10%.

The internal quantum efficiency of our device was estimated to be $\sim 17\%$ with taking into account the effect of oscillator force acceleration in quantum dot nanostructures. Large values of the oscillator strength and the exciton binding energy for type-II quantum dots with finite offsets was predicted by Rorison [4] and observed in the same Ge/Si QD structure [5] for $\sim$ 750-850 meV exciton transitions that corresponds to the excitation wavelengh used in this work. Strong oscillator strength of the Ge QDs was explained by electron leakage into the dots. However, a small thickness of Ge layer, absorbing the 1.55 μm wavelength light yields only $\sim 10^{-3}\%$ of external efficiency. To increase this value, the many-layers Ge structure and waveguide system is proposed to be used.

Acknowledgments

We give thanks to A. V. Nenashev for useful discussion. This work was supported by RFBR (Grants 10-02-00618).

References

[1] A.I. Yakimov, A.V. Dvurechenskii, V.V. Kirienko, Yu.I. Yakovlev, A.I. Nikiforov, C.J. Adkins, *Phys. Rev.* **B 61**, 10868 (2000).
[2] Yakimov A.I. et.al., JETP Letters, **77**, 376 (2003)
[3] N.P. Stepina, A.I. Yakimov, A.V. Nenashev, A.V. Dvurechenskii, N.A. Sobolev, J.P. Leitao, V.V. Kirienko, A.I. Nikiforov, E.S. Koptev, L. Pereira, Journal of Experimental and Theoretical Physics **103** 269 (2006).
[4] J. M. Rorison, Phys. Rev. **B48**, 4643 (1993).
[5] A.I. Yakimov, N.P. Stepina, A.V. Dvurechenskii, A.I. Nikiforov, A.V. Nenashev, Phys. Rev. **B63**, 45312 (2001).

Phonon-mediated relaxation in doped quantum dot molecules

Anna Grodecka-Grad[1,2], Jens Förstner[2]

1. QUANTOP, Danish National Research Foundation Center for Quantum Optics, Niels Bohr Institute, University of Copenhagen, DK-2100 Copenhagen Ø, Denmark
2. Computational Nanophotonics Group, Theoretical Physics, University of Paderborn, 33098 Paderborn, Germany

E-mail: anna.grodecka-grad@nbi.dk

Abstract. We study a single quantum dot molecule doped with one electron in the presence of electron-phonon coupling. Both diagonal and off-diagonal interactions representing real and virtual processes with acoustic phonons via deformation potential and piezoelectric coupling are taken into account. We employ a non-perturbative quantum kinetic theory and show that the phonon-mediated relaxation is dominated by an electron tunneling on a picosecond time scale. A dependence of the relaxation on the temperature and the strength of the tunneling coupling is analyzed.

1. Introduction

Coupled quantum dots (QDs) called quantum dot molecules (QDMs) have attracted much attention due to their proposed application in various quantum computation schemes. The two ground states of a single confined electron can be used as the logical qubit states and their electrical [1] and optical control [2, 3] have already been demonstrated. However, not only the electron-electron interaction [4] but also phonon-assisted relaxation [5, 6] can affect the coherent control to a large extent. In particular, the energy difference between the two electron ground states, which is of order of a few meV, can lead to both virtual pure dephasing processes [7, 8, 9] as well as efficient electron tunneling [10, 11].

In this contribution, a fully quantum kinetic description of the phonon-mediated relaxation in doped quantum dot molecules including non-Markovian effects is presented. It is shown that the coupling to the phonon reservoir in QDMs can lead to a fast electron tunneling on a picosecond timescale, which strongly affects the coherent electron evolution. We employ non-Markovian correlation expansion technique [12, 13] including up to three particle correlations. Both diagonal and usually neglected off-diagonal electron-phonon couplings representing virtual and real phonon-assisted processes are included. We study the temperature dependence of the phonon-mediated relaxation as well as the influence of the tunneling coupling strength.

2. Model system

We study a QDM doped with a single electron and consider the relaxation between the two energetically lowest states of the electron $|1\rangle$ in the left and $|2\rangle$ in the right quantum dot (see Fig. 1 with schematic plot of the energy levels in the QDM). The free Hamiltonian of the electron reads

$$H_c = \epsilon \left(|2\rangle\langle 2| - |1\rangle\langle 1|\right) + \Gamma \left(|1\rangle\langle 2| + |2\rangle\langle 1|\right). \tag{1}$$

Here, Γ is the tunneling coupling between the QDs and 2ϵ is the energy difference between the ground states in both QDs.

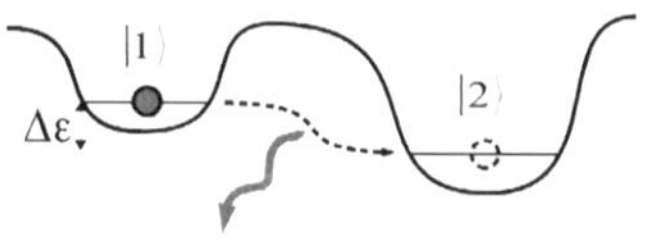

Figure 1. A schematic plot of two coupled quantum dots with two electron states $|1\rangle$ and $|2\rangle$ with the energy difference of $\Delta\epsilon = 2\epsilon$. The dashed and curled arrows depict electron tunneling with simultaneous phonon emission.

The free phonon Hamiltonian is $H_{\rm ph} = \sum_{\mathbf{k},s} \hbar\omega_{\mathbf{k},s} b^\dagger_{\mathbf{k},s} b_{\mathbf{k},s}$, where $b^\dagger_{\mathbf{k},s}$ and $b_{\mathbf{k},s}$ are phonon creation and annihilation operators, respectively, $\omega_{\mathbf{k},s}$ denotes phonon frequencies with phonon wave vector $\mathbf{k}$, where different phonon branches are labeled by s (longitudinal l and two transverse $t1$ and $t2$). The carrier-phonon interaction is described by the following Hamiltonian

$$H_{\rm int} = \sum_{i,j=1,2} \sum_{\mathbf{k},s} \left[g_{ij,s}(\mathbf{k}) |i\rangle\langle j| b_{\mathbf{k},s} + g^*_{ij,s}(\mathbf{k}) |j\rangle\langle i| b^\dagger_{\mathbf{k},s} \right]. \tag{2}$$

The relevant coupling elements for the acoustic phonons coupled via both piezoelectric coupling and deformation potential read:

$$g^{\rm PE}_{ij,s}(\mathbf{k}) = -i\sqrt{\frac{\hbar}{2\rho c_s k}} \frac{d_P e}{\varepsilon_0 \varepsilon_r} M_s(\hat{k}) \int d^3r\, \psi^*_i(\mathbf{r}) e^{i\mathbf{k}\cdot\mathbf{r}} \psi_j(\mathbf{r}), \tag{3}$$

$$g^{\rm DP}_{ij,l}(\mathbf{k}) = \sqrt{\frac{\hbar k}{2\rho V c_l}} D_e \int d^3r\, \psi^*_i(\mathbf{r}) e^{i\mathbf{k}\cdot\mathbf{r}} \psi_j(\mathbf{r}) \tag{4}$$

with Gaussian electron wave functions

$$\psi_i(\mathbf{r}) = \frac{1}{\pi^{3/4} l_i^{3/2}} \exp\left(-\frac{\mathbf{r}^2}{2 l_i^2}\right). \tag{5}$$

We have numerically simulated the self-assembled GaAs QDs with the following parameters: $l_1 = 4$ nm, $l_2 = 4.1$ nm, and $d = 6$ nm, where l_i is the electron wave function size of the i-th QD, and the distance between the dots is d. Here, $\rho = 5360$ kg/m^3 is the crystal density, V is the normalization volume of the phonon modes, c_s is the speed of sound (longitudinal $c_l = 5150$ m/s or transverse $c_t = 2800$ m/s, depending on the phonon branch), $d_P = 0.16$ C/m^2 is the piezoelectric constant, $D_e = -8$ eV is the deformation potential constant for the electrons and $\varepsilon_r = 13.2$ is the static dielectric constant. The function $M_s(\hat{k})$ in the piezoelectric coupling element reads: $M_l(\hat{k}) = \frac{3}{2}\sin(\theta)\sin(2\theta)\sin(2\varphi)$, $M_{t1}(\hat{k}) = -\sin(2\theta)\cos(2\varphi)$, and $M_{t2}(\hat{k}) = \sin(\theta)[3\cos^2(\varphi) - 1]\sin(2\varphi)$.

The dynamics of the electron confined in a QDM in the presence of electron-phonon interaction is studied by means of a second order correlation expansion method [12, 13], which is based on the assumption that correlations involving an increasing number of particles are of decreasing importance. This is a non-perturbative technique commonly used to describe the microscopic quantum dynamics, which covers the memory effects in the non-Markovian regime. The equations of motion for the quantities of interest: electron occupation $f = \langle |2\rangle\langle 2| \rangle$ and coherence $p = \langle|1\rangle\langle 2|\rangle$ are derived with the help of Heisenberg equation. The equations for the time evolution of the electron density and the coherence

$$
\begin{aligned}
\dot{f} &= \frac{2}{\hbar}\Gamma\,\mathrm{Im}(p) - \frac{i}{\hbar}\sum_{\mathbf{k}} g_{12}(\mathbf{k})\left[s_{\mathbf{k}} - s_{\mathbf{k}}^{(+)*} + 2iB_{\mathbf{k}}\mathrm{Im}(p)\right] - \frac{i}{\hbar}\sum_{\mathbf{k}} g_{12}^{*}(\mathbf{k})\left[s_{\mathbf{k}}^{(+)} - s_{\mathbf{k}}^{*} + 2iB_{\mathbf{k}}^{*}\mathrm{Im}(p)\right] \\
\dot{p} &= -\frac{i}{\hbar}2\epsilon p - \frac{i}{\hbar}T(1-2f) - \frac{i}{\hbar}\sum_{\mathbf{k}} g_{12}(\mathbf{k})\left[2t_{\mathbf{k}} + (1-2f)B_{\mathbf{k}}\right] - \frac{i}{\hbar}\sum_{\mathbf{k}} g_{12}^{*}(\mathbf{k})\left[2t_{\mathbf{k}}^{*} + (1-2f)B_{\mathbf{k}}^{*}\right] \\
&\quad -\frac{i}{\hbar}\sum_{\mathbf{k}}\left[g_{22}(\mathbf{k}) - g_{11}(\mathbf{k})\right](s_{\mathbf{k}} + pB_{\mathbf{k}}) - \frac{i}{\hbar}\sum_{\mathbf{k}}\left[g_{22}^{*}(\mathbf{k}) - g_{11}^{*}(\mathbf{k})\right]\left(s_{\mathbf{k}}^{(+)} + pB_{\mathbf{k}}^{*}\right)
\end{aligned}
$$

couple to each other, to $B_{\mathbf{k}} = \langle b_{\mathbf{k}}\rangle$, and to the phonon-assisted correlations: $s_{\mathbf{k}} = \langle|1\rangle\langle 2|b_{\mathbf{k}}\rangle^{corr}$, $s_{\mathbf{k}}^{(+)} = \langle|1\rangle\langle 2|b_{\mathbf{k}}^{\dagger}\rangle^{corr}$, and $t_{\mathbf{k}} = \langle|1\rangle\langle 1|b_{\mathbf{k}}\rangle^{corr}$. The factorization scheme has been used above with $\langle|1\rangle\langle 2|b_{\mathbf{k}}\rangle = \langle|1\rangle\langle 2|\rangle\langle b_{\mathbf{k}}\rangle + \langle|1\rangle\langle 2|b_{\mathbf{k}}\rangle^{corr}$, where the quantities have been decomposed into all possible lower-order factorizations. The next step is to derive the equations of motion for all the phonon-assisted correlation quantities, which will couple to three particle correlations, e.g. $\langle|1\rangle\langle 2|b_{\mathbf{q},s}^{\dagger}b_{\mathbf{k},s'}\rangle^{corr}$. These couple to up to four particle correlations, etc. Thus, in order to get a closed set of equations, one needs to truncate the hierarchy by neglecting higher order correlations. In the present paper, we included up to three particle correlations.

3. Phonon-induced relaxation

Initially, the electron is injected into a quantum dot with higher energy, $f = \langle|2\rangle\langle 2|\rangle = 1$, where the injection time is much shorter than the reservoir response (all correlation terms can be initially set to zero). The time evolution of the occupation of the QD with higher energy in the presence of the phonon reservoir at three different temperatures is shown in Fig. 2. If the electron is not coupled to phonons (ideal evolution, solid black line in the inset of Fig. 2), it tunnels back and forth between the two QDs with a period determined by the tunneling coupling Γ and the energy difference $\Delta\epsilon = 2\epsilon$ and is proportional to $\Gamma/(\Delta\epsilon)$. Using this parameter set with $\Delta\epsilon = 0.5$ meV and $\Gamma = 0.1$ meV, the electron mostly stays in the QD with higher energy since the tunneling coupling is too small to let the electron fully tunnel between the dots.

The dynamics of the electron is strongly affected if we include the electron-phonon interaction, which leads to the electron tunneling to the neighboring dot already on the picosecond time scale. At $T = 0$ K, only the emission processes are available, thus, the electron tunnels completely to the QD with lower energy. The probability of a phonon to be absorbed increases with temperatures, thus the final electron state shifts from the quantum dot with the lower energy towards an equal superposition of electron confined in the left $|1\rangle$ and right $|2\rangle$ QD. It results from the fact, that the probabilities of phonon absorption and emission are equal at high temperatures. Similarly, the damping gets stronger at higher temperatures, thus the oscillation amplitude decreases and the initial decay gets faster.

The fast relaxation processes result mainly from the direct electron tunneling assisted by phonons, which is described by the off-diagonal electron-phonon interaction. It can be seen in the inset of Fig. 2, where the time evolution of the QD occupation is modeled using only the diagonal part of the interaction Hamiltonian. In this case, the relaxation is considerably slower up to two orders of magnitude at all considered temperatures. Thus the proper description of the electron dynamics in a QDM should include the off-diagonal coupling to the phonon reservoir. We compared the results from the non-Markovian quantum kinetic model including up to two-phonon assisted processes to those with up to one phonon and we found out that the former leads only to a small correction. Thus, one can properly describe the electron dynamics by means of the first order correlation expansion technique.

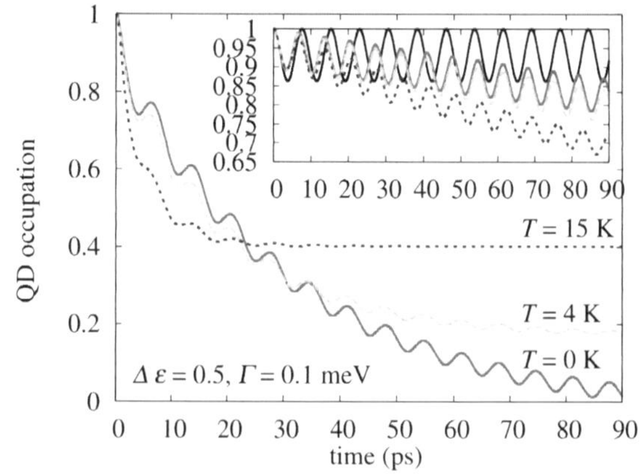

Figure 2. The time evolution of the occupation of the QD with the higher energy at different temperatures with $\Delta\epsilon = 0.5$ meV, and $\Gamma = 0.1$ meV. Inset: The ideal electron evolution without coupling to phonons (black solid line) and the evolution including only the diagonal electron-phonon coupling.

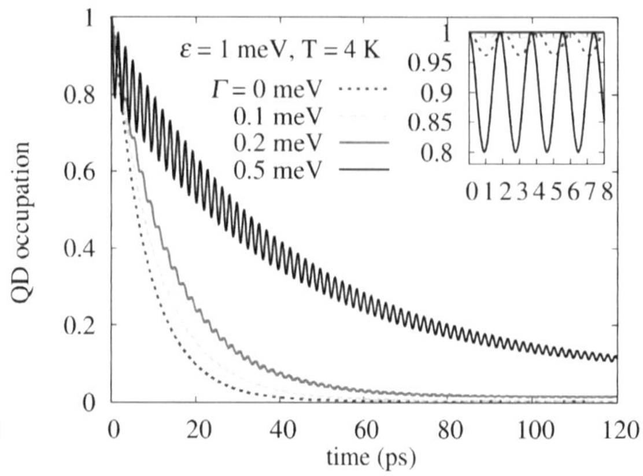

Figure 3. The QD occupation probability as a function of time for four different tunneling couplings with the energy difference $\Delta\epsilon = 2$ meV and at the temperature of $T = 4$ K. Inset: The corresponding ideal evolution of the electron in a QD without electron-phonon interaction.

The phonon-mediated relaxation in quantum dot molecules depends on the parameters of the system, in particular on the strength of the tunneling coupling Γ between the dots. In Fig. 3, we show the time evolution of the quantum dot occupation for four different values of the tunneling coupling and for considerably large energy difference, $\Delta\epsilon = 2$ meV. As one can see from the inset in Fig. 3 with ideal time evolution of the electron, the amplitude of the oscillations grows when the direct tunneling coupling is increased. For small values of the coupling, the phonon-induced relaxation is very fast, the electron tunnels to the neighboring dot within 20 ps. If the tunneling element grows, the relaxation slows down, however the increased oscillations in the case of large Γ are damped by the phonon reservoir even at low temperature of $T = 4$K.

4. Conclusion

A full description of the electron dynamics in the presence of the electron-phonon coupling in a quantum dot molecule doped with a single electron has been presented. The two couplings to the relevant acoustic phonons via deformation potential and piezoelectric coupling with diagonal and off-diagonal interactions have been taken into account. It has been shown that the phonon-mediated relaxation is a fast process on a picosecond timescale strongly modifying the coherent evolution of the electron. It is dominated by the usually neglected off-diagonal electron-phonon coupling two orders of magnitude faster then the relaxation due to pure dephasing. We analyzed the dependence of the relaxation on the temperature as well as on the strength of the tunneling coupling. The presented results are important for nanoscience, in particular for quantum information processing and for the currently performed experiments in the area of quantum transport and optics of quantum dots.

References

[1] J. R. Petta, A. C. Johnson, J. M. Taylor, E. A. Laird, A. Yacoby, M. D. Lukin, C. M. Marcus, M. P. Hanson, and A. C. Gossard, Science **309**, 2180 (2005).
[2] M. Bayer, P. Hawrylak, K. Hinzer, S. Fafard, M. Korkusinski, Z. R. Wasilewski, O. Stern, and A. Forchel, Science **291**, 451 (2001).

[3] H. E. Türeci, J. M. Taylor, and A. Imamoglu, Phys. Rev. B **75**, 235313 (2007).
[4] D. S. Golubev and A. D. Zaikin, Phys. Rev. Lett. **81**, 1074 (1998).
[5] T. Brandes and B. Kramer, Phys. Rev. Lett. **83**, 3021 (1999).
[6] S. Debald, T. Brandes, and B. Kramer, Phys. Rev. B **66**, 041301 (2002).
[7] A. Grodecka, C. Weber, P. Machnikowski, and A. Knorr, Phys. Rev. B **76**, 205305 (2007).
[8] P. Machnikowski and L. Jacak, Phys. Rev. B **69**, 193302 (2004).
[9] A. Grodecka, P. Machnikowski, and J. Förstner, Phys. Rev. A **79**, 042331 (2009).
[10] A. Grodecka, P. Machnikowski, and J. Förstner, Phys. Rev. B **78**, 085302 (2008).
[11] A. Grodecka-Grad and J. Förstner, Phys. Rev. B **81**, 115395 (2010).
[12] F. Rossi and T. Kuhn, Rev. Mod. Phys. **74**, 895 (2002).
[13] J. Förstner, C. Weber, J. Danckwerts, and A. Knorr, Phys. Rev. Lett. **91**, 127401 (2003).

Journal of Physics: Conference Series **245** (2010) 012036 doi:10.1088/1742-6596/245/1/012036

Effect of magnetic field on optical anisotropy of CdZnSe quantum dots

Jitendra Kumar[1] **and Sheetal Kapoor**[2]

[1]Department of Electronics and Instrumentation, Indian School of Mines, Dhanbad, 826004, India

[2]Department of Physics, University of York, Heslington, York, YO10 5DD, UK

E-mail: `jitenkg@rediffmail.com`

Abstract. The effect of magnetic field on the electronic structure and optical anisotropy of $Zn_{1-x}Cd_xSe/ZnSe$ quantum dots (QDs) has been analyzed for varying geometrical confinement related to the in-plane asymmetry. The disk-shaped QD is modeled by anisotropic parabolic potential with the magnetic field considered in Faraday geometry. The multiple band Hamiltonian in presence of magnetic field has been numerically diagonalized using appropriate basis functions and coordinate transformations for fast convergence. The eigenvalues and eigenvectors thus obtained are utilized for obtaining the dipole matrix elements. Hence, the photoluminescence spectra and the degree of linear polarization have been studied for the anisotropic QDs. The polarization degree is found to increase with magnetic field and anisotropy parameter.

1. Introduction

Spintronic devices offer a possibility to inject, modulate and detect spin of charge carriers In the new area of spintronics aimed at quantum computing and quantum communication, self-assembled semiconductor quantum dots (QDs) are a promising systems for spin-based quantum computers. Quantum computation applications based on spins of electons or holes as quantum bits require the preparation of the system in definite spin state and the ability to readout the spin of the output state. These tasks can be achieved through resonant excitation of excitons or trions using polarized optical pulse and analysis of the polarized photoluminescence (PL) spectrum of the output radiation [1].

The precise information about the the spin polarization of the system can be obtained from the study of optical anisotropy of the PL spectra of QDs which depend on the geometrical symmetry of the QDs. Magnetic field plays an important role in a spin based application by lifting the spin degeneracy and controlling the separation between the states of opposite spins.

The objective of this work is to analyze the effect of magnetic field on the electronic structure and optical anisotropy of $Zn_{1-x}Cd_xSe/ZnSe$ QDs for varying geometrical confinement related to the in-plane asymmetry. The disk-shaped QD is modeled by anisotropic parabolic potential with the magnetic field considered in Faraday geometry. The multiple band Hamiltonian in presence of magnetic field can be solved by constructing the Hamiltonian matrix in the Hilbert space of Hermite Gaussian functions for QDs with significant in-plane anisotropy [2, 3]. However, in order to achieve faster convergence, we have worked with the canonical transformations suggested by Madhav and Chakraborty [4] for conduction band electrons in anisotropic quantum dots, and generalized to the case of multiple valence band structure. The Hamiltonian calculated in the transformed coordinates has been diagonalized numerically. The eigenvalues and eigenvectors thus obtained are utilized for obtaining the dipole matrix elements. Hence, the degree of linear polarization (P_L) have been studied for the anisotropic QDs. The polarization degree is found to increase with magnetic field and anisotropy parameter A.

2. Theoretical formulations

The anisotropic $Zn_{1-x}Cd_xSe/ZnSe$ QD is modeled using a quantum well confinement potential, $V_z(z)$ along the z-axis (growth direction) and a parabolic confinement $V_{xy}(x, y)$ along the x- and y-axes, respectively. Further, we consider magnetic field applied in the Faraday configuration, i.e. along z-axis. The potential energy functions $V_z(z)$ and $V_{xy}(x, y)$, are determined by various physical parameters including the extent of interdiffusion of the barrier ions in the QD, valence band offsets, hydrostatic and biaxial strain, size and the shape anisotropy of the QD. The variation of the composition x of the QD material modifies the energy band gap as $E_g(\mathrm{x}) = E_{g\mathrm{CdSe}} + (E_{g\mathrm{ZnSe}} - E_{g\mathrm{CdSe}} - b)\mathrm{x} + b\mathrm{x}^2$ where b is the bowing parameter [5]. The other parameters for $Zn_{1-x}Cd_xSe$ are obtained from a linear interpolation of the ZnSe and the CdSe parameters.

The electronic structure of the QD can be obtained by solving the Luttinger Hamiltonian including the strain effects [6, 7]. Due to the s-type symmetry of the conduction band, the strain simply leads to the shifting of the conduction band edge caused by the hydrostatic deformation potential. The effect of strain on the valence band is more complicated due to the p-type symmetry. The hydrostatic strain shifts the heavy-hole (hh) and light-hole (lh) bands by the same amount, so that, in the strained QD structure the valence band shifts towards higher energy as compared to the unstrained case. In addition to this, in our system $Zn_{1-x}Cd_xSe/ZnSe$, the biaxial strain component has the effect of increasing the energy splitting between the two valence subbands [8]. We have considered the magnetic field in the Faraday geometry and adopted Coulomb gauge whereby the vector potential is given by $\mathbf{A} = (-By/2, Bx/2, 0)$. The in-plane confinement V_{xy} is defined by the anisotropic parabolic potential $V_{xy}(x, y) = (\alpha_x x^2 + \alpha_y y^2)/2$. Here, $\alpha_{x,y}$ is dependent on the confinement energies along x and y axes. The Hamiltonian for z variable can be solved in terms of usual quantum well wavefunctions. For the in-plane confinement potential V_{xy} the standard solution would be the Hermite-Gaussian functions as in the case of Harmonic oscillator. Further, the $Zn_{1-x}Cd_xSe/ZnSe$ QDs tend to show a high degree of in-plane anisotropy. This effect is quantified in our model as the anisotropy parameter A which is defined through the difference in the harmonic oscillator frequencies along x and y directions resulting in the expression $A = (\omega_x - \omega_y)/(\omega_x + \omega_y)$ [3]. Thus, an isotropic QD is denoted by $A = 0$ whereas $A = \pm 1$ would mean a very high degree of in-plane anisotropy approaching the case of a quantum wire. Here, $\omega_{x,y} = \sqrt{(\alpha_{x,y}/m_j)}$, m_j being the mass of electron, hh or lh depending on the jth band under consideration.

With the inclusion of magnetic field there is coupling of the terms in x and y coordinates in the Hamiltonian, thus complicating the solution. The equations can be uncoupled through the canonical transformation of the Hamiltonian [4] so that the diagonal terms reduce to a system of uncoupled Harmonic oscillators. The off-diagonal terms which are coupled in all four variables of the transformed coordinate system can be calculated algebraically. The Hamiltonian is diagonalized numerically using the product of harmonic oscillator wavefunctions in the transformed coordinates for the basis. The off-diagonal terms in this basis lead to a reasonably fast convergence as they just provide a correction factor due to mixing of valence subbands. This transformation introduces an approximation due to presence of two hole species, however, from the numerical estimates the error was found to be less than 2%.

The energy eigenvalues and wavefunctions obtained after the numerical diagonalization of the Hamiltonian are used for the analysis of various optical properties of the QDs. The intensity for a particular transition can be obtained from the square of the absolute value of the corresponding dipole matrix element, $|\langle\psi^c_{n_c}|\mathbf{e}.\mathbf{p}|\psi^v_{n_v}\rangle|^2$ where $\mathbf{e}$ gives the direction of linear polarization of radiation, $\mathbf{p}$ is the momentum operator, and $\psi^c_{n_c}$ ($\psi^v_{n_v}$) is the conduction (valence) band state under consideration [3]. The degree of linear polarization is obtained using the relation $P_L = (I_{max} - I_{min})/(I_{max} + I_{min})$ where, I_{max} and I_{min} are the maximum and minimum intensities obtained by varying the polarization angle of the electric field vector $\mathbf{e}$.

3. Results and discussion

In this section, we study the effect of magnetic field on the energy eigenvalues and the degree of linear polarization in a $Zn_{1-x}Cd_xSe/ZnSe$ QD using the formulation developed in Section 2. The material parameters used in the present numerical analysis are given in Table 1. We have considered the QDs to have Cd content of 40% which yields results in close conformity with the experimental observations [10, 11]. The dimension of the QD is taken to be 3.5 nm along the growth axis. The characteristic lateral size of the QD is estimated through the relation $r_0 = \sqrt{(\hbar/m\omega_0)}$ and gives an order of 8 nm for QD diameter at the typical value of $\hbar\omega_0 = 5$ meV. The effect of z-confinement has been calculated through a finite quantum well model with confinement potential determined by the band offsets and strain effects. The band offsets for x=0.4 calculated through linear interpolation from the pure ZnSe and CdSe parameters are found to be 387 meV and 108 meV for the conduction band and valence band, respectively. In case of valence band, the biaxial compressive strain shifts the energy of hh subband by 40 meV closer to the conduction band thus increasing the confinement potential for hh to 148 meV. The effect is just the opposite for the lh subband, thus decreasing the well potential by about 96 meV so that the lh remains weakly confined. Using the above values for the band offsets and the confinement potential along z-direction, the hole energy eigenvalues and eigenfunctions

Table 1. Material parameters for CdSe, ZnSe [5, 9] and $Zn_{0.6}Cd_{0.4}Se$ (linearly interpolated from data of CdSe and ZnSe).

	CdSe	ZnSe	$Zn_{0.6}Cd_{0.4}Se$
$E_g(eV)$	1.765	2.821	2.326
γ_1	2.1	3.77	3.10
γ_2	0.55	1.24	0.96
γ_3	0.55	1.67	1.22
m_c	0.13	0.16	0.15
C_{11}(GPa)	66.7	82.6	76.2
C_{12}(GPa)	46.3	49.8	48.4
$D_c(eV)$	-2.44	-2.83	-2.68
$D_{v1}(eV)$	-1.22	-1.41	-1.34
$D_{v2}(eV)$	-0.8	-1.2	-1.04
a_L (Å)	6.077	5.668	5.831

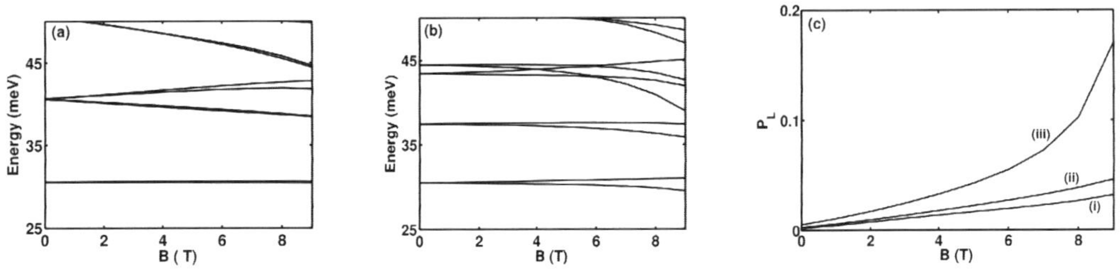

Figure 1. Plots (a) and (b) show the variation of hole energy eigenvalues of $Zn_{0.6}Cd_{0.4}Se$ QD with magnetic field for $A = 0.0$ and 0.3, respectively. Plot (c) shows the Variation of P_L with magnetic field for $Zn_{0.6}Cd_{0.4}Se$ QD at (i) A =0.1, (ii) A =0.3, and (iii) A =0.5.

are calculated. Fig.1 (a) and (b) show the variation of energy eigenvalues with respect to the magnetic field for the isotropic ($A = 0.0$) and anisotropic QDs ($A = 0.3$), respectively. The zero of the energy is taken to be the valence band edge for strained CdZnSe without confinement effects. In case of isotropic QDs, the mixing of hole states results in an extremely small splitting of lowest pair of states and the first excited states. On the other hand, in the anisotropic QDs there is an enhancement in splitting energy with red-shifting of the states [12]. This anomalous red-shifting of the lowest states has been experimentally observed by Miura et al [13].

We have also plotted the degree of linear polarization, P_L of the lowest state as a function of magnetic field for different values of A (Fig. 1(c)). We find that it increases with the magnetic field which is a clear signature of hh and lh mixing[3]. It can be seen that the P_L enhances as A is changed from 0.1 to 0.3 and 0.5 which can be attributed to the fact that the anisotropy increases the valence subband mixing effects leading to enhanced degree of linear polarization.

4. Conclusions

We have studied the effect of magnetic field on the electronic structure and optical properties of strained $Zn_{1-x}Cd_xSe$ QDs. Zeeman splitting of energy levels is seen which enhances with increase in magnetic field. This splitting is found to be more for anisotropic QDs. The degree of linear polarization is also seen to increase with magnetic field and is higher for the anisotropic QDs. These features are explained in terms of valence subband mixing which gets accentuated by the magnetic field.

References

[1] Henneberger F and Benson O (Eds) 2008 *Semiconductor Quantum Bits* (Singapore: Pan Stanford Publishing)
[2] Gupta S K, Kapoor S, Kumar J and Sen P K 2007 *Nanotechnology* **18** 325402
[3] Kumar J, Kapoor S, Gupta S K and Sen P K 2006 *Phys. Rev. B* **74** 115326
[4] Madhav A V and Chakraborty T 1994 *Phys. Rev. B* **49** 8163
[5] Lozykowski H J and Shastri V K 1991 *J. Appl. Phys.* **69** 3235
[6] Luttinger J M and Kohn W 1955 *Phys. Rev.* **97** 869
[7] Lü C, Cheng J L and Wu M W 2005 *Phys. Rev. B.* **71** 075308
[8] Kelly M J 1995 *Low Dimensional Semiconducors : Materials, Physics, Technology, Devices* (Oxford: Oxford Univ. Press)
[9] Cingolani R, Prete P, Greco D, Giugno P V, Lomascolo M, Rinaldi R, Calcagnile L, Vanzetti L, Sorba L and Franciosi A 1995 *Phys. Rev. B* **51** 5176
[10] Gu Y, Kuskovsky I L, Fung J, Robinson R, Herman I P, Neumark G F, Zhou X, Guo S P and Tamargo M C 2003 *Appl. Phys. Lett.* **83** 3779
[11] Koudinov A V, Akimov I A, Kusrayev Y G and Henneberger F 2004 *Phys. Rev. B* **70** 241305(R)
[12] Kusrayev Y G, Namozov B R, Sedova I V and Ivanov S V 2007 *Phys. Rev. B.* **76** 153307
[13] Miura N, Uchida K, Yasuhira T, Kurtz E, Klingshirn C, Nakashima H, Issiki F and Shiraki Y 2002 *Physica E* **13** 263

Temperature dependence of the emission spectra of individual self-assembled quantum dots

A. Zora[1], **C. Simserides**[1,2] **and G. P. Triberis**[1]

[1] University of Athens, Physics Department, Panepistimiopolis, Zografos, GR-15784, Athens, Greece

[2] Institute of Materials Science, NCSR Demokritos, GR-15310, Athens, Greece

E-mail: azora@phys.uoa.gr

Abstract. We have developed a quantum-mechanical theory for the interaction of light and electron-hole excitations in semiconductor quantum dots. Our theoretical analysis results in an expression for the photoluminescence intensity in the non-linear regime. The validity of the theoretical results is tested analyzing experimental data reported for the temperature dependence of the emission spectra of an individual lens-shaped $In_{0.4}Ga_{0.6}As$ self-assembled quantum dot in a wide temperature range up to 300 K. Our theoretical predictions for the redshift of the emission peak with increasing temperature, in the range 2-300 K, agree with the experiment.

1. Introduction

Semiconductor quantum dots (QDs) are intensively studied due to their wide optoelectronic applications such as low threshold lasers and optoelectronic detectors. Single or double QDs are excellent candidates for the implementation of quantum bits [1], as single or entangled photons emitters [2, 3], as well as for biological imaging, detection and targeting [4]. Therefore, a detailed understanding of the electronic and optical properties of single QDs and their temperature dependence is of importance.

In a previous work [5] we studied the photoluminescence intensity of a single quantum dot at room temperature and interpreted the experimental findings reported by Matsuda et al. [6]. For the single-particle distribution functions of the single quantum dot we used Fermi-Dirac statistics, as electrons and holes in all the quantum dots of the assembly were at the same equilibrium state, with a uniform Fermi level in the middle of the band gap (for low excitation conditions) and energetic separation of two successive levels lower than the thermal energy. In this case, the calculated sum of the single-particle distribution functions was negligible ($1 - f^{\mu} - f^{\nu} \simeq 1$, i.e. linear regime). μ (ν) denotes the different electron (hole) states.

In this short article, we present our theoretical findings for the temperature dependence of the photoluminescence intensity of an individual self-assembled quantum dot in the non-linear response regime (where the sum of the single-particle distribution functions $f^{\mu} + f^{\nu}$ obtained using Gibbs statistics, is not negligible compared to unity), and we compare our results with experiment [7].

2. Theory

The total Hamiltonian describing the physical system comprises four terms: the Hamiltonian of the non-interacting electrons and holes, the free-photons Hamiltonian, the electron-hole interaction Hamiltonian, and the interaction of carriers with the background photon-field [8]. Based on the semiclassical density formalism and applying the Heisenberg equation of motion to the photon number expectation values, to the field-matter correlation term, and to the electron (hole) distribution functions f^μ (f^ν), we obtain a set of four luminescence equations in the non-linear regime, as we did before in the linear regime [5]. Under quasi-equilibrium conditions, this set is reduced to a closed set of two photoluminescence equations, the analytical solution of which leads to the theoretical expression of the PL intensity in the non-linear regime [9].

Within our theoretical framework, (i) we calculate the single-particle energies and wave functions within the effective mass and envelope function approximations [10]. For each temperature, using an expansion -within a periodicity box- of the electron (or hole) envelope functions into the orthonormal plane-wave basis $|\varphi^\mu\rangle = \sum_{\mathbf{k}} C^\mu_{\mathbf{k}}|\mathbf{k}\rangle$ [10], we obtain the eigenenergies E^μ as well as the expansion coefficients $C^\mu_{\mathbf{k}}$ by solving the following eigenvalue problem

$$\sum_{\mathbf{k}'}[\langle\mathbf{k}|(\frac{\mathbf{p}^2}{2m^*} + V(\mathbf{r}))|\mathbf{k}'\rangle - E^\mu\delta_{\mathbf{k},\mathbf{k}'}]\langle\mathbf{k}'|\varphi^\mu\rangle = 0. \quad (1)$$

(ii) We compute the Coulomb matrix elements between the carriers $V^{cc}_{\kappa_1\kappa_2\kappa_3\kappa_4}$, where κ_i denote the different electron or hole states, as in [10].

(iii) We evaluate the eigenenergies E^λ and eigenfunctions $\Phi^\lambda(\mathbf{r},\mathbf{r})$ of the non Hermitian -in the non-linear regime matrix [9]

$$S_{\ell\ell'} = E^{e,renorm}_{\mu\mu'}\delta_{\nu\nu'} + E^{h,renorm}_{\nu\nu'}\delta_{\mu\mu'} - V^{eh}_{\mu\nu\nu'\mu'}(1 - f^\mu - f^\nu). \quad (2)$$

Here, we have determined the single-particle renormalized energies as

$$E^{e,renorm}_{\mu\mu'} \equiv E^\mu\delta_{\mu\mu'} - \sum_{\mu_1} V^{ee}_{\mu\mu_1\mu'\mu_1} f^{\mu_1} \quad (3)$$

$$E^{h,renorm}_{\nu\nu'} \equiv E^\nu\delta_{\nu\nu'} - \sum_{\nu_1} V^{hh}_{\nu\nu_1\nu'\nu_1} f^{\nu_1}. \quad (4)$$

(iv) Finally, we calculate the PL spectra according to:

$$I_{em}(\omega_q) \sim Im[\sum_\lambda \frac{\int d\mathbf{r}\Phi^\lambda(\mathbf{r},\mathbf{r})\int d\mathbf{r}'\sum_{\ell'} C^{\lambda *}_{\ell'}\varphi^{\mu' *}(\mathbf{r}')\varphi^{\nu' *}(\mathbf{r}')f^{\mu'}f^{\nu'}}{E^\lambda - i\gamma - \hbar\omega_q}]. \quad (5)$$

C^λ_ℓ are the components of the expansion of the excitonic eigenfunction Φ^λ in terms of the non-interacting single-particle basis [5], $\hbar\omega_q$ is the photon energy, $\ell \equiv \mu\nu$, and γ is a small damping constant that accounts for the dephasing due to phonon scattering [5].

3. Results and discussion

We use our theoretical results to interpret the experimentally observed PL peak positions as a function of temperature, reported by Bayer and Forchel [7]. The QD under study is an individual lens-shaped $In_{0.4}Ga_{0.6}As$ self-assembled quantum dot, with diameter at the base 24 nm, height 3.8 nm, and thickness of the wetting-layer 0.5 nm, containing one electron-hole pair (low excitation conditions). The material parameters used for $In_{0.4}Ga_{0.6}As$ quantum dot at 4 K and 300 K are listed in Table 1. The $In_{0.4}Ga_{0.6}As$ and GaAs band gaps shrinkage can be roughly described by the Varshni law

$$E_g(T) = E_g(0) - \frac{\alpha T^2}{T + \beta} \quad (6)$$

Table 1. Material parameters used for the $In_{0.4}Ga_{0.6}As$ quantum dot. The effective masses are expressed in units of the free-electron mass m_0. ε is the static dielectric constant. The low and high temperature values of the effective masses and of the dielectric constant are obtained by Ref. [11].

T (K)	m_e	m_{hh}	ε
4	0.067	0.34	12.38
300	0.0632	0.5	13.18

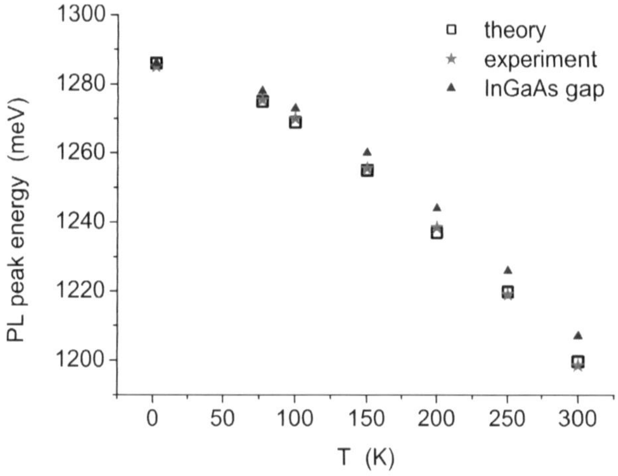

Figure 1. Comparison of the calculated PL peak position as a function of temperature, with the experimental one [7]. The temperature dependence of the $In_{0.4}Ga_{0.6}As$ band gap according to Varshni law is also depicted. To make the comparison clear, the values of $E_g(T)$ have been shifted so that all PL peaks for $T = 2$ K coincide.

The Varshni coefficients of GaAs and InAs were taken from Ref. [12], while for $In_{0.4}Ga_{0.6}As$ we use the interpolated values. The temperature dependence of the conduction and valence band-offsets is obtained from the temperature dependence of the band-gap discontinuity according to $V_{BO} = Q[E_g^{GaAs}(T) - E_g^{InGaAs}(T)]$ with $Q_c = 0.67$ and $Q_v = 0.33$ for the conduction and the valence band-offset respectively. The 3D confining potential is zero inside the wetting layer and the self-assembled dot, and V_{CBO} (or V_{VBO}) inside the barrier. We find that three confined shells exist, as it has also been reported by Ref. [7]. For the derivation of the spectra, we set $E_g^{InGaAs}(0) = 1.132$ eV, so that the theoretical and experimental PL peaks at $T = 2$ K coincide. Moreover, we fit the γ values in order to have the experimentally observed FWHM's.

Figure 1 depicts the theoretical PL peak energies as a function of temperature, compared with the experimentally observed. The theoretically derived PL peak energies are in agreement with the experimental ones. We find that the PL peak position energy follows the band gap temperature variation for low temperatures, while for higher temperatures, there is a small deviation, due to the single particle energies variation and the carriers occupation probabilities variation with temperature.

4. Conclusions

In summary, based on the theoretically derived expression for the photoluminescence intensity of individual quantum dots in the non-linear regime, we investigate the temperature dependence of the PL peak position energy. The theoretical results interpret the experimental data consistently. The PL peak position energy follows the band gap temperature variation for low temperatures. However, for higher temperatures, corrections should be made, due to the temperature variation of the single particle energies and of the term $(1 - f^{\mu} - f^{\nu})$.

References

[1] Moskal S, Bednarek S and Adamowski J 2007 *Phys. Rev.* A **76** 032302
[2] Michler P 2003 *Single Quantum Dots. Fundamentals, Applications and New Concepts* (Berlin: Springer)
[3] Young R J, Stevenson M R, Atkinson P, Cooper K, Ritchie D A and Shields A J 2006 *New J. Phys.* **8** 29
[4] Ghasemi Y, Peymani P and Afifi S 2009 *Acta Biomed.* **80** 156
[5] Zora A, Simserides C, and Triberis G 2007 *J. Phys: Condens. Matter* **19** 406201
[6] Matsuda K, Saiki T, Saito H and Nishi K 2000 *Appl. Phys. Lett.* **76** 73
[7] Bayer M and Forchel A 2002 *Phys. Rev.* B **65** 041308
[8] Kuhn T and Rossi F 1992 *Phys. Rev.* B **46** 7496
[9] Zora A, Simserides C, and Triberis G, to be submitted
[10] Simserides C, Zora A and Triberis G 2006, *Phys. Rev.* B **73** 155313
[11] E. H. Li 2000 *Physica* E **5** 215
[12] Paul S, Roy J B and Basu P K 1990 *J. Appl. Phys.* **69** 827

Design of columnar quantum dots for polarization -independent emission using 8–band k·p method

J Andrzejewski[1], G Sęk[1], E O'Reilly[2], A Fiore[3] and J Misiewicz[1]

[1]Institute of Physics, Wrocław University of Technology, Wybrzeże Wyspiańskiego 27, 50-370 Wrocław, Poland

[2]Department of Physics, University College Cork and Tyndall National Institute, Lee Maltings, Cork, Irleand

[3]CORBA Research Institute, Eindhoven University of Technology, P.O. Box 513, 5600 MB Eindhoven, The Netherlands

E-mail: janusz.andrzejewski@pwr.wroc.pl

Abstract. Control of the polarization of the emitted light can be highly beneficial for certain optoelectronic applications such as optical amplifiers. It has been recently demonstrated experimentally that semiconductor quantum dots with large height to base length aspect ratio are able to emit polarization–independent light from the edge of the wafer. However, analysis of the physics responsible for the observed polarization properties of such nano–objects (like columnar quantum dots or quantum rods) is still rather limited. In particular, the role of the material surrounding the columnar QD on the strain and thus on the polarization properties has not been considered previously. We report here, based on original software, the results of eight–band k·p calculations of the electronic and polarization properties of columnar $In_yGa_{1-y}As$ quantum dots (CQD) with high aspect ratio (up to 6) embedded in an $In_xGa_{1-x}As/GaAs$ quantum well. We calculate the relative intensities of transverse-magnetic (TM) and transverse-electric (TE) linear polarized light emitted from the edge of the semiconductor wafer as a function of the two main factors affecting the heavy hole – light hole valence band mixing and hence the polarization dependent selection rules for the optical transitions, namely i) the composition contrast y/x between the dot material and the surrounding well, and ii) the dot aspect ratio. Our numerical results show, in contrast to the previously reported expectations, that the former is the main driving parameter for tuning the polarization properties. This is explained analyzing the biaxial strain in the CQD, based on which it is possible to predict on the TM to TE intensity ratio.

1. Introduction

Quantum dots (QDs) have a potential for application in semiconductor optical amplifiers (SOAs), due to their high saturation power related to the low differential gain, fast gain recovery and wide gain spectrum compared to quantum wells (QWs) [1]. For in-line amplifier applications polarization-independence is a key requirement. Due to their flat shape and quasi-biaxial compressive strain, self-assembled quantum dots have a valence-band ground state of the heavy-hole (hh) type, which only couples to the in-plane light polarization (transverse-electric, TE) and not to the polarization along the growth axis (transverse-magnetic, TM). By varying the QD aspect ratio, both the shape of the confinement potential and the strain distribution can be changed. As a result, semiconductor quantum dots with large height to base length aspect ratio are able to emit polarization–independent light from the edge of the wafer [2], [3] and [4]. Such "columnar" quantum dots (CQDs) [5] with large aspect ratio can for example be obtained by cycled submonolayer deposition, and are thus promising candidates for amplifier applications. On the other hand, the analysis of the physics responsible for the observed polarization properties of such nano-objects is still rather limited and has considered only the influence of the dot geometry within quite a narrow aspect ratio range [6]. In particular, the role of the material surrounding the CQD on the strain and thus on the polarization properties has not been considered.

2. Computational approach

For the calculations of the electronic states and optical transitions in columnar quantum dots we have developed a three–dimensional strain–dependent eight–band **k·p** model. The model is implemented and all physical equations are numerically solved using the finite difference method. The model includes strain fields, piezoelectric effects and the spin–orbit interaction. The linear strain field has been calculated using continuum mechanical elastic theory. The detailed description of such calculations is given for instance in [7]. Based on strain field, the linear piezoelectric potential [8] and the potential resulting from deformation potential have been derived. Afterwards, the energy levels were calculated by using the multi band approach developed by Bahder [9]. In order to calculate the TE and TM mode transition intensities in the columnar QDs, we first calculate the oscillator strength for the electron-hole transition between the initial and final states, according [10], where the matrix elements between these states are calculated according to the scheme in [11].

3. Results

The geometrical model of the columnar QD with 2D surrounding (immersion) layer is presented in figure 1, including an illustration of the linear polarization directions. The TE-mode polarization vector lies in the X-Y plane, while the TM-mode is polarized along the growth (Z) direction. Light is considered to be emitted from the CQD along either the [110] or [1-10] direction. The CQD is placed in the centre of the box and the height of the 2D immersion layer is assumed the same as the height of the CQD (see the TEM images in [5]). The border between CQD and the faces of the numerical box used for calculating the strain field is 45 nm with a fixed boundary condition at the base of the box and with free-standing boundary conditions at the other faces of the box. For the numerical box for calculating of energy levels the border is 15 nm with fixed boundary conditions at the all faces of the box.

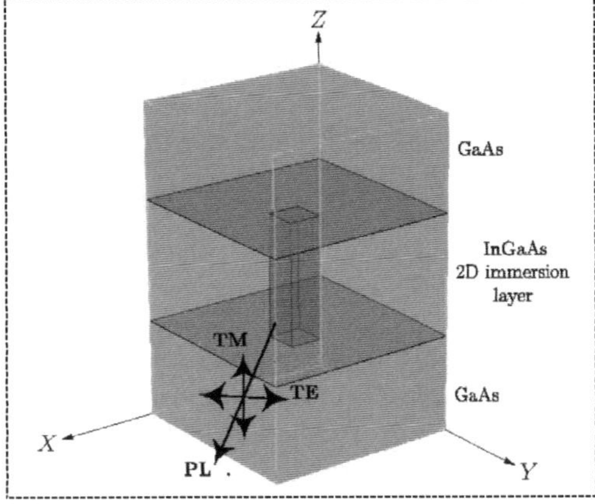

Figure 1. Schematic image of an InGaAs columnar QD in centre of image (red) with InGaAs 2D surrounding (immersion) layer in the plane (blue) and GaAs host material above and below (grey). The arrows show the direction of the considered in-plane emission and the respective linear light polarization directions (TE and TM).

All the simulations presented assume a cuboid QD with a square base of the diagonal length of 20 nm (14.14 nm base length), and with the aspect ratio (height to base length) changed from 3 to 6 (height between 42.42 and 84.84 nm). All the material parameters used in the calculations are taken from [12], assuming room temperature values for the lattice constants and energy band gaps.

In figures 2 and 3 there are presented the calculated oscillator strength of the optical transitions and the fraction of the intensity in the TM polarization for a CQD with an aspect ratio 3 and 6 respectively, and In content of 45% inside the dot. The In content in the immersion layer is changed from 15% down to 9 % in figures 2 (a) to (c) and figures 3 (a) to (c), respectively (which corresponds to the compositional contrast between the dot and the 2D surrounding changing from 3 to 5). For clarity, only those transitions which have a significant intensity are shown. In addition, we also plot on the right hand axes in figure 2 and figure 3 the light hole contribution to each valence band wave function calculated as in [6].

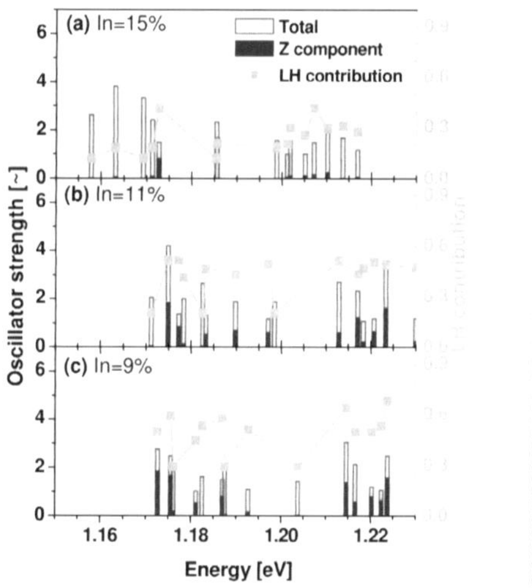

Figure 2. Total (red empty bars) and the Z (TM) component (blue filled bars) of the transition oscillator strength (left axis) of a cuboidal QD with aspect ratio equal to 3 and with QD In content equal to 45% for three different In contents in the 2D surrounding layer. The green squares joined by lines show the light hole contribution to the total valence wave function (right axis).

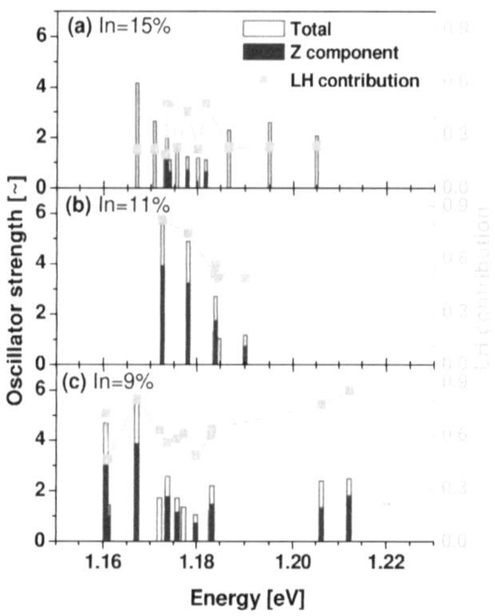

Figure 3. Total (red empty bars) and the Z (TM) component (blue filled bars) of the transition oscillator strength (left axis) of a cuboidal QD with aspect ratio equal to 6 and with QD In content equal to 45% for three different In contents in the 2D surrounding layer. The green squares joined by lines show the light hole contribution to the total valence wave function (right axis).

Several conclusions can be drawn directly from figures 2 and 3. We can see that there is a significant TM polarization contribution when the contents contrast equals 3 but this contribution is mainly found in the higher order transitions for both aspect ratios equal 3 and 6 in figure 2a and figure 3a, respectively. Increasing the contents contrast from 3 to 4 or further to 5 can dramatically enhance the TM polarization even for the lowest energy transitions. But still for high compositional contrast of 5 the TM will not dominate over TE if the aspect ratio is low (e.g. 3).

Based on these calculations, we have obtained that the strong TM polarization is directly connected with the value of the biaxial strain inside the quantum dot and that the minimum requirement is that it needs to be negative. Figure 4 presents biaxial strain calculated in the middle of the CQD for In content equal to 45% inside dot and for various In concentrations in the immersion layer. It is seen, that negative biaxial strain (and hence high intensity in TM polarization) can be obtained for high contents contrast being then weakly dependent on the aspect ratio. On the other hand, the TM mode will never be strong for small composition contrast (like 3, i.e. the case of 15 % In in the immersion layer in figure 4) for even very high aspect ratios.

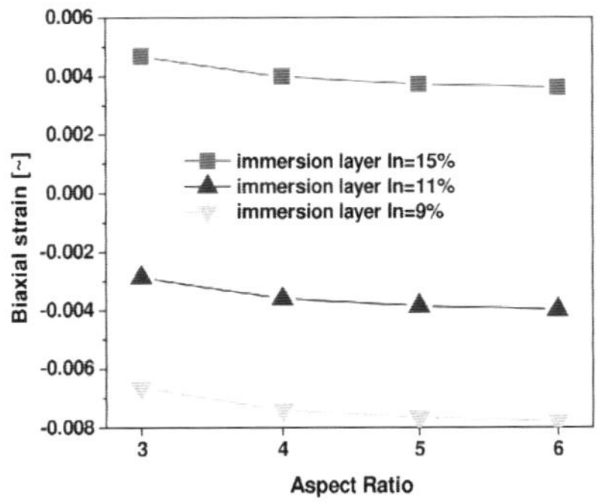

Figure 4. Calculated biaxial strain in the middle of the CQD for In content of 45% in the dot and for various In contents in the immersion layer versus the dot aspect ratio.

4. Conclusions

We have theoretically studied the optical properties of $In_yGa_{1-y}As/GaAs$ columnar QDs from the point of view of their possible application in polarization independent optical amplifiers. We have taken into consideration and shown the importance of a 2D $In_xGa_{1-x}As$ layer of lower composition ($x < y$) which surrounds such dots in–plane in real structures. We have shown how this immersion layer can strongly affect the TM to TE intensity ratio of the optical transitions. Eventually, we have found that if biaxial strain in the quantum dot has a negative value, then there exist conditions for strong TM intensity, which can be further enhanced by increasing the aspect ratio of the columnar dot.

Acknowledgments

Most of the computations were performed in Wrocław Centre for Networking and Supercomputing, Wrocław University of Technology, Poland and some of them in the Computer Facilities of the Ecole Polytechnique Fédérale de Lausanne, Switzerland.

References

[1] Akiyama T, Ekawa M, Sugawara M, Kawaguchi K, Sudo H, Kuramata A, Ebe H and Arakawa Y 2005 *IEEE Photon. Tech. Lett.* **17** 1614
[2] Kawaguchi K, Yasuoka N, Ekawa M, Ebe H, Akiyama T, Sugawara M and Arakawa Y 2008 *App. Phys. Lett.* **93** 121908
[3] Podemski P, Sęk G, Ryczko K, Misiewicz J, Hein S, Höfling S, Forchel A and Patriarche G 2008 *Appl. Phys. Lett.* **93** 171910
[4] Ridha P, Li L, Fiore A, Patriarche G, Mexis M and Smowton P M 2007 *App. Phys. Lett.* **91** 191123
[5] Li L H, Patriarche G, Chauvin N, Ridha P, Rossetti M, Andrzejewski J, Sek G, Misiewicz J and Fiore A 2008 *IEEE J. Sel. Top. Quantum Electron.* **14** 1204
[6] Saito T, Ebe H, Arakawa Y, Kakitsuka T and Sugawara M 2008 *Phys. Rev.* B **77** 195318
[7] Stier O, Grundmann M and Bimberg D 1999 *Phys. Rev.* B **59** 5688
[8] Grundmann M, Stier O and Bimberg D 1995 *Phys. Rev.* B **52** 11969
[9] Bahder T B 1990 *Phys. Rev.* B **41** 11992
erratum: 1992 *Phys. Rev.* B **46** 9913
[10] Sheng W and Leburton J-P 2001 *Phys. Rev.* B **64** 153302
[11] Szmulowicz F 1995 *Phys. Rev.* B **51** 1613
[12] Vurgaftman I, Meyer J R and Ram-Mohan L R 2001 *J. App. Phys.* **89** 5815

Tunnel migration in ensembles of silicon nanocrystals doped with phosphorus.

V A Belyakov, K V Sydorenko, A A Konakov, N V Kurova, V A Burdov

University of Nizhniy Novgorod, Gagarin ave. 23, Nizhniy Novgorod, Russia, 603950

E-mail: belyakovva@nifti.unn.ru

Abstract. A comparative study of radiative recombination and tunnel migration in ensembles of P-doped Si nanocrystals embedded in wide-band dielectric matrix is performed. We have found the migration process to be strongly suppressed at donor concentration exceeding some critical value. At the same time, the radiative recombination rate gradually rises as the concentration increases. This essentially enhances the radiative-channel efficiency in the photoluminescence process.

1. Introduction

Improvement of luminescent properties of various silicon-based structures remains up to now a challenge for modern optoelectronics. Because of indirect band gap of silicon, light emission turns out to be suppressed or completely forbidden in such structures. This drawback is partially overcome in Si nanocrystals due to the Heisenberg uncertainty relations and phonon assistance. An observation of visible-range emission from silicon nanocrystals [1] confirms this fact.

As a means to improve the emittance of silicon crystallites, doping with phosphorus was proposed. It has been shown experimentally that the doping can increase photoluminescence intensity by several times at certain conditions [2]. This rise of the photoluminescence intensity is a consequence of interplay between radiative and nonradiative processes. In the present paper, we analyze an effect of one of the most efficient nonradiative processes – tunneling of excited carriers from one quantum dot to an adjacent one – on light emission of an ensemble of highly and homogeneously P-doped Si nanocrystals. For this purpose, we estimate the rates of the radiative interband recombination and electron tunneling in Si nanocrystals doped with phosphorus. The calculations were carried out within the framework of envelope function approximation for spherical quantum dots.

2. Results and discussion

Coulomb interaction of the excited electron with the phosphorus ion plays a crucial role in the recombination process. The Coulomb field in a nanocrystal consists of two parts [3]: the long-range part; and the short-range microscopic part usually named the central-cell field [4]. The short-range part was determined by Pantelides and Sah [5] using microscopic dielectric function.

Determination of the electronic structure of a P-doped silicon crystallite also exhibits a key role of the short-range field in the electron state formation [6]. In particular, in the conduction band, the spinless ground state (being sixfold degenerate in an undoped dot) splits due to the short-range field similarly to bulk Si into a singlet with the lowest energy, and doublet and triplet.

It should be noted that, phonon-assisted and no-phonon donor-induced transitions individually contributes to the total recombination rate τ_{R}^{-1}. The phonon-assisted radiative transitions are, in fact, independent of doping [7]. Their rate τ_0^{-1} remains almost constant as function of phosphorus concentration n_{D} in the dot. On the contrary, the donor-induced transitions are essentially intensified as n_{D} increases. As was shown earlier [8], the central-cell potential of the phosphorus ions embedded in Si nanocrystal efficiently mixes electronic states of X and Γ bands. Such a mixing produces a partially direct radiative transition enhanced by the quantum confinement effect, which causes rising the rate.

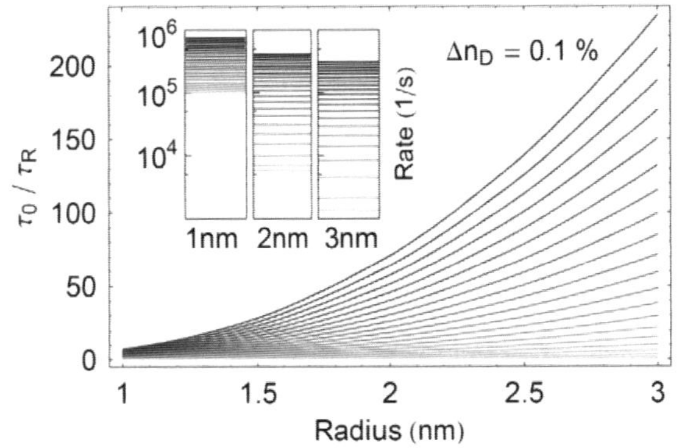

Figure 1. Relative rate of the radiative recombination as a function of the dot radius at donor concentration increasing with the constant step 0.1% from 0 to 2%. Inset shows absolute values of the rate for three Si nanocrystals with different radii indicated in the inset.

The results of our calculations of radiative rates [8] are presented in Fig. 1. As seen, at donor concentrations n_{D} greater than approximately 1% (1 P atom per 99 Si atoms), τ_{R}^{-1} is much greater than the radiative rate τ_0^{-1} in undoped nanocrystal ($n_{\mathrm{D}}=0$) for all dot sizes. This means that the donor-induced transitions become much faster than the phonon-assisted ones. This effect is more pronounced for bigger crystallites. For instance, at $n_{\mathrm{D}}=2\%$, the donor-induced transitions have the rates $1-2$ orders of magnitude greater than the rates of the phonon-assisted transitions as the dot radius increases from 1 to 3 nm, respectively.

Let us now consider the tunnel migration of excited carriers. After laser pumping, electrons from the lowest energy level in the conduction band of one quantum dot can either relax into the valence band of this dot with photon emission or tunnel into the adjacent nanocrystal with phonon assistance occupying there the level with close energy. Similarly, the holes can tunnel from the upper level in the valence band onto the energy level in the valence band of a neighboring nanocrystal.

We consider two nanocrystals with radii R_1 and R_2, and assume the first nanocrystal to be less than the second one, i.e. $R_1 < R_2$. The distance between the dots, equal to the minimal distance between their edges, is L. Denote the wave functions of kth stationary state in the first and the second isolated nanocrystals as $\Psi_k^{(1)}$ and $\Psi_k^{(2)}$, respectively. Then, taking into account a weak tunnel connection between the nanocrystals, it is possible to construct the wave functions of the carriers in the double-dot system as a superposition of $\Psi_k^{(1)}$ and $\Psi_k^{(2)}$. In particular, one can formally write the wave functions of the initial, and the final electron (or hole) states in the form:

$$\Psi_{I,F} = A_k^{(I,F)}\Psi_k^{(1)} + B_k^{(I,F)}\Psi_k^{(2)}. \tag{1}$$

Here, $A_k^{(I,F)}$ and $B_k^{(I,F)}$ are the expansion coefficients, and the Einstein convention was used when summing over k. Suppose, for definiteness, that initially (before tunneling), the electron mainly populates the first nanocrystal (i.e. $A_k^{(I)} \gg B_k^{(I)}$), and has the energy approximately equal to E_1 that is the energy of the ground state in the first isolated nanocrystal. On the contrary, in its final state (after tunneling), the electron mainly occupies the second nanocrystal. Correspondingly, in this case $A_k^{(F)} \ll B_k^{(F)}$, and the electron energy is close to the energy E_{2n} of some n^{th} stationary state in the conduction band of the second nanocrystal. Note that, we suppose E_1 to be greater than E_{2n}. In the opposite case the tunneling is strongly suppressed, as was also pointed out by Roman and Pavesi [9]. Evidently, similar treatment can be performed for hole tunneling.

Tunneling with phonon emission is defined in the first order of the perturbation theory by the rate

$$\tau_{tun}^{-1} = \frac{2\pi}{\hbar} \sum_{s=1,2} \sum_{\ell,\mathbf{q}} \left| \langle \Psi_F | U_{\ell\mathbf{q}}^{(s)} | \Psi_I \rangle \right|^2 \delta_L \left(E_1 - E_{2n} - \hbar \nu_\ell(\mathbf{q}) \right). \quad (2)$$

Here, $U_{\ell\mathbf{q}}^{(1,2)}$ stands for the operator of the electron interaction with ℓ^{th} and $\mathbf{q}^{\text{th}}$ phonon mode in the first, or in the second, crystallite. Because of homogeneous broadening, we have replaced standard Dirac delta-function by the Lorentz function

$$\delta_L(x) = \frac{1}{\pi} \frac{\gamma}{x^2 + \gamma^2} \quad (3)$$

with semi-width γ which is about 10 meV according to Allan and Delerue [10]. It is worth noting that, the Lorentz function allows one to violate the energy conservation under the tunnel transition. Nevertheless, the difference $E_1 - E_{2n}$ for the nanocrystals with radii ranging within 1 – 3 nm can achieve 0.2 – 0.3 eV. In this case, the single-phonon process turns out to be inefficient for compensation this difference to within γ. As a result, instead of the Fermi golden rule in the form given by Eq. (2) we have to treat the electron-phonon interaction in the second order of the non-stationary perturbation theory. Results of our calculations for undoped and doped nanocrystals are presented in Fig. 2 and Fig. 3, respectively.

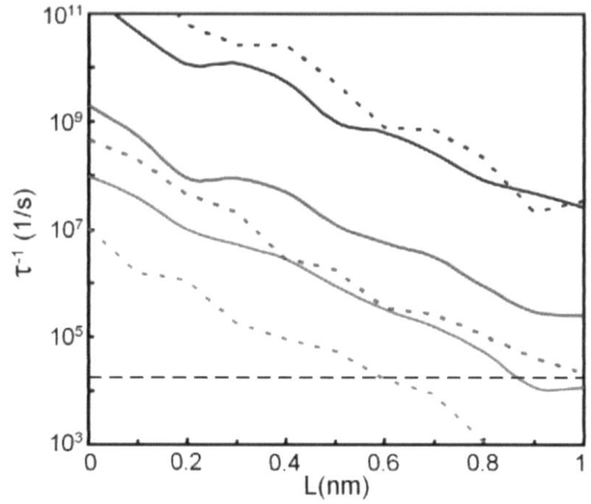

Figure 2. Tunnel rates of the electrons (solid lines) and holes (dots) in undoped nanocrystal as functions of the distance between the edges of the nanocrystals. The emitting nanocrystal has the radius $R_1 = 1.5$ nm. The radius R_2 of the accepting nanocrystal equals (from top to bottom): 1.6 nm; 1.9 nm; 2.2 nm. Dashed line represents the radiative recombination rate in the emitting nanocrystal [7].

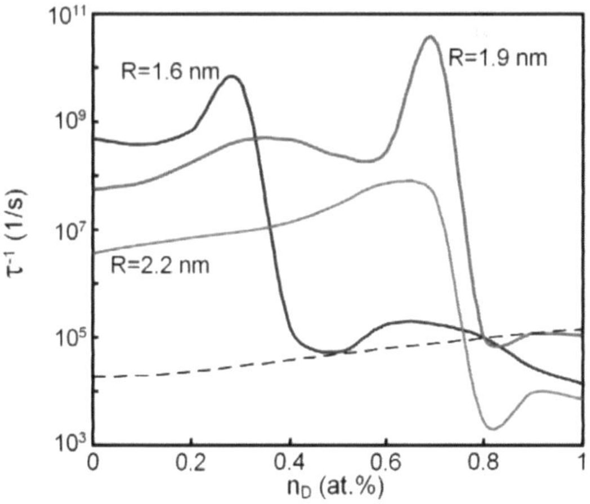

Figure 3. Electron tunnel rates as functions of the donor concentration n_D at the distance $L = 0.5$ nm between the edges of the emitting and accepting P-doped nanocrystals. Radius of the emitting nanocrystal $R_1 = 1.5$ nm. Radii of the accepting nanocrystals are indicated in the figure. Dashed line represents the radiative recombination rate in the emitting nanocrystal according to [8].

It is seen in Fig. 2, that in the mean, after averaging oscillations caused by the resonant tunneling, the migration rate quickly (exponentially) drops as the distance between the nanocrystals increases. However, if L is less than some distance L_0 depending mainly on the difference $R_2 - R_1$, the tunnelling turns out to be much faster than the radiative electron-hole recombination (dashed horizontal line) in the first nanocrystal.

As has been already mentioned, phosphorous doping leads to the partial degeneracy removal in the electron spectrum. As a result, the energy level has more homogenous distribution in the conduction band that effectively reduces the difference $E_1 - E_{2n}$. This, in turn, leads to some initial rise of the tunnel rate with increasing n_D, as shown in Fig. 3. However, further increase of the phosphorus concentration sharply reduces the tunnel rate. In the accepting nanocrystal, electrons (emitted from the donors) populate the lowest levels in the conduction band and block tunneling of the ground-state electron (which could participate in the radiative transition) from the emitting nanocrystal if the Fermi level in the accepting nanocrystal turns out to be higher than the ground-state energy level in the emitting nanocrystal. Tunneling with increasing energy is strongly suppressed by temperature factor like $\exp(-\Delta E/kT)$ while more probable tunneling with decreasing energy is forbidden due to the Pauli principle. As a consequence, the tunnel rate sharply drops when the phosphorus concentration achieves some critical value defined by the difference in sizes of the emitting and accepting nanocrystals.

For comparison, we have depicted in Fig. 3 the radiative recombination rate in Si nanocrystal with radius 1.5 nm as a function of the phosphorus concentration. It is seen that, at concentrations exceeding a certain value, the radiative recombination rate becomes greater than the migration rate.

The tunnel rates for holes are rather less than, or of the same order as, the ones for electrons, which is explained by the higher potential barrier for holes compared to the one for electrons [11]. In the valence band, phosphorus doping results, in fact, only in a common shift of the energy levels without their splitting and degeneracy removal. Therefore, the migration rates of holes shown in Fig. 2 remain almost invariable with varying phosphorus concentration.

3. Conclusion

Thus, doping with phosphorus can be an essentially positive factor from the point of view of improving emittance of silicon nanocrystals. First, doping accelerates the radiative interband recombination. Second, doping blocks one of the most efficient nonradiative channels – tunnel migration of excited electrons. Both these factors should enhance the photoluminescence intensity of Si nanocrystal ensembles.

This work was in part financially supported by the Russian Ministry of Education and Science through the project NK-129P/64 and the RNP-program, RFBR (grant # 10-02-00995), and by the President of the Russian Federation (grant MK-675.2009.2).

References

[1] Pavesi L, DalNegro L, Mazzoleni C, Franzo G and Priolo F 2000 *Nature* **408** 440
[2] Fujii M, Mimura A, Hayashi S, Yamamoto Y and Murakami K 2002 *Phys. Rev. Lett.* **89** 206805
[3] Belyakov V A and Burdov V A 2007 *Phys. Rev. B* **76**, 045335
[4] Kohn W and Luttinger J M 1955 *Phys. Rev.* **97** 1721
[5] Pantelides S T and Sah C T 1974 *Phys. Rev. B* **10** 621
[6] Melnikov D V and Chelikowsky J R 2004 *Phys. Rev. Lett.* **92** 046802
[7] Belyakov V A, Burdov V A, Lockwood R and Meldrum A 2008 *Adv. Opt. Technol.* **2008** 279502
[8] Belyakov V A and Burdov V A 2009 *Phys. Rev. B* **79**, 035302
[9] Roman H E and Pavesi L J. Phys.: Condens. Matter **8**, 5161 (1996)
[10] Allan G and Delerue C, 2007 *Phys. Rev. B* **75**, 195311
[11] Wallace R and Wilk G, 2001 *Semicond. Int.* **8**, 227

Scanning near-field optical microscopy of quantum dots in photonic crystal cavities.

Matthias Skacel[1], Marco Francardi[2], Annamaria Gerardino[2], Blandine Alloing[3], Lianhe Li[3] and Andrea Fiore[1]

[1] COBRA Research Institute, Technical University Eindhoven, Den Dolech 2, 5600 MB Eindhoven, The Netherlands

[2] Institute of Photonics and Nanotechnology, CNR, via del Cineto Romano 42, 00156 Roma, Italy

[3] Institute of Photonics and Quantum Electronics, EPFL, CH-1015 Lausanne, Switzerland

E-mail: m.s.skacel@tue.nl

Abstract. Nanophotonic devices are of major interest for research and future quantum communication applications. Due to their nanometer feature size the resolution limit of far-field microscopy poses a limitation on the characterization of their optical properties. A method to overcome the resolution limit is the Scanning Near-Field Optical Microscope (SNOM). By approaching a fiber tip into the close vicinity of the sample the optical emission in the near-field regime is collected. This way of collecting the light is not affected by the diffraction limit. We employ a low temperature SNOM to investigate the photoluminescence of InAs QDs emitting at 1300nm wavelength embedded in photonic crystal cavities. At each location of an image scan the tip is stopped and a spectrum is acquired. We then plot maps of the photoluminescence for each wavelength. With this instrument it is now possible to directly observe the coupling of QDs to photonic crystal cavities both spectrally and spatially. We show first results of photoluminescence mapping of InAs QDs in photonic crystal cavities.

Quantum dots (QDs) in photonic crystal cavities are of major interest for future quantum communication and quantum key distribution applications. The spatial alignment of quantum dots to photonic crystal cavities (PhCCs) is still an unresolved issue. As yet, little experimental data is available on how the spatial position of the quantum dot within the cavity affects the coupling of the QD's excitons to the cavity mode. In order to determine the requirements necessary for proper coupling more information is needed about the dependence on the spatial alignment. To obtain such information an instrument is needed with high spatial resolution. Mostly, solid state cavity QED systems are studied by micro-photoluminescence (μPL). However, μPL relies on the use of lenses, which are limited by the Abbe limit. Thus, for cavity QED systems operating in the telecom wavelength range the achievable optical resolution is in the order of 1μm. The typical QD size is in the range of a few nanometers. Likewise, the mode pattern of a PhCC has spatial features well below 1μm [1, 2]. Comparing this to the resolution of μPL it is evident that the spatial resolution of μPL is far from providing the relevant details of the QD-PhC coupling.

In SNOM a sharp fiber tip collects the non-propagating (spatial) high-frequency components of light, enabling a spatial resolution well below the diffraction limit. We set up a SNOM working at low temperatures (5K) to investigate InAs QDs in PhCCs. It employs a shear-force tuning-fork feedback for distance regulation. The laser for photoluminescence (PL) excitation as well

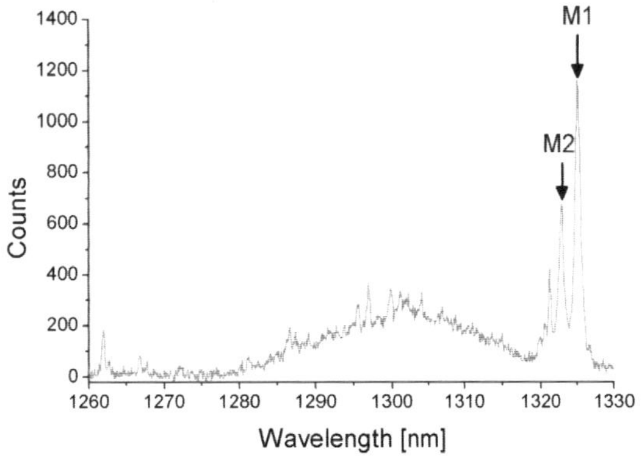

Figure 1. μPL spectrum acquired at 7K under 150μW pulsed excitation at 750nm.

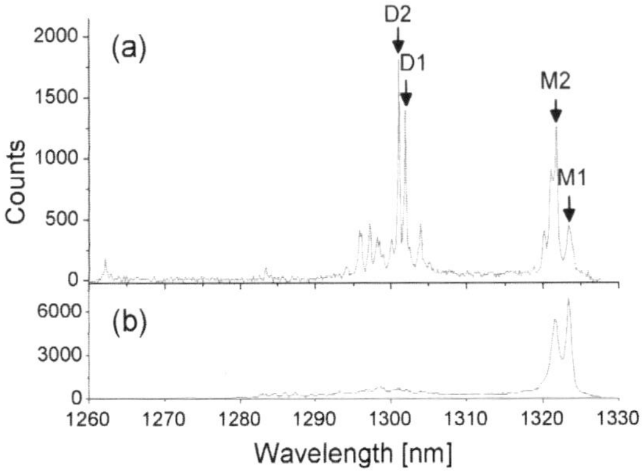

Figure 2. (a) Near-field PL spectrum acquired at 7K under 330nW excitation by a 635nm cw laser. (b) Near-field PL spectrum under 4.7μW excitation.

as the detected PL is coupled through the fiber tip. We use an uncoated fiber tip for optimal throughput and minimal disturbance of the dielectric environment of the photonic crystal [2]. The sample is placed onto a piezo scanner which is mounted, together with tuning fork and tip, onto a dipstick. The dipstick is inserted into a cryostat, in the flow of cold Helium gas. An operating temperature of 7K is chosen with a stability of $\pm$20mK.

The sample under investigation contains low density InAs QDs (approx. 5-10 QDs/μm^2), emitting at 1300nm at low temperature, in a 320nm thick membrane on top of a 1.5μm AlGaAs sacrificial layer [3]. PhCCs are fabricated by electron beam lithography, dry etching of the PhC holes and selective wet etching of the sacrificial layer. The addressed cavities are of a modified H1 geometry. In a triangular photonic crystal of air holes with a lattice constant of 350nm a hole is removed to form a cavity. The inner row of holes around the cavity is slightly shifted in position and reduced in size. Figure 3 shows the topography signal of the SNOM of the studied cavity acquired at 7K.

Figure 1 shows the μPL spectrum of such an H1 cavity. It shows three distinct peaks between 1320nm and 1330nm. We attribute the two peaks marked with M1 and M2 to dipole modes of the H1 cavity [4]. We attribute the third one to an excitonic emission of a QD, as will be shown later. Several sharp lines are superimposed onto a broad background between 1280nm and 1320nm, which presumably originate from the emission of several QDs. For comparison, a near-field PL spectrum where the tip is positioned in the center of the same cavity is plotted in figure 2 (a). In this case, the undistinguished broad emission observed in μPL breaks up into a small number of sharp lines, indicative of the emission of single QDs. Whereas the sharp lines at 1301nm (denoted as D2) and 1302nm (denoted as D1) are barely apparent in μPL, they are remarkably strong in the near-field spectrum, as it shows much less background emission. This can be explained by the much smaller collection and excitation area of near-field microscopy as compared to far-field collection. In order to confirm the origin of the emission lines we investigated their power dependence. Figure 4 shows the power dependence of the most prominent lines of the near-field spectrum. Whereas, the lines at 1301nm (D2) and 1302nm (D1) show a saturation behaviour, the lines at 1321.7nm (M2) and 1323.4nm (M1) do not saturate with higher pump power, and become dominant at high powers, as shown in figure 2(b). We thus conclude that M1 and M2 are cavity modes while D1 and D2 are related to QD (multi-)excitons. Indeed, while the emission of a QD exciton saturates and decreases for increasing QD population, the cavity resonance can be fed by multiexcitonic transitions involving a large number of carriers in the wetting layer [5]

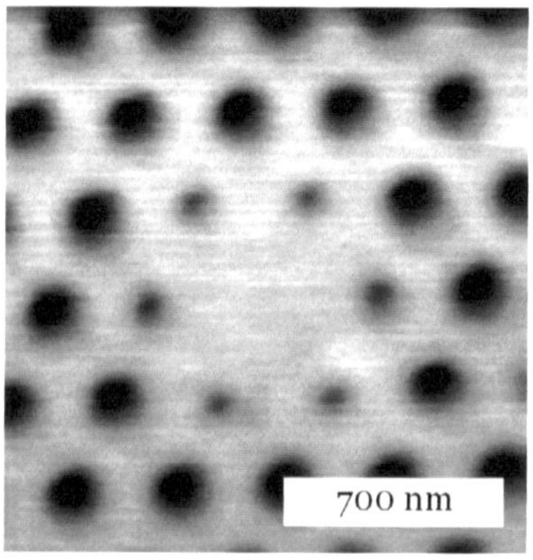

Figure 3. Topography of the investigated cavity.

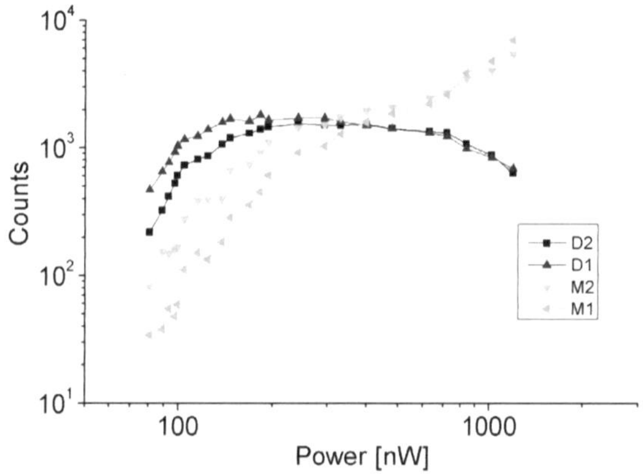

Figure 4. Power dependence of the brightest lines in figure 2.

and is therefore dominant. While the absolute position of the lines is slightly offset in the μPL and SNOM spectra due to different environmental conditions (e.g. different cryostat, sample temperature, cryogetting of surface layers [6]), the M1-M2 energy spacing is the same (2nm). The quality factors are most easily deduced from the high power SNOM spectrum in figure 2 (b) as $Q(M1) = 1800$ and $Q(M2) = 800$.

Besides spectroscopy, the SNOM can also be used to spatially map the PL by raster-scanning the tip over the surface. At regularly spaced locations during the scan (here at 32×32 locations) the tip is stopped and a near-field spectrum with an integration time of 1 second is acquired. After the scan has finished the acquired 1024 recorded spectra are rearranged to show the emission maps for each wavelength (one map for each pixel of the CCD of spectrometer). A selection of the most interesting maps is presented in figures 5, 6, 7 and 8. They correspond to the M1, M2, D1 and D2 lines. All maps are taken under cw excitation at 635nm with a power of $P = 400$nW.

Figures 5 and 6 show the cavity modes M1 and M2. M2 shows a distinct spatial pattern which resembles a dipole mode [4]. On the other hand, M1 has only one lobe. Both modes are displaced to the right side of the cavity center, as compared to the topography (the holes of the PhC are drawn as white circles to guide the eye). We attribute these unexpected features to an asymmetric geometry of the tip [1] and manufacturing imperfections of the PhCC.

The maps of the two sharp lines D1 and D2 are shown in figures 7 and 8, respectively. The round-shaped emission pattern together further confirms that these stem from QDs. Since the spots in figure 7 and 8 both come from approximately the same location in the PhCC we conclude that they are coming from the same QD.

In conclusion we have studied QDs in photonic crystal cavities by scanning near-field optical microscopy. We are able to image the mode profile of photonic crystal cavities with sub-wavelength resolution and can identify individual QDs. Future work will consist of spatially resolved mapping of the PL decay time and of the second-order correlation function, as a function of the QD-mode spectral tuning.

Acknowledgments
We want to thank Silvia Vignolini, Francesca Intonti, Mehmet Dündar, Rob van der Heijden and Fouad Karouta for many helpful suggestions and fruitful discussions.

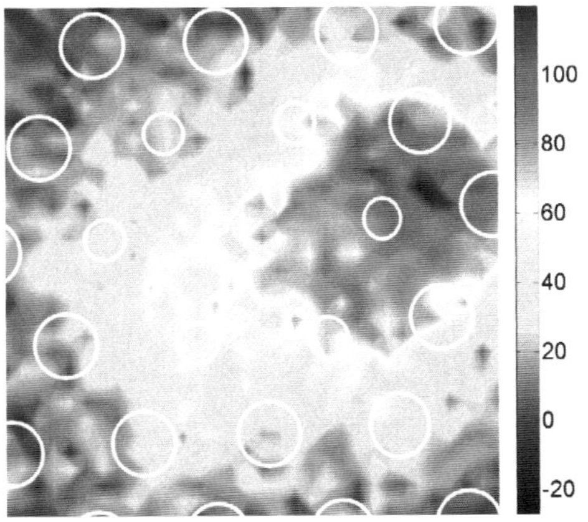

Figure 5. Cavity mode M1 at 1323nm.

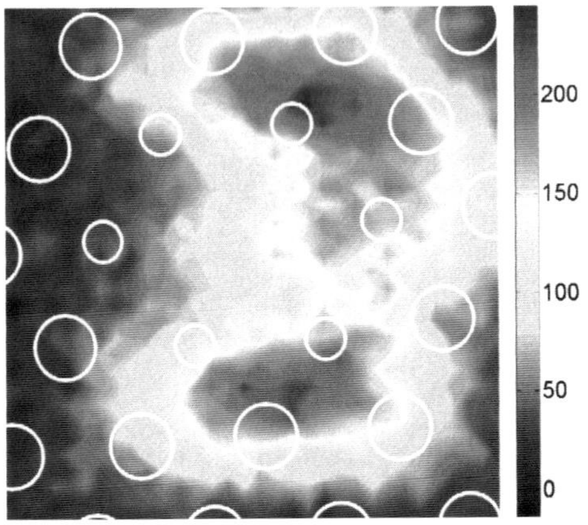

Figure 6. Cavity mode M2 at 1321.7nm

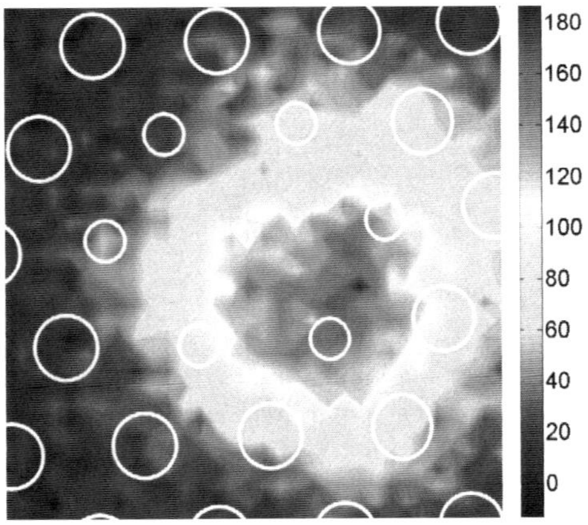

Figure 7. Map of D1 at 1301.96nm.

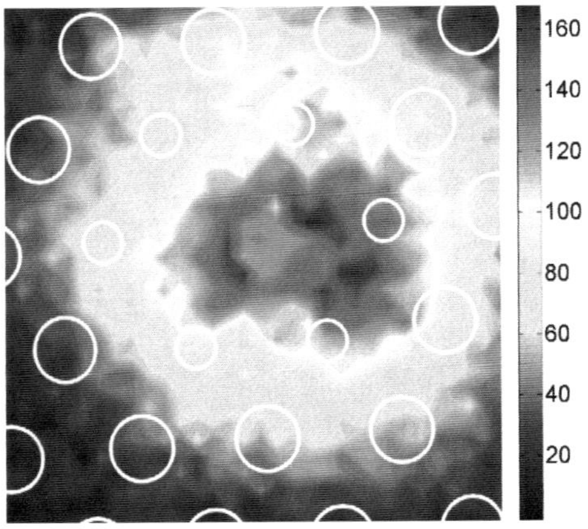

Figure 8. Map of D2 at 1301.12nm.

References

[1] Vignolini S, Intonti F, Riboli F, Wiersma D S, Balet L, Li L H, Francardi M, Gerardino A, Fiore A and Gurioli M 2009 *Applied Physics Letters* **94** 163102–3 URL `http://link.aip.org/link/?APL/94/163102/1`

[2] Intonti F, Vignolini S, Riboli F, Vinattieri A, Wiersma D S, Colocci M, Balet L, Monat C, Zinoni C, Li L H, Houdre R, Francardi M, Gerardino A, Fiore A and Gurioli M 2008 *Physical Review B (Condensed Matter and Materials Physics)* **78** 041401–4 URL `http://link.aps.org/abstract/PRB/v78/e041401`

[3] Alloing B, Zinoni C, Zwiller V, Li L H, Monat C, Gobet M, Buchs G, Fiore A, Pelucchi E and Kapon E 2005 *Applied Physics Letters* **86** 101908–101908–3 ISSN 10773118

[4] Shirane M, Kono S, Ushida J, Ohkouchi S, Ikeda N, Sugimoto Y and Tomita A 2007 *Journal of Applied Physics* **101** 073107 ISSN 00218979

[5] Chauvin N, Zinoni C, Francardi M, Gerardino A, Balet L, Alloing B, Li L H and Fiore A 2009 *Physical Review B* **80** 241306 URL `http://link.aps.org/doi/10.1103/PhysRevB.80.241306`

[6] Mosor S, Hendrickson J, Richards B C, Sweet J, Khitrova G, Gibbs H M, Yoshie T, Scherer A, Shchekin O B and Deppe D G 2005 *Applied Physics Letters* **87** 141105 ISSN 00036951

Quantum Hall regime in emission spectra of single self-organized InP/GaInP quantum dots

A M Mintairov[1], N A Kalugnyy[2], V M Lantratov[2], S A Mintairov[2] and J L Merz[1]
[1] Department of Electrical Engineering, University of Notre Dame, Notre Dame, IN, USA
[2] Ioffe Physical Technical Institute, St. Petersburg, Russia

E-mail: Mintairov.1@nd.edu

Abstract. We used high spatial resolution near-field scanning optical microscopy (NSOM) to measure low-temperature (10K) emission spectra of single InP/GaInP QDs (lateral size ~100 nm) at magnetic fields up to 10 T. A multi-electronic structure having energy splitting 3-5 meV, resulting from charging of the dot up to 20 electrons, has been observed for these QDs. We measured a strong variation of the near-field emission intensity on a length scale ~30 nm related to Wigner localization. Magnetic-field-induced shifts of the multi-shell peaks for some dots follow remarkably well the energy levels of a Fock-Darwin Hamiltonian, which allow the observation of a Wigner-molecular-droplet transition.

A Wigner phase is a strongly correlated state of an electron system, in which electrons occupy separate sites forming a regular lattice. The possibility of crystallization of an electron gas at densities below a certain critical value (n_s) was predicted by Wigner in 1934 [1]. Experimentally such crystallization has been observed in two dimensional electron systems on the surface of liquid He [2], in a GaAs/GaAlAs heterojunction [3] and in Si [4].

The electrons confined in traps having volume $>1/n_s$ form Wigner Molecules (WMs). Wigner localization (WL) of electrons in such traps formed by interface fluctuations is responsible for the quantum Hall effect in high mobility semiconductor heterostructures [5,6]. The WL regime has been realized in single electron transistors (SETs) using GaInAs/AlGaAs quantum dots (QDs) nano-fabricated from modulation doped quantum well structures [7]. Coulomb blockade measurements [7] and theoretical analysis [8] have shown that a WM in such SETs can reveal a rich set of electron arrangements and spin states, which are controlled by the number of electrons present, and by an applied magnetic field.

In the present contribution we report the observation of the signatures related to WL in emission spectra of single self-organized InP/GaInP QDs measured using high-spatial-resolution near-field scanning optical microscope (NSOM). To the best of our knowledge our results are the first observation of WL in QDs using emission spectra [9], which opens the way to study WMs by nano-optical methods.

1. Experimental details

InP/GaInP QD structures having GaInP cap layer thickness from 0 to 60 nm were grown by Metal-Organic Chemical Vapor Deposition epitaxy as described in [10, 11]. The dot density ($1\text{-}5 \times 10^9$ cm^{-2}) and their sizes (base ~100 nm, height ~15 nm) were estimated using transmission electron microscopy (TEM) for capped samples (see below) and atomic force microscopy (AFM) for uncapped samples.

The details of the NSOM experiments were described elsewhere [12]. We measure NSOM images and spectra in the collection-illumination mode. Emission was excited by the 488 nm line of an Ar ion laser and dispersed using a 300 or 1200 gr/mm grating in a 270 mm focal length monochromator. The spectra were measured using a nitrogen-cooled CCD detector and 1200 gr/mm

[1] To whom any correspondence should be addressed.

grating. Monochromatic NSOM images were measured using a 300 gr/mm grating and a GaAs photomultiplier working in the photon-counting regime with an accumulation time of 20 ms per pixel. Excitation power density was ~20-100 and 0.1-1 W/cm^{-2} for 300 and 10 K, respectively.

2. Single dot spectroscopy and imaging

Fig.1a-f shows the results of the NSOM measurements of InP/GaInP QD structures (a, d-f) together with TEM images (b - cross section and c - plan view). Room temperature NSOM images (spatial resolution ~150 nm) presented in the inserts in Fig.1a show a set of bright spots having density ~20 μm^{-2} (see also the TEM image in Fig.1c) and size ~200 nm related to single QDs. Clearly-resolved single QD images are observed for 5 and 20 nm capped QDs as well as for the uncapped QDs. For a 60 nm cap the single dot image contrast strongly decreases, which demonstrates the expected reduction of spatial resolution due to increased tip-dot separation. It is important to note the strong emission intensity for the uncapped QDs at room temperature. We believe that this is the first observation of such strong emission from uncapped QDs.

The spectra of single InP QDs at room temperature (Fig.1a) consists of a single band having wavelength in the range of 710-790 nm (1.75-1.56 eV) and halfwidth γ~60 nm (~150 meV). At 10K the emission spectra of a single InP QD results from the filling of several electron shells (see Fig.1d), which occurs as a result of residual doping (~10^{16} cm^{-3}) of the GaInP matrix [13]. For most of the dots three peaks (s, p and d shells) having width γ~3 meV and energy splitting 3-5 meV are observed, indicating charging of the QD by 7-16 electrons. For some dots a fine structure of the shells consisting of a several ultra-narrow lines with γ=0.2 meV is revealed as was shown in detail in ref [12]. Spectra of three selected QDs denoted by QD1, QD2 and QD3 in Fig.1d show a progressive decrease of the shell energy splitting ΔE from ~5 to ~4 and to ~3 meV, which is directly related to a change of the confinement potential, i.e. "effective" dot sizes, from ~90 to ~100 and to ~115 nm; this is in good agreement with TEM measurements (see Fig.1b and c). For the ~100 nm dot, one can expect WL of electrons [8] and thus each of these InP/GaInP QDs represents a natural WM.

In Fig.1e and f we present high-spatial-resolution monochromatic 10K-NSOM images taken for the same area (size 500x500 nm^2) and detection wavelength (690 nm) in two separate scans. In addition to the bright resonant QD denoted Qa, two weaker non-resonant QDs denoted Qb and Qc

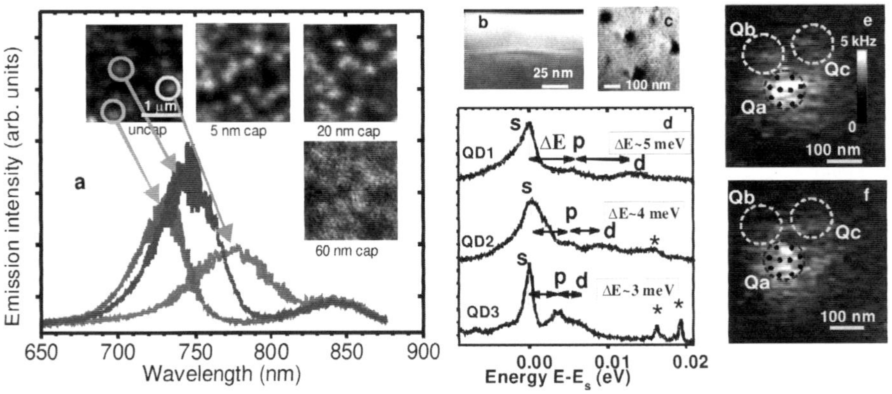

Figure 1. Single InP/GaInP QD emission spectra taken at temperatures: a – 300K and d – 10K, transmission electron microscopy (b and c), and 10K NSOM images (e and f). Inserts in (a) show room temperature NSOM images of InP/GaInP QD structures with different cap layer thickness. In (d) the s-peak energy (E$_s$) is 1.7064, 1.6992, and 1.6994 eV for QD1, QD2 and QD3, respectively; star symbols mark emission bands of neighbouring dots. In (e) and (f) Qa, Qb and Qc dashed large circles outline different QDs, while the small circles in dot Qa represent possible WM electron positions [8].

appear in the images. The image of Qa has size ~120 nm and it reveals a strong (up to 50%) intensity fluctuation on a length scale ~30 nm, which is of the order of the single electron separation for a WL, as can be seen from the possible configuration of an ten-electron-WM [8], indicated by small circles superimposed on the Qa image. We determined that the photon counting detection noise only partly contributes to the spatial intensity fluctuation in Fig.1e and f (note that photons are counted only for 20 ms in monochromatic imaging); this provides evidence that the observed spatial variation of the near-field intensity is related to a WL. Indeed, the measurements of the spectra using a CCD with accumulation time 1 s revealed similar spatial fluctuations of the emission intensity. Spectral measurements also show that the intensity variations are accompanied by spectral diffusion of a few meV. We found that intensity and spectral fluctuations are suppressed by decreasing spatial resolution (near-field defocusing). Spectral diffusion indicates the effects of rearrangement of the charge distribution inside and/or around the QDs under near-field excitation; this will require further investigations, which are in progress.

3. Magnetic field spectra

The multi-shell structure observed in Fig.1d makes it possible to observe a magnetic-field-induced phase transition of a WM using magneto-NSOM spectroscopy (see Fig.2a-d), which is similar to that observed in Coulomb blockade measurements using nano-lithographically-defined GaInAs/AlGaAs QDs in a SET [7]. In Fig.2a and b we show the emission spectra versus magnetic field of two InP/GaInP QDs denoted QD2 and QD4 having ΔE=4.5 and 3.5 meV, respectively, and in Fig.2c and d we compare the magnetic-field-induced shifts of the shell peaks for these QDs with the energy levels of the Fock-Darwin (FD) Hamiltonian. From Fig.2b and d one can see that for QD4 having larger size (i.e. smaller ΔE, see also Fig.1d) the f-shell peak is observed, which indicates the accumulation up to 20 electrons. Here the FD levels poorly describe the experimentally-observed shifts and no distinct assignments of the shell peaks at high fields can be done.

In contrast, for QD2 (Fig.2a and c) the observed magnetic-field-induced shell shifts follow the FD energy levels remarkably well. Here two magnetic-field induced transitions are observed. The first transition occurs at B=2T. It arises from optical pumping of the nuclear spins [14] inducing an internal magnetic field (B_{int}) of 2T which is anti-parallel to the external field. The existence of such a field was established using measurements of circular polarization [15]. In the unpolarized spectra presented in Fig.2a B_{int} is responsible for the shift to lower energy of the s- and d-shell peaks at B<2. For B>2

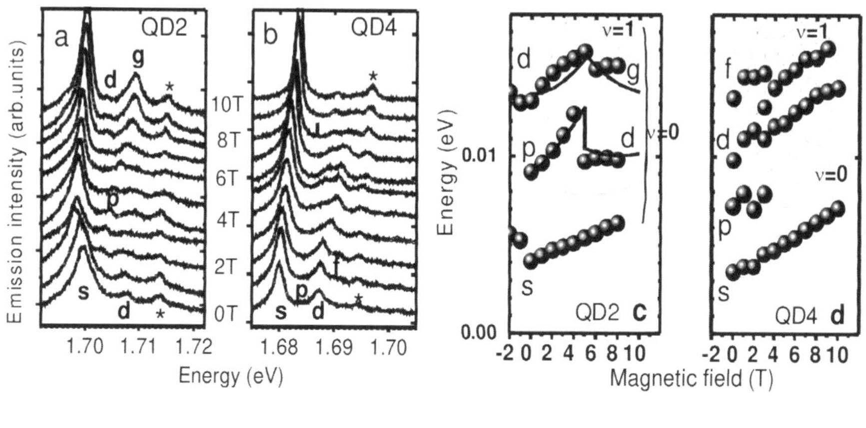

Figure 2. Near-field spectra at 10 K representing shell structure (peaks s-g) of single InP QDs at magnetic fields 0, 1,..,10T (a and b); energy shift of emission peaks of these QDs versus magnetic field (solid circles) together with calculated Fock-Darwin energy levels (solid curves) and Landau levels (ν=0 and 1) dashed lines (c and d). Peaks * in (a and b) are contributions of neighbouring QDs. In c and d the abscissa is the net *internal* magnetic field.

(B_{int}>0) the peak positions shift to increasing energy (see Fig.2c). In the range B=2-7 T (B_{int}=0-5 T) the p-shell peak appears. At B>7 T (second transition) the p-shell peak transforms into a d-shell, and d into g. As can be seen from Fig.2c the second transition (at 7T) corresponds to transition of electrons from the second (ν=1, p/d-shells) to the first Landau level (ν=0, d/g-shells). Such a transition was previously observed in the magnetic field dependence of Coulomb blockade and can be described as a transition from the WM phase to the so-called maximum-density droplet phase [16].

In conclusion, we used high-spatial-resolution NSOM to demonstrate the effect of Wigner localization of electrons (quantum Hall regime) in the emission spectra of self-organized InP/GaInP QDs. Our results open the way to study such localization by nano-optical methods.

4. References

[1] Wigner E P 1934 *Phys. Rev* **46** 1002
[2] Grimes C C and Adams G 1979 *Phys. Rev. Lett.* **42** 795
[3] Andrei E Y, Deville G, Williams F, Paris I B and Etienne B. 1988 *Phys. Rev. Lett.* **60** 2765
[4] Pudalov V M, D'Iotio M, Kravchenko S V and Campbell J W 1993 *Phys. Rev. Lett.* **70** 1866
[5] Laughlin R B 1988 *Phys. Rev. B* **27** 3383
[6] Ilani S, Martin J, Teitelbaum E, Smet J H, Mahalu D, Umansky V, Yacoby A 2004 *Nature* **427** 328
[7] Kastner M A 1993 *Physics Today*, **24**; Ashoori R C 1996 *Nature* **379** 413; Tarucha S, Austing D G, Honda T,. Van der Hage R. J, Kouwenhoven L P 1996 *Phys. Rev. Lett.* **77**, 3613
[8] Maksym P A, Imamura H, Mallon G P and Aoki H 2000 *J. Phys.: Condens. Matter* **12** R299; Reimann S M, Manninen M 2002 *Rev. Mod. Phys.* **74**, 1283; Szafran B, Bednarek S and Adamowski J 2003 *J. Phys. Condens. Matter* **15** 4189
[9] Patel S, Plaut A S, Hawrylak P, Lage H, Grambow P, Heitman D, von Klitzing K, Harbison J P, Florez L T 1997 *Sol. St. Communic.* **101**, 895; Zang Y H, Plaut A S, Weis J, Harbisom J P, Florez L T, Holland M C, Stanley C R 2003 *Phys. Rev. B* **68** 073302; Stallard W G, Plaut A S, Thoms S, Holland M C, Beamont S P, Stanley C R., Hopkinson M 1998 *Appl. Phys. Lett.* **73** 1898; Nomura S, Nakanishi T, Aoyagi Y 2001 *Phys. Rev. B* **63** 165330; Kalliakos S, Garcia C P, Pellegrini V, Zamfirescu M, Cavigli L, Gurioli M, Vinattieri A, Pinczuk A, Dennis B S, Pfeiffer L N, West K W 2007 *Appl. Phys. Lett.* **90** 181902
[10] Vinokurov D A, Kapitonov V A, Nikolaev D N, Sokolova Z N, and Tarasov I S 2001 *Semiconductors* **35** 235
[11] Chu Y, Mintairov A M, He Y, Merz J L, Kalyuzhnyy N A, Lantratov V M, Mintairov S A 2009 *Physics Letters A,* **373** 1185
[12] Mintairov A M, Sun K, Merz J L, Li C, Vlasov A S, Vinokurov D A, Kovalenkov O V, TokranovV, Oktyabrsky S, 2004 *Phys. Rev. B* **69** 15530
[13] Hessman D, Persson J, Pistol M-E, Pryor C, and Samuelson L 2001 *Phys. Rev. B* **64** 233308
[14] Tratakovskii A I, Wright T, Russell A, Fal'ko V I, Van'kov A B, Skiba-Szymanska J, Drouzas I, Kolodka R. S, Skolnik M S, Fry P W, Taharaoui A, Lui H-Y and Hopkinson M 2007 *Phys. Rev. Lett.* **98**, 026806 ; Maleinsky P, Lai C W, Bodalato A, and Imamoglu A 2007*Phys. Rev. B* **75**, 035409; Kaji R, Adachi S, Sasakura H, Muto S, Kumano H, Suemune I 2007 *J. Supercond. Nov. Magn.* **20** 447
[15] Mintairov A M, Vlasov A S, Merz J L 2004 *Quantum Dots, Nanoparticles and Nanowires, (Material Research Society Symposium Proceeding)* **789** 169
[16] Oosterkamp T H, Janssen J W, Kouwenhoven L P, Austing D G, Honda T, and Tarucha S 1999 *Phys. Rev. Lett.* **82**, 2931

Acknowledgments

Work was supported by the National Science Foundation via NSF-NIRT grant No. ECS-0609249.

Temperature stable positive magnetic susceptibility of semiconductor wobbled nano rings

L. M. Thu, W. T. Chiu and O. Voskoboynikov

Department of Electronics Engineering and Institute of Electronics, National Chiao Tung University, 1001 Ta Hsueh Rd., Hsinchu 30010, Taiwan, R.O.C

E-mail: vam@faculty.nctu.edu.tw

Abstract. We theoretically investigate the magnetic response of ensembles of semiconductor nano rings. Using our mapping method we simulated geometrical and material dispersion of the magnetic susceptibility of ensembles of the rings. Unlike the susceptibility of an individual ring the averaged susceptibility shows small temperature effect.

1. Introduction

Recent experimental studies on self-assembled *InAs/GaAs* quantum nano-scale rings (NRs) demonstrate controllable flexibility of geometrical and material characteristics of those semiconductor nano-objects [1,2]. Three-dimensional topological quantum effects (similar to the Aharonov-Bohm effect) for electrons confined in a NR result in unusual behavior of the differential magnetic susceptibility (DMS) of the ring containing only a single electron [1,2,3]. The appearance of the positive peak in the DMS at low temperatures has to be addressed to the Aharonov-Bohm crossing between two lowest energy states of the electron confined in the ring. The actual experimental height of the peak and its temperature dependence were found in contradiction to the theoretical description performed in [1,2]. Unlike in the theoretical simulation results the experimental peak demonstrated a negligible temperature effect. It was proposed in [1] that the temperature stability is due to the rings' ensemble averaging, but no further theoretical or experimental considerations have been done to proof this point. At the same time the inherent property of the contemporary semiconductor self-assembled nano-objects is their geometrical and material parameters' dispersions. The general approach to the description of ensembles of semiconductor nano-objects is to consider a multi-parameter distribution (multi-dimensional) function including dispersions of all appropriate parameters' within wide ranges of their changes. This requires for a computational method which can optimize those extensive simulations of ensembles of semiconductor nano objects. Our mapping method (proposed in [4]) allows us to efficiently simulate quantum mechanical properties of semiconductor nano systems such as NRs. In this study using the method we address the issue of the temperature dependence of the magnetic response of ensembles of three-dimensional torus-shaped wobbled NRs.

2. Theoretical method and simulation results

To determine the magnetization M (total magnetic moment) and differential magnetic susceptibility χ for an isolated nano ring, the standard approach is to calculate the total electronic energy of the ring in the presence of the external magnetic field **B**. The magnetization and DMS of the ring are [3]

$$M = \sum_n \left(-\frac{\partial E_n}{\partial B} \right) f(E_n - \mu) \text{ and } \chi = \frac{\partial M}{\partial B} \tag{1}$$

where n stands for the electronic state with corresponding energy E_n, $f(x)$ is the Fermi distribution function, and μ is the chemical potential of the system determined from the number of the electrons confined in the NR: $N = \sum_n f(E_n - \mu)$.

The magnetic field strength's change results in crossings between lowest subsequent energy levels of NRs and oscillations in the magnetization (the Aharonov–Bohm oscillations). At low temperatures the aperiodic oscillations of magnetization generate delta-like positive peaks in the ring's DMS [3]. At the same time the peaks' positions and amplitudes strongly depend on the actual geometrical and material parameters of NRs such as effective radii, heights, material content, etc. In this paper we concentrate only at the first peak (which is the most visible in experiments). This peak corresponds to the crossing between energy states E_0 and E_1. One should notice that the electronic energies E_n of NRs can be characterized by a set of parameters: $\{R,h,\xi,\ldots\}$, where R, h, ξ represent characteristic radius, height, anisotropy coefficient (see equation 3), etc. The geometrical and material parameters' dispersions in ensembles of NRs are described by a multidimensional distribution function $P(R,h,\xi,\ldots)$ which gives the number of NRs dN with the values of $\{R,h,\xi,\ldots\}$ inside the domain $\{[R,R+dR]$, $[h,h+dh]$, $[\xi,\xi+d\xi],\ldots\}$ as $dN = P(R,h,\xi,\ldots)\, dRdhd\xi\ldots$In this study we assume that the NR parameters in typical structures are distributed according to the non-correlated normal distribution $P(R,h,\xi,\ldots) = P_G(R)P_G(h)P_G(\xi)\ldots$, where the standard normal distributions are presented by the probability density function $P_G(x) = A\cdot G[(x-x^0)/\Delta x]$, A stands for the normalization coefficient, x^0 is an mean value and Δx is the standard deviation of the parameter x, and G[y] is the Gaussian function. With including the parameters' variations in the ensemble of the rings we can write the following expression for the experimental meaningful (average) DMS:

$$\overline{\chi} = \int P(\{x_i\})\chi(\{x_i\})\prod_i dx_i \tag{2}$$

where $\{x_i\}$ represents set of the appropriate parameters.

To bridge NR parameters' variations and DMS changes in this paper we adopt the mapping procedure in constructing a three-dimensional model of a NR as it was described in Ref. [4]. For an asymmetric (wobbled) *InAs/GaAs* NR we first assume that the ring was grown on a substrate parallel to the x-y plane (see inset in figure 1). Using known experimental data obtained from AFM (atomic force microscopy) and X-STM (cross-sectional scanning tunneling microscopy) [1,2] we model the geometry of the ring by mapping the height of the ring $h(x,y)$ with the following function:

$$h(x,y) = h_0 + \gamma_0^2 R_r^{-2}\left[h_r\left(1+\xi\frac{x^2-y^2}{x^2+y^2}\right)-h_0\right]\cdot\frac{R_r^2-\left(\sqrt{x^2+y^2}-R_r\right)^2}{\left(\sqrt{x^2+y^2}-R_r\right)^2+\gamma_0^2};\sqrt{x^2+y^2}\le R_r$$

$$h(x,y) = h_\infty + \gamma_\infty^2\left[h_r\left(1+\xi\frac{x^2-y^2}{x^2+y^2}\right)-h_\infty\right]\left[\left(\sqrt{x^2+y^2}-R_r\right)^2+\gamma_\infty^2\right]^{-1};\sqrt{x^2+y^2}>R_r \tag{3}$$

where h_0, h_r, h_∞ stand correspondingly for the height at the center of the ring, at the rim of the ring (when $x^2+y^2=R_r^2$), and far away from the center of the ring; γ_0 and γ_∞ respectively determine the inner and outer slope near the rim. The parameter ξ defines the anisotropy of the ring height in the (x, y) plane. Three-dimensional confinement potential for electrons can be found from the composition and geometry dependent conducting band profile: $V(x,y,z) = \Delta E_C\cdot\{1\text{-}1/4[1+\tanh(z\text{-}z_0/a)][1\text{-}\tanh(z\text{-}h(x,y)/a)]\}$, where ΔE_C is the electronic band offset for the ring *InAs/GaAs* structures. The slope of the potential and the range of the potential change at the boundaries of the object is controlled by the parameter a. The potential is used to define the mapping function $M(x,y,z) = 1-V(x,y,z)/\Delta E_C$. The mapping function reproduces experimental information about geometry (shape) and position dependent composition of the ring. Using $M(x,y,z)$ we model the position dependent band gap $E_g(x,y,z)$, spin-orbit interaction splitting $\Delta(x,y,z)$, and the effective mass at the bottom of the conducting band $m(x,y,z)$. Those functions we conventionally use for the effective one-electronic-band Hamiltonian (energy and position-dependent electron effective mass and g-factor approximation) [5,6]. The realistic semiconductor material parameters for the *InAs/GaAs* heterostructure with complex strained composition are used according to [4]. To determine the single electron magnetic response of an isolated NR we calculate the energy states of the ring with a predefined set of parameters $\{R_r,h_0,h_r,h_\infty,\xi,\gamma_0,\gamma_\infty,a,\Delta E_C\}$. The energy states are found by the nonlinear iterative method (see for instance [6] and references therein) using the COMSOL Multiphysics package (www.comsol.com). As an example of our simulation results we present in figure 1 two lowest energy states and DMS of the single electron *InAs/GaAs* NR when: $\{R_r = 9$ nm, $h_0 = 1.6$ nm, $h_r = 3.6$ nm, $h_\infty = 0.0$ nm, $\xi = 0.15$, $\gamma_0 = 3.0$ nm, $\gamma_\infty = 5$ nm, $a = 0.5$ nm, $\Delta E_C = 0.44$ eV$\}$.

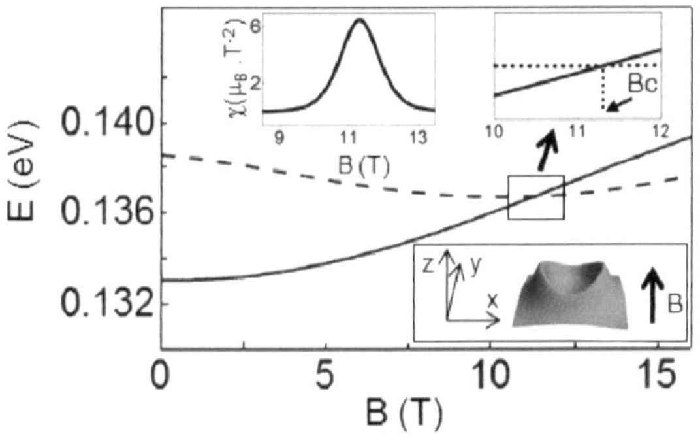

Figure 1. Energy states and DMS (upper left inset, μ_B stands for the Bohr magneton) of a wobbled nano ring. All explanations see in the text.

We stress that at low temperatures the DMS peak position and amplitude are obviously defined by the properties of two lowest energy states as functions on the magnetic field. Near the crossing point (B_C) we can approximate the electronic energies: $E_{0,1}(B) \approx E_0(B_C) + C_{0,1}(B - B_C)$ (see the upper right inset of figure 1). With this approximation the DMS of a single electron NR may be written as

$$\chi(B,T) \approx \frac{(C_1 - C_0)}{2k_BT}\left\{\frac{C_1\exp(\delta)}{[1+\exp(\delta)]^2} - \frac{C_0\exp(-\delta)}{[1+\exp(-\delta)]^2}\right\} \tag{4}$$

where $\delta = (C_1\text{-}C_0)\cdot(B\text{-}B_C)/2k_BT$, T stands for the temperature, and k_B is the Boltzmann constant.

Our mapping method allows us efficiently and economically to simulate energy states of NRs by varying parameters within a wide range. For reason of clarity in this short paper we restrict our consideration only to variations of the DMS peak caused by the rim radius R_r dispersion. Other parameters we assume to be fixed as for the individual ring above. To connect the DMS peak variations and changes of R_r on the base of our simulation experience and results we propose to use the following type of functions to describe the crossing point B_C and coefficients $C_{0,1}$ dependencies on R_r:

$$B_C = a_B + b_B \cdot R_r^{\beta_B};\ C_0 = a_0 + b_0 \cdot R_r^{\beta_0};\ C_1 = a_1 + b_1 \cdot (R_r - \beta_1)^{-1} \tag{5}$$

where $a_{B,0,1}$, $b_{B,0,1}$, and $\beta_{B,0,1}$ are parameters to be fitted by use of our simulation results when only R_r has been varied within the interval 6.5 ÷ 12.5 nm. The parameters have been found to be (in appropriate SI units): a_B = 1683.9, b_B = 0.92, β_B = - 2.24; a_0 = 9.6×10^{-3}, b_0 = 7.4×10^{-3}, β_0 = 9.3×10^{-2}; a_1 = - 1.9 ×10^{-5}, b_1 = 7.3×10^{-5}, β_1 = 6.15. The fitted expressions (5) and equation (4) we use to simulate the average DMS:

$$\overline{\chi}_R(B,T) = \int P(R_r)\chi(B,T;R_r)dR_r \qquad (7)$$

Figure 2 shows the simulation results for the DMS of an individual ring with R_r = 9 nm and the DMS averaged within the ensemble of rings with the mean value R^0_r = 9 nm and standard deviation ΔR_r = 0.5 nm. The averaged DMS peak is much lower than the individual ring's DMS. Without including the radius variations, the calculated amplitude of the individual NR DMS's peak sharply increases with decreasing temperature. At the same time the averaged within the ensemble of NRs DMS's demonstrates temperature stable behaviour. This was observed experimentally in [1].

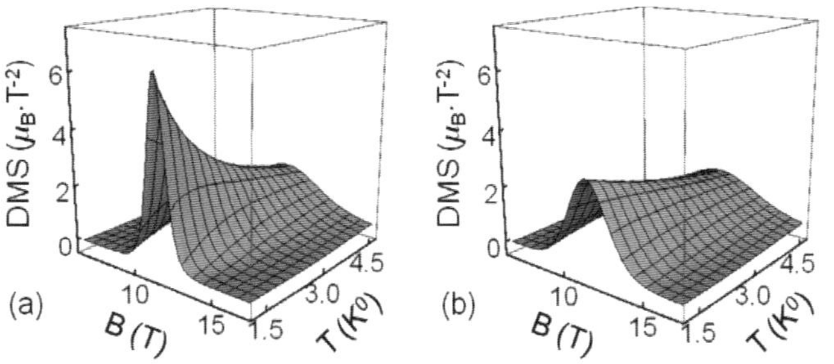

Figure 2. Dependence of the differential magnetic susceptibility on magnetic field and temperature: (a) a single ring; (b) DMS averaged within an ensemble of the rings.

In short conclusions the simulations showed that the DMS amplitude is sensitive to the exact geometry and composition of semiconductor nano rings. We theoretically demonstrated a small temperature effect on averaged DMS of ensembles of the rings. It follows from this study that experimental investigations of the magnetic response of specially designed ensembles of NRs can be potentially useful for further fabrication of systems with principally new magnetic properties.

Acknowledgements
This work is supported by the National Science Council of the Republic of China under Contracts no. 97-2112-M-009-012-MY3 and NSC 97-2120-M-009-004 and by the Aim for the Top University Plan of the National Chiao Tung University and Ministry of Education of Taiwan, ROC.

References

[1] Kleemans N A J M, Bominaar-Silkens I M A, Fomin V M, Gladilin V N, Granados D, Taboada A G, García J M, Offermans P, Zeitler U, Christianen P C M, Maan J C, Devreese J T, and Koenraad P M 2007 *Phys. Rev. Lett.* **99** 146808
[2] Fomin V M, Gladilin V N, Klimin S N, Devreese J T, Kleemans N A J M, and Koenraad P M 2007 *Phys. Rev.* **76**, 235320
[3] Voskoboynikov O and Lee C P 2004 *Physica* E **20** 278
[4] Thu L M and Voskoboynikov O 2009 *Physics Procedia* **3** 1133
[5] Voskoboynikov O, Li Yiming, Lu Hsiao-Mei, Shih Cheng-Feng, and Lee C P 2002 *Phys. Rev.* B **66** 155306
[6] Li Yiming, Voskoboynikov O, Lee C P, and Sze S M 2002 *Comp. Phys. Comm.* **147** 209

Journal of Physics: Conference Series **245** (2010) 012043 doi:10.1088/1742-6596/245/1/012043

Capacitance-voltage spectroscopy on InAs quantum dot valence band states in tilted magnetic fields

D Reuter[1], R Roescu[1], U Zeitler[2], J C Maan[2], and A D Wieck[1]

[1]Lehrstuhl für Angewandte Festkörperphysik, Ruhr-Universität Bochum, Universitätsstraße 150, D-44801 Bochum, Germany
[2]High Field Magnet Laboratory, Institute for Molecules and Materials, Radboud University Nijmegen, Toernooiveld 7, 6525 ED Nijmegen, the Netherlands

E-mail: dirk.reuter@rub.de

Abstract. We have performed capacitance-voltage (C-V) spectroscopy on the valence band states of InAs quantum dots (QDs) in magnetic fields tilted with respect to the sample plane. We observe an unexpected behaviour for the height of the capacitance signal as function of the in-plane magnetic field, when varying the perpendicular field component, for the first two charging peaks corresponding both to an s-like ground state. With increasing perpendicular field component, the signal height for the first peak decreases strongly and for the second peak the maximum signal height is observed at finite in-plane field values. These observations are not in agreement with the usual assumption that the height-vs.-in-plane-field traces reflect the in-plane k-space probability distribution of the charged QD level because an s-like state should not vary its general shape with perpendicular field. The qualitatively different behaviour of the two peaks is attributed tentatively to correlation effects when the second electron is charged.

1. Introduction

InAs quantum dots (QDs) have been intensively studied in the past decades because they are model systems for strong three-dimensional carrier confinement as well as building blocks for novel optoelectronic devices, e.g. for quantum information technology [1]. One point of interest is the carrier-carrier interaction in QDs, which have been intensively studied by charging spectroscopy employing capacitance-voltage (C-V) measurements [2,3]. Extending the capabilities of C-V spectroscopy, it was recently possible to map the in-plane k-space probability distribution corresponding to the individual charging peaks for conduction as well as valence band states of InAs QDs [4,5]. In these experiments, based on an approach of Patanè and co-workers for magneto-tunneling spectroscopy [6], C-V measurements at certain frequencies (for which the charging dynamics becomes relevant) were employed and the k-space mapping was realized by a variable in-plane magnetic field.

In this paper, we present C-V experiments on the valence band states of InAs QDs in tilted magnetic fields intended to take the wave function mapping one step further. The idea of the work was to manipulate the wave function by the perpendicular component of the magnetic field and to map the altered probability distribution by the in-plane field component. We will see that this straight forward interpretation is not sufficient to explain our experimental finding.

2. Experimental Details

The InAs QD samples were prepared by solid source molecular beam epitaxy on a GaAs(100) substrate. The active part of the layer sequence consists of a 300 nm thick, carbon doped (3×10^{18} cm^{-3}) GaAs back-contact, a 19 nm thick GaAs tunnelling barrier, an InAs QD layer, 30 nm GaAs and 32 periods of a (3 nm AlAs)/(1 nm GaAs) superlattice, followed by a 10 nm thick GaAs cap layer. The InAs QDs were prepared by depositing a nominal coverage of 2.0 ML InAs at a substrate temperature of 510 °C. The ground state photoluminescence for these samples is between 1250 and 1270 nm at 300 K. From these samples, Schottky diodes were prepared using Cr-Au gates.

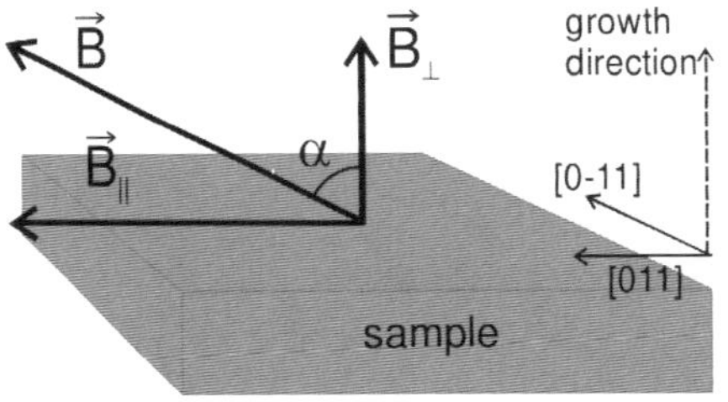

Figure 1: Measurement geometry for the experiments in tilted magnetic field.

The sample was mounted on a rotatable holder so that, for a fixed spatial B-field direction, the B-field orientation with respect to the sample could be varied from in the sample plane (in-plane field) to perpendicular to the sample plane (perpendicular field). The samples could be mounted in two orientations so that the in-plane component of the field is either along the [0-11] or the [011] surface direction. In the experiments, the absolute value of the magnetic field B and the tilt angle α (see Fig. 1 for a sketch of the geometry) were changed simultaneously so that the perpendicular component stayed constant and the in-plane component was varied. For each configuration a C-V measurement was performed at 8 kHz for the first two charging peaks (see Fig. 2 for the peak labeling).

The height of the charging peaks, which should be for an appropriate measurement frequency approximately proportional to the tunneling rate [4,5,7], was determined by subtracting first a background due to the capacitance of the Schottky diode and then fitting Gaussians to the individual charging peaks. The heights of the capacitance signal were normalized to their corresponding values in the absence of the magnetic field and plotted as a function of the in-plane field for various values of the perpendicular component.

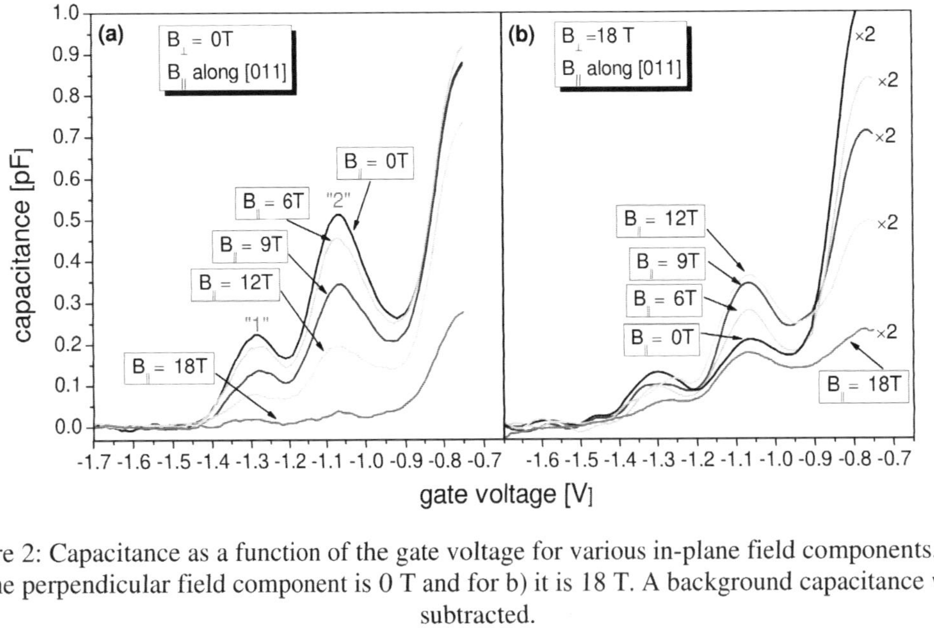

Figure 2: Capacitance as a function of the gate voltage for various in-plane field components. For a) the perpendicular field component is 0 T and for b) it is 18 T. A background capacitance was subtracted.

3. Results and Discussion

Fig. 2 shows the first two charging peaks for various in-plane magnetic fields oriented along [011]. Dispersion measurements supported the assignment that these two peaks belong to the charging of a two-fold degenerated s-like ground state [3,5]. For $B_{\perp} = 0$ T (Fig. 2a), the height of the capacitance signal decreases monotonically with increasing in-plane field as reported before [5] and expected for an s-like ground state. For $B_{\perp} = 18$ T (Fig. 2b), the situation has changed: The height of the capacitance signal in general has decreased significantly for both peaks, which indicates that the perpendicular field hampers somehow the tunnelling into the ground state. The most significant change occurs for peak 2, whose maximum value is not measured for zero in-plane field, but for fields around 12 T, which is totally unexpected for tunnelling into an s-like ground state.

To display the changes induced by the magnetic field in more detail, Fig. 3 shows the normalized height of the capacitance signal as function of the in-plane field for various values of $B_{\perp}$ for the first two charging peaks. At $B_{\perp} = 0$ T, for both peaks and both high symmetry directions a bell-like shape with a maximum at $B_{\|} = 0$ T is observed. This is expected for an s-like ground state under the assumption that the normalized height of the capacitance signal reflects quite well the in-plane k-space probability distribution of the charged energy level [4,5]. The extension in k-space is different for the two high symmetry directions pointing to an elliptically shaped QD as discussed in detail in [5].

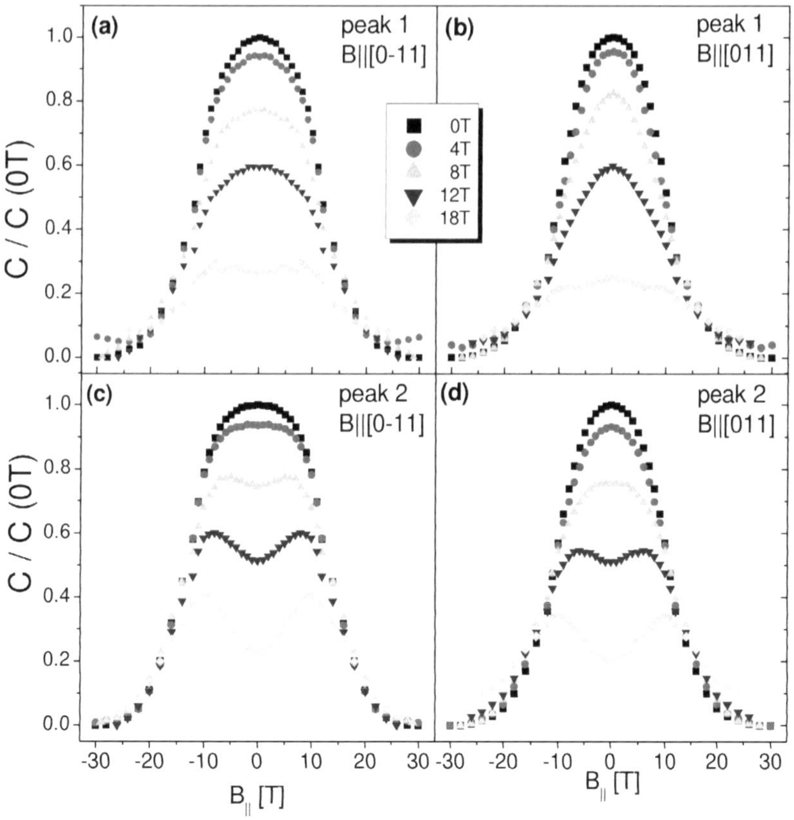

Figure 3: Normalized height of the capacitance signal for the first two charging peaks as function of the in-plane magnetic field component for various values of $B_{\perp}$. The orientation of the in-plane component is indicated in the panels. The trace for negative in-plane fields was generated be mirroring the values measured for positive in-plane fields.

If $B_\perp$ is increased up to 12 T, the bell shape is nearly conserved for peak 1 but the maximum height decreases monotonically. For $B_\perp$ = 18 T almost a plateau is observed between -10 T and 10 T (see Fig. 3a and 3b). The area under the trace decreases significantly with increasing $B_\perp$. This is surprising and not understandable with the interpretation that the trace reflects the in-plane probability distribution, because then the area under the trace should stay constant. It seems that the perpendicular field alters the tunnelling rate beyond changing the QD wave functions.

For charging peak 2, the changes induced by the perpendicular field are even more drastic (see Fig. 3c and d). For $B_\perp$ = 8 T and higher, a dip at $B_\parallel$ = 0 T develops and the largest signal height is found around ± 10 T. This shape becomes more pronounced with increasing $B_\perp$ and for 18 T the signal in the maximum is approximately a factor 2 higher than at $B_\parallel$ = 0 T. The development of the bone-like structure with increasing $B_\perp$ cannot be understood with the simple assumption that the perpendicular field component modifies the k-space probability distribution and the parallel field component is just used to map it because an s-like ground state wave function is expected to retain its bell-like shape under the influence of a perpendicular magnetic field.

The results are quite surprising and difficult to interpret but two remarks can be made:

a) Although the signal height is determined by fitting, the first two charging peaks are well resolved and the non-monotonic bone-like shape for peak 2 can be directly seen in the C-V traces (see Fig. 2). Therefore, we are quite sure that this shape is not a fitting artefact.
b) It is interesting that the two charging peaks show significantly different behaviour, although both arise from tunnelling into the same state. This means that the observations for peak 2 are connected to correlation effects between the two holes in the QD.

In our opinion, to understand the experimental observations, it is necessary to perform calculations of the tunnelling probability taking into account the influence of the field on the whole heterostructure, especially the p-doped back-contact. A chirality induced here by $B_\perp$ might be reflected in the measured tunnelling probabilities.

4. Summary

We have investigated the tunnelling dynamics into InAs QDs under the influence of a tilted B-field by employing C-V spectroscopy. For the first charging peak the tunnelling probability decreases monotonically with increasing in-plane field component, resulting in a bell-shaped trace. This holds for perpendicular fields up to 18 T, but the area under the trace decreases significantly with increasing $B_\perp$. For the second peak, the bell-like trace at $B_\perp$ = 0 T is transformed to a bone-like shape for increasing $B_\perp$. This behaviour cannot be explained by simply assuming that the tunnelling probability reflects only the in-plane k-space probability distribution for the corresponding energy level, because for an s-like ground state a bell-like trace should be observed even under perpendicular B-field.

Acknowledgment

We gratefully acknowledge financial support by the Research School of the Ruhr-University Bochum and by EuroMagNET II under the E. U. contract No. 228043.

References

[1] for a review see: D. Bimberg, M. Grundmann, and N. N. Ledentsov, Quantum Dot Heterostructures (Wiley, New York, 1999),
Self-Assembled Quantum Dots, Ed. Z. M. Wang, (Springer, Berlin, 2008).
[2] H. Drexler et al., Phys. Rev. Lett. **73**, 2252 (1994).
[3] D. Reuter et al., Phys. Rev. Lett. **94**, 26808 (2005).
[4] O. S. Wibbelhoff, A. Lorke, D. Reuter, and A. D. Wieck, Appl. Phys. Lett. **86**, 92104 (2005).
[5] G. Bester et al., Phys. Rev. B **76**, 75338 (2007).
[6] A. Patanè et al., Phys. Rev. B **65**, 165308 (2002).
[7] R. J. Luyken, et al., Appl. Phys. Lett. **74**, 2486 (1999).

Journal of Physics: Conference Series **245** (2010) 012044 doi:10.1088/1742-6596/245/1/012044

Efficiency of spin injection in novel InAs quantum dot structures: exciton vs. free carrier injection

J Beyer[1], I A Buyanova[1], S Suraprapapich[2], C W Tu[2] and W M Chen[1]

[1] Department of Physics, Chemistry and Biology IFM, Linköping university, SE-581 83 Linköping, Sweden

[2] Department of Electrical and Computer Engineering, University of California at San Diego, La Jolla, USA

E-mail: beyer@ifm.liu.se

Abstract. Unambiguous experimental evidence for a significant difference in efficiency of excitonic vs. free carrier spin injection is provided in novel laterally arranged self-assembled InAs/GaAs quantum dot structures, from optical orientation and tunable laser spectroscopy. A lower efficiency of exciton spin injection as compared to free carrier spin injection from wetting layers into QDs results in a distinct feature in luminescence polarization of the QDs as a function of excitation photon energy. It is shown that this difference is not related to carrier density and state-filling effects arising from the difference in optical absorption efficiency between the excitons and free carriers. Rather, it is a genuine property for exciton spin injection that suffers stronger spin relaxation due to Coulomb exchange interaction.

1. Introduction

Three-dimensionally confined carriers in quantum dots (QDs) are a topic of great research interest at present due to their extremely long spin lifetimes[1]. In(Ga)As QDs are a model system for spin-functionality-enabled quantum devices as they can be controllably positioned[2] and embedded into active device structures[3]. At the basis of these spintronic applications is the ability to prepare the confined carriers in a defined spin state. Regardless of whether this is accomplished electrically or optically, the carriers are usually injected non-resonantly, i.e. in one of the surrounding layers and energetically well above their final ground state. During their transfer and energy relaxation, they are subject to various spin relaxation processes which depend on the exact injection mechanism, being either individual carrier or exciton injection[4]. Previously, we showed results evidencing lower spin injection efficiency under resonant optical orientation of free excitons (FE) compared to free carriers of the wetting layer (WL) and interpreted this as a result of accelerated spin relaxation due to exchange interaction inside the injected exciton[5]. The aim of the present work is to investigate some aspects of injected carrier density on the spin injection efficiency into the QDs.

2. Materials and Methods

Several InAs/GaAs-based QD structures have been studied, such as self-assembled single QDs (denoted by SQD) and lateral QD rings (denoted by QDR), consisting of five to seven dots per ring. The nominally undoped structures were grown in Stranski-Krastanov growth mode by molecular beam epitaxy on (001) semi-insulating GaAs substrate. Details on the growth process can be found in [6].

To investigate spin injection into the QDs from the InGaAs WL, we used optical pumping of spin-polarized carriers in the WL and monitored the circular polarization of QD luminescence. A tunable Ti:sapphire laser was used as an excitation source, whose circular polarization could be controlled by a rotatable quarter-wave plate. The samples were kept at a temperature of either 6 K or 10 K for the measurements to be presented here. Polarization-resolved detection of the emitted photoluminescence (PL) was achieved by a rotatable quarter-wave plate in conjunction with a linear polarizer in front of a grating monochromator, equipped either with a liquid-nitrogen-cooled Ge-detector in case of power-dependent PL measurements or with a liquid-nitrogen-cooled infra-red-enhanced Si-CCD camera for all other measurements. The degree of luminescence circular polarization was defined as $P_\alpha = (I_\alpha^{\sigma+} - I_\alpha^{\sigma-})/(I_\alpha^{\sigma+} + I_\alpha^{\sigma-})$, where $I_\alpha^{\sigma+}$ and $I_\alpha^{\sigma-}$ denote the PL intensities of σ^+ and σ^- polarized luminescence, respectively. The excitation polarization state α was adjusted to either σ^+ or σ^-. Excitation power was varied by inserting neutral density filters in the excitation light path.

Optical selection rules in the WL, where both strain and quantum confinement split heavy- and light-hole valence band states, allow creation of spin-polarized carriers/excitons by circularly polarized optical excitation along the direction normal to the WL plane. The degree of luminescence circular polarization can be used as a measure of spin conservation during injection and energy relaxation within the QD as well as spin relaxation in the QD ground state, due to the validity of the same optical selection rules inside the QDs.

3. Results and Discussion

Figure 1 shows typical PL spectra from the studied QD structures, which are dominated by the exciton emission associated with the QD ground state. With increasing excitation power, PL emissions from the QD excited states gain strength due to the well-known state-filling effect in QDs. A close-up PL/PLE map of the ground state PL under excitation nearby the WL absorption edge, taken the QDR sample as an example, is presented in figure 2(a). Similar PL/PLE maps were observed in all studied samples. A PLE cross-section through these maps at the QD ground state PL maximum, shown in figure 2(b), exhibits a distinct peak arising from the heavy-hole (hh) FE absorption in the WL (denoted by XH), followed at shorter excitation wavelengths by an absorption continuum due to band-to-band transitions between the hh valence band and the conduction band. Figures 2(c) and 2(d) show the corresponding PL polarization under circularly polarized excitation over the same excitation and detection wavelength ranges as that in figures 2(a) and 2(b). Excitation of free carriers from the WL hh VB results in QD ground state PL polarization of up to almost 15%. On tuning the excitation wavelength to the WL hh FE resonance, the PL polarization noticeably drops down to 7%.

As shown by experiments[7,8] and theory[9], the ground state of neutral excitons in QDs at low temperature is characterized by two linearly polarized eigenstates, due to an anisotropic electron-hole exchange interaction. A likely explanation of the observed circular polarization is the observation of trion (charged exciton) luminescence[10], originating from unintentional doping during MBE growth. In trions, exchange interaction is cancelled[11], leading to circularly polarized ground states. As trion and neutral exciton luminescence cannot be distinguished in ensemble measurements due to their small energetic difference[12], the ground state PL band may contain PL of both species and thus

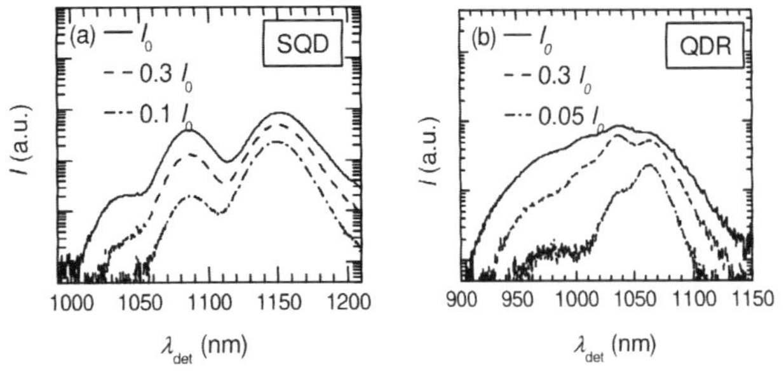

Figure 1. PL spectra (PL intensity I over detection wavelength λ_{det}) of the studied samples as a function of excitation power: SQD (a) and QDR (b), taken at 6 K under above-barrier excitation (514 nm).

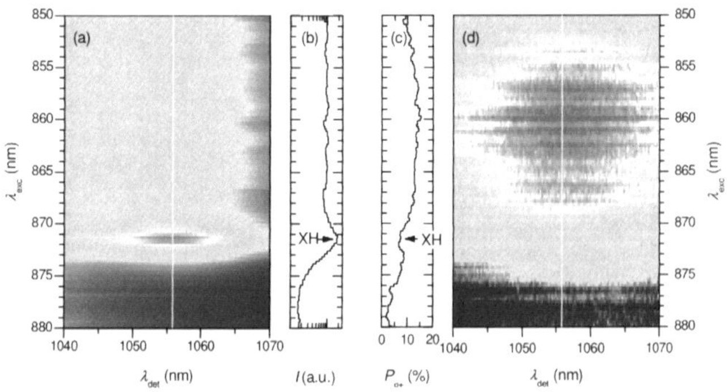

Figure 2. PL/PLE intensity (a) and polarization (d) map of the QD ground state PL (sample QDR) for excitation wavelengths λ_{exc} nearby the WL edge at a low excitation power (< 1 mW). (b) and (c) show PLE intensity I and polarization $P_{\sigma+}$ cross-sections through the respective maps at the ground state PL peak maximum, indicated by vertical yellow lines in the maps. XH is the WL hh FE absorption peak.

show a non-vanishing circular polarization. Furthermore, the presence of trions opens the possibility of dynamic nuclear polarization (DNP) of the lattice atoms forming the QDs via the hyperfine interaction between the spin-polarized electrons in trions and the nuclei[13,14]. The long-lived DNP will act as an effective magnetic (Overhauser) field, possibly suppressing the effect of anisotropic exchange interaction, which will lead to restoration of circularly polarized eigenstates for the neutral exciton and the observation of corresponding circularly polarized luminescence. Additionally the Overhauser field may also contribute to an increase in carrier spin relaxation time, again assisting the circular polarization of QD luminescence for both neutral and charged excitons. It must be pointed out that all of the aforementioned mechanisms for the observation of the circularly polarized QD PL originate from electron spin polarization in the QDs that is generated by spin injection from, in our case, the WL. Therefore, the degree of the QD PL polarization can be taken as a measure of spin injection efficiency from the WL. The observed decrease in QD PL polarization upon spin injection from the WL hh excitons, shown in figure 2, is thus an indication that the efficiency of exciton spin injection is lower than that of uncorrelated free-carrier spin injection.

It should be noted that the observed XH peak in PLE implies a higher concentration of injected carriers, leading to a higher concentration of QD ground state excitons. Such a change in the QD exciton density can, in principle, affect the QD PL polarization via several mechanisms. First, due to Pauli blocking, the QD ground state can only be occupied by two carriers of opposite spin[15] at the same time. Thus, with increasing injected carrier density, we can expect the QD ground state luminescence to show weaker polarization, as the second (counter-oriented) spin-state starts to be filled by carriers having undergone spin flips during injection and energy relaxation into the QD ground state. Second, an increase in spin-polarized electron injection into the QDs may lead to an increase in dynamic nuclear polarization in the QDs[14] and thus to an increasing degree of luminescence circular polarization. The observed increasing carrier density at resonant excitation of the WL hh exciton may be expected to lead to either rising or falling PL polarization, depending on the relative importance of state filling vs. DNP effects. To exclude such carrier-density-related effects from our measurements, we conducted PL polarization measurements at several excitation wavelengths around the WL XH resonance, while adjusting the incident laser power to ensure identical PL intensity of the QD ground state. This way, the carrier density in the QDs was kept constant and the corresponding effects on luminescence polarization were eliminated. The results of these measurements are presented in figure 3.

From the PL/PLE polarization map in figure 3(d) and its corresponding cross-section (PLE polarization) at the ground state PL peak position in figure 3(c), it is clear that QD ground state luminescence polarization decreases at resonant XH excitation. As carrier density effects no longer play a role here, we therefore conclude that this effect must be due to a lower efficiency in exciton spin injection into the QDs as compared with spin injection of free carriers. We attribute the origin of this reduction to more efficient spin relaxation of correlated carriers within the excitons, promoted by the Coulomb exchange interaction. As a direct consequence, the spin polarization of the electrons in the QDs should decrease upon exciton spin injection from the WL. The lower electron spin

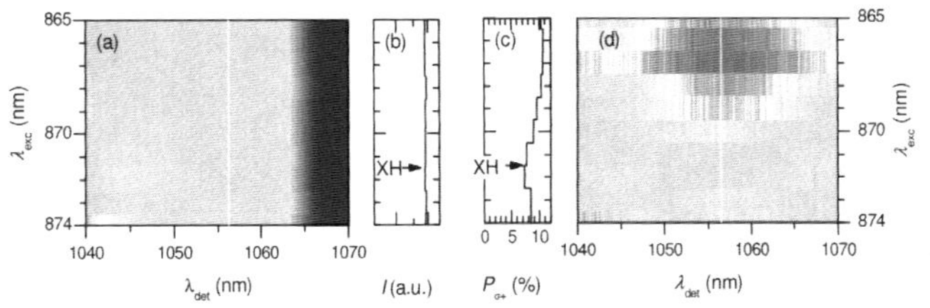

Figure 3. PL/PLE intensity (a) and polarization (d) map of the QD ground state PL (sample QDR) for λ_{exc} nearby the XH resonance. Constant QD ground state PL intensity was achieved by adjusting excitation power at each λ_{exc} step. The PL intensity has been regulated to within ±1% of its absolute value. (b) and (c) as in figure 2.

polarization may subsequently lead to a weaker DNP effect and a reduced Overhauser field, resulting in increasing importance of the anisotropic exchange interaction of the neutral excitons and stronger electron spin relaxation in the QDs. All these effects lead to a decreasing PL polarization of the trions and neutral excitons in the QD, as observed in our case.

4. Conclusion

In conclusion, we have demonstrated the significance of a distinction between excitonic and free carrier spin injection from an InGaAs WL into novel InAs multiple-QD structures. Reduced PL polarization of the QD ground state under injection of resonantly created heavy hole excitons in the WL is confirmed to be a specific property of exciton injection and is not related to the differing absorption efficiency of WL free carrier continuum and WL free exciton resonance leading to differing carrier densities in the QD structures. Thus we attribute this effect to accelerated spin relaxation of the excitons promoted by the Coulomb exchange interaction between the carriers.

References

[1] For a recent review, see e.g. Dyakonov M I 2008 *Spin Physics of Semiconductors* (Berlin: Springer)
[2] Heidemeyer H, Denker U, Müller C and Schmidt O G 2003 *Phys. Rev. Lett.* **91** 196103
[3] Holub M and Bhattacharya P 2007 *J. Phys. D: Appl. Phys.* **40** R179
[4] Dawson P, Rubel O, Baranovskii S D, Pierz K, Thomas P and Göbel E O 2005 *Phys. Rev. B* **72** 235301
[5] Beyer J, Buyanova I A, Suraprapapich S, Tu C W and Chen W M 2009 *Nanotechnology* **20** 375401
[6] Suraprapapich S, Panyakeow S and Tu C W 2007 *Appl. Phys. Lett.* **90** 183112
[7] Paillard M, Marie X, Renucci P, Amand T, Jbeli A and Gérard J M 2001 *Phys. Rev. Lett.* **86** 1634
[8] Bayer M, Kuther A, Forchel A, Gorbunov A, Timofeev V B, Schäfer F, Reithmaier J P, Reinecke T L and Walck S N 1999 *Phys. Rev. Lett.* **82** 1748
[9] Bester G, Nair S and Zunger A 2003 *Phys. Rev. B* **67** 161306
[10] Tartakovskii A I *et al.* 2004 *Phys. Rev. Lett.* **93** 057401
[11] Bayer M *et al.* 2002 *Phys. Rev. B* **65** 195315
[12] Landin L, Miller M S, Pistol M-E, Pryor C E and Samuelson L 1998 *Science* **280** 262
[13] Lai C W, Maletinsky P, Badolato A and Imamoglu A 2006 *Phys. Rev. Lett.* **96** 167403
[14] Belhadj T, Simon C-M, Amand T, Renucci P, Chatel B, Krebs O, Lemaître A, Voisin P, Marie X and Urbaszek B 2009 *Phys. Rev. Lett.* **103** 086601
[15] Kalevich V K *et al.* 2001 *Phys. Rev. B* **64** 045309

Resonance effects in Raman scattering of quantum dots formed by the Langmuir-Blodgett method

A G Milekhin[1,2], L L Sveshnikova[1], T A Duda[1], N V Surovtsev[3], S V Adichtchev[3], Yu M Azhniuk[4], C Himcinschi[5], M Kehr[6] and D R T Zahn[6]

[1]Institute of Semiconductor Physics, Lavrentjev av.13, 630090, Novosibirsk, Russia
[2]Novosibirsk State University, Pirogov av.2, 630090, Novosibirsk, Russia
[3]Institute of Automation and Electrometry, Koptyug av.1, 630090, Novosibirsk, Russia
[4]Institute of Electron Physics, Universytetska Str. 21, 88017, Uzhhorod, Ukraine
[5]Institut für Theoretische Physik, TU Bergakademie Freiberg, Leipziger Str. 23, 09596, Freiberg, Germany
[6]Semiconductor Physics, Chemnitz University of Technology, Chemnitz, Germany

E-mail: milekhin@thermo.isp.nsc.ru

Abstract. The enhancement of Raman scattering by optical phonon modes in quantum dots was achieved in resonant and surface-enhanced Raman scattering experiments by approaching the laser energy to the energy of either the interband transitions or the localized surface plasmons in silver nanoclusters deposited onto the nanostructures. Resonant Raman scattering by TO, LO, and SO phonons as well as their overtones was observed for PbS, ZnS, and ZnO quantum dots while enhancement for LO and SO modes in CdS quantum dots with a factor of about 700 was measured in surface enhanced Raman scattering experiments. Multiple phonon Raman scattering observed up to 5th and 7th order for CdS and ZnO, respectively, confirms the high crystalline quality of the grown QDs.

1. Introduction

The Langmuir-Blodgett (LB) technique is considered as a versatile, relatively simple, fast, and low-cost method for the fabrication of model quantum dots of a variety of materials with controllable areal density on different substrates. Most of the published data on the phonon spectra of QDs were obtained by measuring Raman scattering in QD arrays since the small spatial volume of a single QD and thus the small Raman cross-section leads to extremely low Raman efficiencies. However, the determination of the phonon spectrum in a single QD is one of the most exciting tasks of solid state spectroscopy. To overcome the challenge of the low signal in QD arrays of a low areal density appropriate approaches for increasing the Raman efficiency are required. One of the possibilities is to employ resonant Raman spectroscopy when the energy of the incident laser light matches that of some electronic transitions (*e.g.* interband transitions) in QDs. The second is to use the conditions of surface enhanced Raman scattering (SERS) when the energy of the incident laser light is close to that of surface plasmon in noble metal clusters intentionally fabricated in the vicinity of QDs. Of course, the combination of both methods (resonant SERS) is considered as the most beneficial approach.

In this paper we present a study of Raman scattering by optical phonons (including transverse, longitudinal, and surface optical (TO, LO, and SO) phonons and their overtones) in nanostructures with (Cd, Zn, Pb)S and ZnO QDs enhanced under various resonant conditions.

2. Experimental

The LB technology was used to prepare high quality films of Cd, Zn, and Pb behenates on quartz and Si substrates covered by a 100 nm thick Au or Pt layer. The thickness of LB films under investigation was 400 monolayers. The reaction of the behenates with gaseous H_2S results in the formation of CdS, ZnS, and PbS QDs. Free standing QDs were obtained after removing the organic matrix by thermal annealing in an inert atmosphere at temperatures between 100 and 200°C. Free standing ZnO QDs were formed as a result of annealing of either Zn behenate films or ZnS QDs at 600°C. Ag nanoclusters films were deposited on the surface of CdS QD also using the LB technology [1].

High-resolution transmission electron (HRTEM) experiments were performed using an electron microscope JEM-400EX (JEOL). Scanning electron microscopy (SEM) experiments were carried out using a scanning electron microscope FIB, 1540XB, Cross Beam, Carl Zeiss. X-ray diffraction with a fixed incidence angle was used for determining the crystallographic structure of the QDs. The measurements were carried out using a Seifert XRD3000 PTS diffractometer. A line focus Cu-X-ray tube operated at 40 kV and a parabolic multilayer mirror were employed to produce a parallel beam of the wavelength 1.54 Å (Cu-K_α). UV-vis Absorption spectra were recorded using a UV spectrometer Shimadzu UV-31000 in the wavelength range of 200-800 nm.

Raman experiments were carried out using Dilor XY800 and T64000 spectrometers in back-scattering geometry analysing non-polarized scattered light. Ar^+ and Kr^+ and HeCd lasers were used as excitation sources in the wavelength range from 752.5 to 325 nm. In the macro geometry the incident laser light with power of 100 mW or less was focused by a cylindrical lens to avoid sample overheating.

3. Results and discussion

The size of the QDs determined from the HRTEM and SEM images is in the range of 2-6 nm. The shape of the QDs was found to be predominantly spherical. These data are in accordance with our previous results obtained for CdS, ZnS, and PbS QDs [2].

All reflection peaks of PbS and ZnS QDs seen in XRD pattern (Fig.1a and b) can be indexed as a cubic structure with cell parameters a=5.9362 Å and a=5.406 Å, respectively, which is in good agreement with previous literature data (JCPDS Cards Nos. 05-0592 and 05-0566). The peaks observed for ZnO QDs (Fig.1c) are assigned to the hexagonal (Card No. 36-1451) structure. The hexagonal structure was also evidenced in our earlier study of CdS QDs [1].

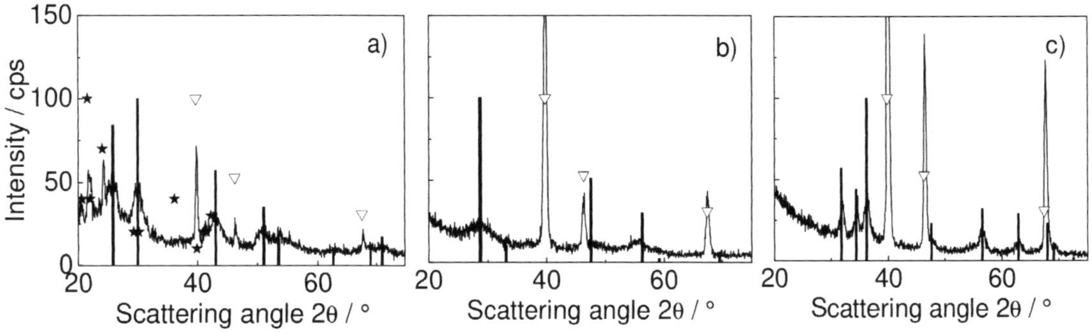

Figure 1. XRD patterns of a) PbS, b) ZnS, c) ZnO QDs. The peak positions of the observed phases are displayed with solid vertical lines while triangles and stars mark the peaks originating from the Pt layer beneath the QDs and the rest of behenic acid, respectively.

Fig.2 shows absorption spectra of Ag nanoclusters and QDs after removal of the organic matrices. The positions of the absorption edge in the QDs indicated by arrows correspond to the interband transition energies and cover the spectral range from red to UV. The transition energies for PbS, CdS, and ZnS QDs exceed the bandgap energies for the corresponding bulk materials (0.4, 2.5, and 3.7 eV, respectively) due to confinement effect. The transition energy for ZnO QDs is close to the bandgap energies of the bulk materials (3.3 eV). According to Fig.2 resonant Raman scattering by optical phonons is expected to be observed from red (for PbS QDs) through blue (for CdS QDs) to UV spectral region (ZnS and ZnO QDs).

Fig.3a shows Raman scattering spectra of PbS QDs measured at 30 K after removal of the organic matrix. The peaks observed at 71 and 212 cm^{-1} correspond to the Raman scattering by TO and LO phonons at Γ point of the Brillouin zone (BZ). The frequency position of the TO phonon line shows no remarkable dependence on the excitation energy while the line shape of LO phonon varies significantly for excitation wavelengths from 514.5 to 647.1 nm (where interband absorption in PbS QDs occurs (Fig.2)) due to the appearance of a new additional peak at 223 cm^{-1}. This feature is assigned to scattering by LO phonons at the Δ point of the BZ which are normally prohibited for PbS but occurs near the interband resonance in PbS QDs. A 5-fold increase of the intensity of first order

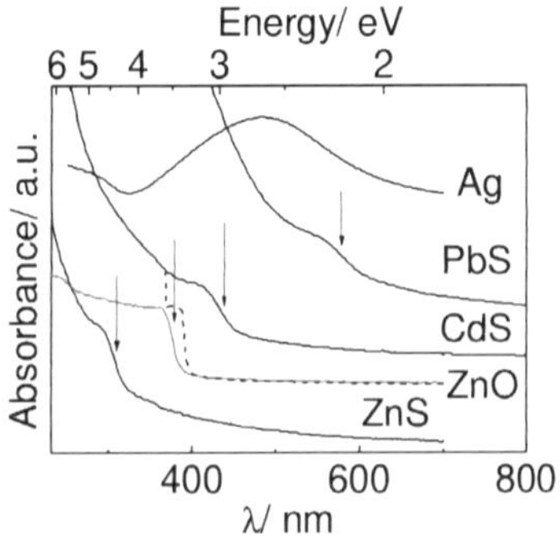

Figure 2. Absorption spectra of the structures with Ag nanoclusters and (Pb,Cd,Zn)S and ZnO QDs. The spectra are shifted and scaled for clarity. The energy of the interband transitions in the QDs is denoted by vertical arrows. For comparison the dashed line shows the spectrum of a ZnO bulk crystal. The maximum of absorption of the Ag nanoclusters corresponds to the energy of localized surface plasmons in the nanoclusters.

Raman scattering by optical phonons is observed in the region of 647.1÷676.4 nm which is below the interband absorption (not shown here). The frequency position of other phonon modes seen in the Raman spectrum and their assignment based on the comparison with literature data [3,4] taking into account temperature shift of 6 cm^{-1} are presented in Table 1. Moreover, the following tentative assignment is performed for the observed multiple-phonon features:

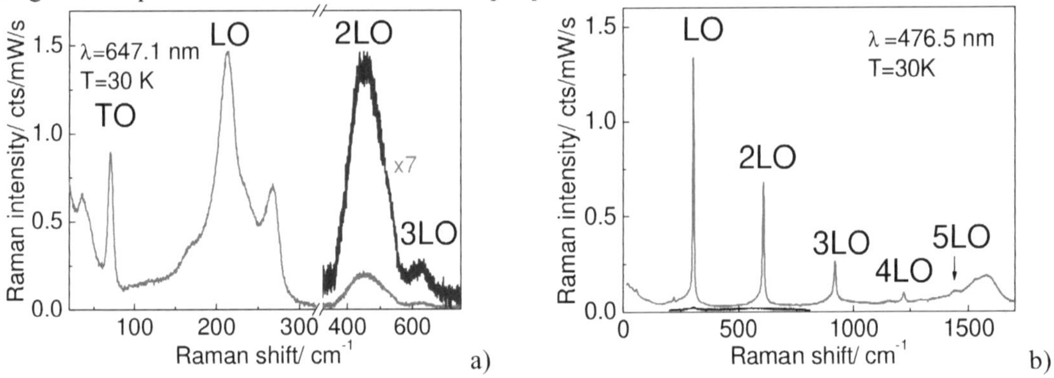

Figure 3. a) Experimental Raman spectrum of PbS QDs. b) Raman and SERS spectra of CdS QDs.

The mode at 173 cm^{-1} is attributed to the LA+TA combinational mode at L (118+55=173) cm^{-1}. The assignment of the mode at 267 cm^{-1} to 2TO phonon at Σ (2x135=270) cm^{-1} is based on the calculation of the phonon density of states [5]. A broad band centered at 450 cm^{-1} can be fitted by three lorentzian curves with maxima at about 425, 450, and 500 cm^{-1} due to 2LO phonon scattering at Γ (2x212=424), Δ (2x223=446), L (2x241=482) cm^{-1} or 2LO+TO at Γ (2x212+71=495) cm^{-1}, respectively. The mode at 636 cm^{-1} is the third-order Raman mode (3LO along Γ (3x212=636) cm^{-1}).

Table 1. Assignment and frequencies of phonon modes observed in Raman spectra of PbS QDs (in cm^{-1}).

phonon	TA	TA	LA	TO	TO	LO	LA	TO	LO	LO	LO
BZ point	X	L	X	Γ	X	X	L	L	Γ	Δ	L
previous data [3]	46	55	59.7	71.4	94.7	98.7	107.7	200.4	210.8	230[4]	243.8
Present data	38	-	-	71	-	-	118	-	212	223	241

Raman spectra of CdS QDs reveal a resonant behaviour at 441 nm (not shown here) and demonstrate 10-fold increase of the intensity of the LO phonon line at 300 cm^{-1} with respect to the non-resonant conditions. Deposition of Ag clusters onto the sample surfaces leads to appearing the multiple LO phonon modes up to 5th order and a strong enhancement of the LO phonon modes with a maximum at the energy position of the surface plasmons localized in Ag clusters (Fig.3b). Details of the discussion on SERS in CdS QDs are reported in [6]. By adjusting the plasmon energy to that of the laser light and the interband electronic transitions in CdS QDs the conditions of resonant SERS were realized.

LO phonons and their overtones (2LO and 3LO) as well as combinational modes (LA+TA along X point of BZ) were observed in Raman spectra of ZnS QDs measured with λ=325 nm which is in the vicinity of the interband transitions in the QDs (Fig.4a). Note that in resonance the intensity of the LO phonon mode of ZnS QDs at 350 cm^{-1} is two orders of magnitude larger than that in the non-resonant conditions (*e.g.* for λ=488 nm as it is shown in Fig.4a).

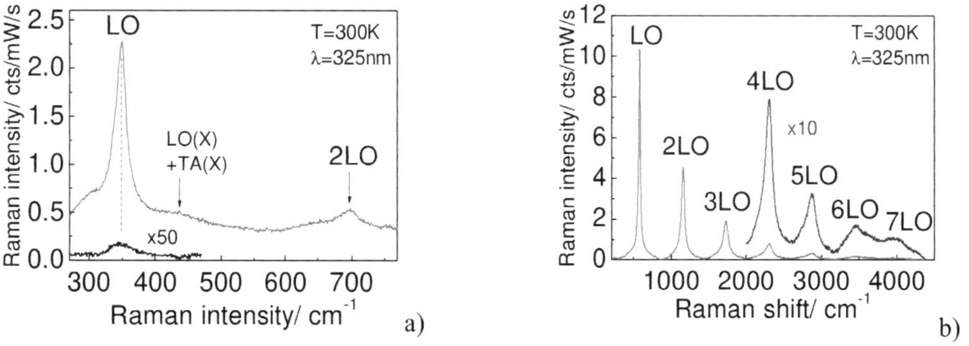

Figure 4. Resonant UV Raman spectra of a) ZnS and b) ZnO QDs measured with λ=325 nm using a microscope (laser power of 3 mW). Non-resonant Raman spectrum of ZnS QDs measured with λ=488 nm is shown for comparison.

In the wurtzite ZnO QDs the resonant and non-resonant Raman spectra differ significantly not only in the sense of the mode intensity but also in the mode symmetry. The most pronounced phonon mode seen in the non-resonant Raman spectra of the QDs is the E_2 phonon mode at 440 cm^{-1} (not shown here) while a mode at 577 cm^{-1} and its overtones (up to 7th order) are observed in the resonance with the interband transitions in the QDs (Fig.4b). This mode has a frequency intermediate between those of A_1(LO) and E_1(LO) phonons (574 and 591 cm^{-1}, respectively) and can be referred to confined LO phonons in the QDs according to theoretical calculations [7]. It is worth mentioning that the heating effect causes no frequency shift of the phonon modes at the laser power used in the experiment.

4. Conclusion

We demonstrate a significant enhancement of Raman scattering by optical phonon modes achieved in resonant and surface-enhanced Raman scattering experiments in semiconductor QD arrays formed by LB technology. Multiple phonon Raman scattering up to seventh order observed at room temperature confirms a high crystalline quality of the grown QDs and evidences a strong electron-phonon interaction in the QDs.

Acknowledgments

This work was supported in part by the Russian Foundation for Basic Research (grants 09-02-00458_a, 09-02-92007-ННС_a, 09-02-92650-ИНД_a), Siberian Branch of Russian Academy of Sciences (joint grant with NAS of the Republic of Belarus №.9), Deutsche Deutsche Forschungsgemeinschaft (grant Za146/21-1) and the International Research Training Group (Internationales Graduiertenkolleg, GRK 1215) "Materials and Concepts for Advanced Interconnects".

5. References

[1] Milekhin A G, Sveshnikova L L, Duda T A, Surovtsev N V, Adichtchev S V and Zahn D R T, 2008, *JETP Letters* **88** 799

[2] Milekhin A G, Sveshnikova L L, Repinski S M, Gutakovski A K, Friedrich M and Zahn D R T 2002 *Physics of the Solid State* **44** 1976

[3] *Landolt-Bornstein Tables* 1999 ed. O Madelung, New Series, Group III, v. 41B (Berlin: Springer-Verlag)

[4] Etchegoin P G, Cardona M, Lauck R, Clark R J H, Serrano J and Romero AH 2008 *Phys. Stat. Sol. (b)* **245** 1125

[5] Romero A H, Cardona M, Kremer R K, Lauck R, Siegle G, Serrano J and Gonze X C 2008 *Phys. Rev. B* **78** 224302

[6] Milekhin A G, Sveshnikova L L, Duda T, Surovtsev N V, Adichtchev SV and Zahn D R T 2009 *Proc. Int. Workshop and Tutorials on Electron Devices and Materials*, Erlagol **5173943** 112

[7] Fonoberov V A and Balandin A A 2004 *Phys. Rev. B* **70** 233205

Ab Initio Study of functionalized 1 nm Silicon Nanoparticles

M. E. Ahmed, J. P. Goss, R. J. Eyre, P. R. Briddon, M. A. Taylforth

School of Electrical, Electronic and Computer Engineering, Newcastle University, Newcastle upon Tyne, NE1 7RU, England, United Kingdom

E-mail: M.E.M.Ahmed@ncl.ac.uk

Abstract. Functionalization of silicon nanoclusters reveals electronic and optical phenomena that can be utilized in a range of applications, including optical sensing, biological imaging and optoelectronic devices. Using density functional theory calculations, light-absorption and luminescence processes are modelled at the quantum mechanical level. In this study, a number of chemical functional groups are attached to the surface of silicon quantum dots (Si-QDs) of ~1 nm diameter via either C_4H_8 or C_8H_{16} alkane chains. In these results, the impact of the functional groups upon the electronic structure and optical absorption spectra of composite systems depend upon the chemical nature of the functional group and the alkane chain length.

1. Introduction

In recent years, silicon nanocrystalline materials have become an interesting area in physics, chemistry, material science and biophysics. Various applications have resulted in biomedical applications, semiconductor nanomaterials and quantum dot devices[1, 2].

According to experimental observations and theoretical predictions, the properties of large Si-QDs are close to bulk silicon material in terms of the energy gap and emission wave length [3]. However, when Si-QDs have a diameter less than 5 nm, PL emission is in the visible range with high intensity. Quantum confinement is a key factor, but because there are up to around 50% of the atoms at the surface, the role of surface states is likely to be important.

Termination using hydrocarbon chains is an effective way of stabilizing the surface against oxidation [4]. Furthermore, functionalization of Si-QDs with different chemically active molecules is currently being investigated for possible applications as luminescent labels in biological applications, since the intense luminescence of Si-QDs are at wavelengths where biological systems do not absorb strongly. Theoretically, the role of surface functional groups with regards to their optical properties is an area still in development. For complete coverage, as the polarity of the Si-QD interface increases, the optical band-gap becomes increasingly affected by charge transfer at the interface rather than by quantum confinement [5].

Previous studies of surface functionalization largely involved functional groups bonded directly to the Si-QD rather than at the end of the long alkane chains. In this paper we present a quantum chemical study into the impact upon the electronic and optical properties of Si-QDs for various functional groups which terminate the alkane chains bonded to the Si-QD surface.

2. Computational Method

Calculations were carried out using the density functional technique, which is implemented in AIMPRO code (ab initio modelling program) [6, 7]. All calculations presented here use the generalized gradient approximation [8] for total energies and forces, and core electrons are treated using norm-conserving pseudo potentials [9].

The Si-QD is modeled starting from tetrahedral, hydrogen-terminated cluster, $Si_{35}H_{36}$, which is illustrated in Fig. 1. The non-functionalized cluster was first structurally relaxed, and the resulting geometry analyzed for its electronic structure and optical absorption. The main features of the electronic structures are the energies of the electronic-states, which represent the highest occupied molecular orbitals (HOMOs) and lowest unoccupied molecular orbitals (LUMOs). Optical absorption spectra are obtained from the calculated complex dielectric function [10]. Each electron state is assigned an width using a polynomial broadening, characterized by a width of 0.1 eV.

For each functional group, one of the surface hydrogen atoms of the Si-QD is substituted by a linear alkane chain, and the functional group is attached to the other end of this chain. In all cases there are either four or eight methylene units (CH_2) between the "functional group" and the Si-QD, as shown in Fig. 1. The functional groups explored in this study are listed in table 1. To determine how significant the impact that functionalization has upon electronic and optical properties of the Si-QDs, in addition to varying the chemical functional group, we change the length of alkane chain from C_4H_8 to C_8H_{16} with every functional group.

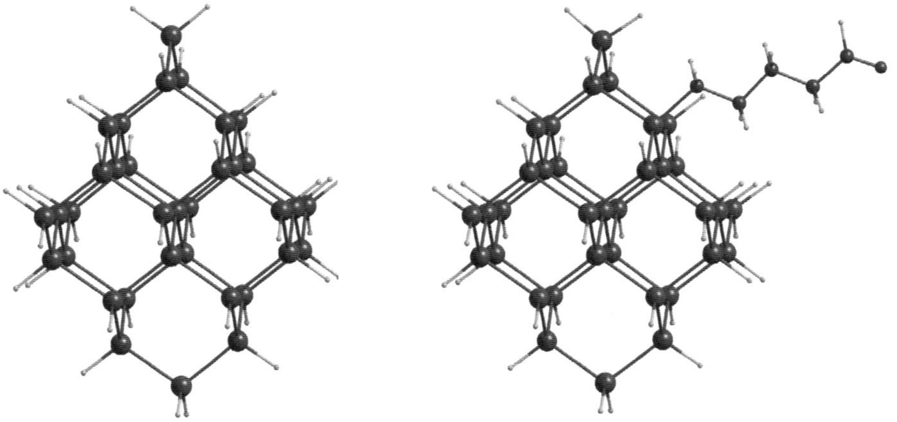

Figure 1. Schematic showing the fully-hydrogenated silicon dot $Si_{35}H_{36}$, The left side is undecorated fully-hydrogenated silicon dot, while the right side is the Si-QD after the functionalization with a single functional group. Pink, white, grey and red atoms are silicon, hydrogen, carbon and oxygen, respectively.

3. Results

Calculations for Kohn-Sham energy levels, and the electronic spectra are given for functionalized and non-functionalized Si-QD systems. They can be viewed from the HOMOs and LUMOs energy levels with the associated optical gap, as illustrated in table 1.

The results divided into three main groups depending upon the electronic states introduced into the optical gap. Firstly, group(A) which contains alcohol, alkane, alkene, nitride, carbonic acid chloride, and halogens functional groups. As in table 1, in this group(A), the attached functional groups to the Si-QDs introduce no new electronic state into the optical gap. Fig. 2

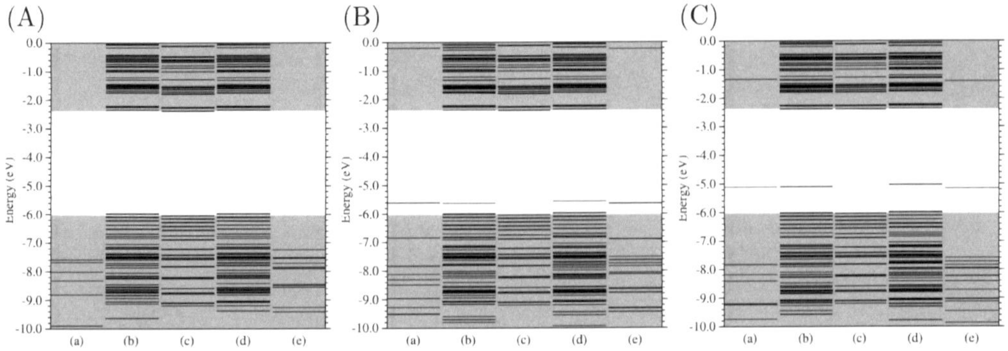

Figure 2. The Kohn-Sham levels for 1 nm Si-QD of (A), (B) and (C) for alkane, carbonic-acid and aldehyde respectively, when (a), molecule with four carbon chain length (b) Si-QD attached to four carbon chain length, (c) undecorated H-terminated Si-QD, (d) molecule with eight carbon chain length and (d) molecule with eight carbon chain length. The shaded regions indicate the position of the bands for undecorated Si-QD (c).

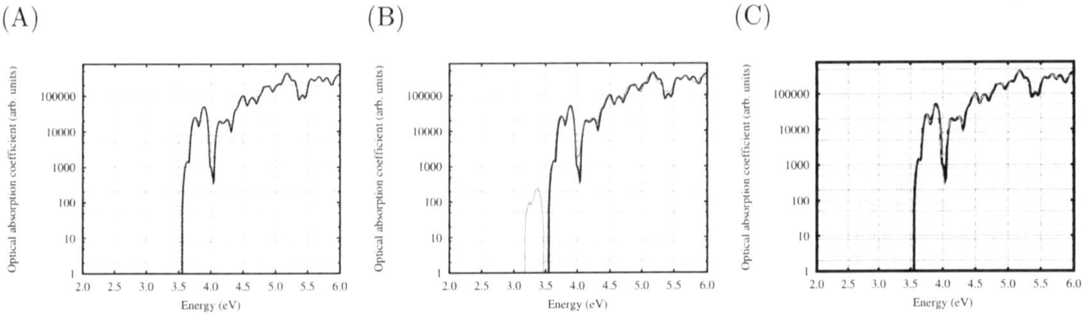

Figure 3. Plot of the calculated optical absorption spectra corresponding to the undecorated hydrogen terminated Si-QD (black full line) in three graphs, (A) alkane functional group on silicon quantum dot (purple line), (B) carboxylic acid functional group (red line) and (C) aldehyde functional group (dashed green line).

(A) is represents alkane functional group Kohn-Sham energy levels as an example of this group. From the plot we can see that the chain of eight alkane chain has no new electronic state in the gap. This confirms that the length of alkane chain does not affect the states in the gap. At the same time, from Fig.3(A), the optical absorption spectrum for the same alkane functional group, produces no difference in terms of optical absorption coefficient compared to the undecorated Si-QD.

Secondly in group (B), which includes alkyne, arene and carboxylic acid. Shallow gap-states are introduced just above the HOMO of the Si-QD. Carboxylic acid functional group (carboxylic acid contains a carbon-oxygen double bond and carbon-oxygen-hydrogen single bonds) is chosen as an illustration to represent this group in Fig. 2 (B). From Fig. 3 (B), we can see a minor shift in absorption spectra.

Finally, group(C) which contains amide, thiole, amine, aldehyde, sulfide and ferrocene. We can see deep electronic states introduced into the optical gap. As a clear example, Fig. 2 (C) shows the electronic structure of aldehyde functionalized Si-QD (aldehyde contains a carbon-oxygen double bond). Here, a electronic state localised on the functional group lies *well above* the HOMO of the Si-QD. In Fig. 2 (C) we can see that the electronic state in HOMO-LUMO gap corresponding with optical energy shift in the optical absorption by eV in Fig. 3 (C).

Table 1. Electronic structure data contain HOMO and LUMO energies for various functional groups attaches to the Si-QD via (C_4H_8 and C_8H_{16}). HOMO and LUMO energies are in eV relative to the vacuum level at zero. The groups are ranked by increasing energy of the HOMO and divided into the three groups as indicated in the text.

			C_4H_8		C_8H_{16}	
Category	Family	Functional group	HOMO	LUMO	HOMO	LUMO
	Pure QD			-2.39	-6.03	
A	Nitrile	(CN)	-6.10	-2.49	-6.05	-2.44
	Carbonic acid chloride	(COCl)	-6.06	-2.44	-6.02	-2.41
	Chloride	(Cl)	-6.04	-2.43	-6.01	-2.39
	Bromide	(Br)	-6.03	-2.42	-6.00	-2.39
	Fluoride	(F)	-6.03	-2.42	-6.00	-2.39
	Alcohol	(OH)	-5.99	-2.38	-5.95	-2.37
	Alkene	(C_2H_3)	-5.98	-2.37	-5.94	-2.37
	Alkane	(H)	-5.97	-2.36	-5.96	-2.35
B	Alkyne	(C_2H_1)	-5.78	-2.38	-5.73	-2.37
	Arene	(C_6H_5))	-5.66	-2.37	-5.61	-2.37
	Carboxylic acid	(C(OH)O)	-5.62	-2.38	-5.55	-2.36
C	Amide	($CONH_2$)	-5.51	-2.40	-5.45	-2.39
	Thiol	(SH)	-5.50	-2.41	-5.45	-2.40
	Amine	(N_2)	-5.44	-2.37	-5.36	-2.35
	Aldehyde	(CHO)	-5.11	-2.39	-5.04	-2.38
	Sulfide	(SCH_3)	-5.02	-2.40	-4.96	-2.39
	Ferrocene	($(C_5H_5)_2Fe$)	-4.07	-2.37	-4.01	-2.37

4. Discussion and Conclusions

The presence and location of electronic states in the band-gap of the 1 nm Si-QD is strongly dependent upon the chemical nature of the functional group. In our calculations the increase in the chain length from four to eight carbon atoms results in only a small perturbation of the electronic structure, but dramatically reduces any sub-band-gap optical absorption due to the associated decrease in the overlap between the HOMO and LUMO molecular orbitals. The role of oxidation is the subject of future investigations.

References

[1] Jiang C and Green MA, 2006 *J. Appl. Phys.* **99** 114902
[2] Jang H, Pell LE, Korgel BA and English DS, 2003 *J. Photochem. Pthoboil. A, Chem.* **158** 111
[3] Wang B, Chou Y. Deng J and Dung Y, 2008 *J. Phys. Chem. A* **112** 6351
[4] Mangolinil L, Jurbergs D, Rogojina E and Kortshagen U, 2006 *Phys. Status Solidi C* **11** 3975
[5] König D, Rudd J, Conibeer G and Green MA, 2009 *Mater. Sci. Eng. B* **159–160** 117–121
[6] Briddon PR and Jones R, 2000 *Phys. Status Solidi B* **217** 131–171
[7] Rayson MJ and Briddon PR, 2008 *Computer Phys. Comm.* **178** 128–134
[8] Perdew JP, Burke K and Ernzerhof M, 1996 *Phys. Rev. Lett.* **77** 3865–3868
[9] Hartwigsen C, Goedecker S and Hutter J, 1998 *Phys. Rev. B* **58** 3641–3662
[10] Fall CJ, Blumenau AT, Jones R, Briddon PR, Frauenheim T, Gutiérrez-Sosa A, Bangert U, Mora AE, Steeds JW and Butler JE, 2002 *Phys. Rev. B* **65** 205206

Journal of Physics: Conference Series **245** (2010) 012047 doi:10.1088/1742-6596/245/1/012047

Tunnel injection structures based on InGaAs/GaAs quantum dots: optical properties and energy structure

W Rudno-Rudziński[1,3], J Andrzejewski[1], G Sęk[1], M Syperek[1], J Misiewicz[1], EM Pavelescu[2], C Gilfert[2] and JP Reithmaier[2]
[1] Institute of Physics, Wroclaw University of Technology, Wybrzeże Wyspiańskiego 27, 50-370 Wrocław, Poland
[2] Institute of Nanostructure Technologies and Analytics, Technische Physik, Universitaet Kassel, Heinrich-Plett-Str. 40, 34132 Kassel, Germany

E-mail: wojciech.rudno-rudzinski@pwr.wroc.pl

Abstract. The limited speed of direct modulation due to the significant population of hot carriers in quantum dots (QD) is a major drawback of the application of QD based lasers in fibre optics telecommunication. One of the most promising methods of alleviating this problem is the design of tunnel injection (TI) structures, where already cold carries are injected by means of tunnelling through a thin barrier from an adjacent quantum well (QW) directly to the QD ground state. We have investigated the properties of TI structures consisting of an $In_xGa_{1-x}As$ quantum well and a layer of self-assembled $In_{0.6}Ga_{0.4}As/GaAs$ quantum dots versus the properties a reference QD sample. Photoreflectance spectroscopy is applied to determine the energies of optical transitions in this complex system. The obtained energies are then used to verify the reliability of the calculations in the 8 band *kp* model, which take into account the realistic geometry of the dots, influence of the strain and the coupling between the dot layer and the injector quantum well. Finally, time resolved photoluminescence (PL) experiment is performed at low temperature and its results are related to the acquired structure of confined levels. The influence of the tunnelling process on PL rise and decay times is explained.

1. Introduction

Layers of self-assembled quantum dots (QD) are already used as active material in semiconductor lasers. Devices that are based on them offer, as was both predicted theoretically and confirmed experimentally, improvements to some of the operating parameters, when compared to still more common quantum well (QW) lasers. This concerns a low threshold current [1, 2] and high temperature insensitivity [3-5]. Due to the large inhomogeneous broadening of QD emission they can also be tuned in a wide range of wavelengths [6]. However, there are still problems that need to be overcome to push the performance characteristics of the QD lasers further, for instance, the existence of the large population of hot carriers occupying QD excited states or wetting layer, which limits the speed of modulation. In order to deal with that issue, tunnel injection (TI) [7-10] system has been proposed, where an additional QW, separated by a thin barrier from QD layer, is put into the structure. If the QW is properly designed, the lowest lying levels confined in it, especially electron levels, should have

[3] To whom any correspondence should be addressed.

energies similar to the energies of electrons confined in QDs. Then the carriers trapped by the QW would very fast thermalise to the QW ground state and tunnel directly to the QD ground state, either resonantly or with the assistance of phonons, usually lateral optical. This process should be much faster than the thermalisation in a QD itself, which can suffer from the phonon bottleneck problem. Moreover, due to a bigger volume, QW can trap more carriers than QDs and wetting layer, thus the TI structure can have higher efficiency of emission.

In this paper we use photoreflectance (PR) spectroscopy, which is one of the absorption-like modulation techniques proven highly effective for the investigations of low dimensional semiconductor structures [11, 12], even complex ones, such as TI structures [10, 13], to study the optical transitions of three samples, two consisting of one layer of $In_{0.6}Ga_{0.4}As$ QDs separated by a thin barrier from $In_{1-x}Ga_xAs$ QW and a reference sample without the injector QW. The measurement results are compared with the results of numerical calculations in an 8-band *kp* model (similarly as in Ref. [14]), which takes into account the realistic geometry of the dots and strain distribution. The calculations are necessary in order to determine the energies of levels confined in the system, since the experiment gives only the total energy of transition. Finally, a time resolved photoluminescence (TRPL) experiment is carried out and the determined rise and decay times of emission are discussed in the light of established energy structure.

2. Experiment

2.1. Investigated structures

The samples are molecular-beam-epitaxy grown on an *n* doped GaAs (100) substrate. The reference sample contains a layer of $In_{0.6}Ga_{0.4}As$ QDs with the nominal thickness of 1.8 nm, surrounded by 15 nm of GaAs and embedded in an AlGaAs/GaAs superlattice. In TI structures A an B, beside the 15 nm GaAs layer and the layer of QDs grown in the same conditions as in the reference structure, an additional injector QW is inserted, separated by a 2 nm GaAs barrier from the QDs layer. In the case of sample A, this QW consists of a 15 nm thick $In_{0.2}Ga_{0.8}As$ material, in the sample B the QW is 7 nm wide, with the 30 % of In.

2.2. Experimental setup

Photoreflectance is measured in a standard bright configuration setup, with the pump beam provided by the 532 nm line of a frequency doubled yttrium-aluminium-garnet laser, the probe beam obtained from a tungsten halogen lamp, and the detection supplied by an InGaAs photodiode combined with a 0.55 m focal length monochromator. Further details can be found in [15]. For the time resolved PL measurements, the 800 nm line from a high power mode-locked Ti:Sapphire laser is used as a nonresonant excitation source and a Hamamatsu streak camera equipped with a Peltier cooled S1 photocathode and combined with a 0.25 m focal length monochromator as a detection system. Low temperature measurements are carried out in a continuous-flow helium cryostat.

3. Experimental results and calculations

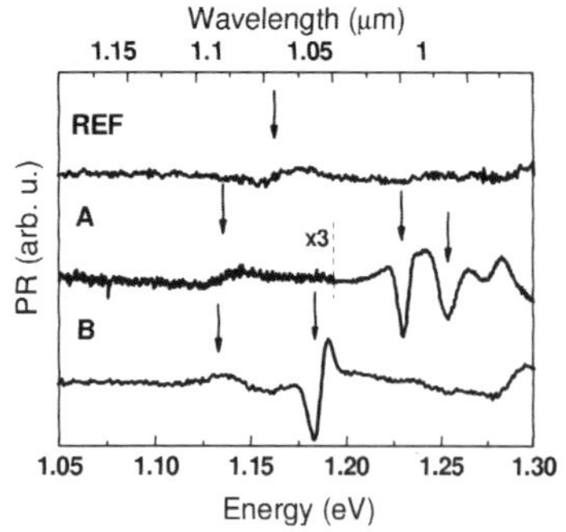

Figure 1. Room temperature PR spectra of reference QD sample and TI samples A and B in the region of lowest energy transitions. Arrows indicate the energies of optical transitions relevant for this paper.

Journal of Physics: Conference Series **245** (2010) 012047 doi:10.1088/1742-6596/245/1/012047

Figure 1 shows PR spectra measured at room temperature for all samples, at the region of lowest energy transition. In the reference sample only one weak (due to small total volume of QDs) and broad (due to the broad distribution of QDs sizes and contents) transition can be seen, attributed to the QD ground state. In TI samples, beside analogical transition, slightly shifted in energy because of a different strain distribution governing Stranski-Krastanov growth mode and coupling with injector QW, more PR resonances at higher energies are visible, attributed to the transitions between levels confined in the complex QW+QD system. In this paper we will concentrate only on two QW+QD transitions in sample A and one in sample B, because they are primarily involved in the tunnelling process.

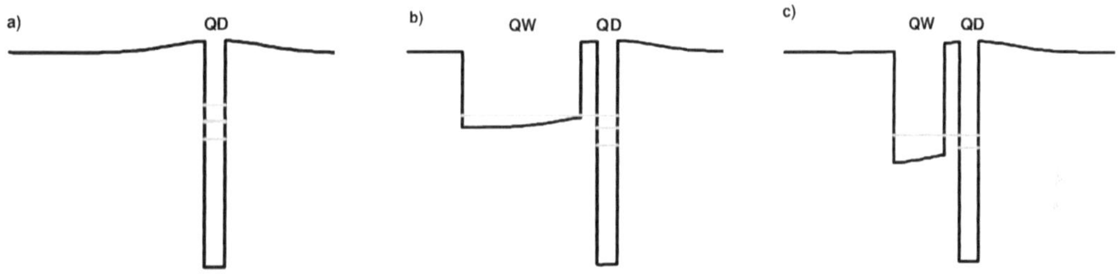

Figure 2. Calculated cross-sections of realistic confining potentials in conduction band (including the effect of strain) in the growth direction, at room temperature, for a) reference sample, b) TI sample A and c) TI sample B. Horizontal lines indicate energies of electron levels discussed in the paper.

In the next step the 8 band *kp* model is used in order to calculate the energy structure of investigated samples. The details of the model can be found in [14], material parameters were taken after [16]. Figure 2 presents calculated energies of a few lowest lying electron levels confined in the conduction band potential, projected onto the cross-section of the potential in the growth direction, with the strain effect included. The reference QDs were used as a test of the model. Good agreement between calculated transition energies and energies obtained from high excitation PL measurements (not shown here) has been obtained. Only electron levels are discussed, since there are many hole levels confined, and separation between consecutive levels is on the order of a single meV, thus at room temperature holes can be treated as having a quasi-continuous band. The results obtained for TI structures indicate that in the sample A first two electron levels are confined in QD, only the third level is smeared over entire QW+QD system, whereas in the sample B it is the case of already the second one.

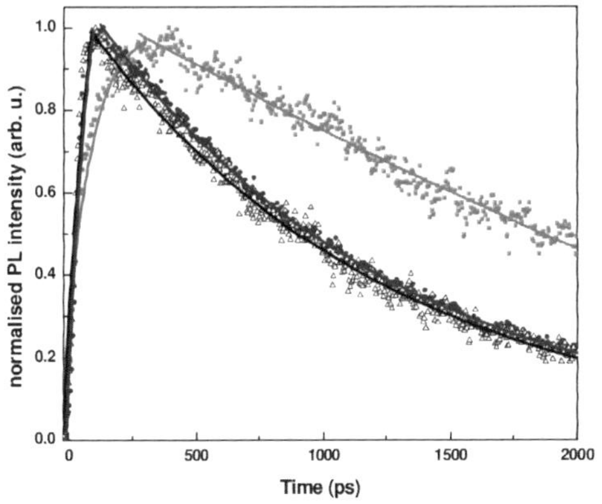

Figure 3. Time-resolved PL spectra for reference sample (open triangles), TI sample A (filled circles) and TI sample A (filled squares), measured at 5 K, with moderate excitation power (100 μW). Solid lines are fits by the exponential decay formulae.

Finally, the time resolved photoluminescence was measured at low temperature, at moderate excitation power density of 80 mW/cm^2. In order to determine the rise and decay times of emission, fits with exponential function were performed. The results for the reference sample are ~40 ps rise time (with the setup response time ~30 ps) and 1.2 ns decay time, which is typical for such QDs. Surprisingly, almost identical times were obtained for TI structure A. On the other hand, both the rise time and the decay time for sample B were found to be longer (~100 ps and ~2 ns, respectively). This difference can be explained by the determined energy structure. In the TI sample B the tunnelling occurs to the QD ground state, thus it determines the relevant times of emission, because QW feeds the carriers directly to the emitting state. In the sample A the tunnelling occurs to higher state in the dot, thus the rise and decay times are governed by the relaxation processes within the dot, which are approximately the same as in the reference sample.

4. Conclusions

Photoreflectance measurement results, followed by the calculations in an 8 band kp model involving realistic dot geometry and strain distribution, lead to the determination of the energy structure of TI samples, containing an InGaAs QW and a layer of self-assembled InAs/GaAs QDs. The obtained diagram of confined energy levels was used to explain the differences in rise and decay times between TI structures with various injector wells and QD reference sample. It was found that the time characteristics of emission depend on the fact, if the tunnelling occurs directly to QDs ground state or to QDs excited state.

Acknowledgments
This work has been supported by the DeLight project no 224366, 7th Framework Programme of the European Commission.

References

[1] Eliseev PG, Li H, Liu T, Newell TC, Lester LF and Malloy KJ 2001 IEEE Journal of Selected Topics in Quantum Electronics 7 135
[2] Ledentsov NN, Kovsh AR, Zhukov AE, Maleev NA, Mikhrin SS, Vasil'ev AP, Semenova ES, Maximov MV, Shernyakov AM, Kryzhanovskaya N et al 2003 Electronics Letters 39 1126
[3] Maksimov MV, Gordeev NY, Zaitsev SV, Kop'ev PS, Kochnev IV, Ledentsov NN, Lunev AV, Ruvimov SS, Sakharov AV, Tsatsul'nikov AF et al 1997 Semiconductors 31 124
[4] Shchekin OB and Deppe DG 2002 Appl. Phys. Lett. 80 3277
[5] Fathpour S, Mi Z, Bhattacharya P, Kovsh AR, Mikhrin SS, Krestnikov IL, Kozhukhov V and Ledentsov NN 2004 Appl. Phys. Lett. 85 5164
[6] Varangis PM, Li H, Liu GT, Newell TC, Stintz A, Fuchs B, Malloy KJ and Lester LF 2000 Electronics Letters 36 1544
[7] Mi Z, Fathpour S and Bhattacharya P 2005 Electronics Letters 41 1282
[8] Mi Z, Bhattacharya P and Fathpour S 2005 Appl. Phys. Lett. 86 153109
[9] Bhattacharya P and Ghosh S 2002 Appl. Phys. Lett. 80 3482
[10] Sek G, Poloczek P, Podemski P, Kudrawiec R, Misiewicz J, Somers A, Hein S, Höfling S and Forchel A 2007 Appl. Phys. Lett. 90 081915
[11] Rudno-Rudzinski W, Kudrawiec R, Sek G, Misiewicz J, Somers A, Schwertberger R, Reithmaier JP and Forchel A 2006 Appl. Phys. Lett. 88 141915
[12] Rudno-Rudzinski W, Sek G, Misiewicz J, Lamas TE and Quivy AA, 2007 J. Appl. Phys. 101 73518
[13] Rudno-Rudzinski W, Sek G, Ryczko K, Syperek M, Misiewicz J, Semenova E S, Lemaitre A and Ramdane A 2009 Appl. Phys. Lett. 94 171906
[14] Andrzejewski J, Sek G, O'Reilly E, Fiore A and Misiewicz J 2010 J. Appl. Phys. 107 073509
[15] Kudrawiec R and Misiewicz J 2009 Review of Scientific Instruments 80 096103
[16] Vurgaftman I, Meyer JR and Ram-Mohan LR, 2001 J. Appl. Phys. 89 5815

Journal of Physics: Conference Series **245** (2010) 012048 doi:10.1088/1742-6596/245/1/012048

Low-lying states of exciton in type-II quantum dot with ring-like geometry

J. Sierra-Ortega[1], **R. A. Escorcia**[1], **and I. D. Mikhailov**[2]
[1]Universidad del Magdalena, A.A. 731, Santa Marta, Colombia
[2]Universidad Industrial de Santander, A. A. 678, Bucaramanga, Colombia

E-mail: jsierraortega@gmail.com

Abstract. A theoretical investigation of the properties of excitons in type-II InP quantum ring deposited on a wetting layer and embedded in a matrix made of GaInP material is presented. The energies of the low-lying states as functions of the strength of a magnetic field applied along the growth axis for rings with different profiles are calculated by means of the diagonalization method. We find that the increase of the wetting layer thickness as the smoothness of the transverse cross section profile lead to a lowering of the effective barrier height for the electron in-plane motion and an increase of the electron tunnelling toward the axis.

1. Introduction

Study of excitons in type-II self-assembled quantum dots (QDs) has attracted considerable attention in recent years. Separate localization of the carriers in type-II single and vertically stacked QDs can lead to interesting effects such as persistent currents, oscillations of the exciton ground state energy, and crossovers of the energy levels, typical for structures with ring-like geometry [1-3]. At the same time, analysis realized recently shows that the modification of the density distribution in two-dimensional structures with broken symmetry produces a partial or total quenching of these effects for low-lying energy levels [4]. Other fact that can lead to similar consequence is the tunneling of the particles through the junctions. According to the results obtained previously for type-I QDs in form of disk [5,6], lens, and pyramid [7], the three-dimensional density distribution of carriers in QDs is modified due to an appreciable tunneling of carriers toward the matrix through the QD's junctions. In contrast to the type-I QDs the tunneling in type-II QDs can decrease the electron-hole separation and quench their properties stipulated by a strong electron-hole separation. Other important factor that may affect essentially the electronic properties of the type-II QDs is the existence of the wetting layer (WL) whose thickness frequently is comparable with the height of the QD. As it was shown previously [8], the electron and the hole tunneling in type-I from QDs toward the WL becomes significant and it is much stronger in QDs with ring-like geometry, where the tunneling toward central hole along the WL is dominant. In this work we analyze the low-lying states of the exciton captured by quantum ring (QR) in order to establish how much the increase of the WL thickness affects their properties.

2. Theoretical model

We consider a model of axially symmetrical QR deposited on a wetting layer made of InP material and embedded in a GaInP matrix, where the electron is mainly confined in the dot, but the hole is located in the barrier material. The correspondent geometry is schematically shown in figure 1. The dot together with the wetting layer forms a thin film with variable thickness d whose dependence on the distance ρ from the symmetry axis is given by a function $d(\rho)$ that tends to d_w (the thickness of

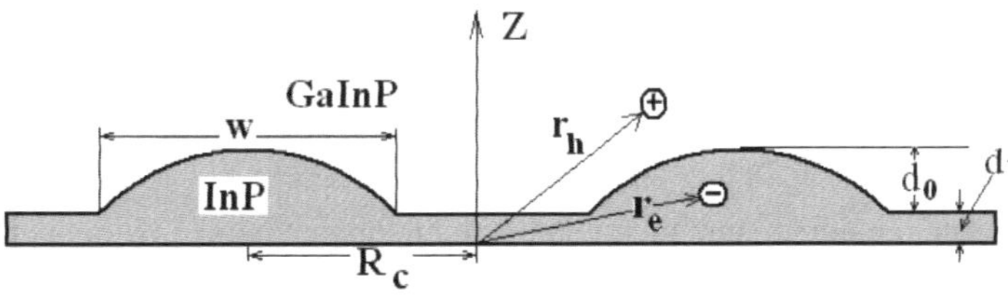

Figure 1. The geometry of the system under consideration with side view

the wetting layer) for large values of ρ and to $d_w + d_0$ (d_0 is the height of the QR over the wetting layer) as $\rho \to 0$. Other two parameters that characterize the QR geometry are the ring width w and the centerline radius R_c. In accordance with the experiment and, besides, for sake of the mathematical convenience, we assume in what follows, that the height-to-base radius aspect ratio in all cases is small, i.e. $(d_0 + d_b)/R_c << 1$. Also we consider a simplified infinite-barrier model, for which the confinement potentials for the electron, V_e and for the hole, V_h are equal to zero inside the InP layer and to infinity in the GaInP matrix for V_e and vice versa for V_h.

Within the framework of the effective-mass approximation and using cylindrical coordinates, the Hamiltonian of the electron-hole system can be written in the following dimensionless form:

$$H = H_0 + U(\mathbf{r}_e, \mathbf{r}_h); \quad H_0 = H_e(\mathbf{r}_e) + H_h(\mathbf{r}_h); \quad U(\mathbf{r}_e, \mathbf{r}_h) = -2/|\mathbf{r}_e - \mathbf{r}_h| + 2Z_i/r_h$$
$$H_e = \eta_e\left(-\Delta_e + \gamma^2\rho_e^2/4 - i\gamma\partial/\partial\varphi_e\right) + V_e; \; H_h = -\eta_h\left(-\Delta_h + \gamma^2\rho_h^2/4 - i\gamma\partial/\partial\varphi_h\right) - 2Z_i/r_h + V_h(\rho_h, z_h) \quad (1)$$

Here Z is an artificially introduced variational parameter that allows us to replace the actual complicate potential the attraction between electron and hole by hydrogenic-like one in order to separate the one-particle part of the Hamiltonian. It is clear that the parameter Z_i should be chosen in a way that the electron-hole separation were of the order one Bohr radius. It is the reason why we use in our calculation the value $Z_i = 1/R_c$. In these relations, additionally are used notations $\eta_e = \mu/m_e = 0.88$, $\eta_h = \mu/m_h = 0.12$ being $m_e = 0.077m_0$, $m_h = 0.6m_0$ the effective masses of the electron and the hole, respectively, and $\mu = m_e m_h/(m_e + m_h) = 0.068$ the reduced mass. Dimensionless units are used in the Hamiltonian (1), the exciton effective Bohr radius $a_0^* = \hbar^2\varepsilon/\mu e^2$, as the unit of length, the effective Rydberg $Ry^* = e^2/2\varepsilon a_0^* = \hbar^2/2\mu a_0^{*2}$, as the unit of energy and $\gamma = e\hbar B/2\mu c Ry^*$ as the unit of the strength of the magnetic field applied along the symmetry axis. Here and above subscripts e and h are used for parameters and variables corresponding to the electron and to the hole, respectively.

One-particle eigenfunctions and eigenvalues of the Hamiltonian H_0 can be found explicitly:

$$\psi^{(0)}_{n_e,l_e,n_h,l_h}(\mathbf{r}_e, \vec{\mathbf{r}}_h) = \psi^{(e)}_{n_e,l_e}(\mathbf{r}_e)\psi^{(h)}_{n_h,l_h}(\vec{\mathbf{r}}_h); \; E^{(0)}(n_e,l_e,n_h,l_h) = E_e(n_e,l_e) + E_h(n_h,l_h) \quad (2a)$$

The electron wave function $\psi^{(e)}_{n_e,l_e}$, that depends on two quantum numbers, radial n_e and angular momentum l_e, one can find in the adiabatic approximation [9] as:

$$\psi^{(e)}_{n_e,l_e}(\mathbf{r}_e) = e^{il_e\varphi_e} \cos\left[\pi\left(z_e - d(\rho_e)/2\right)/d(\rho_e)\right] f^{(e)}_{n_e,l_e}(\rho_e) \quad (2b)$$

where the radial part $f^{(e)}_{n_e,l_e}(\rho_e)$ for in-plane motion is a solution of one-dimensional wave equation:

$$\eta_e\left(-\frac{1}{\rho_e}\frac{\partial}{\partial\rho_e}\rho_e\frac{\partial}{\partial\rho_e} + \frac{l_e^2}{\rho_e^2} + \frac{\gamma^2\rho_e^2}{4} + \gamma l_e\right) f^{(e)}_{n_e,l_e}(\rho_e) + \frac{\pi^2}{d^2(\rho_e)} f^{(e)}_{n_e,l_e}(\rho_e) = E_e(n_e,l_e) f^{(e)}_{n_e,l_e}(\rho_e) \quad (2b)$$

In order to find the one-particle wave functions $\psi^{(h)}_{n_h,l_h}(\vec{\mathbf{r}}_h)$ we use the following approximation. Due to the infinite-barrier repulsion the hole may be localized only over or under the very narrow InP layer. As the scale of the hole displacement is significantly larger than thickness of this layer, we in our calculations of the hole wave function for sake of the mathematical convenience depreciate the layer thickness and consider a simple boundary condition: the hole waves functions should be equal to zero at all points of the plane Z=0. Additionally, the paramagnetic term in H_h gives contribution in the total energy of the order $\mu_h^2 \approx 0.01$ and in the first order of the theory of the perturbation may be depreciated. Therefore the eigenfunctions of H_h are given by the hydrogen-like atom wave functions that satisfy the above condition and in spherical coordinates corresponding them energies have the following form:

$$\psi^{(h)}_{n_h,l_h}(\vec{\mathbf{r}}_h)=e^{il_h\varphi_h}R_{n_h,l_h+1}(r_hZ/\eta_h)P_{l_h+1,l_h}(\vartheta_h);\quad E_h(n_h,l_h)=-\eta_h\left[(Z_i/\eta_h n_h)^2+\gamma^2\left\langle\rho_h^2\right\rangle/4+\gamma l_h\right] \quad (2c)$$

where n_h and l_h are radial and angular momentum quantum numbers, $R_{n_h,l_h+1}(r)$ and $P_{l_h+1,l_h}(\vartheta_h)$ are the Laguere and Legendre polynomials, respectively and $\left\langle\rho_h^2\right\rangle$ is the averaged distance of the hole from Z-axis calculated analytically by using the hydrogenic-like wave functions.

The exciton wave functions with fixed total angular momentum $L=l_e+l_h$ (a unique "good" quantum number for this system) are constructed as linear combinations of these one-electron wave functions:

$$\Psi^{(L)}(\mathbf{r}_e,\mathbf{r}_h)=\sum_{n_e=0}^{n_m}\sum_{n_h=0}^{n_m}\sum_{l_e=-l_m}^{l_m}C^{l_h}_{n_e,n_h}\psi^{(e)}_{n_e,l_e}(\mathbf{r}_e)\psi^{(h)}_{n_h,L-l_e}(\mathbf{r}_h) \quad (3)$$

The matrix elements of Hamiltonian (1) are found using solutions of the one particle problem as the basis. Then eigenenergies and eigenvectors are calculated using the Hausholder technique, for different values of the total momentum L. For most of our numerical results, we used $n_m=10$, $l_m=10$ which leads to an accuracy for the ground state energy better than 5%.

3. Results

In order to analyze the effect of the variation of the ring width, WL thickness and shape of the transverse cross-section profile on the properties of a heterostructure with ring-like geometry, we have performed numerical calculations of exciton energies of some low-lying states as functions of the magnetic field strength for rings with different geometric parameters.

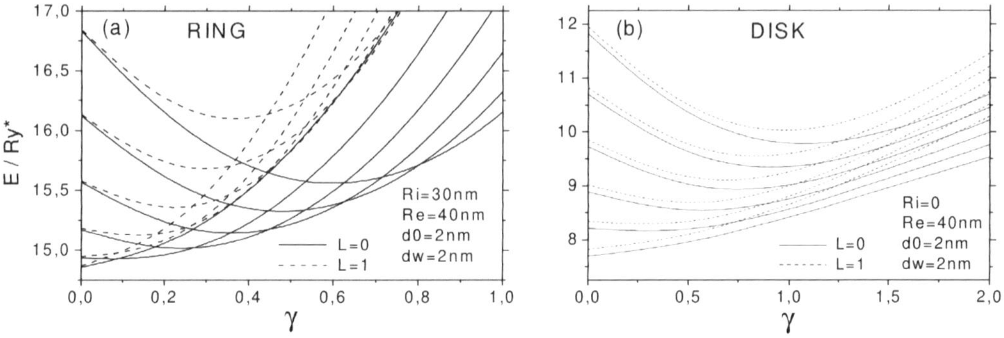

Figure 2. Energy levels as a function of the magnetic field for exciton in (a) QR, (b) quantum disk. Parameters are indicated in figure

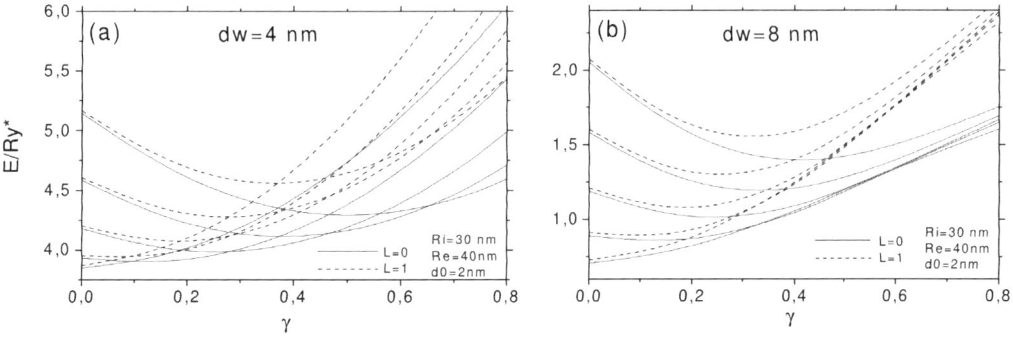

Figure 3. The same as Figure 2a but now for QRs with different WL

In figure 2 are shown dependencies for 12 lowest energy levels, corresponding to angular momenta along the *Z*-axis *L*=0 and *L*=1, on the magnetic field strength for exciton trapped in (a) ring and (b) disk with the same exterior radii and thicknesses of the WL and the ring. A similar transformation of the curves but associated to the increase of the WL thickness one can observe in figure 3. As the WL thickness increases from 4nm to 8nm the crossovers between curves with the same exciton angular momentum disappear and the oscillation of the lowest energy level quenches.

Finally, in figure 4 we compare curves for 42 lowest energy levels in two QRs with different profiles of the transverse cross sections and with very small interior radii. It is seen that the curves for structure with rectangular profile have a form typical for quantum dots with disk-like geometry while for structure with soft-edge profile there are crossovers between curves with the same angular momentum typical for rings. It is due to the fact that the confinement in the QR with soft-edge profile is stronger and the tunneling toward the axis is weaker than in QR with rectangular profile

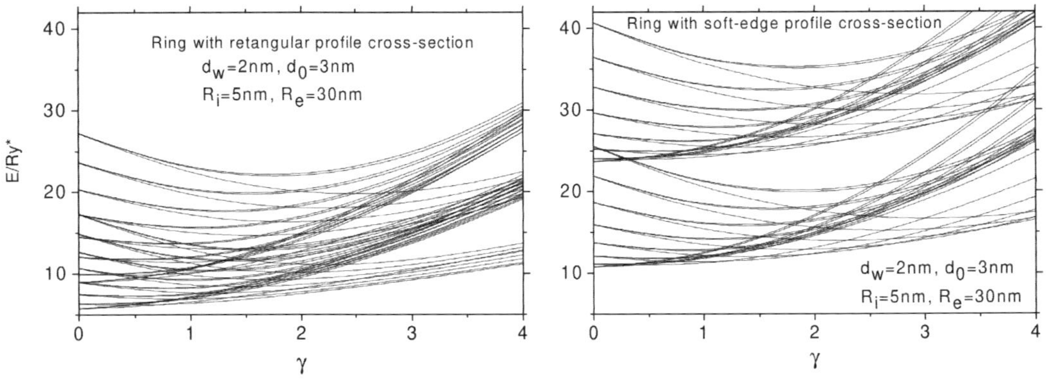

Figure 4. The same as Figure 2a but now for QRs with different profiles of transverse cross-sections

In short we propose a new simple method for calculating the energies of the low-lying states of the exciton trapped in type-II quantum ring in the presence of the magnetic field applied along the symmetry axis. We show that the increase of the wetting layer thickness and the smoothing of the transverse cross-section profile lead to quenching of the oscillations of the low-lying energy curves.

Acknowledgements
This work was financed by the Industrial University of Santander (UIS) through the Dirección General de Investigaciones (DIF Ciencias, Cod. 5124) and the Excellence Centre of Novel Materials –ECNM, under Contract No. 043-2005 and the Cod. No. 1102-05-16923 subscribed with COLCIENCIAS.

References
[1] Govorov A O, Kalameitsev A V, Warburton R, Karrai K, Ulloa S E 2002 *Physica E* **13** 297
[2] Mlinar V, Tadić M and Peeters F M 2006 *Phys. Rev. B* **73**, 235336
[3] Kuskovsky I L, MacDonald W, Govorov A O, Mourokh L, Wei X, Tamargo M C, Tadic M and Peeters F M 2007 *Phys. Rev. B* **76**, 035342
[4] Degani M H, Maialle M Z, Medeiros-Ribeiro G and Ribeiro E 2008 *Phys. Rev. B* **78** 75322
[5] Janssens K L, Peeters F M and Schweigert V A 2001 *Phys. Rev. B* **63** 205311
[6] Janssens K L, Patroens B and Peeters F M, 2001 *Phys. Rev. B* **64**, 155324
[7] Marín J H, Mikhailov I D and García L F 2007 *Physica B* **398** 135
[8] Mikhailov I D, García L F and Marín J H 2006 *Phys.: Condens. Matter* **18** 9493
[9] Mikhailov I D, Marín J H, and García F, 2005 *Phys. Stat. Sol. (b)* **242**,1636

Fine structure splitting reduction in droplet epitaxy GaAs quantum dots grown on (111)A surface

M. Abbarchi, T. Kuroda, T. Mano and K. Sakoda

National Institute for Materials Science, 1-2-1 Sengen, Tsukuba 305-0047, Japan

E-mail: ABBARCHI.Marco@nims.go.jp

Abstract. In this contribution we compare the geometrical symmetry of self–assembled GaAs quantum dots (QDs) grown by a *refined* droplet epitaxy (DE) method on (100) and (111)A AlGaAs substrate. By atomic force microscopy (AFM) we study the morphology of the QDs and by micro–photoluminescence spectroscopy (μPL) we compare the dependence of the fine structure splitting on QD size and shape. We show that: i) the *refined* DE enhances the elongation of QDs grown on (100) AlGaAs substrate increasing the fine structure splitting; ii) by growing the QDs on the (111)A AlGaAs surface it is possible to recover a good geometrical circular symmetry obtaining a small value of the fine structure splitting .

1. Introduction

The control of the anisotropic exchange Coulomb interaction between electron and hole spin in QDs (usually said fine structure splitting, FSS) is a relevant issue for quantum information applications. Single QDs in fact represent a promising building block for the realization of polarization–entangled photons [1] from the biexciton–exciton–ground state cascade (XX-X-gs) provided that the FSS is tuned to zero [2].

Generally speaking, the FSS originates from asymmetries in the QD confinement potential [3, 4] and brings to a polarization splitting in the X photoluminescence (PL) which complicates the opportunity of realizing entangled photon pairs. In Stranski Krastanov (SK) QDs, anisotropies came from several sources (shape, strain, atomistic effects and piezoelectricity) and claims of the major role of strain induced effects on the FSS have been reported [3].

Taking advantage of the DE growth method [5, 6, 7], we can realize strain–free GaAs/AlGaAs QDs in order to investigate only the effect of geometrical asymmetry. In previous studies on DE QDs grown with a *conventional* low–temperature method [5, 6, 7] on (100) AlGaAs surface, a reduction in the FSS was shown for small circular–symmetric nanoemitters [8]. Nevertheless, a significant drawback of the low temperature used during the growth was a large PL linewidth of the excitonic recombination from single QDs ($\sim$100 μeV) ascribed to the fluctuating charged environment [9]. In this paper we demonstrate that a *refined* high–temperature DE [10] on AlGaAs (100) surface improves the QDs quality in terms of spectral linewidth [9, 10] but increases the geometrical elongation and the FSS [11] respect to the *conventional* low–temperature growth [8]. Moreover, the same *refined* DE on the AlGaAs (111)A surface allows also a reduction in the FSS.

The paper is organized as follows: in section 2 we describe the sample growth and the method of investigation, in section 3 we show the analysis of AFM measurements on uncapped samples and μPL measurements on single QDs, finally, in section 4, we summarize the results.

2. Experimental method

Two GaAs QD samples were grown on AlGaAs (100) and (111)A (respectively indicated as sample A and B) in a solid-source molecular beam epitaxy (MBE) system with a *refined* DE procedure [10] which is here summarized. After the growth of a AlGaAs barrier layer at 580 °C, the temperature was set to 350 °C. At this stage Ga droplet were deposited and then irradiated by As flux for the crystallization of GaAs QDs at 200 °C. The QDs density, determined by substrate temperature, was as small as 10^8-10^9 QDs/cm^{-2}. The GaAs QDs were annealed at 400 °C and than covered by 40 nm AlGaAs capping layer grown by standard MBE at the same 400 °C plus 20 nm thick AlGaAs layer at 580 °C. Than a 10 nm thick GaAs capping layer was grown on top by standard MBE at 580 °C and finally the samples were annealed at 800 °C.

The main difference between this *refined* procedure and the *conventional* DE is the annealing step after the QDs crystallization. The increased temperature allows to grow a high quality top barrier which is the key factor ruling the PL quality in terms of spectral linewidth [9, 10].

Low-temperature μ–PL experiments were performed by using a confocal setup in illumination–collection configuration allowing for a lateral resolution better than 1 μm. The small QDs areal density allowed collecting the PL from few nanoemitters and it was enough for accessing the single nanostructure emission after spectral filtering. The PL polarization was detected by using a half-wave plate and a polarizer. The collected PL was fed into a grating monochromator and detected by a silicon-based CCD camera allowing to detect lineshift as small as ~10 μeV and line broadening as small as ~50 μeV after a fitting procedure. For illumination we used a CW solid–state laser emitting at a wavelength of 532 nm to excite the high energy states of the AlGaAs barrier.

3. Results and discussion

Figure 1 (a) and (b) shows AFM profiles of QDs from samples A and B before capping. We note that the QD shape in sample A is elongated along [1-10] direction. We ascribe this behavior to the different adatom migration along the orthogonal in–plane directions [110] and [1-10][10]. Furthermore we observe a increase in the geometrical elongation respect to QDs grown with the *conventional* DE on (100) surface [10, 11, 8, 12]. We relate this effect to the increased temperature during the *refined* growth process which enhances the material migration along [1-10] direction [10]. Differently from A the case of sample B, despite the high temperature of the growth, doesn't show any preferential elongation axis in the QDs shape. We ascribe this feature to the three fold rotational symmetry of the (111)A substrate.

Single QD PL of samples A has been studied in previous works and emission from neutral exciton (X) charged exciton (T) and neutral biexciton (XX) was demonstrated by FSS measurements [10]. In addition, the *refined* DE growth allowed for a drastic reduction in the inhomogeneous broadening (in the high excitation regime the linewidth was as small as ~30 μeV, about 4 times smaller respect to samples grown with *conventional* DE)[10, 9]. Moreover, in the low excitation regime, we can achieve the very small linewidth of 4 μeV as measured by scanning Fabry–Perot interferometry (not shown). This accounts for the improvement achieved thanks to the *refined* DE in terms of line broadening.

Typical PL spectra of QDs from sample B are shown in Fig. 3 (a) and (d) for different excitation power. The power dependence of X and XX lines is reported in Fig. 3 (b) for the QD of graph 3 (a). Power dependence of X and T of the QD shown in graph 3 (d) are displayed in graph 3 (e). On the basis of first and second order Poissonian or quasi–Poissonian dependence of the peak intensity on the excitation power we interpret these lines as neutral exciton X,

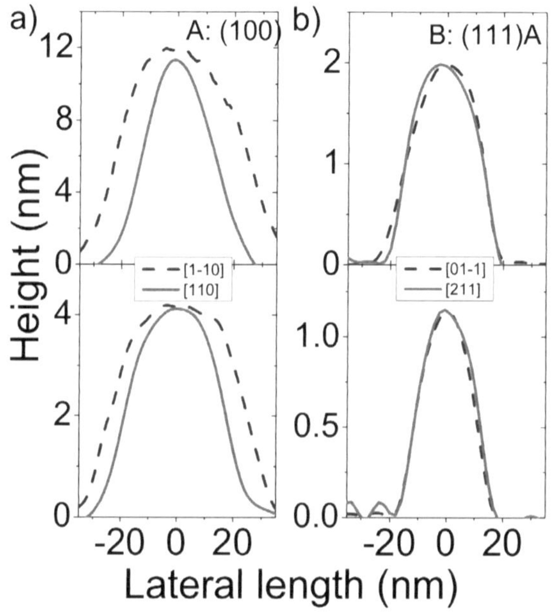

Figure 1. (a) and (b): AFM profiles of uncapped samples in orthogonal directions for a large (top panel) and a small (bottom panel) QD grown on AlGaAs (100) (sample A) and (111)A (sample B) respectively. Dashed (continuous) lines refer to scan in [1-10] ([110]) for sample A and [01-1] ([211]) for sample B.

neutral biexciton XX and charged exciton T recombination[13]. This attribution is confirmed by polarization resolved PL reported in graphs 3 (c) for the QD of graph 3 (a) and in graph 3 (f) for the QD of graph 3 (d). While X and XX show the typical mirror–symmetric polarization splitting, the charged complex T is not split. The measured FSS are $\simeq$22 μeV for the QD in 3 (a) and $\simeq$26 μeV for the QD in 3 (d). We note also that the anisotropy axes of the two different QDs are noticeably different [8, 12]. In the merit of this we note that a random orientation of the polarization axes was found for all the QDs grown on (111)A surface (not shown) and in small, circular–symmetric QDs grown on (100) surface [8, 12]. The linewidth, which cannot be resolved by this experimental setup, is below 50 μeV.

Similar μPL investigation was performed on several QDs from sample A and B. The results of the measured FSS are reported as a function of the X energy in Fig. 3. A comparison between the FSS in the present sample A and previous samples grown by *conventional* DE on (100) highlights the enhanced geometrical elongation of the QDs shape due to the different diffusion process which takes place in [100] and [1-10] directions. In fact the FSS is in average larger than the previous reported results [11, 8, 12]: the new *refined* growth has the relevant drawback of increasing the FSS from $\sim$90 μeV up to $\sim$140 μeV. This points out that the annealing step, after the QDs crystallization, rules the enhanced geometrical elongation. On the contrary the FSS in sample B is as small as $\simeq$17 μeV (note that, due to the poor resolution of the experimental setup, this value should be considered as an upper limit).

This results account for the overall quality improvement of the DE QDs grown on (111)A surface in terms of linewidth and FSS. In fact, by keeping a high growth temperature we obtain a small PL broadening which is the signature of the improved material quality. At the same time, thanks to the choice of the (111)A plane for the droplet deposition on which any diffusion process is equivalent every 120 degrees, we can realize circular symmetric QDs.

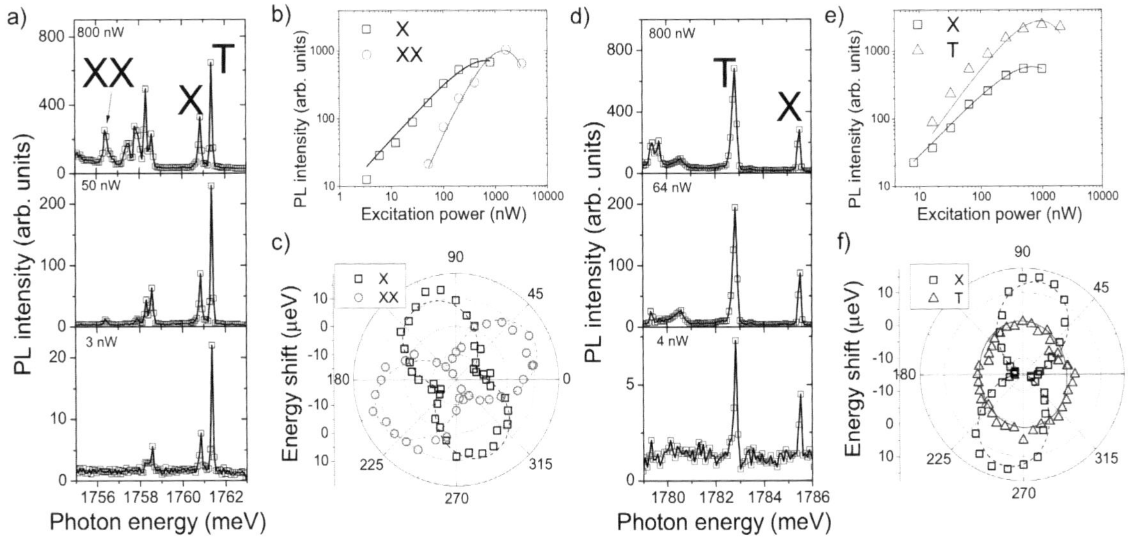

Figure 2. (a) and (d): PL spectra of a QD grown on (111)A at different excitation power. (b): power dependence of X and XX peak intensity of the corresponding QD in (a) as a function of the incident power. (e): power dependence of X and T peak intensity of the corresponding QD in (d) as a function of the incident power. The symbols represent the experimental data. Solid lines represent first and second order Poissonian fit for X and XX (graph (b)), first order Poissonian and quasi–Poissonian fit for X and T (graph (e)). (c): energy shift of X and XX of the QD in (a) as a function of the detected polarization angle represented in a polar plot. (f): energy shift of X and T of QD in (d) as a function of the detected polarization angle represented in a polar plot. Symbols are the experimental data while solid lines are sinusoidal fit to data.

4. Conclusions

In conclusion we have studied two samples of GaAs QDs grown on AlGaAs (100) and (111)A by a *refined* DE technique. By PL spectroscopy we accessed the single nanoemitter discriminating between X, T and XX recombination. By polarization resolved PL analysis we investigated the FSS of several QDs allowing a comparison of the geometrical properties in different samples. The effect of the *refined* DE technique on QDs grown on (100) AlGaAs surface is an enhanced elongation which results in a larger FSS. On the contrary in the case of AlGaAs (111)A surface, due to the lack of preferential direction for the material diffusion, a more circular–symmetric shape of the QDs is obtained leading to a reduction in the FSS. We believe that the present achievement represent a relevant step toward the realization of solid–state sources of

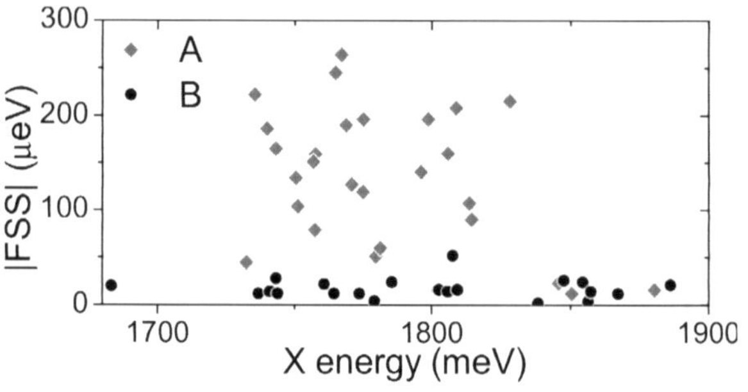

Figure 3. Absolute value of the FSS for sample A (diamonds) and B (circles).

polarization–entangled photon pairs

References

[1] D. Bouwmeester, A. K. Ekert, and A. Zeilinger, The Physics of Quantum Information (Springer, Berlin, 2000).
[2] R. M. Stevenson, R. J. Young, P. Atkinson, K. Cooper, D. A. Ritchie, and A. J. Shields, Nat. **439**, 179 (2006).
[3] R. Seguin, A. Schliwa, S. Rodt, K. Pötschke, U.W. Pohl, and D. Bimberg, Phys. Rev. Lett. **95**, 257402 (2005).
[4] A. Schliwa, M. Winkelnkemper, and D. Bimberg, Phys. Rev. B **76**, 205324 (2007).
[5] N. Koguchi, S. Takahashi, and T. Chikyow, J. Cryst. Growth **111**, 688 (1991).
[6] K. Watanabe, N. Koguchi, and Y. Gotoh, Jpn. J. Appl. Phys. Part 2, **39**, L79 (2000).
[7] V. Mantovani, S. Sanguinetti, M. Guzzi, E. Grilli, M. Gurioli, K. Watanabe, and N. Koguchi, J. Appl. Phys. **96**, 4416 (2004).
[8] M. Abbarchi, C. A. Mastrandrea, T. Kuroda, T. Mano, K. Sakoda, N. Koguchi, S. Sanguinetti, A. Vinattieri, and M. Gurioli, Phys. Rev. B **78**, 125321 (2008).
[9] M. Abbarchi, F. Troiani, C. Mastrandrea, G. Goldoni, T. Kuroda, T. Mano, K. Sakoda, N. Koguchi, S. Sanguinetti, A. Vinattieri, and M. Gurioli, Appl. Phys. Lett. **93**, 162101 (2008).
[10] T. Mano, M. Abbarchi, T. Kuroda, C. A. Mastrandrea, A. Vinattieri, S. Sanguinetti, K. Sakoda and M. Gurioli, Nanotechnology **20**, 395601 (2009).
[11] M. Abbarchi, T.Kuroda, C. Mastrandrea, A. Vinattieri, S. Sanguinetti, T. Mano, K. Sakoda, M. Gurioli, Physica E **42**, 881 (2010).
[12] M. Abbarchi, T. Kuroda, T. Mano, K. Sakoda, and M. Gurioli, Phys. Rev. B **81**, 035334 (2010).
[13] M. Abbarchi, C. Mastrandrea, T. Kuroda, T. Mano, A. Vinattieri, K. Sakoda, and M. Gurioli J. Appl. Phys. **106**, 053504 (2009).

Evidence of correlated electron hole pairs in dots in asymmetric quantum well structures

M.H.Abdellatif [1,2], Jin Dong Song[1], Won Jun Choi[1], Nam Ki Cho[1] and Jung Il Lee[1]*

[1]Nano Convergence Devices Center, Korea Institute of Science and Technology, Seoul 136-791, Korea

[2]University of Science and Technology, 113 Gwahangno, Yuseong-gu, Daejeon, 305-333, Korea

Abstract. We introduce the study of InAs quantum dots embedded in an InGaAs quantum well. The quantum dot potential strongly modifies the energy states of the quantum well structure. Hence, the carrier thermodynamics is also influenced. The dependence of the integrated photoluminescence on the excitation power density shows linear dependence with very little deviation from linearity at a high temperature. That result contributes to the strong correlation between the electron and hole pairs, which is attributed to the enhanced confinement provided by the asymmetric quantum well structure.

1. Introduction

It was shown earlier that the InAs quantum dot (QD) density can be increased essentially if the dot is grown inside the InGaAs/GaAs quantum well (QW) (dot in a well, DWELL) structure [1]. The most important aspect for light emitting device based on InAs QD and operated at room temperature is the understanding of excitation power and temperature dependence photoluminescence emission (PL)[3]. The problem of how to reach the 1.3μm emission at room temperature has been discussed before [4], however the problem is that the total InAs amount in active region is relatively high which leads to high probability of dislocation formation, which in return affects the PL emission [6]. DWELL structure based on InAs QD has been effective in improving the performance of emitting device in the long wavelengths regime since it helps decrease the thermal escape rate of the carriers from QDs at high temperature [7, 8]. Among the proposed mechanisms of long wavelength emission from InAs QDs, it is believed that residual strain plays a key role. The asymmetric DWELL was reported to have enhanced uniformity in dot morphology, narrower line width of emission from ground state transition [9]. The fundamental problem is to understand the temperature dependence of QD PL. Two main ways: (a) thermal escape (capture) of carriers (electron and hole) in/from QDs takes place as an exciton or correlated electron hole pair [10]; or (b) the excitons dissociate and single electrons (holes) thermally escape or capture independently [11]. In Asymmetric DWELL structure the introduction of surrounding QW changes the potential barrier height for exciton thermal escape from QDs and can increase the density of nonradiative (NR) centers. In this work, the parameters of the QW alters the thermal quenching of the PL intensity, the study of PL, its temperature and power dependence in InAs QDs embedded in asymmetric InGaAs/GaAs DWELL structures are presented.

2. Experimental details

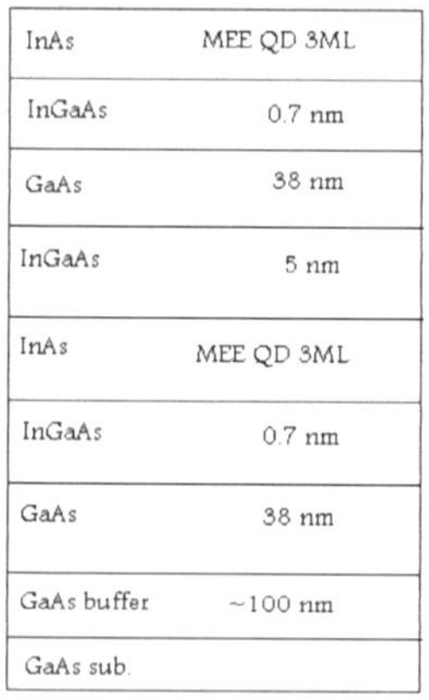

Figure 1. Schematic diagram of InAs QD in InGaAs/GaAs (asymmetric well) structure.

The schematic diagram of the sample structure, InAs QD embedded in asymmetric InGaAs/GaAs QW, is shown in figure 1. Details of the growth procedure are described elsewhere [8]. The sample is excited with 532 nm from solid state laser. The PL spectra were collected using a grating monochromator with Peltier cooled detector. The monochromator is ACTON-SPECTRA PRO-23001. The sample was mounted in a closed-cycle He cryostat where the temperature is varied from 16 to 300K. The power dependence of the PL is measured in the range 1 to 30 mW-measured using power meter directed to the laser beam. Laser power is changed using neutral density filter. Spot size of laser on the sample is ~ 1 mm^2.

3. Results and discussion

The PL spectra of InAs QD in asymmetric InGaAs/GaAs QW at 16K are shown in figure 2(a) for different exciting laser power, namely 1mW to 30 mW. The graph shows that PL intensity increase with increasing laser power. Four peaks are found, the low energy peak around 850 nm is assigned to GaAs emission, 940 nm assigned to emission from wetting layer (WL), while the other two peaks at 1100, 1200 nm are assigned to QW and QD emission, respectively. The PL spectrum represents recombination of excitons localized at the QD ground state as well as InGaAs QW.

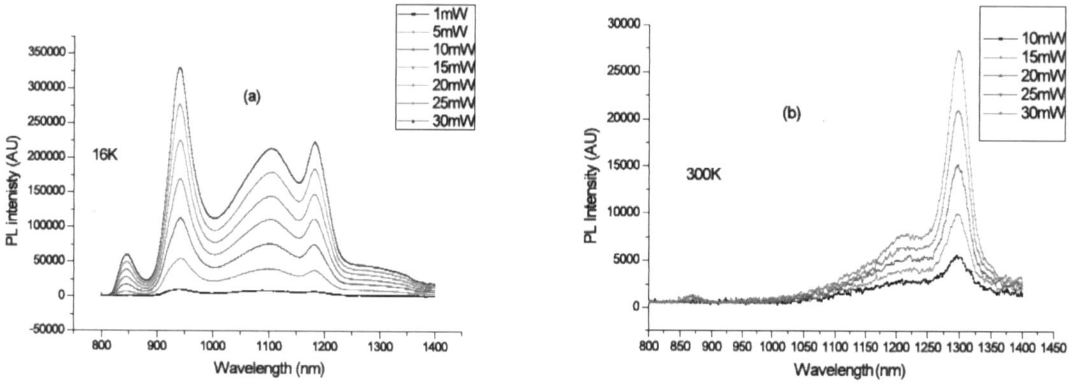

Figure 2. Photoluminescence spectra of asymmetric DWELL structure (a) at 16 K (b) at 300K.

The PL spectra of the asymsmetric DWELL structure in Figure 2(b) shows that the long wavelength emission near 1.3 μm could be achieved at room temperature. The PL intensity is found to decrease as the temperature increase, due to thermal quenching. In general thermal quenching of the PL intensity can be explained by thermal escape of excitons from QDs into the GaAs barrier or WL with subsequent nonradiative recombination as well as by thermal activation of exciton capture by nonradiative (NR) centers. Our data enhance the picture of the two stage process of exciton capture (thermal escape) from GaAs barrier to QW (I stage) and from QW to QD (II stage) as shown before [12].

The power dependence of the integrated intensity over the entire PL spectra for QD peak is shown in figure 3 (b). The fitted value of the integrated intensity of the ground state QD with the power law $I_{PL} = \xi P_{exc}^{\alpha}$ where I_{PL} is the integrated intensity ξ is the proportionality constant and the exponent α determines the linearity or the super-linearity. The plot for the exponent α is shown in figure 3 (a). The

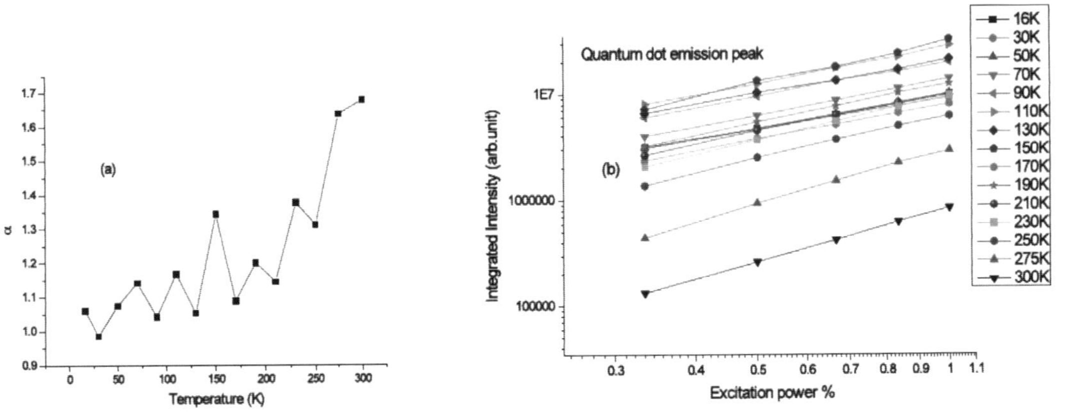

Figure 3. (a) Temperature dependence of the exponent α for the QD PL Peak emission. (b) Dependence of the integrated PL intensity of the QD peak emission on the excitation power ratio measured at different temperature.

exponent α is found to be directly proportional to excitation power up to 210K and then switch to super linear, it has been shown [13, 14] that the power dependence show linear behavior in a temperature region where carrier dynamics are controlled by correlated electron hole pair and then switches to super linear in a region where electrons and holes behaves independently, with some degree of correlation.

The temperature dependence of peak position for QD peak emission is fitted with Varshni relation $E(T)=E_o-aT^2/(T+b)$, (graph is not shown here). The fitted parameters E_o, a and b estimated from the peak position of the PL are presented in Table 1. Comparison of the corresponding parameters for the variation of the energy band gap versus temperature in the bulk of InAs, GaAs are presented as well. The parameters of QD are found to be bigger than the corresponding value in the InAs bulk crystals that permit the assumption that the studied asymmetric DWELL has a material composition that partially changed due to Ga/In interdiffusion [15].

Table 1: Comparison of Varshni Parameters with bulk values

		Eo (eV)	a(meV/K)	b(K)	References
QD-Peak		1.05394	0.5409	202.7	This work
InAs	Bulk	0.418	0.308	65	15
GaAs	Bulk	1.519	0.540	204	16, 17

4. Conclusion

We investigated PL temperature dependence of QD embedded in InGaAs asymmetric DWELL structure. The sample showed 1.3 μm PL emission at room temperature Analysis of power dependence shows linear dependence in the temperature region 16K up to 210K enhancing the idea that carrier dynamics are controlled by correlated electron-hole pairs. The temperature dependence of the peak position for QD is measured and fitted with Varshni relation, and the bigger values of Varshni parameters are attributed to In/Ga inter diffusion during growth.

5. Acknowledgement

This work is supported by MEST by GRL Program.

References

[1] A. Stintz, G.T. Liu, L.Gray, R. Spillers, S.M. Delgado, K.J. Malloy, J. Vac.Sci. Technol. B. 18 (**3**) (2000)1496.
[2] G.T.Liu, A. Stintz, H. Li, Kj. Malloy, L.E. Lester, Electron. Lett. **35** (1999) 1163.
[3] D. Bimberg, M. Grundman, N.N. Ledentsov, Quantum Dot heterostructures, Ed., Wiley & Sons, 2001, 328.

[4] M. Knodow, K. Uomi, A. Niwa, T.Kitatani, S. Watahiki, Y. Yazawa, Jpn. J. Appl. Phys. **35**, (1996). 11273.
[5] G.E. Girlin, V.G. Durbovskii, V.N.Petrov, N.K. Polyakov, N.P. Korneeva, V.N. Demidov, A.O. Golubok, S.A. Masalov, D.V.Kurochkin, O.M. Gorbenko, N.I. Komyak, V.M. Ustinov, A.Yu. Egorov, A.R. Kovsh, M.V. Maximov, A.F. Tsatsul'nikov, B.V. Volovik, A.E.Zhukov, P.S. Kop'ev, Zh.I. Alferov, N.N. Ledentsov, M. Grundmann, D. Bimberg, Semicond.Sci, Technol. **13** (1998) 1262.
[6] V.A. Egorov, N.K. Pluakov, A.A. Tonkikh, V.N. petrov, G.E.Cirlin, B.V. Volovik, A..E. Zhukov, Yu.G.Musikhin, N.A. Cherkashin, V.M. Ustinov Applied Surface Science,175-176, (2001) 243-248.
[7] I.R. Seller, H.Y. Liu, T.J. Badcock, K.M. Groom, D.J. Nowbray, M.Gutierrez, M. Hopkinson, M.S. Skolnick, Physica E.**26** 382 (2005).
[8] Kwang Woong Kim, Nam Ki Cho, Sung Phil Ryu, Jin Dong Song, Won Jun Choi, Jung Il Lee and Jung Ho Park. Japanese Journal of Applied Physics, Vol. **45**, No. 10A, 2006, pp. 8010–8013.
[9] V. D. Dasika, J. D. Song, W. J. Choi, N. K. Cho, J. I. Lee, and R. S. Goldman, Applied Physics Letter **95**, 163114 (2009).
[10] R. Heitz, F. Guffarth, K. Potschke, A. Schliwa, D. Bimberg, N. D. Zakharov and P. Werner, Phys. Rev. B 71, 045325 (2005).
[11] S. Sanguinetti, M. Henini, M. Grassi Alessi, M. Capizzi, P. Frigeri, S. Franchi, Phys.Rev. B. **60** (1999) 8276.
[12] S. P. Ryu and Y. T. Lee, N. K. Cho, W. J. Choi, J. D. Song, and J. I. Lee, H. S. Kwack and Y. H. Cho, Journal of Applied Physics **102**, 023105 (2007).
[13] T.V. Torchynska, J. L. Casas Espinola, L.V. Borkovska, S. Ostapenko, M. Dybic, O. Polupan, N. O. Korsunska, A. Stintz, P.G. Eliseev, K.J. Malloy, J. Appl. Phys. **101** (2007) 024323 & references therein.
[14] S. Sanguinetti, D. Colombo, M. Guzzi, E. Grilli, M. Gurioli, L. Seravalli, D. frigeri and S. Fanchi, Phys. Rev.B **74**, 205302 (2006).
[15] J.L. Casas-Espinola, T.V. Torchynska, G.P. Polupan, E. Velazquez-Lozada, Materials Science and Engineering B **165** (2009) 115-117.

Journal of Physics: Conference Series **245** (2010) 012051 doi:10.1088/1742-6596/245/1/012051

Entanglement variations in model core-shell quantum dots

P. P. Marchisio, J. P. Coe, I. D'Amico

Department of Physics, University of York, York YO10 5DD, UK

E-mail: ppm501@york.ac.uk,jpc503@york.ac.uk,ida500@york.ac.uk

Abstract. We model a core-shell quantum dot using a one dimensional system of two interacting electrons with a potential that can vary from one corresponding to a dot with comparable width of core and shell to one where the core well width becomes very small, but is always non-zero, while the shell becomes wide. We calculate the entanglement due to the spatial degrees of freedom and also the site or local entanglement found in a partition of the potential accessible to measurement. We quantify both using the linear entropy of the reduced density matrix. A sharp variation of the entanglement had been seen in earlier work for the transformation from a one dimensional model core-shell dot to a double dot [1]. We investigate whether the spatial entanglement and energy derivatives display any discontinuity as the core well width becomes small and if the two-electron state is always bound to the core well.

1. Introduction

Quantum dots are considered a good choice as a possible platform for the realisation of quantum information processing, which in principle would lead to significant benefits over classical information processing. This is partly due to the ability of quantum states to exhibit entanglement. We investigate the entanglement in a quantum dot model using an interacting two-electron one-dimensional system. We consider potential shapes which could correspond to core-shell dots or electrostatically gated dots. We compare the spatial entanglement with the site or local entanglement of the accessible partitions. For the latter we divide the system into three but assume that only two partitions are accessible to measurement. Furthermore, systems like this may also offer an interesting test-bed for investigating entanglement transitions in the few-particle regime [1]. Entanglement transitions—in the thermodynamic limit—have recently been linked with quantum phase transitions and shown to be able to identify them [2, 3]: in this case a non-analyticity in the entanglement as the control parameter is varied could be associated with a non-analyticity in the energy.

2. The system

The one-dimensional Hamiltonian in effective atomic units for our system reads

$$H = \sum_{i=i}^{2} \left[-\frac{1}{2}\frac{d^2}{dx_i^2} + V(x_i, R) \right] + \delta(x_1 - x_2). \quad (1)$$

Here we model the Coulomb repulsion in one-dimension by using $\delta(x_1 - x_2)$. $V(x, R)$, is the potential characterised by an outer well of depth -10 Hartree and a deeper inner well of depth

-20 Hartree corresponding respectively to shell and core. The former increases whilst the latter decreases as the control parameter R increases. However, the inner well never disappears. The outer well width is $W_{OW}(R) = 5 + R$. The inner well width varies as $W_{IW}(R) = \left(\frac{4}{R}\right)^{\gamma}$ where $\gamma = 13$. See Fig. 1 for an illustration of the potential when $R = 4\ a$ where a is the Bohr radius.

We solve this system for various values of R using 'exact' diagonalisation with a basis comprising of the first 50 solutions to the 1D harmonic oscillator. With ω chosen appropriately for the length scale of the system; we use $\omega = 2$. We then calculate quantities as a function of R.

3. Bound state

We note that although a one-dimensional square well always has a bound state for one particle this may not be the case for an interacting system. We find that when we have two electrons interacting via a delta function then for a small enough inner well ($R = 5.136\ a$) the ground-state energy per particle of the system is greater than the depth of the outer well (see Fig. 2). Hence the repulsion means that the state is no longer bound to the inner well.

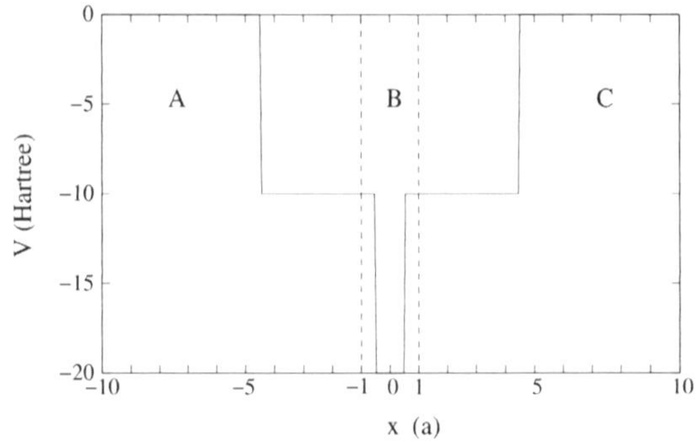

Figure 1. Potential for $R = 4\ a$ (solid line) and the partitions for the local entanglement calculation (dashed line).

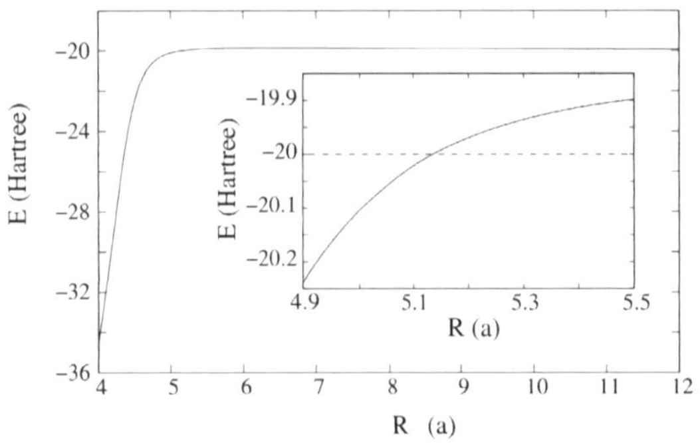

Figure 2. Ground-state energy of the system as a function of the control parameter R. Inset: comparison with twice the depth of the outer well (dashed line).

4. Entanglement

We calculate the entanglement using the linear entropy of the reduced density matrix $L = 1 - Tr\rho_{red}^2$. We first consider the spatial entanglement [4], where the reduced density matrix is found by tracing out the spatial degrees of freedom of one of the particles $\rho_{red} = Tr_1|\Psi\rangle\langle\Psi|$ which is then used to give L_{spatial}. This entanglement quantifies how much extra information one would gain on measuring the position of one particle about the other particle's position.

The spatial entanglement initially decreases (Fig. 3) as the inner well narrows due to the electrons being more confined and their spatial state becoming more similar to a non-entangled product state. However as the inner well continues to narrow more and more of the wavefunction spills into the outer well until the electrons become no longer bound to the inner well. In this parameter region the entanglement increases quickly as the electrons become much less confined in the wider outer well. Then the rate at which the entanglement increases diminishes and finally becomes driven by the (linear) expansion of the outer well (see also Fig. 5).

We next calculate the local or site entanglement [5, 6] which can be thought of as a measure of how much information we gain, upon measuring the occupation of a site, about the other sites. We move to a site occupation basis and partition the system into three 'sites': A $[X_{min}, -1]$, B $[-1, 1]$ and C $[1, X_{max}]$ (see Fig. 1). Here $X_{max} = -X_{min}, X_{max} \sim 36\ a$. For the minimum value

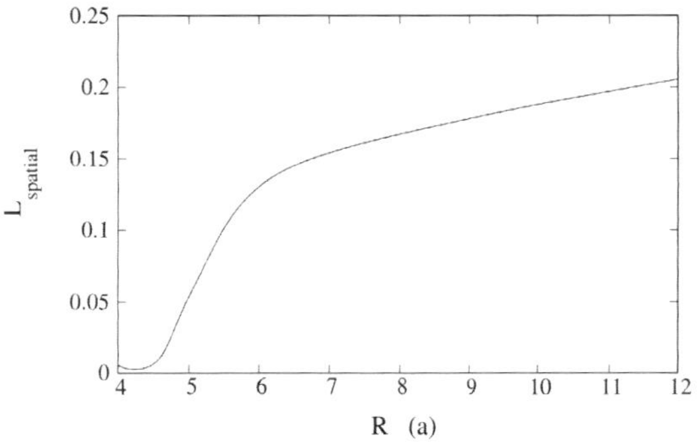

Figure 3. Spatial entanglement quantified using the linear entropy of the reduced density matrix as a function of the control parameter R.

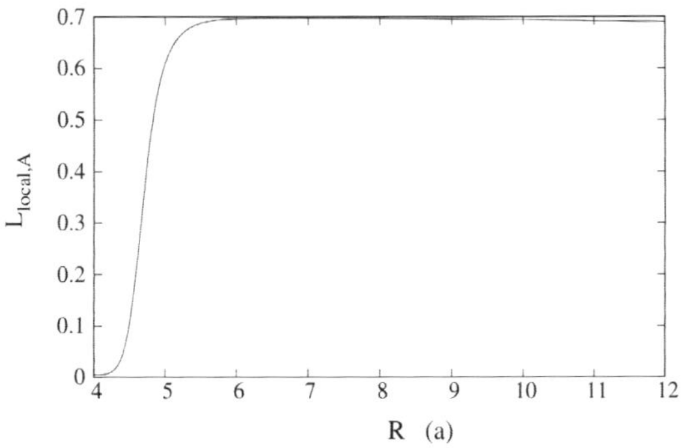

Figure 4. Local entanglement for site A using the linear entropy of the reduced density matrix ($L_{\text{local,A}}$) as a function of the control parameter R.

of R we consider ($R = 4\ a$), the inner well width is at its maximum of 1. This well is centred at the origin so our centre partition (B) always contains the inner well.

We may calculate the local entanglement using the linear entropy of the reduced density matrix to give L_{local}. Now the reduced density matrix is found by tracing out the other two partitions. This reduced density matrix is conveniently calculated in the site occupation basis as $diag(P(\uparrow), P(\downarrow), P(\uparrow\downarrow), P(0))$ where $P(\chi)$ is the probability of occupation χ in the partition of interest.

Rather than using the average local entanglement we calculate the mean local entanglement of sites A and C thereby neglecting the local entanglement of site B. We use this to model the local entanglement available if only these two parts of the system are accessible to measurement, thereby introducing a spatial separation between the measurements. As the system is symmetric about 0 the average 'accessible' local entanglement of partitions A and C, is just the local entanglement of partition A, i.e., $L_{\text{local}}^{\text{accessible}} = (L_{\text{local,A}} + L_{\text{local,B}})/2 = L_{\text{local,A}}$. We see in Fig. 4 that this entanglement, similarly to the spatial entanglement, generally increases with increasing R once the inner well is small enough. We attribute this to the wavefunction spreading out and thereby increasing the possibilities on measurement in a partition beyond simply finding it empty all the time. So although an increase in spatial entanglement may suggest an increase in the interaction of the system, and stronger interactions tend to reduce the local entanglement (see for example [7]) by limiting the possibilities on measurement, here the spatial entanglement and local entanglement of the accessible partitions behave similarly, both quickly increasing as the wavefunction spreads into the outer well. We note that this local entanglement appears to decrease very slightly after $R \sim 8\ a$. This may be a small numerical inaccuracy due to the widening outer well but finite X_{max} in the definition of the partitions. Or, and potentially more interesting, the decrease in 'accessible' local entanglement could signify that now that the wavefunction has spread out to completely occupy the outer well, increasing R combined with the Coulomb interaction results in a reduction in the probability of the measurement of double occupation in a partition. This may be worthy of further investigation.

The behaviour of this 'accessible' local entanglement suggests that it could be possible to experimentally begin with a localised state which has little or no spatial entanglement nor local entanglement accessible to measurement then adiabatically decrease the inner well and expand the outer well thereby increasing the spatial entanglement and additionally the amount of available local entanglement.

5. Derivatives

Previous work [1] showed that when the potential of a system similar to the one considered here splits into two wells the behaviour of the derivatives of the energy and the entanglement were suggestive of something akin to an analogue of a quantum phase transition in the few-particle regime. In this work we find that the narrowing of the well alone appears to not be enough to cause a sharp transition as both of the derivatives are smooth (see Figs. 5 and 6). However it seems that we can associate the region around the second stationary point in the entanglement derivative (Fig. 5) at $R \sim 5.1\ a$ with the particles no longer being bound to the inner well (see the inset of Fig. 2).

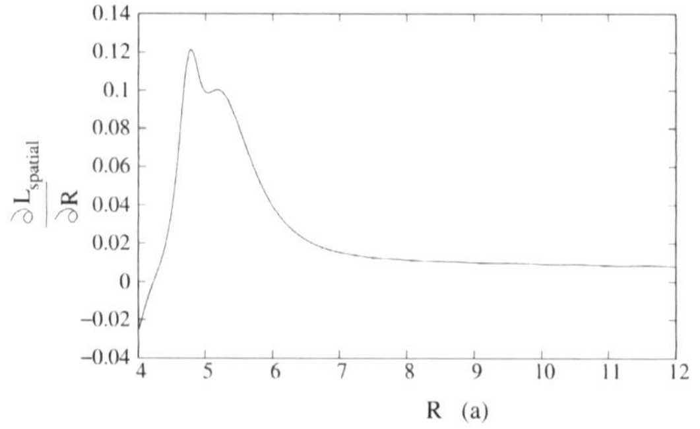

Figure 5. Derivative of the linear entropy of the reduced density matrix (spatial entanglement) with respect to the control parameter R.

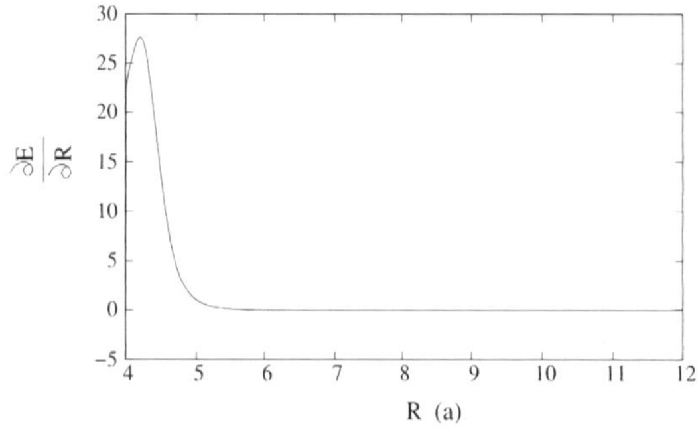

Figure 6. Derivative of the ground-state energy with respect to the control parameter R.

6. Summary

We saw that the two interacting electrons are not always bound to the inner well as its width becomes narrow. This results in the spread of the wavefunction and an increase in entanglement.

We have shown that both the spatial and local entanglement of the 'accessible' partitions of the system can be varied by narrowing the width of the inner while increasing the outer well width. Interestingly the 'accessible' local entanglement we defined behaved similarly to the spatial entanglement with respect to the potential control parameter R.

We noted that for this system the derivatives of the spatial entanglement and energy did not show any discontinuity demonstrating that the entanglement and energy change with respect to R is a smooth process in this system. However the derivative of the entanglement did appear to signal when the electrons were no longer bound to the inner well.

References

[1] Abdullah S, Coe J P and D'Amico I 2009 *Phys. Rev. B* **80** 235302
[2] Wu L A, Sarandy M S, Lidar D A and Sham L J 2006 *Phys. Rev. A* **74** 052335
[3] Wu L A, Sarandy M S and Lidar D A 2004 *Phys. Rev. Lett.* **93** 250404
[4] Coe J P, Sudbery A and D'Amico I 2008 *Phys. Rev. B* **77** 205122
[5] Zanardi P 2002 *Phys. Rev. A* **65** 042101
[6] Larsson D and Johannesson H 2005 *Phys. Rev. Lett.* **95** 196406
[7] Coe J P, França V V and D'Amico I 2010 arXiv:1003.3562, Phys. Rev. A in press

Spin dynamics in two-dimensional arrays of quantum dots with different shapes

A.F. Zinovieva[1], A. S. Lyubin[1], N. P. Stepina[1], E. S. Koptev[1], A. I. Nikiforov[1], A. V. Dvurechenskii[1], N. A. Sobolev[2], M. C. Carmo[2]

[1] Institute of Semiconductor Physics, 630090 Novosibirsk, Russia
[2] Departamento de Fisica e I3N, Universidade de Aveiro, Aveiro, Portugal

E-mail: aigul@isp.nsc.ru

Abstract. Electron paramagnetic resonance and spin echo methods are used to probe the spin dynamics in two-dimensional quantum dot (QD) arrays with different shape of nanoclusters. Two types of QD structures were investigated: 1) with single shaped QDs (hut-clusters, having the ratio of height h to lateral size l, $h/l = 0.1$), and 2) with two groups of QDs, *hut*- and *dome*-clusters ($h/l = 0.2$). Both types of structures demonstrate the EPR signals from electrons localized in QD layers.The orientation dependence of EPR line width for first type structures is well described by model of spin relaxation through precession in the effective magnetic field, arising during tunneling between QDs due to structure-inversion-asymmetry. In the experiments on structures with dome-clusters the additional peculiarity, which can not be explained in the framework of precession model, is observed. The different orientation dependencies can be explained by different localization degree of electrons in investigated structures. Spin echo measurements give the longest spin decoherence time for structures with single shaped QDs.

1. Introduction

A dimensionality reduction leads to the appearance of new effects, many of them are governed by symmetry of nanostructures. In particular, the symmetry of nanostructure has a crucial impact on the spin dynamics. The low symmetry of nanostructures can lead to appearance of additional spin relaxation mechanisms, which takes place in the arrays of tunnel-coupled quantum dots (QDs) with structure-inversion-asymmetry. At high density of QDs the overlapping between localized states is sufficient for coming in force the most efficient Dyakonov-Perel mechanism of spin relaxation [1]. In this case the spin relaxes during series of random tunneling events (in hopping transport) through precession in the effective magnetic field whose direction can be changed after each tunneling event. There are two ways to increase the spin relaxation time. The first follows from origin of the effective magnetic field. It relates to spin-orbit interaction and can vanish in more symmetrical structures. The second way is related to the suppression of probability of tunneling between QDs, that can be realized by the creation of well-separated QD array.

The present paper is aimed to realize the increase of spin relaxation time in Si/Ge structures with quantum dots. To reach this goal we examine the electron spin coherence in QD structures with different shapes of nanoclusters.

2. Samples and Experiment

Samples were grown by molecular-beam epitaxy on n-Si(001) substrates with a resistivity of 1000 Ω cm. The QD structure with single shaped QDs (hut-clusters) was optimized with the aim to enlarge the electron binding energy in strain-induced potential well in Si near Ge QD. The vertical stack of four Ge QDs layers was inserted into the 0.6 μm epitaxial n-Si layer (Sb concentration $4 \cdot 10^{16}$cm^{-3}) at the distance of 0.3 μm from the substrate. The space charge spectroscopy confirms the localization of electrons in Si near QDs with binding energy $\sim$ 50 meV [2]. The test structures uncovered by Si capping layer was investigated by scanning tunneling microscopy (STM). It was shown that the Ge islands have the shape of *hut*-clusters with the average lateral size l=20 nm and height h=2 nm. The density of QDs is $\sim 10^{11}$ cm^{-2}.

The QD structure with *dome*-clusters contains 6 layers of QDs separated by 30 nm Si spacer layers. Each QD layer are formed by deposition of 7 ML Ge at the temperature $T = 550°$. On top of the structure a 0.3 μm epitaxial n-Si layer (Sb concentration $4 \cdot 10^{16}$cm^{-3}) is grown, the same layer is grown under QDs layers. The STM of structure with single QD layer uncovered by Si shows the bimodal distribution of QDs (*huts* and *domes*). The average lateral size of *dome*-clusters is l=50 nm and height h=10 nm. The density of *dome*-clusters is 10^9 cm^{-2}. The *hut*-clusters are distributed between *dome*-clusters with density 10^{11} cm^{-2}.

Measurements were performed with a Bruker X-band EPR spectrometer at the temperatures T = 4.5 − 20 K. To avoid the needless EPR signal from dangling bonds (g = 2.0055) the passivation with atomic hydrogen was done before measurements. To increase the number of registrable spins the sandwiched sample was prepared.

The spin echo measurements were carried out at temperature 4.5 K in resonance magnetic field H = 3470 G (can be slightly varied ±5 G in dependence on resonance conditions) with direction corresponding to the narrowest EPR line width, for the first type structure $\mathbf{H} \parallel Z$, where Z is the [001] growth direction of the structure, for the second type structure the magnetic field deviates from Z on $\theta = 30°$. A two-pulse Hahn echo experiment ($\pi/2-\tau-\pi-\tau-$ echo) was used to measure T_2 (a detailed explanation can be found in Ref. 3). In order to observe a longitudinal spin relaxation (corresponding time T_1), a different pulse sequence is applied ($\pi-\tau-\pi/2-T-\pi-T-$ echo). In the first and second type of experiments, the durations of $\pi/2$ and π pulses were 60 ns and 120 ns, respectively; the interpulse time in the second experiment was kept T = 200 ns.

3. Results and Discussion

Both types of structures demonstrate the EPR signals from electrons localized in QD layers. The characteristic properties of EPR line (g-factor and EPR line width) depends on the type of the structure. The narrowest EPR lines (width $\Delta H \approx$ 0.8 G) was detected for the first type structures. The principal values of g-factor are very close to g-factor values in uniaxially deformed Si (g_{zz}=1.9995 ± 0.0001 and g_{xx}=g_{yy}=1.9984 ± 0.0001). These EPR lines have a strong orientation dependence of EPR line width. When external magnetic field deviates from growth direction of the structure the EPR lines become broader and weaker, and for in-plane magnetic field they have the maximal width ($\Delta H \approx$ 3 G). The second type structures show the wider EPR lines, having a weaker orientation dependence of g-factor and EPR line width. In contrast to the single shaped QD samples the maximum of EPR line width for the structures with *dome*-clusters is observed at $\theta = 60°$. Here the additional peculiarity is observed: at the deviation of external magnetic field on small angles (up to $\theta = 30°$) the EPR line width sharply decreases from 2.3 G to 1.5 G (Fig. 1). The g-factor behavior completely correlates with EPR line width change (Fig. 2). In the magnetic field perpendicular to the plane of QD array the g-factor value is g = 1.9991 ± 0.0001. At the deviation of magnetic field from Z-axis the g-factor value slightly grows up to g = 1.9992 ± 0.0001, then at $\Theta \approx 60°$ the minimal g-factor value g = 1.9987 ± 0.0001 is observed, and in the end $\Theta = 90°$ the g-factor value increases and practically returns to its initial value $g = 1.9990 \pm 0.0001$.

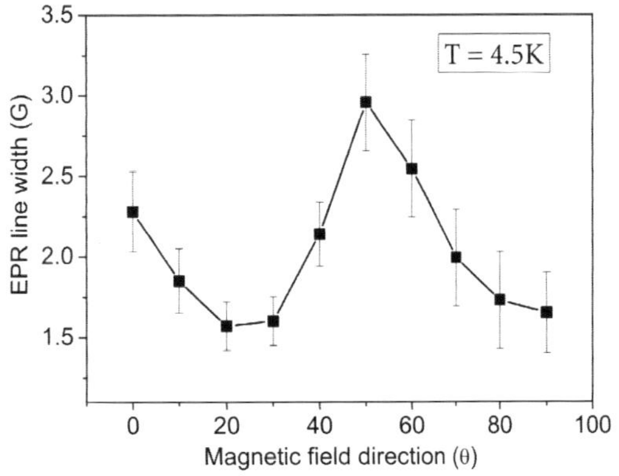

Figure 1. The angular dependence of EPR-line width for heterostructure with dome-clusters. For $\Theta = 0$ the magnetic field is parallel to the growth direction of the structure.

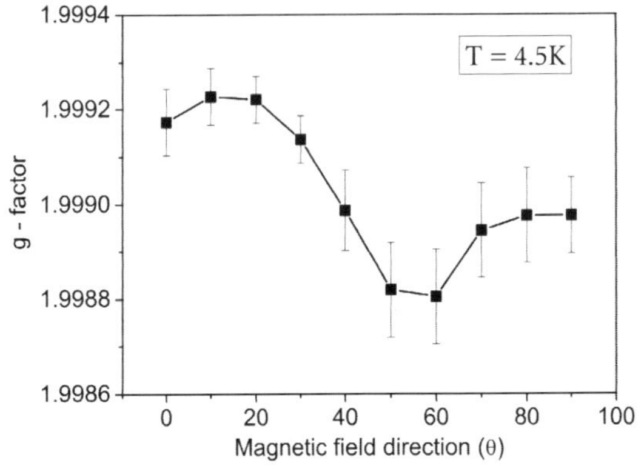

Figure 2. The angular dependence of electron g-factor for heterostructure with dome-clusters. For $\Theta = 0$ the magnetic field is parallel to the growth direction of the structure.

According to results of two-pulse Hahn echo experiment for both types structures the spin echo behavior can be described by superposition of two exponentially decaying functions (Fig. 3). The decay parameters give two times of spin dephasing: for the first type structure $T_2^{(1)} \approx 0.9\,\mu$ s and $T_2^{(2)} \approx 20$ μs; for the second type structure $T_2^{(1)} \approx 0.28$ μs and $T_2^{(2)} \approx 3$ μs.

The analysis of an inversion signal recovery, measured in three-pulse echo experiments, shows a non-exponential behavior for both single-shaped and two-shaped QD structures (Fig. 4). The experimental curve can be described by the superposition of two functions:

$$M(t) = M_{0z} - M_z^{(1)} exp(-\tau/T_1^{(1)}) - M_z^{(2)} exp(-\tau/T_1^{(2)}),$$

where M_{0z} is the equilibrium magnetization, $M_{0z} = M_{0z}^{(1)} + M_{0z}^{(2)}$, $M_z^{(1,2)} = M_{0z}^{(1,2)} - M_z^{(1,2)}(0)$, $M_z(0) = M_z^1(0) + M_z^2(0)$ is the magnetization just after applying of an inverting π-pulse. The characteristic times obtained by fitting for the first type structure is about $T_1^{(1)} \approx 400$ ns and $T_1^{(2)} \approx 10\,\mu s$. For the second structure $T_1^{(2)} \approx 2\,\mu s$ and $T_1^{(1)} \approx 20 \mu s$. For interpretation of results the existence of two electron groups with different spin relaxation times is suggested. For the first type structure both groups are formed by electrons in QD layers. These groups located in Si spacers with different width and have the different probability of electron hopping between Ge QDs [4]. In both groups of carriers the special relation between T_2 and T_1, $T_2 \approx 2T_1$, is observed. The unusual relation $T_2 > T_1$ confirms that the spin relaxation is caused by the interaction with the effective magnetic field arising due to the structure-inversion-asymmetry. For the second structure the spin echo signal is also formed by two groups of electrons. There are the free electrons in Sb-doped Si layers with isotropic g-factor 1.9987 and electrons localized in QD layers (g_{zz}=1.9995 and g_{xx}=g_{yy}=1.9984). For first group the relation $T_2 < T_1$ is observed, while for QD electrons the inverse relation ($T_2 < T_1$) is obtained.

The orientation dependence of EPR line width for first type structures is well described by model of spin relaxation through precession in the effective magnetic field, arising during tunneling between QDs due to structure-inversion-asymmetry [5]. For the second type structure the orientation dependence of EPR line width is more complicated and can not be explained by simple model. For explanation of this orientation dependence we consider two mechanisms,

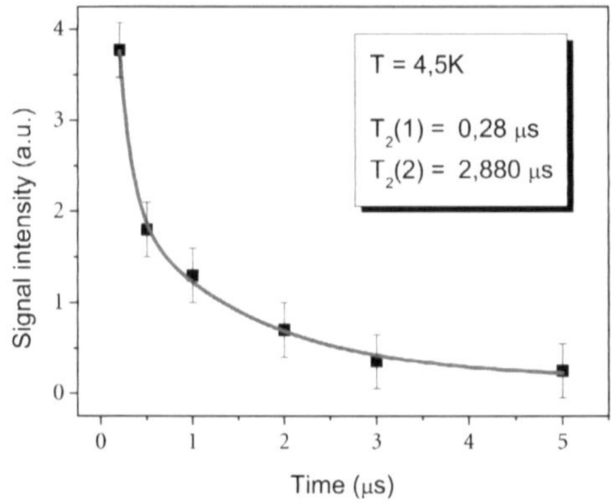

Figure 3. Results of two-pulse spin echo experiments (points) and approximation by superposition of two exponential functions (line). Time of transversal relaxation.

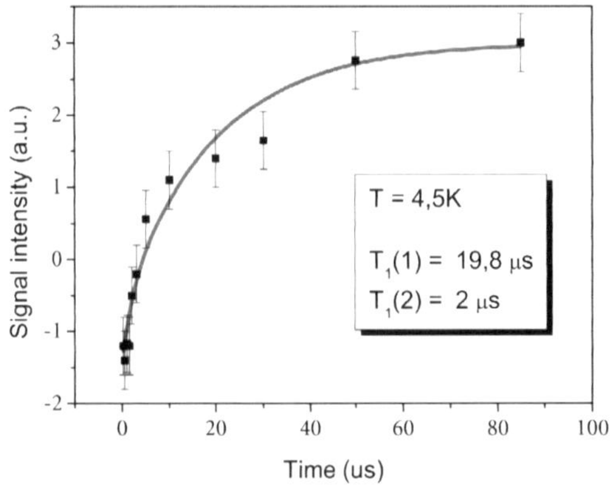

Figure 4. Results of three-pulse spin echo experiments (points) and theoretical approximation (line). Time of longitudinal relaxation.

which control the EPR line width. The first mechanism is the motional narrowing of EPR line width. It controls the EPR line width at the initial stage of orientation dependence, up to $\theta = 30°$. At $\theta = 0°$ the tunneling transitions between QDs is suppressed due to wave function shrinking. With deviation of magnetic field from Z axis the effect of wave function shrinking becomes weaker, that leads to increasing probability of tunneling transitions between QDs. As result, we observe the narrowing of EPR line down to 1.5 G ($\theta = 30°$). After this point the probability of tunneling transitions between QDs is sufficiently high for coming in force the second mechanism. It is the spin relaxation due to interaction with the spin-orbit fields that works also in the first type structures. Then at larger deviation angles θ the EPR line width behavior can be described in framework of the same model of spin precession, but with another parameters of model. The main difference is the dependence of characteristic fluctuation time (hopping time) τ_c on the external magnetic field $\tau_c = \tau_0 \exp(\alpha H^2)$. The theoretical approximation of the orientation dependence of EPR line width for second type structures in the range $\theta = 30° - 90°$ allows to estimate the effective radius of electron localization in second type structures, as 80 nm. While for the first type structures it is about 10-15 nm that is several times smaller than magnetic length at given external magnetic field ($H = 3470$ G) $\lambda \approx 50$ nm. Namely the last leads to independence of τ_c on the external magnetic field for the first type structures and more simple orientation dependence of EPR line width. The different localization degree leads to different spin dynamics in the structures under study and longer spin relaxation times (up to 20 μs) in first type structures.

This work was supported by RFBR (Grants 08-02-00121, 09-02-90480).

References

[1] Shklovskii B I 2006 *Phys. Rev. B* **73**, 193201
[2] Yakimov A I, Dvurechenskii A V, Nikiforov A I, Bloshkin A A, Nenashev A V, Volodin V A 2006 *Phys. Rev. B* **73**, 115333
[3] Schweiger A and Jeschke G 2001 *Principles of Pulse Electron Paramagnetic Resonance* (Oxford: Oxford University Press)
[4] Zinovieva A F, Dvurechenskii A V, Stepina N P, Nikiforov A I, Lyubin A S, Kulik L V 2010 *Phys. Rev. B* **81** 113303
[5] Zinovieva A F, Dvurechenskii A V, Stepina N P, Deryabin A S, Nikiforov A I, Rubinger R M, Sobolev N A, Leitao J P, Carmo M C 2008 *Phys. Rev. B* **77**, 115319

Journal of Physics: Conference Series **245** (2010) 012053 doi:10.1088/1742-6596/245/1/012053

Decoherence of a resonantly driven exciton in a single quantum dot

M Ravaro[1], A Enderlin[1], C Tonin[1], V Voliotis [1,2], R Grousson[1], A Lemaitre[3] and A Martinez[3]

[1]Institut des NanoSciences de Paris, CNRS, Université Pierre et Marie Curie, Campus Boucicaut, 140 Rue de Lourmel, 75015 Paris, France
[2]Université d'Evry Val d'Essonne, Boulevard F. Mitterand, 91025 Evry, France
[3]Laboratoire de Photonique et Nanostructures, CNRS, Route de Nozay, 91460 Marcoussis, France

E-mail: voliotis@insp.jussieu.fr

Abstract. We report on coherent control of the resonantly driven exciton state in a single quantum dot in the strong coupling regime. With picosecond laser pulses, the two-level system undergoes Rabi oscillations and a given coherent superposition of states can be addressed. Using a pair of phase-locked pulses, optical manipulation is achieved during the coherence time of the qu-bit. The dephasing processes are also studied and we show that in typical quantum dots, the coherence time T_2 can be of the same order of magnitude as the effective lifetime of the exciton T_1 (a few hundreds of ps) although not reaching the upper theoretical limit of $2T_1$. We conclude that energy relaxation and pure dephasing processes due to virtual scattering with phonons contribute on an equal footing to the loss of coherence.

1. Introduction

Semiconductor quantum dots (QDs) have attracted considerable interest since they appear as promising candidates to realize quantum bits for future quantum information processing. To this aim, a fundamental requirement is the coherent manipulation of QDs, which is severely limited by the decoherence mechanisms arising from the strong interaction with their environment and thus limiting the number of possible quantum operations. Apart from radiative recombination, these decoherence mechanisms are due to interaction with phonons and/or fluctuations of the electrostatic background.

In order to take full advantage of the longest possible coherence time, it is more interesting to drive the lowest lying level of the exciton, since it is the least interacting state with its surrounding. However, resonant optical spectroscopy remains a real experimental challenge. Only recently, resonant luminescence has been investigated by inserting the QDs in microcavities [1]. Photocurrent measurements have also been performed under resonant excitation of single QDs in Schottky diodes [2]. We choose an alternative approach by embedding the QDs in a one-dimensional waveguide for the confinement of the driving-field, so the resonance luminescence can be detected from the waveguide top surface [3]. This experimental configuration combines micro-photoluminescence (μPL) and waveguiding geometry that allows an all-optical control of the driven state. Moreover, the reduced volume of the optical mode in the one-dimensional waveguide, together with a large absorption cross-section of the semiconductor QD, yields an enhanced light-matter interaction which reduces the effective lifetime of the exciton.

Coherent control (CC) of a single dot has also been achieved [4] using a pair of phase-tailored laser pulses. By addressing different excitation regimes, depending on pulse delay and power, we show that CC provides a useful tool not only to manipulate the excitonic state but also to measure its relaxation-time constants.

2. Resonant Rabi regime

The QDs of interest consist of InAs self-assembled QDs grown by molecular beam epitaxy on a GaAs substrate. The InAs wetting layer and the QDs are embedded in a 1 μm GaAs layer, between two AlGaAs layers. This structure forms a 2D waveguide, the GaAs being its core and the AlGaAs its barrier. 1 μm ridges are then wet-etched on the surface of the sample in order to obtain a 1D waveguide (figure 1). The density of QDs is small enough to be spectrally resolved.

The sample is fixed on the cold finger of a three optical access cryostat and cooled down to 10 K. The excitation is provided by a tunable Ti-Sapphire laser with 1 ps pulses. The pump beam is focused with a large numerical aperture microscope objective, in front of the sample edge. Then, it propagates along the 1D waveguide and interacts with the QDs. The luminescence is then collected from the top surface of the sample by another microscope objective. Since the laser light is confined in the guided mode, the scattered light is greatly suppressed and the resonant luminescence is almost background free, at low pump power. Experimental details can be found elsewhere [3,4].

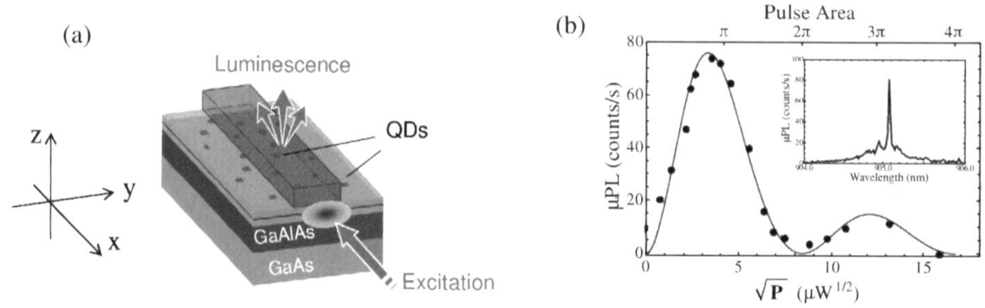

Figure 1. (a) Schematic experimental guided-wave configuration showing the 1D optical mode and the polarization axes (x and y). (b) μPL maximum intensity extracted from spectra like shown in the inset, for on-resonance excitation in an InAs/GaAs QD demonstrating the Rabi oscillations versus square-root of incident power. The solid curve is a data simulation using the optical Bloch equations.

The optical eigenmodes of the waveguide are linearly polarized (y for TE mode and z for TM) and the QD eigenstates are linearly polarized along x and y. Thus, for low pump power resonant experiments, the QD μPL and the laser should be in principle, copolarized. However, in the case of self-assembled QDs we have very often observed luminescence in both polarization directions, suggesting that the QDs are tilted from the crystallographic axis [5]. The splitting of the exciton fine structure is of the order of 10 μeV and has been measured using a Fabry-Pérot interferometer placed in the luminescence path, because the spectral resolution of our set-up (40 μeV) does not allow spectrally resolving the splitting.

When the pump is resonant with the lowest lying exciton state in the dot and as the pump power is increased, we observe oscillations of the luminescence, demonstrating strong coupling between the QD and the driving field (figure 1 b). The number of periods is limited by the laser scattering as the pump power is increased, but we also observe that the oscillations are damped. Very recently, in [6] interaction with acoustic phonons has been proposed as the main source for the intensity damping of Rabi oscillations (RO). However, in our case the damped RO can be well reproduced only by

Journal of Physics: Conference Series **245** (2010) 012053 doi:10.1088/1742-6596/245/1/012053

introducing a power-dependent phenomenological parameter in the population term of the Bloch equations [3]. This means that there is a very fast relaxation of the population as power is increased. It has also to be related to the increase of the spontaneous emission rate observed in the resonant time-resolved experiments. Indeed, the exciton effective radiative lifetime is reduced by about 50% for InAs dots. We ascribe this effect to an enhanced coupling of the emitted photons to the guided 1D mode arising from a modification of the refractive index that leads to a reduction of the propagating pulses group velocity. This phenomenon of "slow light" is due to self-induced transparency of the dipole coupled to the one-dimensional optical mode, a similar effect observed in QDs coupled to photonic crystals [7]. Recently, slow light effects have been investigated in QDs coupled to photonic crystals waveguides exploiting the waveguide dispersion [8]. In our case, the waveguide dispersion leads to a very small modification of the refractive index, only a few %.

Rabi oscillations experiments allow addressing one qu-bit, i.e. to choose the proper superposition of the two quantum levels. The final state $|\psi\rangle = \cos\left(\frac{\theta}{2}\right)|0\rangle + \sin\left(\frac{\theta}{2}\right)|1\rangle$ is completely determined by the pulse area θ, which is the time integrated Rabi frequency: $\theta = \int_{-\infty}^{\infty} \Omega(t)dt = \int_{-\infty}^{\infty} \frac{\mu E(t)}{\hbar} dt$, μ being the dipole moment of the transition and E(t) the pulse envelope. For typical InAs QDs, the transition dipole moment is 25 D and the Rabi coupling constant is of the order of one meV.

3. Coherent control of a single qu-bit

The optical manipulation of the prepared state can then be achieved using a pair of phase-locked pulses. The delay between the pulses δ, and the relative phase ϕ are well controlled using a stabilized Michelson interferometer on a scale of λ/ 100. The most interesting case for investigating the coherence properties of the system is when π/2 pulses excite the QD. The first pulse creates a state with the maximum coherence and the second one, depending on its relative phase, can interfere constructively or destructively, thus enhancing or annihilating the dipole oscillation [4]. Measuring the resonant luminescence as a function ϕ and δ yields the total coherence time T_2 of the state. Figure 2 shows an example of coherent control experiments, where the population oscillation is observed as a function the phase ϕ, for a given delay. The broad peak on which is sitting the sharp μPL line is due to the spectrally broad laser line (1 ps laser pulse corresponds to a 0.5 meV wide spectral line) .

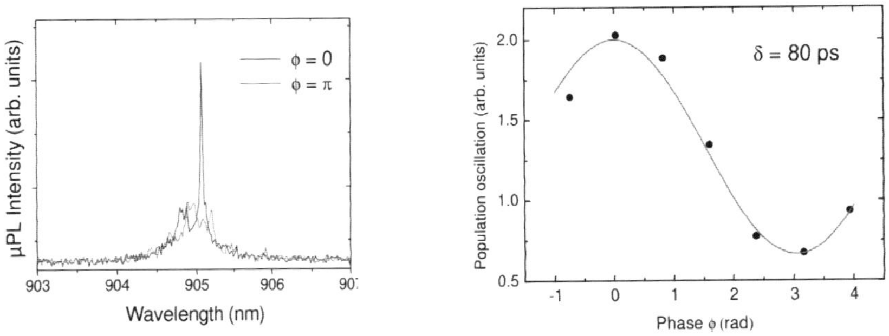

Figure 2 (a). Resonant luminescence spectra of a QD pumped by two π/2 pulses in phase (ϕ = 0) and out of phase (ϕ = π) delayed by 10 ps. **(b)** The population oscillates as a function of ϕ at a given delay δ.

The contrast between maximum (ϕ = 0) and minimum (ϕ = π) luminescence as a function of delay δ, is shown in figure 3. The exponential decay depends on both the state coherence time T_2 and its effective lifetime T_1 and has the following expression: $C(\delta) = \frac{e^{-\delta/T_2}}{2 - e^{-\delta/T_1}}$. Independent time-resolved measurements allow fixing the effective lifetime T_1, thus the only adjustable parameter is the coherence time T_2. In several dots, we find that the coherence time is of the same order of magnitude as the effective lifetime.

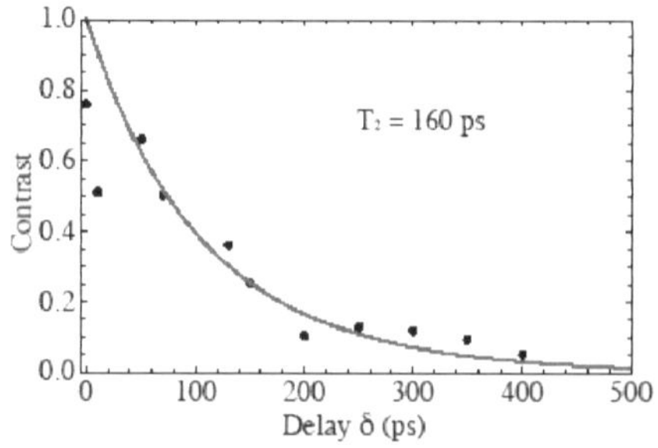

Figure 3. Quantum interference contrast as a function of delay δ (solid circles). The solid line corresponds to a fit with fixed T_1=400 ps (see text). The dephasing time obtained is T_2 =160 ps.

The decoherence probability rate, $1/T_2$ can be written as: $\frac{1}{T_2} = \frac{1}{2T_1} + \frac{1}{T^*}$, T_1 represents the exciton effective lifetime (radiative recombination and inelastic scattering with phonons) and T^* represents the elastic pure dephasing processes (virtual phonons, and environmental charge fluctuations). For the ground exciton state that we are probing, phonon absorption at low temperature has been calculated to be very inefficient [4], typically of the order of 1 ns. Thus T_1 is limited only by emission of photons. As T_2 is less than the theoretical upper limit of $2T_1$, then pure dephasing processes are also relevant. In the example given above, T^* is of the order of 250 ps. Due to low enough temperature, polarons are not formed, as confirmed by the absence of lateral wings in the emission line. Therefore, the pure dephasing contribution is due to virtual phonon scattering mechanisms or environmental charge fluctuations if they exist.

4. Conclusion

Coherent control in the strong coupling regime has been achieved on the ground exciton state in a single QD. Resonant excitation in the 1D guided-wave geometry allow to observe long coherence time of the ground exciton state and faster exciton radiative lifetime. These are basic requirements for realizing single indistinguishable photon sources with high generation rate. The role of acoustic phonons in the dephasing processes has to be further investigated and CC experiments as a function of temperature are currently carried out. For future applications, the collection efficiency remains a challenge and new improvements of the structures design have to be made.

References

[1] Muller A *et al*, 2007 Phys. Rev. Lett. **99** 187402
[2] Stufler S *et al*, 2005 Phys. Rev. B **72** 121301 (R)
[3] Melet R *et al*, 2008 Phys. Rev. B **78** 073301
[4] Enderlin A *et al*, 2009 Phys. Rev. B **80** 085301
[5] Kowalik K *et al*, 2008 Phys. Rev. B **77** 161305 (R)
[6] Ramsay A J *et al*, 2010 Phys. Rev. Lett. **104** 017402
[7] Faraon A *et al*, 2008 Opt. Express **16** 12154
[8] Nielsen T R *et al*, 2009 Appl. Phys. Lett. **94** 113111; Dewhurst S J *et al*, 2010 Appl. Phys. Lett. **96** 031109

Journal of Physics: Conference Series **245** (2010) 012054 doi:10.1088/1742-6596/245/1/012054

Excitonic complexes in InGaAs/GaAs quantum dash structures

A Musiał[1], G Sęk[1], P Podemski[1], M Syperek[1], J Misiewicz[1], A Löffler[2], S Höfling[2] and A Forchel[2]

[1]Institute of Physics, Wrocław University of Technology, Wybrzeże Wyspiańskiego 27, 50-370 Wrocław, Poland

[2]Technische Physik, Universität Würzburg, Wilhelm-Conrad-Röntgen-Research Center for Complex Material Systems, Am Hubland, D-97074 Würzburg, Germany

E-mail: anna.musial@pwr.wroc.pl

Abstract. Optical properties of single self-assembled InGaAs/GaAs quantum dashes have been investigated. The excitation power dependent measurements together with the rate equation model calculations enable to identify the excitonic and biexcitonic emission from one quantum dash. Based on that the biexciton binding energy and the relative exciton to biexciton lifetime ratio have been estimated. The emission spectra have also revealed a fingerprint of Coulomb interaction between excitons confined in the quantum dash and adjacent wetting layer. The conditions required for observing this phenomena and the influence of the wetting layer on the emission of individual quantum dashes have been discussed.

1. Introduction

Semiconductor quantum dash technology development is mostly driven by their optoelectronic applications. Quantum dash ensemble used as an active region in lasers or optical amplifiers exceeds in some cases the standard quantum-dot emission properties. This has already been proven for InAs dashes on InP and the improved device performance has been demonstrated [1].

Recently, the possible applications of quantum dashes have evolved toward the single-dash ones and quantum electrodynamics experiments (e.g. single photon sources or quantum bits for quantum cryptography and computation) as both the weak and strong coupling between excitons confined in single dashes and the three-dimensionally confined electromagnetic field was proven in InGaAs/GaAs material system [2]. This is possible due to relatively big sizes (from tens up to 100 nm in the elongation direction) and weakening of the quantum confinement resulting in enhanced oscillator strength of optical transitions. Those prospective results demand further and more comprehensive investigations of optical properties of individual quantum dashes, especially that, there is still a lack of both experimental data and theoretical calculations regarding such InGaAs dashes on GaAs. The existing reports have rather preliminary and qualitative character especially in excitonic complexes theme [3]. In this paper more systematic and quantitative investigations of InGaAs/GaAs quantum-dash structures is presented.

2. Experimental details

2.1. Investigated structures

All presented results concern the epitaxially (MBE) grown self-assembled structure with quantum dashes formed of nominally 4.5 nm $In_{0.3}Ga_{0.7}As$ layer deposited on GaAs. Further growth details [4] and ensemble properties [5] can be found elsewhere. The indium content of about 30% enables exceeding the critical thickness for 3D island formation. Low-strain growth conditions results in formation of (5 x 30 x 100) nm structures preferentially elongated in one of the lateral directions ([1-10]). The surface density of quantum dashes is rather low - (6 – 9) 10^9 cm^{-2} resulting in very few quantum dashes to be excited within the laser spot. To further increase the spatial resolution a submicron mesa structures (from 1 μm down to 100 nm) were etched on the sample surface, providing 10 quantum dashes within smaller mesas, or less. The emission of individual objects have been additionally spectrally resolved due to size nonuniformity of self-assembled structures.

2.2. Experimental setup

Optical measurements have been realized in microphotoluminescence setup composed of microscope objective with high numerical aperture (0.4) and long working distance (20 mm), 0.5 m focal single grating spectrometer and InGaAs nitrogen-cooled CCD linear detector. Investigated structures were nonresonantly excited by a semiconductor laser emitting at 660 nm. Described setup characteristics (1 μm spatial and below 100 μeV spectral resolution) are sufficient for single-dot spectroscopy. The samples were mounted in a continues-flow microscopy-type cryostat and cooled with liquid helium down to the temperature of approximately 5K. For time-resolved photoluminescence setup details see ref. 6.

3. Results and discussion

3.1. Excitonic complexes

For single-dot investigations mesa structures sizes of hundreds nanometers have been chosen. They contain number of quantum dashes small enough to resolve single emission lines and at once the mesa sizes are sufficiently big to decrease the influence of their edges (e.g. stronger local electric field fluctuations due to the charges at the mesa sidewalls) broadening the emission. Exemplary microphotoluminescence spectra have been presented in figure 1. The individual emission lines both with the gaussian-type background (characteristic for the nonuniform quantum objects ensemble) due to relatively big mesa size (700 nm) can be clearly seen.

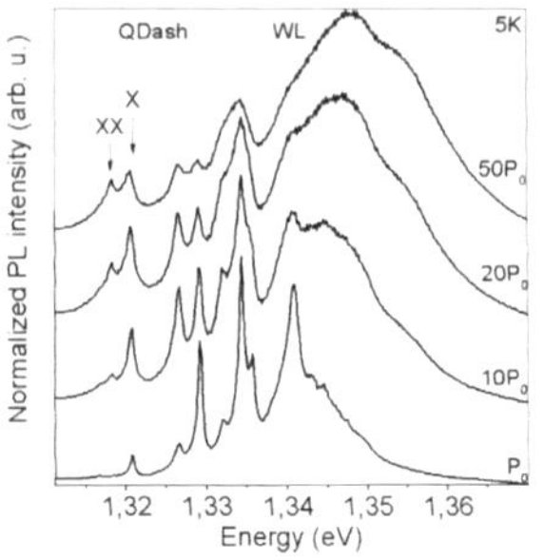

Figure 1. Low temperature normalized microphotoluminescence spectra of InGaAs/GaAs 30% In content quantum dashes for different excitation power values.

Observed lines are relatively broad (hundreds of μeV) compared to the setup resolution due to spectral diffusion effects – fluctuations of the charge environment in the time scale much smaller than the time of the measurements. Those effects are stronger for quantum dashes because of their relatively big sizes with respect to the mesa size causing the location of a structure in the vicinity of mesa edge more probable and hence stronger influence of the mesa sidewalls surface charge fluctuations. Choosing emission lines from the low energy tail of the spectra provides better separated lines originating from ground state emission. Higher energy part of the spectra revealed a wetting layer emission centered at 1.345 eV. Excitation power dependent measurements allow identification of excitonic and biexcitonic emission which intensity dependence is in low excitation conditions linear and quadratic, respectively. As the excitation power increases (above a critical value sufficient for biexciton formation) the excitonic emission saturates and the biexciton starts to dominate the spectra. To describe theoretically the emission intensity as a function of excitation power in the whole range of measured parameters three level rate equation model has been used [7,8]. The agreement between the modeled and measured dependences supports the origin of observed emission from one quantum dash. This allowed biexciton binding energy determination (as a spectral distance between excitonic and biexcitonic emission lines). For investigated structures it is in the range of 1 – 3 meV, which is consistent with our expectations regarding the strength of quantum confinement – due to their relatively big sizes they are rather in weak confinement regime. The best fit has been obtained for the exciton to biexciton lifetime ratio (being the only free parameter of this procedure) lower than 1 – see figure 2. This value is an additional fingerprint of the weak quantum confinement [9].

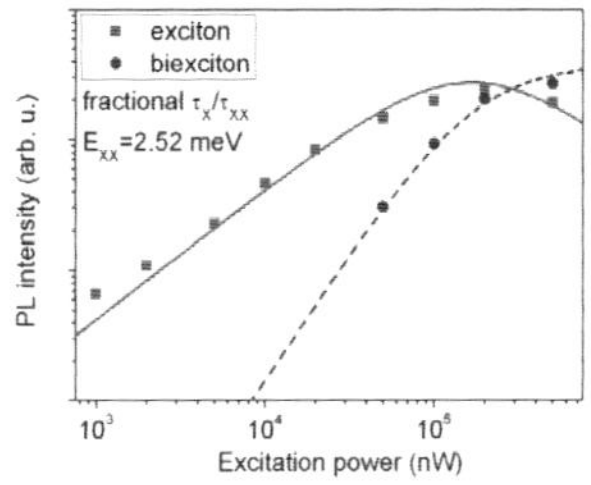

Figure 2. Excitonic and biexcitonic emission lines intensity vs. excitation power analysis in double logarithmic scale: experimental points and theoretical curves obtained from rate equation model for fractional exciton to biexciton lifetime ratio.

3.2. Sideband feature

Another very interesting feature in the emission spectra of individual quantum dashes under investigation is an undoubted asymmetry of biexcitonic line denoted as 'sideband' in figure 3.

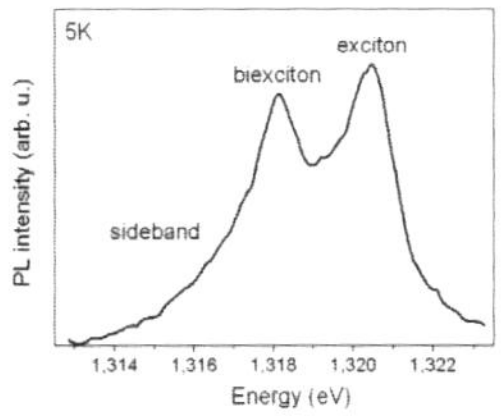

Figure 3. Experimental emission spectra of single quantum dash for high enough excitation power to provide equal emission intensity of exciton and biexciton and appearance of low-energy sideband accompanying the biexcitonic line.

This low-energy emission band accompanying the biexcitonic line has been so far observed for monolayer fluctuation GaAs/AlGaAs quantum dots [10]. The sideband can be interpreted as a fingerprint of the Coulomb interaction between the biexciton and the excitons confined in the neighboring environment (e.g. wetting layer) which lowers biexciton emission energy. The energy difference between the 'bare' biexciton and interacting one is increasing with increasing number of interacting excitons. The sideband is composed of biexcitonic emission in different environmental conditions (different wetting layer occupation). As a result the microphotoluminescence measurements are probing not only the state of the quantum object but also the state of its surrounding, because the interaction with the environment cannot be separated or neglected. This phenomena can be observed only when the quantum dash (bi)excitons have lifetimes comparable with the wetting layer excitons. This condition guarantees that the Coulomb interaction is efficient. There are two possible cases: i) the emission lifetime of quantum dashes is very short or ii) the wetting layer states have localized character (like in [11]) and their lifetimes are longer than for typical quantum well. Time-resolved measurements confirmed that for investigated structures both is true, revealing relatively long radiative lifetime of excitons confined in the wetting layer (caused probably by localization effects) and short biexciton lifetimes (in agreement with predicted enhanced oscillator strength). Furthermore obtained results have shown that observed cases of biexcitonic emission accompanied by the sideband are limited to the higher energy part of the emission band (figure 4).

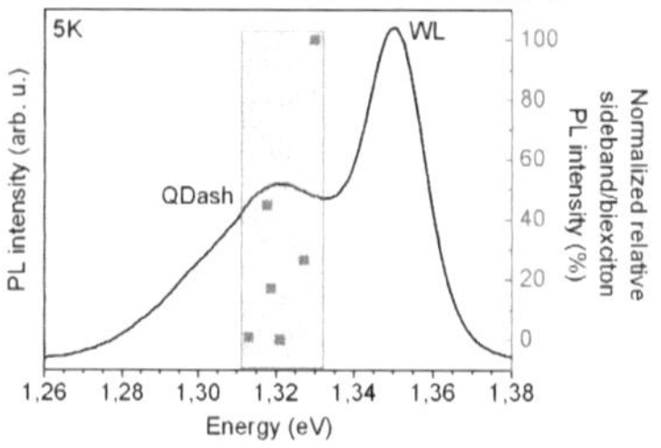

Figure 4. Normalized relative sideband to biexciton emission intensity for different biexciton cases (defined by their emission energy) against the background of the photoluminescence spectrum of the quantum-dash structure.

Moreover they differs in the intensity of the sideband with respect to the intensity of biexcitonic emission line. To compare those values for different quantum dashes integrated relative intensity has been obtained in rather high excitation regime providing almost equal excitonic and biexcitonic emission intensity. This could be a result of difference in (bi)exciton lifetimes or the efficiency of carrier capture between quantum dashes within the ensemble which tunes the experimental conditions out of the optimum for efficient Coulomb interaction.

4. Conclusions

Optical properties of single InGaAs/GaAs quantum dashes have been examined. Biexciton binding energy of about 1 – 3 meV has been obtained. The comparison of measured and modeled by rate equations excitation power emission intensity dependence enabled to indirectly estimate the kinetic properties of carriers confined in quantum dashes. Determined fractional exciton to biexction lifetime ratio is an evidence of weak confinement regime in these structures.

The effect of lowering the biexciton emission due to Coulomb interaction with excitons confined in their neighborhood has been observed. The condition of comparable lifetimes of those excitations fulfillment has been confirmed by time-resolved measurements. As it has been shown in photoluminescence measurements one cannot separate the optical properties of investigated structure from its environment. In this sense this structure has to be treated as an open system.

Acknowledgements

The work has been supported by the Polish Ministry of Science and Higher Education within Grant No. N N202 181238 and project: "National Laboratory of Quantum Technologies" – NLTK POIG.02.02.00-00-003/08-00. Authors acknowledge also the financial support from the Foundation for Polish Science and within the Fellowship co-financed by EU within European Social Fund.

References

[1] Reithmaier J P, Eisenstein G and Forchel A 2007 *Proc. Of the IEEE* **95** 9
[2] Reithmaier J P, Sęk G, Löffler A, Hofmann C, Kuhn S, Reitzenstein S, Keldysh L V, Kulakovski V D, Reinecke T L and Forchel A 2004 *Nature* **432** 197
[3] Mensing T, Reizenstein S, Löffler A, Reithmaier J P and Forchel A 2006 *Physica E* **32** 131
[4] Löffler A, Reithmaier J P, Forchel A, Sauerwald A, Peskes D, Kümmell T and Bacher G 2006 *J. Cryst. Growth* **286** 6
[5] Poloczek P, Sęk G, Löffler A, Reithmaier J P and Forchel A 2006 *J. Appl. Phys.* **100** 013503
[6] Syperek M, Kudrawiec R, Baranowski M, Sęk G, Misiewicz J, Bisping D, Marquardt B, Fischer M, and Forchel A 2010 *Appl. Phys. Lett.* **96** 041911
[7] Grundmann M and Bimberg D 1997 *Phys. Rev. B* **55** 9740
[8] Dekel E, Gershoni D, Ehrenfreund E, Garcia J M and Petroff P M 2000 *Phys. Rev. B* **61** 11009
[9] Wimmer M, Nair S V and Shumway J 2006 *Phys. Rev.* B **73** 165305
[10] Sennelart P, Peter E, Hours J, Cavanna A and Bloch J 2005 *Phys. Rev.* B **72** 115302
[11] Babiński A, Borysiuk J, Kret S, Czyż M, Golnik A, Raymond S and Wasilewski Z R 2008 *Appl. Phys. Lett.* **92** 171104

Time-resolved Hanle effect in (In,Ga)As/GaAs quantum dots

R. V. Cherbunin,[2] S. Yu. Verbin,[2] K. Flisinski,[1] I. Ya. Gerlovin,[2] I. V. Ignatiev,[2] D. V. Vishnevsky,[2] D. Reuter,[3] A. D. Wieck,[3] D. R. Yakovlev,[1] and M. Bayer[1]

[1] Experimentelle Physik 2, Technische Universität Dortmund, 44221 Dortmund, Germany
[2] Physics Department, St.Petersburg State University, 198504 St.Petersburg, Russia
[3] Angewandte Festkörperphysik, Ruhr-Universität Bochum, 44780 Bochum, Germany

E-mail: r.chebunin@gmail.com

Abstract. The circular polarization degree of (In,Ga)As/GaAs quantum dot photoluminescence is experimentally studied as function of magnetic field applied perpendicular to the optical excitation direction (Voigt geometry). The measurements are performed using various modulation timing protocols for the excitation polarization and intensity. The experimental data obtained allow us to estimate the characteristic times of polarization and relaxation processes for the nuclear spin system.

Introduction

The hyperfine interaction of electron and nuclear spins in semiconductor quantum dots (QDs) has recently become a matter of extensive theoretical and experimental research [1]. The electron spin dynamics is found to depend sensitively on the state of nuclear system. On the other hand, the oriented electron spin can generate a dynamical nuclear polarization (DNP) through the hyperfine interaction. The steady-state properties of DNP in semiconductors are under study for at least three decades [2], however experimental and theoretical studies of the DNP dynamics in QDs have been developed only during the last several years [3, 5, 6]. In particular, it has been predicted theoretically that the dynamics in QDs should be quite complicated [7, 8]. Experiments reveal a large variety of nuclear spin relaxation times from milliseconds [3, 6] to seconds [4, 5].

Most experimental data about the DNP in QDs have been obtained in magnetic field applied along the axis of optical excitation and signal detection (Faraday geometry), see, e. g., Ref. [1] and references therein. In this case the magnetic field prevents the nuclear spin system from angular momentum leakage due to the dipole-dipole interaction between the nuclear spins. Such experiments provide results, which allow a straightforward analysis of the processes studied.

In magnetic field perpendicular to the optical axis (Voigt geometry), the polarization of the electron spin decreases due to its precession about the magnetic field (Hanle effect). Strong optical pumping accompanied by relaxation processes in the nuclear spin system gives rise to an effective DNP field aligned along the external magnetic field and, correspondingly, normal to the direction of optical orientation. The presence of this DNP field may considerably modify the Hanle curve [9, 10].

In the present work, we report on experimental results of the time-resolved Hanle effect for a (In,Ga)As/GaAs quantum dot ensemble. These results allow us to estimate the characteristic times of processes occurring in the nuclear spin system.

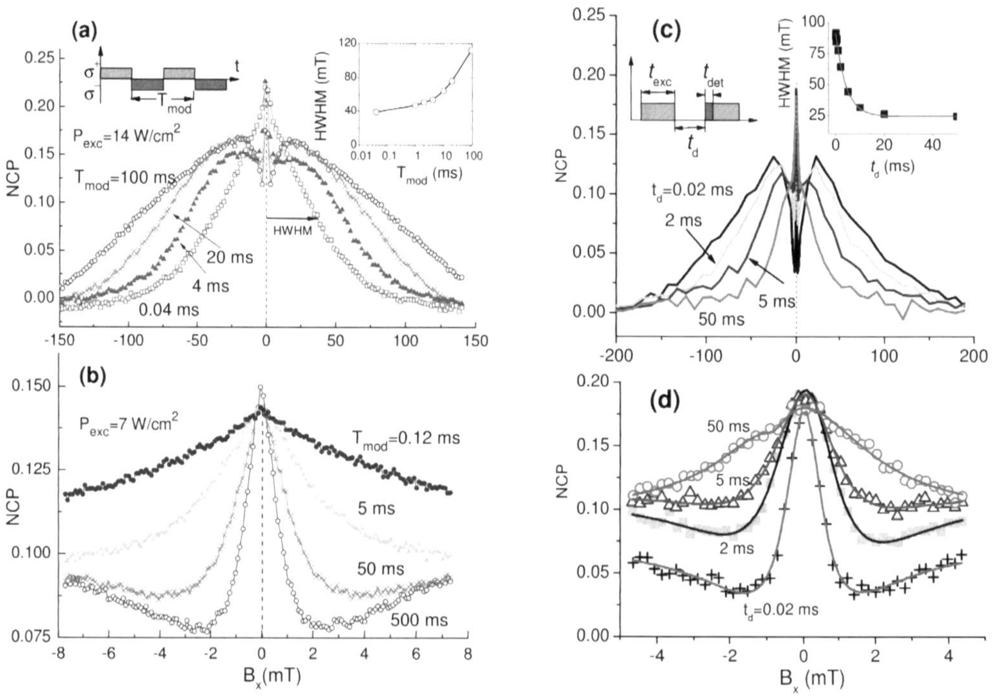

Figure 1. Hanle curves in relatively strong (a) and weak (b) magnetic fields for excitation with constant intensity but modulated circular polarization. The left inset in panel (a) shows the modulation protocol and the right inset shows the change of HWHM of the Hanle curves with rise of modulation period T_{mod}. Hanle curves measured in relatively strong (c) and weak (d) magnetic fields for excitation with constant circular polarization and modulated intensity are shown to the right. The left inset in panel (c) shows the excitation protocol and the right inset shows the change of HWHM with rise of dark time t_d.

1. Experimental details

A heterostructure containing 20 layers of self-assembled (In,Ga)As QDs sandwiched between n-delta modulation doped GaAs barriers has been studied. Donor ionization supplies every QD with on average a single resident electron.

Photoluminescence (PL) was excited by a continuous-wave Ti:Sapphire laser. Special timing protocols for the optical excitation which are schematically shown in Fig. 1 were developed to study the dynamics of nuclear polarization. An acousto-optical modulator was used to form the excitation pulse sequences. Modulation of excitation polarization was realized by an electro-optical modulator. The PL polarization was measured using a photoelastic modulator operated at a frequency of 50 kHz and a multi-channel photon counting system. The experiments were done at temperature T = 1.8 K. The electron spin polarization was determined from the degree of circular polarization of the QD PL excited at the energy of the wetting layer optical transition. We analyzed the degree of the negative circular polarization (NCP) effect, which is provided by the spin orientation of the resident electrons in singly-charged QDs. Further experimental details can be found in Ref. [6].

2. Experimental results

To study the dynamics of the nuclear polarization rise we have measured a set of Hanle curves at modulated polarization of the excitation. The modulation period was varied over a wide range. For large modulation period (T_{mod} > 100 ms), the Hanle curve is found to coincide almost with that for continuous-wave (CW) measurements. It consists of a wide Lorentzian-like part and a w-shaped peculiarity in its central part. Decrease of the modulation period is accompanied by a strong decrease of the half width at half maximum (HWHM) of the Hanle curve. Besides, a remarkable modification of the curves is observed in the range of small magnetic fields. As seen in Fig. 1(b), the width of the central peak increases, the symmetrical dips around the peak are smoothed, and the curve approaches the standard form of a Hanle curve. According to Ref. [9], the w-shape of Hanle curve with narrow central peak is caused by DNP whose effective magnetic field is directed along the external magnetic field. Our results presented in Fig. 1 show that for the excitation conditions used DNP develops in the time range of tens of milliseconds.

To study the relaxation of DNP, we used amplitude modulation of the optical excitation keeping the helicity of polarization as shown in the inset of Fig. 1(c). We used relatively long and intensive pump pulses to excite the electron-nuclear spin system up to a stationary state, which does not depend on the history of excitation, and in particular on the dark interval duration. The dark interval, t_d, was varied in the time range from 0.02 ms to 50 ms. The NCP signal and, correspondingly, the spin polarization were detected during the first millisecond just after the dark interval when the nuclear spin polarization was still not restored by the pumping. Results of these experiments are shown in Fig. 1(c,d). The Hanle curve measured after a very short dark interval is similar to that measured in CW regime. In particular, the central peak related to DNP is very pronounced. However already at t_d > 2 ms, a noticeable deformation of the peak becomes observable. This means that the nuclear spin relaxation time is units of milliseconds. This conclusion agrees well with the results on DNP dynamics measured in longitudinal magnetic field [6].

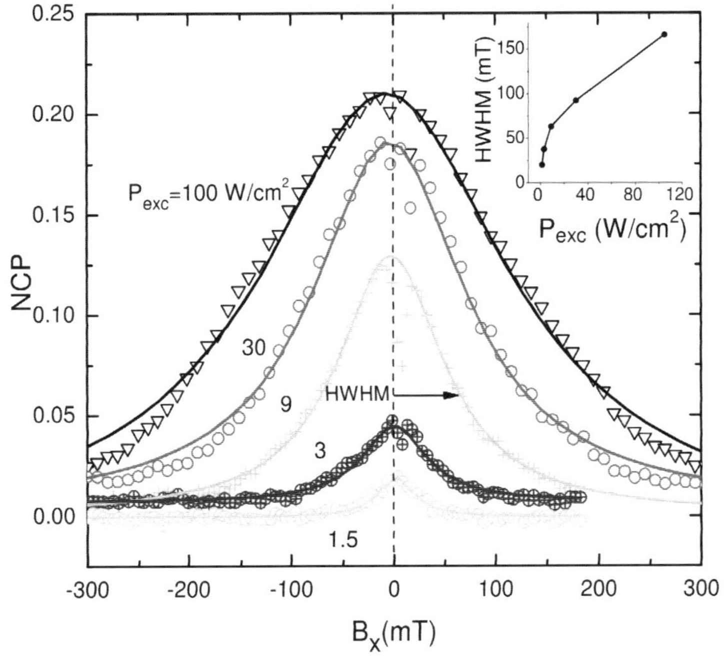

Figure 2. Hanle curves measured in CW regime at different excitation densities. Inset shows the dependence of HWHM of Hanle curve on excitation density.

In Figure 2, Hanle curves measured in CW regime at different pump powers are shown. A pump power increase is accompanied by increase of both the amplitude and the HWHM of the curves. At the maximal excitation power P_{exc} = 100 W/cm^2 used in the experiment, HWHM exceeds 150 mT and is almost four times larger than the one obtained in the absence of DNP [compare with Fig. 1(a), T_{mod} = 0.04 ms]. We believe that the HWHM increase is related to the increase of DNP field and its value exceeds 100 mT at the largest power density used. Recently, a very large broadening of the Hanle curves was observed for single QDs and was assigned to the DNP effects [10].

3. Conclusion

The time resolved measurement of the Hanle effect in (In,Ga)As QDs gives two new results. Firstly, the rise and decay times of nuclear spin polarization are estimated for the weak magnetic field range where DNP appearance may be explained by a classical mechanism. Secondly, a sharp rise of the Hanle curve width with excitation power increase has been found. We can conclude that the broadening of the Hanle curve is a result of DNP formation.

The authors thank V. K. Kalevich, K. V. Kavokin, and V. L. Korenev for fruitful discussions. This work has been supported by the Deutsche Forschungsgemeinschaft, by the Russian Foundation for Basic Research, and by the Ministry of Education and Science of the Russian Federation.

[1] Kalevich V K, Kavokin K V, and Merkulov I A, in *Spin Physics in Semiconductors* 2008 ed Dyakonov M I (Berlin: Springer) chapter 11, p 309
[2] *Optical Orientation* 1984 ed Zakharchenya B P and Meier F (Amsterdam: North-Holland)
[3] Maletinsky P, Badolato A, and Imamoglu A 2007 *Phys. Rev. Lett.* **99** 056804
[4] Oulton R, Greilich A, Verbin S Yu, Cherbunin R V, Auer T, Yakovlev D R, Bayer M, Stavarache V, Reuter D, and Wieck A 2007 *Phys. Rev. Lett.* **98** 107401
[5] Makhonin M N, Tartakovskii A I, Van'kov A B, Drouzas I, Wright T, Skiba-Szymanska J, Russell A, Fal'ko V I, Skolnick M S, Liu H-Y, and Hopkinson M 2008 *Phys. Rev. B* **77** 125307
[6] Cherbunin R V, Verbin S Yu, Auer T, Yakovlev D R, Reuter D, Wieck A D, Gerlovin I Ya, Ignatiev I V, Vishnevsky D V and Bayer M 2009 *Phys. Rev. B* **80** 035326
[7] Christ H, Cirac J I, and Giedke G 2007 *Phys. Rev. B* **75** 155324
[8] Petrov M Yu, Kozlov G G, Ignatiev I V, Cherbunin R V, Yakovlev D R, and Bayer M 2009 *Phys. Rev. B* **80** 125318
[9] Paget D, Lampel G, Sapoval B, Safarov V I 1977 *Phys. Rev. B* **15** 5780
[10] Krebs O, Maletinsky P, Amand T, Urbaszek B, Lemaître A, Voisin P, Marie X, and Imamoglu A 2010 *Phys. Rev. Lett.* **104** 056603

Journal of Physics: Conference Series **245** (2010) 012056 doi:10.1088/1742-6596/245/1/012056

Dynamical nuclear polarization and nuclear magnetic resonance in a (In,Ga)As/GaAs quantum dot ensemble

K. Flisinski,[1] I. Ya. Gerlovin,[2] I. V. Ignatiev,[2] M. Yu. Petrov,[2] S. Yu. Verbin,[2] D. R. Yakovlev,[1,3] and M. Bayer[1]

[1] Experimentelle Physik 2, Technische Universität Dortmund, 44221 Dortmund, Germany
[2] Physics Department, St.Petersburg State University, 198504 St.-Petersburg, Russia
[3] A. F. Ioffe Physical-Technical Institute, Russian Academy of Sciences, 194021 St.Petersburg, Russia

E-mail: Elkarol@gmx.de

Abstract. Peculiarities in the dependence of the photoluminescence polarization of (In,Ga)As/GaAs quantum dots on the transverse magnetic field were observed under radio frequency (RF) field application. We show that these peculiarities result from the suppression of dynamical polarization of the ^{71}Ga and As nuclei by the RF field. The effect of quadrupole splittings on the nuclear spin dynamics has been evaluated.

Optically detected nuclear magnetic resonance (ODNMR) has been used as an effective tool for studying the dynamics of nuclear polarization in bulk semiconductors [1, 2] and semiconductor quantum wells [2, 3, 4]. The high sensitivity of optical detection makes ODNMR promising for studying semiconductor quantum dots (QDs), which contain a relatively small number of nuclei. At the same time, there are very few experimental data on ODNMR in such structures [5, 6].

In this paper we present experimental studies of nuclear magnetic resonances in (In,Ga)As/GaAs QDs detected in the polarization of their PL in transverse magnetic field (Voigt geometry). Analysis of the experimental results allows us to identify the nuclear species responsible for DNP formation in the magnetic field range 0–10 mT.

We studied two samples, #1 and #2, with self-assembled (In,Ga)As/GaAs QDs annealed at temperatures 900 °C and 980 °C, respectively. The annealing causes relaxation of the internal strain in the QDs, which is one of the important sources of the quadrupole splitting of the nuclear spin states, which considerably affects the nuclear spin polarization. In the structures studied, each QD contains on average one resident electron, which is supplied by delta-doped barrier layers. The optically oriented electron spin interacts with the nuclear spins thereby efficiently polarizing the nuclei. At the same time, the electron spin polarization degree can be used as an indicator of the nuclear spin polarization.

The electron spin polarization was detected via the negative circular polarization (NCP) of the photoluminescence (PL) [7]. The PL was excited by a cw Ti-Sapphire laser, whose wavelength was tuned to the wetting layer optical transition. The PL polarization was detected using a photoelastic modulator operating at a frequency of 50 KHz and a photon counting system. All experiments were performed at a temperature $T = 2$ K. A split-coil superconducting magnet was used for the magnetic field (along the x-axis) to the samples. A radio frequency (RF) field was generated by a coil with its axis (y-axis) oriented perpendicular to both the optical axis and the external magnetic field.

We studied the PL polarization degree as function of the magnetic field applied perpendicular to the optical pumping direction parallel to z-axis (Hanle effect). We found that the Hanle curves have a W-shape, which can be explained by the development of a nuclear field along the total field acting on the

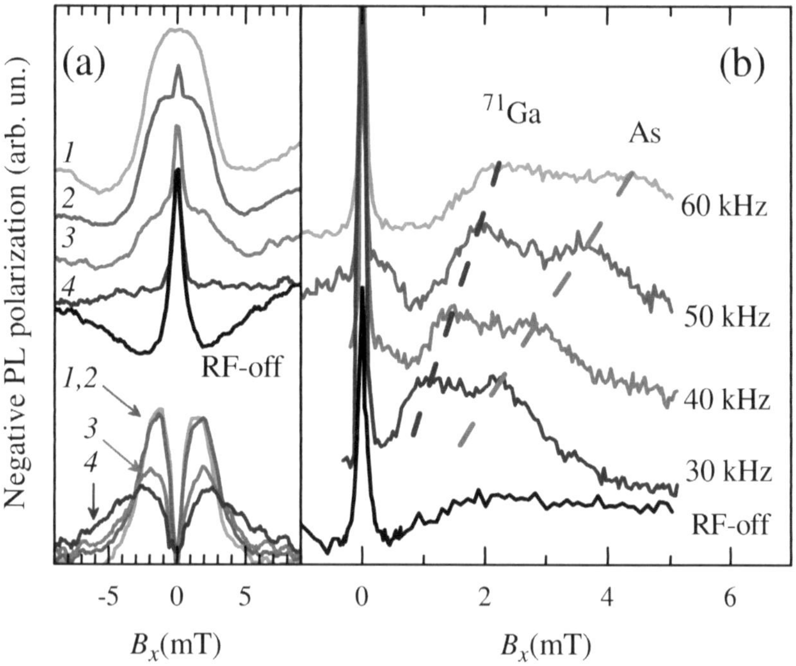

Figure 1. Effect of RF field on the Hanle curves in the range of the W-shape for sample #1 [top part of panel (a)] and for sample # 2 (b). The RF frequencies are given at each curve with "RF-off" indicating the RF field switched off. Numbers 1, 2, 3, and 4 label the curves measured at RF = 30, 50, 75, and 100 kHz, respectively. The curves are shifted vertically for clarity of presentation. The bottom part of panel (a) shows the difference of the Hanle curves measured for "RF-on" and "RF-off". Dashed lines in panel (b) indicate the magnetic field positions of NMR for the ^{71}Ga and As nuclei.

nuclei [1, 8]. This total field is the sum of the external magnetic field and the Knight field of the spin polarized electron. For sufficiently strong optical excitation the nuclear field can be considerably larger than the external field and, therefore, is able to depolarize the electron spin. The efficiency of nuclear spin polarization along the total field gradually goes down with the magnetic field rise thus forming the W-shape of Hanle curve [8].

We studied the modification of the Hanle curves by RF field application. The effect of RF field on sample #1 is shown in Fig. 1(a), revealing an *increase* of spin polarization. The differences of Hanle curves measured in presence and absence of RF excitation are shown at the bottom of Fig. 1(a) and consists of wide peaks, which shift towards larger external magnetic field with increasing RF frequency. The large width of these peaks does not allow us to uniquely identify, which nuclei are responsible for the observed resonances.

Figure 1(b) shows the Hanle curves measured for sample #2. The stronger annealing of this sample changes the Hanle curves considerably. In this case, application of RF field is accompanied by the appearance of pronounced resonances. The resonance position is found to be depend linearly on RF frequency. We suggest that the two observed resonances correspond to the $|1/2\rangle \leftrightarrow |-1/2\rangle$ transitions of the ^{71}Ga and As nuclei which are subject of a change of their gyromagnetic ratios due to the quadrupole splitting of the nuclear states. The source of this splitting is the strain-induced electric field gradient, ∇F, which lifts the degeneracy of nuclear spin states with different absolute values of the projections of nuclear spin on the principal axis of the tensor ∇F [9].

The Hamiltonian of the nuclear spin $\hat{\boldsymbol{I}}$ interaction with an external magnetic field $\boldsymbol{B}$ in presence of a quadrupole splitting along the z axis can be written as [9]

$$\hat{\mathcal{H}} = -\gamma\hbar\left(\hat{\boldsymbol{I}}\cdot\boldsymbol{B}\right) + \frac{h\nu_Q}{2}\left(\hat{I}_z^2 - I(I+1)/3\right), \tag{1}$$

where γ is the nuclear gyromagnetic ratio, and $\hbar$ is the Planck constant. The quadrupole splitting is determined solely by the frequency

$$\nu_Q = \frac{3eV_{zz}Q}{2I(2I-1)h}. \tag{2}$$

Here e is the elementary charge, Q is the nuclear quadrupole momentum, and $V_{zz} = S_{11}\varepsilon_{zz}$ is the strain-induced electric field gradient at the nuclear site. The values of quadrupole momentum Q and of S-tensor for different isotopes are extracted from experimental data on nuclear acoustic resonance [10, 11]. We also estimated the value of strain ε_{zz} to be about 0.01 (one percent) for the QD annealed at 980 °C using a cylindrically symmetric model for the QD [12] and the transversal isotropic approximation in the frame of continuum elasticity theory [13]. The magnitude of ν_Q obtained for gallium, arsenic, and indium nuclei are summarized in Table 1.

Isotope	I	Q (mbar)	S_{11} (statC/cm^3)	ν_Q (kHz)
^{71}Ga	3/2	107	9.1×10^{15}	353
^{69}Ga	3/2	171	9.1×10^{15}	564
^{75}As	3/2	314	1.31×10^{16}	1490
^{113}In	9/2	759	1.67×10^{16}	388
^{115}In	9/2	770	1.67×10^{16}	383

Table 1. Nuclear quadrupole parameters. The values of Q are taken from Ref. [10] and the values of S_{11} are taken from Ref. [11]. Quadrupole frequencies ν_Q are calculated for the value of strain $\varepsilon_{zz} = 0.01$.

Calculated Zeeman splittings of different levels of As and In nuclei obtained by direct diagonalization of Hamiltonian (1) are shown in Fig. 2. The Zeeman splittings strongly depend on the angle between the principal quadrupole axis (vertical axis) and the magnetic field direction for the case of strong quadrupole splitting (or weak magnetic field). In particular, when the magnetic field is applied along the quadrupole axis, there is a normal Zeeman splitting for which $\Delta E_{\pm n/2} = n\Delta E_{\pm 1/2}$ with $n = 3$ for As and $n = 3, 5, 7, 9$ for In. When the magnetic field is weak enough and applied perpendicular to the quadrupole axis, the $|\pm n/2\rangle$ states almost are not split for $n \geq 3$. At the same time, the splitting of the $|\pm 1/2\rangle$ states is larger by a factor of two for the As nuclei and a factor of five for the In nuclei as compared to the absence of a quadrupole splitting. A similar behavior of Zeeman splittings is also obtained for the Ga nuclei. Taking into account the two-fold increase of the gyromagnetic ratio for the $|\pm 1/2\rangle$ states, the resonances observed in Fig. 1(b) well correspond to the RF transitions $|1/2\rangle \leftrightarrow |-1/2\rangle$ for the ^{71}Ga and As nuclei.

Figures 2(c) and 2(d) show the dependence of Zeeman splittings of the As and In nuclei on magnetic field applied perpendicular to the principal quadrupole axis. The quadrupole splittings, $h\nu_Q$, of the nuclei were calculated for a strain $\varepsilon_{zz} = 0.01$, as estimated for sample #2. Even such small strain gives rise to quadrupole splittings, which are larger than the Zeeman splitting of the $|\pm 1/2\rangle$ states up to magnetic fields of about 10 mT for the In and 100 mT for the As nuclei. A similar analysis for the Ga nuclei gives a comparable Zeeman splitting at about $B = 20$ mT for the ^{69}Ga and $B = 15$ mT for the ^{71}Ga nuclei. In all cases, the approximation of large quadrupole splittings in the analysis of resonances shown in Fig 1(b) is valid.

In conclusion, we have optically detected nuclear magnetic resonances in an ensemble of (In,Ga)As/GaAs quantum dots, which is a direct proof of DNP formation. The resonances are detected as relatively wide peaks in the magnetic field dependence of the PL polarization degree, which appear when a radio-frequency field is applied. The origin of this effect is the suppression of dynamical nuclear polarization by the resonant field.

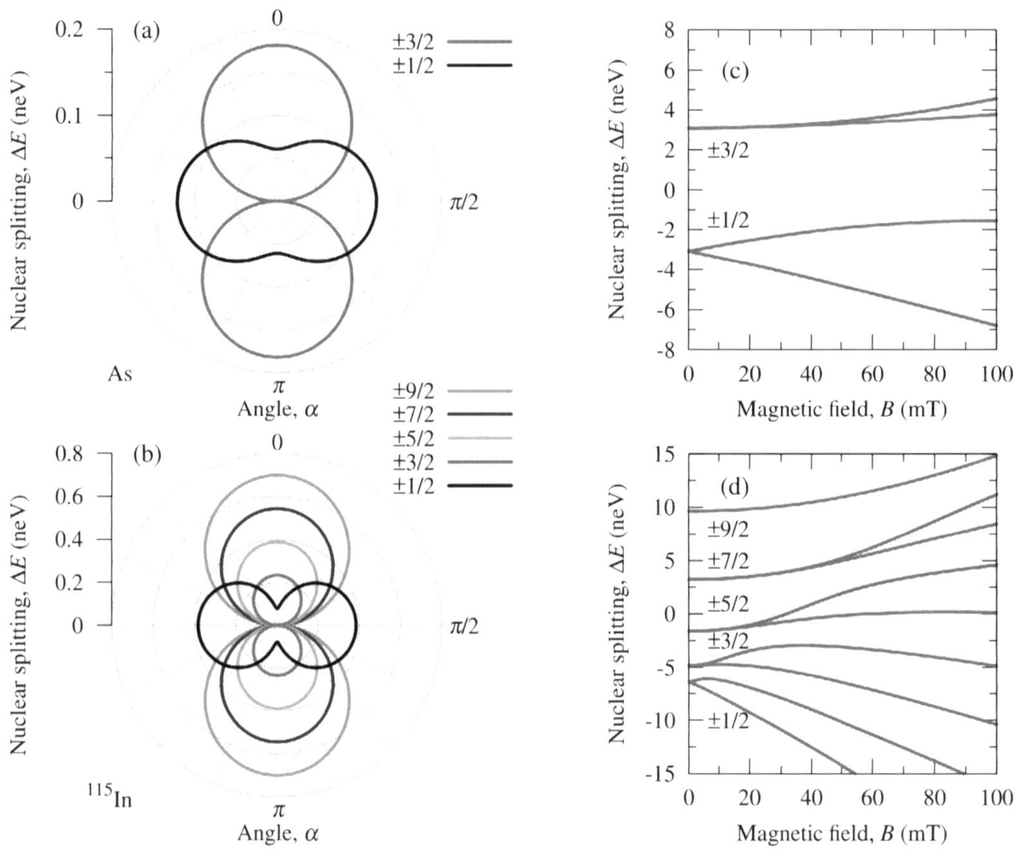

Figure 2. (a) Angular dependence of Zeeman splittings for As nuclei for large quadrupole splittings. The vertical axis corresponds to the principal axis of the quadrupole tensor. (b) The same for In nuclei. (c) Zeeman splittings for As nuclei in magnetic field perpendicular to quadrupole axis. (d) The same for In nuclei.

The authors thank V. K. Kalevich, K. V. Kavokin, and V. L. Korenev for fruitful discussions. This work has been supported by the Deutsche Forschungsgemeinschaft, by the Russian Foundation for Basic Research, by the Russian Ministry of Education, and by the "Dynasty" foundation (MYP).

[1] *Optical Orientation* 1984 ed Zakharchenya B P and Meier F (Amsterdam: North-Holland).
[2] Kalevich V K, Kavokin K V and Merkulov I A in *Spin Physics in Semiconductors* ed Dyakonov M I 2008 (Berlin: Springer) chapter 11, p 309
[3] *Semiconductor Spintronics and Quantum Computation* 2002 ed Awschalom D D, Loss D and Samarath N (Berlin: Springer)
[4] Salis G, Fuchs D T, Kikkawa J M, Awschalom D D, Ohno Y and Ohno H 2001 *Phys. Rev. Lett.* **86** 2677
[5] Gammon D, Brown S W, Snow E S, Kennedy T A, Katzer D S, and Park D 1997 *Science* **277** 85
[6] Makhonin M N, Chekhovich E A, Senellart P, Lemaître A, Skolnick M S, Tartakovskii A I 2010 *arXiv*:1002.0523v2
[7] Cherbunin R V, Verbin S Yu, Auer T, Yakovlev D R, Reuter D, Wieck A D, Gerlovin I Ya, Ignatiev I V, Vishnevsky D V and Bayer M 2009 *Phys. Rev. B* **80** 035326
[8] Paget D, Lampel G, Sapoval B and Safarov V I 1977 *Phys. Rev. B* **15** 5780
[9] Abragam A 1961 *The Principles of Nuclear Magnetism* (Oxford UK: Oxford Science Publications)
[10] Pyykkö P 2008 *J. Mol. Phys.* **106** 1965
[11] Sundfors R K 1974 *Phys. Rev. B* **10** 4244
[12] Petrov M Yu, Ignatiev I V, Poltavtsev S V, Greilich A, Bauschulte A, Yakovlev D R and Bayer M 2008 *Phys. Rev. B* **78** 045315
[13] Even J, Doré F, Cornet C, Pedesseau L, Schliwa A and Bimberg D 2007 *Appl. Phys. Lett.* **91** 122112

Journal of Physics: Conference Series **245** (2010) 012057 doi:10.1088/1742-6596/245/1/012057

Spectroscopy of the two Lowest Exciton Zero-Phonon Lines in Single CdSe/ZnS Nanocrystals

Y Louyer, L Biadala, Ph Tamarat and B Lounis

Centre de Physique Moléculaire Optique et Hertzienne, University of Bordeaux and CNRS, 351 cours de la Libération, Talence, F-33405, FRANCE

E-mail: y.louyer@cpmoh.u-bordeaux1.fr

Abstract. We study the optical properties of the lowest-energy exciton states in highly photostable individual CdSe/ZnS nanocrystals at low temperatures. We observe two sharp zero-phonon lines which we attribute to the radiative recombination from the two lowest-energy levels of the band-edge exciton fine structure. By using resonant laser excitation on these two lines we measure a spectral broadening of 10 μeV over integration times of 100 ms.

1. Introduction

Colloidal semiconductor nanocrystals (NCs) quantum dots have remarkable optical properties, such as tunable emission wavelengths [1] and high quantum efficiencies at room temperature [2]. Due to these properties, NCs are very promising for a broad range of applications [3-5]. The origin of the energy spectrum of the *single exciton* characterized by a fine structure is still not clear at low temperature [6]. In contrast with self-assembled quantum dots, the dark exciton radiative recombination is not completely forbidden. Furthermore, a few meV higher in energy, the optically active (bright-exciton) state is often missing in the low-temperature photoluminescence (PL) spectra of spherical NCs. In this work, we investigate the recombination dynamics of the two-lowest energy exciton states by reporting a detailed spectroscopic study of single CdSe/ZnS NCs as a function of temperature (see also [7]).

2. Samples and experimental setup

The samples are prepared by spin coating clean glass coverslips with a nanomolar solution of NCs in polyvinyl alcohol, ensuring a density <0.5 NC/μm^2. We mainly study NCs having a mean core radius of 4 nm (peak emission at 655 nm at room temperature). For comparison, smaller NCs were also used. The samples are mounted in a helium cryostat containing a xyz piezoscanner with a 0.95 NA objective as being part of a scanning confocal microscope used to image individual NCs. The NCs are excited by the 532 nm line of a cw frequency doubled Nd:YAG laser for measuring the single dot spectra, and by a pulsed laser source (optical parametric oscillator at 570 nm, 250 fs pulse width, 2.5 MHz repetition rate) for studying the emission dynamics. The high photostability of the experimental microscopy setup allows the study of the same single NC over a wide range of temperatures and over several days.

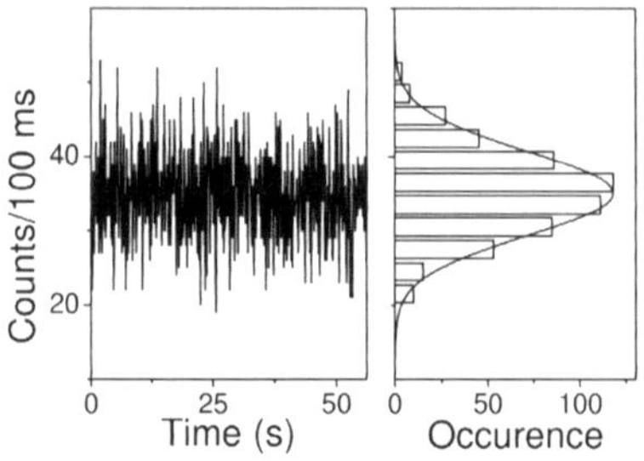

Figure 1. Time trace of the integrated PL intensity of a single NC, showing that the intensity fluctuations are shot-noise limited.

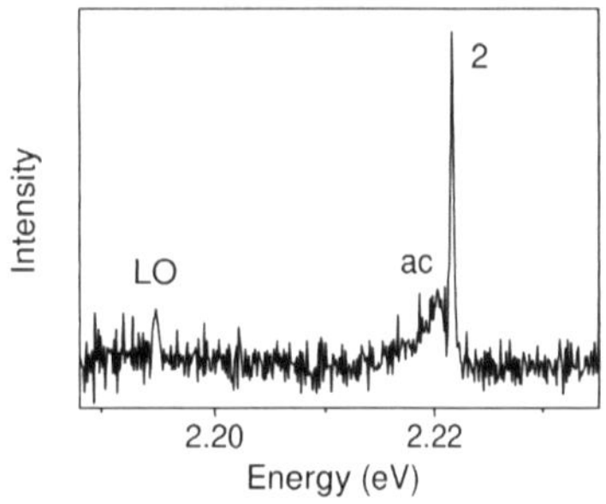

Figure 2. PL spectrum at 2K for a 2.2 nm single NC, where 2, ac and LO stand for the dark state ZPL and the confined acoustic and longitudinal optical phonon bands, respectively.

3. Results and discussion

At 2 K, the single NCs exhibit a blinking-free behaviour and a remarkable stability, as exemplified by the time trace of a single NC presented in figure 1. Figure 2 shows a PL spectrum (at 2 K with resolution-limited linewidths of 300 μeV) of a 2.2 nm radius single NC (peak emission at 585 nm at room temperature). We thus assign [8] the highest intensity peak to the dark zero phonon line (ZPL), while the red-shifted emission band is characteristic of a confined acoustic phonon sideband [9]. This assignment is also supported by the integrated intensity of the ZPL, which is weaker than that of the phonon sidebands (LO and LA phonons): the radiative recombination from the dark state occurs preferentially with emission of phonons. Figure 3 shows the PL spectrum of a 4 nm radius single NC recorded at weak excitation intensity. Surprisingly, the spectrum (at 2K) displays two peaks 1.6 meV apart around 2.005 eV. Their intensities evolve linearly with the excitation power, excluding any biexcitonic origin for this structure. Due to the electron-hole exchange interactions, the theoretical analysis [10] predicts a separation between the bright and dark levels of ~ 1 meV for 4 nm NCs, i.e., a value in agreement with our observations. These two peaks are attributed to the bright and dark exciton radiative recombination ZPLs. The peak around 1.979 eV (26 meV red-shifted from the main peak, at 2K) corresponds to the longitudinal optical phonon replica of the dark state ZPL.

A confirmation of the attribution of these peaks is realized by studying the temperature dependence of the emission lines. As shown in figure 3, we can clearly see that the relative intensity of the higher-energy peak strongly increases with temperature, in agreement with the picture of a phonon mediated thermal mixing between the dark and the bright excitons. Such an increase rules out that these peaks can be attributed to two bright states (as reported previously [11]) and clearly supports the emission of two excitonic states with very different oscillator strengths.

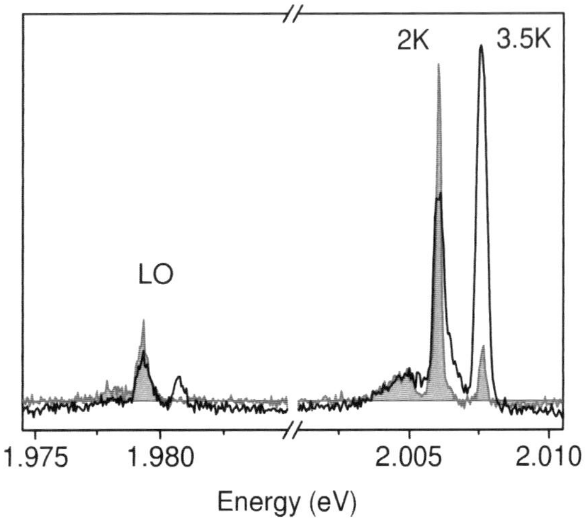

Figure 3. PL spectrum at 2 K and 3.5 K of a 4 nm single NC, showing the thermal mixing between the two lowest-energy exciton states.

In the samples of 4 nm NCs, we observe single dots exhibiting both two ZPLs (in the PL spectra) and a biexponential PL decay curve (when performed at 2K and at low excitation power). We simply use a three-level model, as illustrated in figure 4 (a), for describing the emission dynamics of the lowest-energy doublet. The three-states model takes into account the zero-exciton ground level, $|G\rangle$, and the two lowest-energy states, the bright $|\pm 1^L\rangle$ and the dark $|\pm 2\rangle$ exciton states, with respective radiative rates Γ_1 and Γ_2. Thermal mixing between these states occurs via emission and absorption of acoustic phonons from a mode whose energy matches the bright-dark energy splitting ΔE, which is deduced from the PL spectrum ($\Delta E = 1.4$ meV) recorded on another single NC (PL spectrum not shown). At zero temperature, the nonradiative phonon-mediated spin-flip rate from $|\pm 1^L\rangle \to |\pm 2\rangle$ is γ_0. Figure 4 (b) shows the PL decay for various temperatures. $\Gamma_2 = 3.7\times10^{-3}$ ns^{-1} is given by the long time component of the decay at very low temperature (a regime where $k_B T << \Delta E$). Relaxation rates (Γ_1 and γ_0) are deduced from the fit

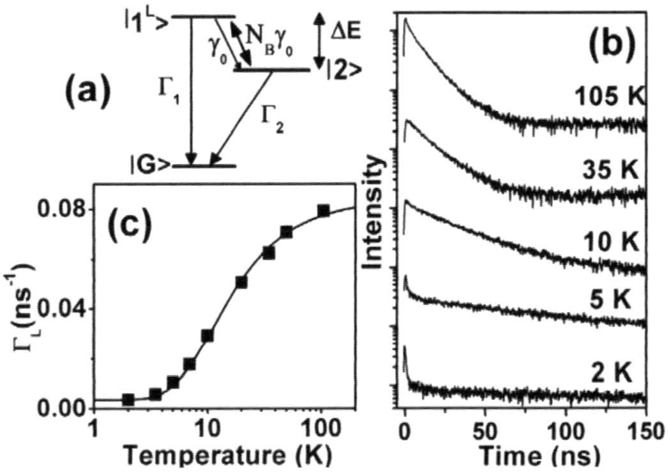

Figure 4. (a) Three-states model: $\gamma_0 N_B$ is the thermal mixing rate between $|\pm 1^L\rangle$ and $|\pm 2\rangle$ due to acoustic phonons. (b) PL decay curves of a single NC for various temperatures (excitation intensity, 20 W/cm^2). (c) Temperature dependence of the slow component decay rate Γ_L. Solid line: fit derived from the three-level model.

of the long time component, Γ_L^{-1}, given by the solutions of rate equations. Figure 4 (c) shows the temperature dependence of Γ_L for a single NC. The solid line well reproduces the data, for which only γ_0 and Γ_1 are used as fitting parameters. We find $\gamma_0 \approx 0.9\,\text{ns}^{-1}$ and $\Gamma_1 \approx 0.16\,\text{ns}^{-1}$. These values are of the same order of magnitude and explain why the bright exciton ZPL is observed. This also suggests a size-dependent spin-flip relaxation rate, which is reduced as the NC size increases. From the three-level model, we can estimate (at low *T*) the relative weight of the emission rate from the state $|\pm 1^L\rangle$. We find $\Gamma_1 / 2(\Gamma_1 + \gamma_0) \approx 0.08$, which is consistent with the relative weight of the bright state ZPL in the PL spectrum.

We also performed ultrahigh resolution resonant PL excitation of the two lowest-energy states of individual NCs at 2K. The method is as follows: The single-frequency (linewidth, 4 neV) of a cw tunable dye laser is swept through the ZPL, while the red-shifted luminescence of the LO-phonon replica is collected through a long pass filter on an avalanche photodiode. Figure 5 shows the resonant PL excitation line of the dark state at an excitation intensity of 1 W/cm^2 (well below saturation). For a single passage (i.e., on a time scale of 100 ms) of the laser through the resonance, the recorded linewidth is in average of the order of 10 μeV (see figure 5). This corresponds to an optical coherence lifetime larger than 100 ps. In the case of the bright state, similar linewidths have been measured. We also recorded resonant PL spectra on longer time scales. As a result of spectral diffusion inherent to NCs, figure 5 (10 scans) shows for a 10-s time scale, a threefold increase in the linewidth.

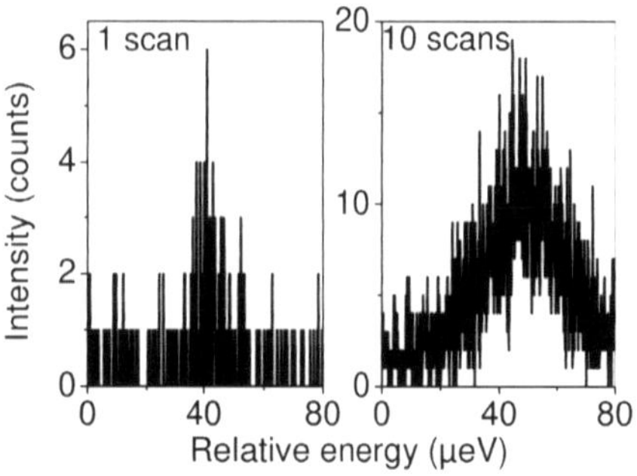

Figure 5. Resonant PL excitation of the dark state ZPL recorded on a single scan (left) and ten frequency-scans (right).

In conclusion, we have directly observed in the low temperature regime and for 4 nm NCs the two lowest-energy exciton recombination ZPLs in PL spectra. For smaller NCs, there is no bright 1^L exciton ZPL. This suggests a size dependence of the bright-to-dark spin-flip rate and a phonon bottleneck effect. We have also found that resonant PL excitation of the two ZPLs leads to an optical coherence lifetime longer than 100 ps over a time scale as long as 100 ms.

References

[1] Alivisatos A P 1996 *Science* **271** 933
[2] Qu L and Peng X 2002 *J. Am. Chem. Soc.* **124** 2049
[3] Colvin V L, Schlamp M C and Alivisatos A P 1994 *Nature* **370** 354
[4] Klimov V I, Mikhailovsky A A, Xu Su, Malko A, Hollingsworth J A, Leatherdale C A, Eisler H J and Bawendi M G 2000 *Science* **290** 314
[5] Michalet X, Pinaud F F, Bentolila L A, Tsay J M, Doose S, Li J J, Sundaresan G, Wu A M, Gambhir S S and Weiss S 2005 *Science* **307** 538
[6] de Mello Donega C, Bode M and Meijerink 2006 *Phys. Rev.* B **74** 085320
[7] Biadala L, Louyer Y, Tamarat Ph and Lounis B 2009 *Phys. Rev. Lett.* **103** 037404
[8] Empedocles S A and Bawendi M G 1997 *Science* **278** 2114
[9] Fernée M J, Littleton B N, Cooper S, Rubinsztein-Dunlop H, Gomez D E and Mulvaney P 2008 *J. Phys. Chem.* C **112** 1878
[10] Efros A L, Rosen M, Kuno M, Nirmal M, Norris D J and Bawendi M 1996 *Phys. Rev.* B **54** 4843
[11] Htoon H, Crooker S A, Furis M, Jeong S, Efros Al L and Klimov V I 2009 *Phys. Rev. Lett.* **102** 017402

Polarized light emission from CdSe/ZnSSe quantum-dot monolithic pillar microcavities

Moritz Seyfried[1], Joachim Kalden[1], Kathrin Sebald[1], Arne Gust[2], Carsten Kruse[2], Detlef Hommel[2], and Jürgen Gutowski[1]

[1]Semiconductor Optics, Institute of Solid state Physics, University of Bremen, Germany
[2]Semiconductor Epitaxy, Institute of Solid state Physics, University of Bremen, Germany

E-mail: `Seyfried@ifp.uni-bremen.de`

Abstract. Small-size II-VI micropillars with asymmetrical cross sections are presented as a way to achieve more than 95% of the emitted light from single quantum dots having one particular linear polarization state. We show that the detected PL intensity of the QD is increased by optimizing the spectral overlap between QD emission and resonator mode. The polarization of the emitted light is defined by the polarization of the resonator mode. Moreover, the internal mode structure in photonic molecules is investigated by studying their far field pattern. The observed field distribution opens the possibility of coupling individual quantum dots to each other via the mediation of the electromagnetic field.

1. Introduction

Future applications in the field of quantum information processing like, e.g., quantum computing or quantum cryptography will be based on robust and reliable single-photon sources [1]. In recent years, self-assembled semiconductor quantum dots (QDs) have been confirmed to act as single-photon sources [2]. By embedding these QDs into small-volume and high-quality pillar microcavities (MCs) the spontaneous emission rate can be enhanced and the collection efficiency increased [3]. However, for quantum information processing it is also crucial to control the polarization state of the emitted photons. This may be achieved by altering the shape of the micropillars, leading to a control of the photonic mode structure to some extent [4]. With this technique a polarized QD emission could be shown for elliptically shaped MCs in the III-V material system [5][6]. Furthermore, the coupling of different QDs is of high interest for the generation of entangled photons which can also be achieved via the electromagnetic field in a cavity. As temperature insensitivity of the QD emission is desirable, the II-VI material system with its remarkable stability of the excitons in the QDs is the material of choice [7][8].
In the following, we show for the II-VI material system that the intensity ratio between different polarizations of the QD emission can be controlled by using asymmetrically shaped MCs and can be changed by temperature detuning of the QD against the cavity mode. Furthermore, photonic molecules (PMs), consisting of two circularly shaped MCs which are interconnected by a small bar, are presented as a possibility to couple individual QDs and discussed with respect to their far field pattern [9].

2. Experimental

Planar MC samples were grown by molecular beam epitaxy on GaAs substrates at a temperature of 280 °C. The ZnSSe λ-cavity contains one sheet of CdSe/ZnSSe QDs grown by migration enhanced epitaxy. For an enhanced temperature stability of the excitonic emission the QDs are embedded in between additional MgS barriers. The cavity is positioned in between an 18.5 period bottom and a 15 period top distributed-Bragg-reflector (DBR) (details in [10]). The structures have been designed such that the spectral position of the longitudinal resonance of the cavity and the low-energy tail of the QD ensemble coincide at 4 K. For investigations of polarized light emission, completely etched elliptically shaped pillar MCs as well as photonic molecules were prepared from the planar sample by focused-ion-beam etching using a FEI Nova NanoLab system [11]. A micro-photoluminescence (µ-PL) setup with a nitrogen cooled charge-coupled camera (CCD) has been used while employing a microscope objective to excite individual pillar structures and to collect their emission. The measurements were performed at variable temperatures using a cw Ar^+ laser at 2.71 eV excitation energy. For polarization dependent measurements a polarizer and a $\lambda/2$ waveplate were inserted into the detection beam path. The far field pattern was obtained with the same setup by measuring sequences of 1D spatial profiles corresponding to a cut through the far field. The final 2D far field pattern was obtained from these individual 1D profiles by using a computational routine.

3. Results

The samples are characterized by Q factors of up to 7860 determined from the fundamental mode (FM) of circularly shaped airpost pillar MCs [10]. In the inset of Fig. 1 the photoluminescence spectrum of an elliptically shaped MC is shown with axis lengths of 1.89 µm and 3.89 µm, respectively. For a better identification of the modes the whole spectrum was measured at an elevated temperature of T=80 K at which the spectrally broadened emission lines of the QDs act as internal light source for the cavity modes. If the measurement is performed for two orthogonal polarizer orientations of 0° and 90°, defined by polarizer settings parallel to the short and long axis of the ellipse, respectively, the polarization degeneracy of the modes is lifted

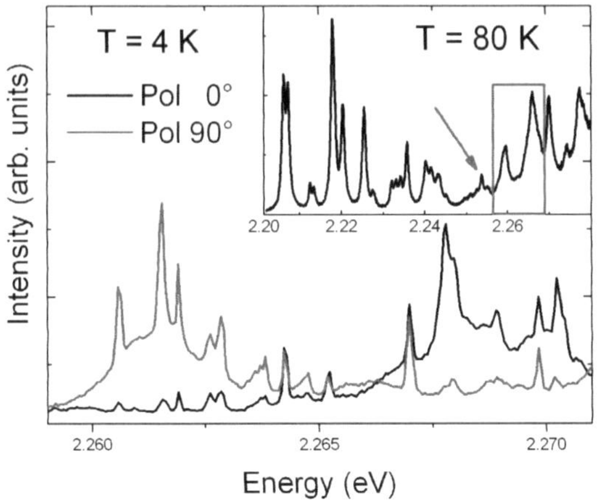

Figure 1. Polarization resolved PL spectra for 0° and 90° polarizer orientation of an elliptically shaped MC with axis lengths of 1.89 µm and 3.89 µm at T = 4 K. In the inset a much extended spectrum is shown at T=80 K.

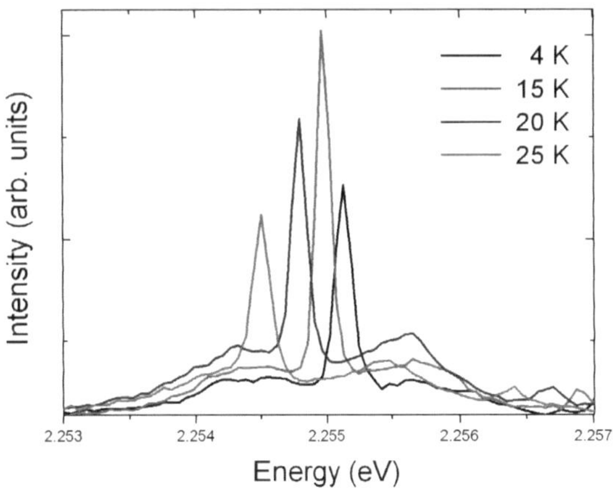

Figure 2. Single line emission of a QD at temperatures between 4 K to 25 K and a polarizer orientation of 0°. The spectral position of the mode is marked with the arrow in the inset of Fig. 1

due to the asymmetric cross section. Thus, an energy splitting between orthogonally polarized mode components becomes visible.

Spectrally sharp and partially resolution limited emission lines from the QDs are visible over the whole spectral range in particular at low temperatures. The collection efficiency of a QD can be dramatically increased by optimizing the spectral matching with a cavity mode [12]. The marked section of the inset was measured polarization resolved at T=4 K and is shown in the main frame of Fig. 1. The polarization orientation of the underlying mode leads to the enhancement of each QD emission in the same polarization direction and to the suppression of the emission into the orthogonal direction and hence to a polarized QD emission. In Fig. 1, QDs emitting in the spectral regime from 2.260 eV to 2.264 eV are more pronounced for the polarizer orientation of 90°, in contrast to another set of dots in the spectral region between 2.267 eV to 2.271 eV which gains intensity for the polarizer orientation of 0°. Comparing the PL intensity of one distinct QD for both polarizer orientations we observe a strong enhancement of the emission into one mode while the emission into the corresponding orthogonally polarized mode is suppressed. This yields, e.g., a polarization degree of 83% for the QD emitting at 2.2615 eV.

The redshift of the modes with increasing temperature due to an increase of the refractive index is small compared to the band gap shrinkage. Thus, the variation of the QD transition energy with temperature can be used to tune the QD emission into resonance with the modes of the cavity and thereby to enhance the QD emission. This is shown in Fig. 2 for a QD and the resonator mode at about 2.255 eV under 0° detection. The QD emission is shifted starting with a small detuning of -90 µeV (T=4 K), going into resonance with the cavity mode at T=15 K, and off resonance again with a positive detuning at T=25 K. For a QD in a spectral region where one polarized mode completely dominates (e.g., the region from 2.260 eV - 2.264 eV in Fig. 1) we can thus systematically change the intensity of the QD emission into this mode. At the same time the emission with orthogonal polarization remains on a constant rather low intensity level

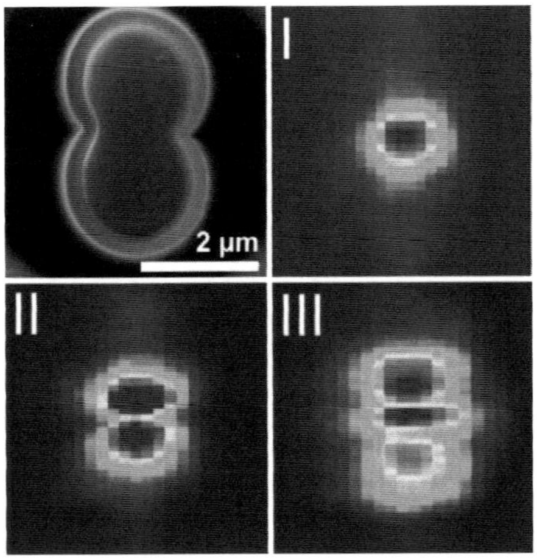

Figure 3. SEM image and two-dimensional spatial distribution of the PL intensity (red highest) of the fundamental mode as well as the first and second higher-order modes of a photonic molecule. The images were taken at the spectral position of the assigned modes in Fig. 4.

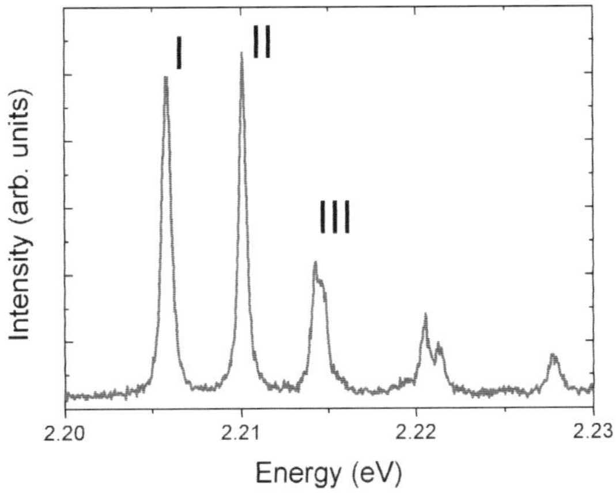

Figure 4. PL spectra of the photonic molecule shown in the SEM image of Fig. 3 measured at T=4 K. The labels on the modes correspond to the displayed far field pattern.

since for this polarization direction no active resonator mode is interacting with this emission. Thus, the polarization degree of the detected photons from the QD is considerably increased. Here, for zero detuning between the QD emission and the cavity mode we thus found 95% of the QD emission to couple into the 90° polarized mode.
Microcavities can also yield a coupling of different QDs by the mediation of the electromagnetic field. To study this effect, we produced several photonic molecules with different distances between the two circular pillars. The asymmetric total shape lifts the polarization degeneracy of the modes and leads to an energy splitting of up to 870 µeV for the FM of the PM with pillar diameters of 1.7 µm each and a bar of 0.64 µm length and 0.53 µm width. In Fig. 3 an SEM image of such a photonic molecule and the corresponding far field images of the fundamental mode (I) and two higher-order modes (II, III) are shown. All measurements were performed under excitation into the center of the PM. For a better understanding of the far field pattern the corresponding PL spectra of the PM with the first three modes numbered is presented in Fig. 4. The FM has its spatial intensity distribution maximum right in the middle of the PM whereas the first higher-order mode has two peaks locally shifted towards the individual pillars. The second higher-order mode possesses three intensity maxima, one in the center of the PM and the two others close to each pillar center. These first and second higher-order modes therefore could offer a possibility to couple two QDs each one located in one of the both pillars via the electromagnetic field [13][14]. The measurements further show the coexistence of localized (fundamental mode) and delocalized modes (next two higher-order modes) in the same microcavity. In combination with a method to tune the QDs' emission on and off such a resonance a switchable coupling of two QDs may be achieved.

4. Conclusions

We successfully fabricated asymmetrically shaped microcavities out of MBE grown planar samples by FIB milling. For the elliptically shaped pillars as well as for the photonic molecules a lifting of the polarization degeneracy of the fundamental cavity mode was observed leading to a more than 95% linear polarization for some QDs. The far field pattern of PMs shows the spread of the electromagnetic field over the whole molecule for some modes making these PMs interesting candidates for investigations concerning the coupling of several QDs.
Financial support by the Deutsche Forschungsgemeinschaft (Se 1846/1-1, Kr 2888/3-1) is gratefully acknowledged.

References

[1] Knill E, Laflamme R and Milburn G J 2001 *Nature* **409** 46
[2] Thompson R M, Stevenson R M, Shields A J, Farrer I, Lobo C J, Ritchie D A, Leadbeater M L and Pepper M 2001 *Phys. Rev. B* **64** 201302
[3] Reitzenstein S and Forchel A 2009 *J. Phys. D: Appl. Phys* **43** 033001
[4] Gayral B, Gérard J M, Legrand B, Costard E and Thierry-Mieg V 1998 *Appl. Phys. Lett.* **72** 1421
[5] Daraei A, Tahraoui A, Sanvitto D, Timpson J A, Fry P W, Hopkinson M, Guimarães P S S, Vinck H, Whittaker D M, Skolnick M S and Fox A M, 2006 *Appl. Phys. Lett.* **88** 051113
[6] Daraei A, Sanvitto D, Timpson J A, Fox A M, Whittaker D M, Skolnick M S, Guimarães P S S, Vinck H, Tahraoui A, Fry P W, Liew S L and Hopkinson M 2007 *J. Appl. Phys.* **102** 043105
[7] Sebald K, Michler P, Passow T, Hommel D, Bacher G and Forchel A 2002 *Appl. Phys. Lett.* **81** 2920
[8] Arians R, Kümmel T, Bacher G, Gust A, Kruse C and Hommel D 2007 *Appl. Phys. Lett.* **90** 101114
[9] Bayer M, Gutbrod T, Reithmaier J P, Forchel A, Reinecke T L, Knipp P A, Dremin A A and Kulakovskii V D 1998 *Phys. Rev. Lett.* **81** 12
[10] Sebald K, Kruse C and Wiersig J 2009 *Phys. Stat. Sol. B* **246** 255
[11] Lohmeyer H, Kalden J, Sebald K, Kruse C, Hommel D and Gutowski J 2008 *Appl. Phys. Lett.* **92** 011116
[12] Gérard J M, Sermage B, Gayral B, Legrand B, Costard E and Thierry-Mieg V 1998 *Phys. Rev. Lett.* **81** 5
[13] Karl M, Li S, Passow T, Löffler W, Kalt H and Hetterich M 2007 *Opt. Expr.* **15** 13
[14] Zhuk V, Regelman D V, Gershoni D, Bayer M, Reithmaier J P, Forchel A, Knipp P A and Reinecke T L 2002 *Phys. Rev. B* **66** 115302

Journal of Physics: Conference Series **245** (2010) 012059 doi:10.1088/1742-6596/245/1/012059

Microcavity quantum-dot systems for non-equilibrium Bose-Einstein condensation

I M Piper[1], P R Eastham[2], M Ediger[1], A M Wilson[1], Y Wu[1], M Hugues[3], M Hopkinson[3] and R T Phillips[1]

[1] Cavendish Laboratory, University of Cambridge, J J Thomson Avenue, Cambridge CB3 0HE, UK

[2] School of Physics, Trinity College Dublin, Dublin 2, Ireland

[3] Department of Electronic and Electrical Engineering, University of Sheffield, Mappin Street, Sheffield, S1 3JD UK

E-mail: imp24@cam.ac.uk

Abstract. We review the practical conditions required to achieve a non-equilibrium BEC driven by quantum dynamics in a system comprising a microcavity field mode and a distribution of localised two-level systems driven to a step-like population inversion profile. A candidate system based on eight 3.8nm layers of $In_{0.23}Ga_{0.77}As$ in GaAs shows promising characteristics with regard to the total dipole strength which can be coupled to the field mode.

1. Introduction

Experimental observation of polariton Bose-Einstein condensates (BEC) has been reported in several different systems [1, 2, 3]. In all of these cases, the ground state is populated by relaxation from an excited state, rather than populating the ground state resonantly. A different approach to creating BEC in a polariton system proposed by Eastham and Phillips [4], does not involve dissipation or inelastic scattering. Instead the condensation occurs as a result of the coherent dynamics of a specific coupled photon-exciton system.

In general the proposed dynamical condensation could be implemented in any system of discrete atomic-like states which are dipole-coupled to a long-lived mode of the electromagnetic field. We consider the practical aspects of implementing this proposal in semiconductor systems.

2. Theoretical requirements

The proposed experiment involves two stages, which are separated in time and can be regarded as independent. The first stage is the creation of a tailored exciton population in the inhomogeneously-broadened exciton line using a controlled pump pulse. For certain population profiles there is then a second stage, in which the photon-mediated interactions between the localized excitons lead to condensation. More specifically, the normal modes of a population of dipole-coupled localized excitons obey an equation [(1)], which is essentially the Cooper equation of BCS theory. It predicts that steps in the energy-dependent exciton population result in new collective modes, leading to condensation. The theoretical requirements for the condensation stage are described by the equation ($\hbar = 1$)

$$\lambda_{\mathbf{k}} - \omega_{\mathbf{k}} + \kappa^2 n \int \frac{\nu(E)[2n(E)-1]}{\lambda_{\mathbf{k}} - E} dE = 0. \tag{1}$$

This equation determines the complex normal-mode frequencies $\lambda_{\mathbf{k}}$ of the electromagnetic field, in a planar microcavity containing an ensemble of localized dipole oscillators. These normal modes are approximately plane waves, with in-plane wavevector $\mathbf{k}$. $\omega_{\mathbf{k}}$ is the frequency of the bare cavity mode, whose imaginary part $-\gamma$ describes the cavity losses. The localized exciton states are assumed to be two-level systems, which can be either unoccupied or occupied. $\nu(E)$ is the normalized distribution of the energies of these transitions, and n is their area density. Their average occupation at energy E is $n(E)$. Adiabatic rapid passage allows the creation of tailored population profiles $n(E)$, with an energy resolution set by the duration of the pump pulse [4].

The coupling κ is the matrix element for a transition from the unoccupied to the occupied state, assumed to be a constant for simplicity. In the dipole gauge its explicit form is

$$\kappa = \langle er \rangle \sqrt{\frac{\hbar\omega}{2\epsilon_0 \epsilon L_{\text{eff}}}}, \tag{2}$$

where $\langle er \rangle$ is the dipole moment of the transition, $\hbar\omega$ its energy, and L_{eff} the effective width of the cavity. A typical value for a semiconductor in the bandgap region is $\langle er \rangle \approx 20$ D, corresponding to an electron moving approximately 0.5 nm [5]. For $\hbar\omega \approx 1$ eV and $L_{\text{eff}} \approx 20$ μm this gives an estimate $\kappa \approx 8 \times 10^{-7}$ meV cm^{-1}.

A solution to (1) with $\lambda''_{\mathbf{k}} = \Im(\lambda_{\mathbf{k}}) > 0$ corresponds to an exponentially growing normal mode, which leads to a large population. Thus the presence of such solutions implies a dynamical phase transition from an exciton population to a non-equilibrium condensate. The character of this condensate depends on that of the unstable normal mode. If it is essentially a pure electromagnetic field mode we have a photon condensate, i.e. a laser. If it has a significant excitonic polarisation component then we have condensation of coupled light-matter states.

Close to a condensation threshold the imaginary part of (1) is a gain-loss equation for the growth rate of the mode: $\lambda''_{\mathbf{k}} = -\gamma + \pi\kappa^2 n \nu(\lambda'_{\mathbf{k}})[2n(E) - 1]$. The energy range of the optically-active localized states is approximately the inhomogeneous linewidth γ_{inh}, so typically $\nu(E) \sim 1/\gamma_{\text{inh}}$. Thus a necessary condition for condensation is, up to numerical factors of order one, $\kappa^2 n > \gamma\gamma_{\text{inh}}$. A second requirement comes from the real part of (1), which determines the shifting of the normal mode energies due to the exciton population. Large shifts indicate a significant exciton component, and hence distinguish exciton-photon condensation from lasing. For realistic $n(E)$ and $\nu(E)$ equation (1) gives shifts of order $\kappa^2 n/\gamma_{\text{inh}}$. Thus if we require energy shifts of order E_{res}, as well as the gain-loss criterion to be satisfied, we have $\kappa^2 n > \gamma_{inh} \max(\gamma, E_{\text{res}})$.

An energy shift greater than the inverse of the exciton's decoherence time implies an excitonic component to the condensate. We hope to achieve somewhat larger shifts, of order 1 meV. This is about one order of magnitude smaller than the vacuum Rabi splittings now routinely observed, and is a reasonable benchmark for a new exciton-photon coupling phenomenon. Assuming an inhomogeneous linewidth also of order 1 meV, and taking the estimate of the coupling strength above, gives an order-of-magnitude estimate of the total area density required near the field antinodes as $n > 10^{12}$cm^{-2}. The previous demonstration of lasing in very similar vertical cavity surface-emitting structures [6] indicates that the required physical régime should be accessible.

3. Candidate semiconductor systems

Stranski-Krastanow (S-K) dots are limited in density by the requirements of the growth mode, and usually are strongly inhomogeneously broadened, which reduces the spectral density. We

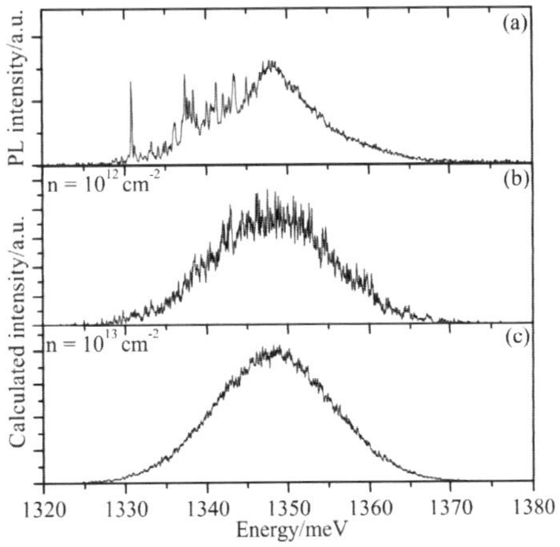

Figure 1. (a) Confocal μPL from a region of area $\approx 1\mu m^2$, showing both resolved and unresolved emission from the dot ensemble. Excitation was at 633nm, at a power density $\approx$ 10Wcm^{-2}. A model density of states of $n = 10^{13}$ cm^{-2} (c) is too high to resolve individual transitions. Localised states can be resolved for an order of magnitude lower density (10^{12} cm^{-2}, (b)), indicating that the density of states in the sample lies between that of (b) and (c).

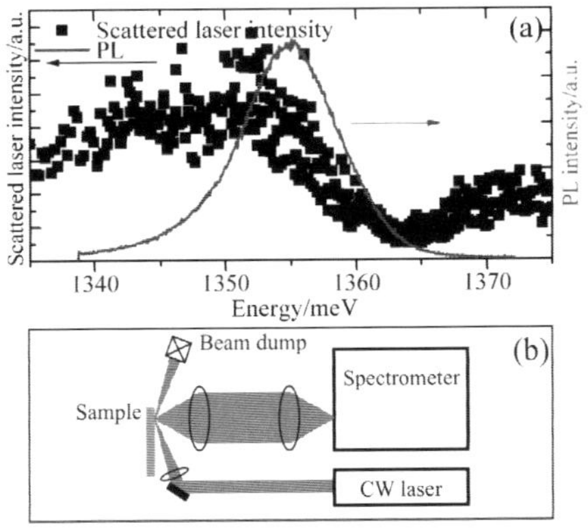

Figure 2. (a) The spectral dependence of the weak RRS enhancement shows a dip on the high energy side of the line associated with reabsorption in the eight-layer stack. The low degree of RRS enhancemnent indicates 0D rather than 2D states.
(b) The sample is excited at an oblique angle and the RRS normal to the sample collected. The scattered intensity as the CW laser is tuned across the transition is recorded.

have therefore considered a system based on a "poor" quantum well, in which alloy and thickness fluctuations localise the nominally 2D states into quantum dots (QD). The key issues here relate to the dipole coupling strength, density of dots, and proof of their behaviour as isolated two-level systems. The sample consists of eight 3.8nm layers of $In_{0.23}Ga_{0.77}As$ with GaAs barriers, grown by molecular beam epitaxy on an insulating GaAs substrate. Microphotoluminescence (μPL) collects PL from an area $< 1\mu m^2$ and resolves emission from individual localised states, as shown in Figure 1. By modelling the PL as a superposition of Lorentzians in a Gaussian distribution we estimate the total density of states (in eight layers) to be at least 10^{12} cm^{-2} (Figure 1(b,c)). This system therefore appears capable of satisfying the density requirement implied in Section 2. A further important feature of the PL evidence is that the total inhomogeneous linewidth of 10meV is very well suited to the proposed cavity BEC experiment, and is narrower than typically found for high density ($n > 10^{10}$cm^{-2}) S-K dot samples.

Resonant Rayleigh scattering (RRS) is sensitive to localisation, and for 2D states shows enhancement by a factor up to 100 on resonance [7, 8]. However, as localisation increases, the intensity of the RRS signal decreases and becomes broader, as the RRS efficiency depends on the ratio between the homogeneous and inhomogeneous broadening of the transitions [7]. The weak enhancement in the present samples (Figure 2(a)) is consistent with a high degree of localisation.

The electron-hole correlation length ρ is given by $\langle\rho^2\rangle = \frac{8\gamma_2\mu}{e^2}$, where $\gamma_2 B^2$ is the diamagnetic shift and μ the reduced exciton mass [9]. The diamagnetic coefficient of one state measured using magneto-optical μPL [10] is found to be 20.1 μeV T^{-2} in the Faraday geometry and 7.6 μeV T^{-2} in Voigt, shown in Figure 3. We estimate the extent of the *e*-*h* correlation length to be approximately 7nm and 4nm respectively. As this is a simple model, these values are consistent with those of localised states in QDs. Electron microscopy of a QD system with

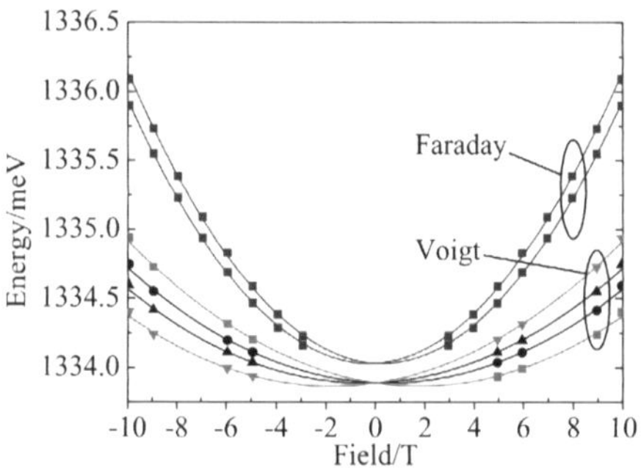

Figure 3. Diamagnetic shifts and spin fine structure of a single localised state in both the Faraday and Voigt geometries, illustrating the change in e-h correlation length as the field is moved from parallel to perpendicular to the growth axis. Markers show every tenth data point; lines show a least-squares minimised regression fit.

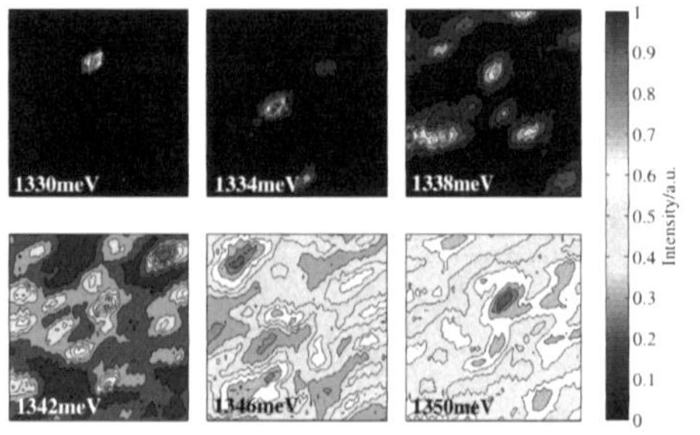

Figure 4. Map of μPL emission from a $15\mu m \times 15\mu m$ region at six different energies.

similar composition shows the QDs are of length 40-100nm and width 20-30nm; these states have similar diamagnetic shifts to those measured in the present work [11].

μPL emission maps at six photon energies are shown in Figure 4 for a $15\mu m \times 15\mu m$ region indicating elongation of the emitting regions in the $[1\bar{1}0]$ direction. The combination of these experiments indicates that the localised states in narrow $In_{0.23}Ga_{0.77}As/GaAs$ quantum wells are promising candidates for the proposed dynamically-driven BEC, as they have a high density of discrete localised states distributed over a suitably narrow energy range.

4. Conclusions

We have outlined the theoretical origin of the requirements for a suitable system in which to realise the proposed condensation scheme. Narrow $In_{0.23}Ga_{0.77}As/GaAs$ quantum wells show a suitably high density of localised states to offer a good system for attempted observation of the dynamically-driven BEC.

Acknowledgments

This work was conducted with financial support from EPSRC grant EP/F040075/1 and the Science Foundation Ireland grant 09/SIRG/I1592. We thank the staff of the Cavendish Laboratory Workshops for their expert technical assistance.

References

[1] Kasprzak J *et al.* 2006 *Nature* **443** 409
[2] Balili R, Hartwell V, Snoke D, Pfeiffer L, West K 2007 *Science* **316** 1007
[3] Bajoni D, Senellart P, Lemaître A and Bloch J 2007 *Phys. Rev* B **76** 201305
[4] Eastham P R and Phillips R T 2009 *Physical Review* **B79** 165303
[5] Asada M and Suematsu Y, 1985, *IEEE J. Quantum Electronics* **21** 434
[6] Saito H, Nishi K, Ogura I, Sugo S and Sugimoto Y 1996 *Appl. Phys. Lett* **69** 3140
[7] Hegarty J, Sturge M D, Weisbuch C, Gossard A C and Wiegmann W 1982 *Phys. Rev. Lett.* **49** 930
[8] Garro N *et al.* 1997 *Phys. Rev.* B **55** 13752
[9] Phillips R T, Steffan A G, Newton S R, Reinecke T L and Kotlyar R 2003 *phys. stat. sol.* (b) **3** 601
[10] Kehoe T, Ediger M, Phillips R T, and Hopkinson M 2010 *Rev. Sci. Instrum.* **81** 013906
[11] Mensing T, Reitzenstein S, Löffler A, Reithmaier J P, and Forchel A 2005 *Physica* E **32** 131

Some reasons of emission variation in InAs quantum dot-in-a-well structures

T V Torchynska[1], J Palacios Gomez[1], G Gómez Gasga[1], A Vivas Hernandez[2], E Velazquez Lozada[2], G Polupan[2] and Ye S Shcherbyna[3]
[1]ESFM– National Polytechnic Institute, México D. F. 07738, México
[2]ESIME– National Polytechnic Institute, México D. F. 07738, México
[3]National Technical University –"KPI" of Ukraine, Kyiv, 03056, Ukraine

E-mail: ttorch@esfm.ipn.mx

Abstract. Photoluminescence (PL) and X ray diffraction have been studied in InAs quantum dots (QDs) embedded in symmetric $In_{0.15}Ga_{1-0.15}As$/GaAs quantum wells (dot-in-a-well, DWELL) with QDs grown at different temperatures. The density of QDs decreases from 1.1×10^{11} down to 1.3×10^{10} cm^{-2} with increasing the QD growth temperatures from 470 to 535°C. The QD density decreasing in DWELLs is accompanied by the non monotonous variation of QD parameters. The PL intensity increases and the PL peak shifts to low energy in structures with QDs grown at 490 and 510°C. On the contrary the structures with QDs grown at 525 and 535°C are characterized by lower PL intensities and PL peak positions shifted to higher energy.

The method of X-ray diffraction has been applied with the aim to study the variation of elastic strain in DWELL structures with QDs grown at different temperatures. It was shown that the minimum of elastic strain corresponds to DWELL with QDs grown at 490-525 °C. For lower (470 °C) and higher (535 °C) QD growth temperatures the level of compressive strain increased in DWELLs. The reasons of strain variation are discussed as well.

1. Introduction

The technology of growth of InAs quantum dots (QDs) has become perfected permitting many different applications of QD structures in optoelectronics such as: an active medium of low-threshold semiconductor lasers for optical fiber communication and infrared photo-detectors [1-3] as well as in memory devices [4]. It was shown earlier [2, 5-7] that the InAs QD density can be increased essentially if QDs are grown inside of InGaAs/GaAs quantum well (dot-in-a-well, DWELL) structures. The optimum composition in symmetric quantum wells is the same composition of capping and buffer layers equal to $In_{0.15}Ga_{0.85}As$ [5]. But even for the optimal capping/buffer layer composition, the InAs/InGaAs/GaAs QD structures are characterized by the essential variation of emission efficiency. In this paper we try to understand some reasons of QD emission variation in DWELLs with QDs grown at different temperatures.

2.Experimental conditions

A set of samples was created using molecular beam epitaxy on (100) oriented 2''diameter semi-insulating GaAs substrates. InAs quantum dots (QDs) were grown inside of the symmetric $In_{0.15}Ga_{0.85}As$/GaAs quantum well (DWELL) structures at QD growth temperatures 470, 490, 510, 525 and 535°C as was described earlier in details [5,6]. At the photoluminescence (PL) study the PL spectra were excited by the 514.5 line of an Ar laser at an excitation power density of 300 W/cm^2. PL spectroscopy setup was described earlier in [8-11]. X-ray diffraction (XRD) experiments were made using an XRD equipment model of D-8 Advance (Bruker Co.) with $K_{\alpha 1}$ line from the Cu source (λ=1.54A).

3. Experimental results and discussion

Figure 1 presents the typical PL spectra measured at different temperatures (120-310K) for the structure with QDs grown at 525 °C. The main ground state (GS) and small excited state (ES) exciton emission bands are clearly visible in PL spectra. Figure 2 presents the dependences of integrated PL intensities and PL peak positions measured in DWELLs with different QD growth temperatures. The integrated PL intensity increases and the PL peak shifts to lower energy in structures with QDs grown at 490 and 510°C. On the contrary the structures with QDs grown at 525 and 535°C are characterized by lower PL intensities and PL peak positions shift to higher energy (Fig.2). Thus the non monotonic behavior of the integrated PL intensities and PL peak positions in DWELLs are revealed experimentally.

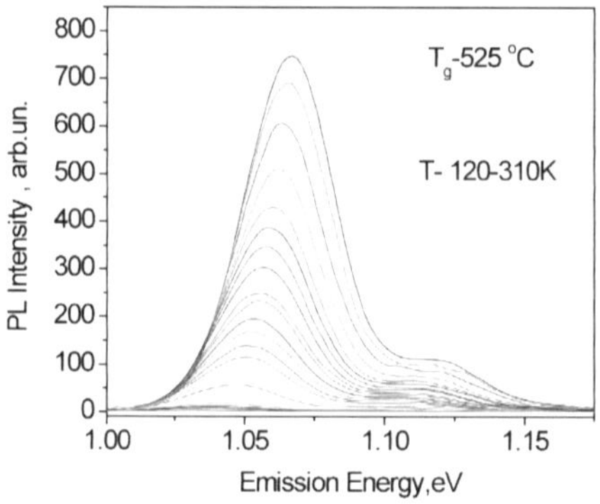

Figure 1. Typical PL spectra measured at different temperatures from the range 120-310K for the structure with QDs grown at 525 °C.

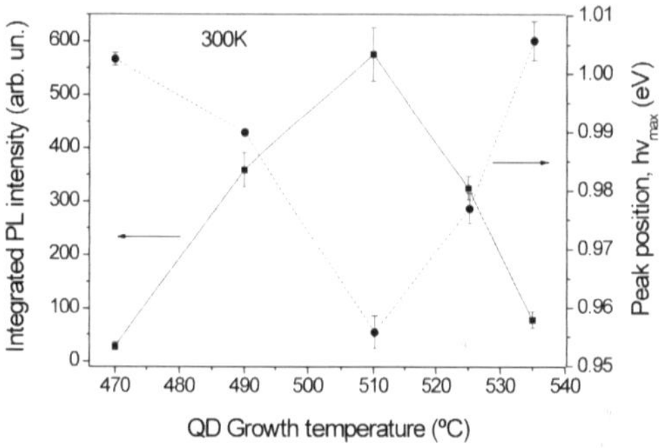

Figure 2. Integrated PL intensity and PL peak positions in DWELLs with QDs grown at 470, 490, 510, 525 and 535 °C.

As it was shown earlier using the AFM method the density of QDs decreases from 1.1×10^{11} down to 1.3×10^{10} cm^{-2} with increasing the QD growth temperatures from 470 to 535°C [5,6]. At the same time the lateral size of QDs increases as well from 12 up to 28 nm. It can be expected that in DWELLS with QD density decreasing and rising their sizes the PL intensity diminishes and the emission peak shifts to lower energy. To understand the reason of the non monotonous variation of PL intensity and peak positions the X ray diffraction has been measured in studied DWELLs.

Figure 3 presents the superposition of XRD peaks related to the diffraction of $K_{\alpha 1}$ and $K_{\alpha 2}$ lines of X-ray beam from the (400) crystal planes of the cubic GaAs substrate and of GaAs layers in studied InGaAs/GaAs QWs. As one can see the peaks related to the $K_{\alpha 1}$ and $K_{\alpha 2}$ line diffraction from the (400) crystal plane of GaAs layers in DWELLs with QDs grown at 490-525 °C (66.05° and 66.24°) locate very close to the corresponding XRD peaks of bulk cubic GaAs for the same (400) crystal plane (Table 1, Fig.3). The last fact testifies that the level of elastic strain in these DWELLs is minimum. In contrary in DWELLs with QDs grown at 470 and 535 °C the corresponding XRD peaks shift to 66.10° and 66.29° (Table 1, Fig.3) testifying the higher levels of compressive strain in these QD structures.

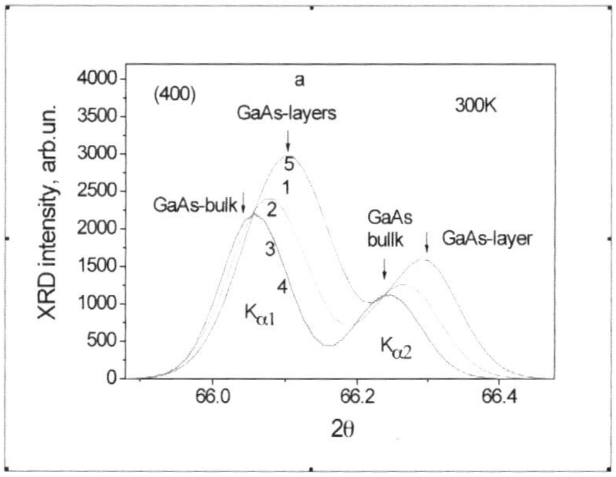

Figure 3. XRD peaks related to the diffraction of $K_{\alpha 1}$ and $K_{\alpha 2}$ lines of the Cu source on (400) cubic GaAs crystal planes in DWELLs with QDs grown at 470 (1), 490 (2), 510 (3), 525(4) and 535 °C (5).

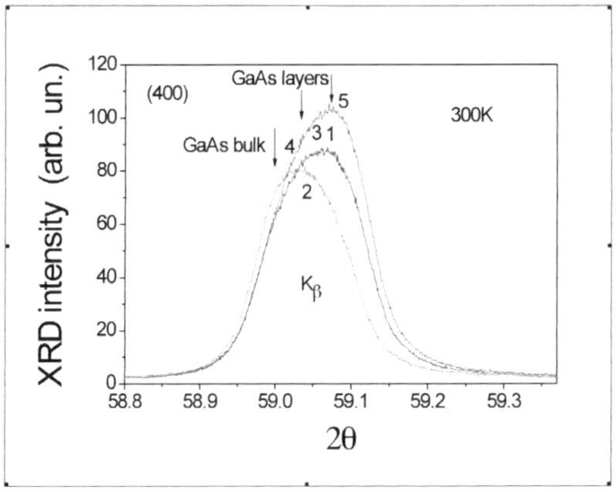

Figure 4. XRD peaks related to the diffraction of K_β line on (400) cubic GaAs crystal planes in DWELLs with QDs grown at 470 (1), 490 (2), 510 (3), 525(4) and 535 °C (5).

Table 1. The values of 2θ angles for the diffraction of $K_{\alpha 1}$, $K_{\alpha 2}$ and K_β lines of the X-ray beam of Cu source from the (400) cubic bulk GaAs crystal plane [12] and from GaAs layers in QWs.

	(400) cristal plane			
Material	$2\theta_1$ (degree) $K_{\alpha 1}$	d_1, A	$2\theta_2$ (degree) $K_{\alpha 2}$	$2\theta_3$ (degree) K_β
GaAs bulk [12]	66.044	1.414	66.225	59.005
GaAs QW layer T_g=470°C	66.10	1.414	66.29	59.07
GaAs QW layer T_g=490°C	66.08	1.414	66.26	59.04
GaAs QW layer T_g=510°C	66.05	1.414	66.24	59.05
GaAs QW layer T_g=525°C	66.05	1.414	66.24	59.04
GaAs QW layer T_g=535°C	66.10	1.414	66.29	59.07

Figure 4 presents the additional confirmation of conclusions mentioned above. It shows the superposition of XRD peaks related to the diffraction of K_β line from the same (400) crystal planes of cubic GaAs substrate and of GaAs layers in InGaAs/GaAs QWs. As one can see the peaks related to the K_β line diffraction from the (400) crystal plan of GaAs layers in DWELLs with QDs grown at 490-525 °C (59.036-59.042°) locate close to the corresponding XRD peak in the bulk GaAs (Table 1, Fig.4) for the same (400) crystal plane. Thus the level of elastic strain in these DWELLs is minimum. In DWELLs with QDs grown at 470 and 535 °C the corresponding XRD peaks shift to 59.07° testifing the higher level of compresive strain. Note that the partial relaxation of elastic strain can stimulate the creation of nonradiative recombination centers and due to this the decrease of the PL intensity in DWELLs with QDs grown at 470 and 535 °C.

It is essential to note that the application of buffer and capping $In_xGa_{1-x}As$ layers coupled with InAs QDs in DWELLs stimulates the lattice mismatch and strain decreasing in the vicinity of QDs, but, simultaneously, enhances the lattice mismatch and strain increasing at the InGaAs /GaAs interface for the surface area between QDs. With reduction of QD density (at the rise of QD growth temperatures [6]) the surface area of QDs reduces and the InGaAs/GaAs interface area between QDs enlarges. The variation non-monotonically of the integrated PL intensity and peak positions versus QD growth temperatures is connected, apparently, with the non-monotonous variation of the compressive strain due to the competition of mentioned above two effects

This work was partially supported by CONACYT (project 058358) as well as by SIP-IPN, Mexico. The authors thank very much Dr. Andreas Stintz from the Center of High Technology Materials at the University of New Mexico, NM, USA for the preparation of the DWELLs studied.

[1]. Bimberg D, Grundman M, Ledentsov NN, Quantum Dot Heterostructures, Wiley & Sons (2001) 328.

[2]. Liu GT, Stintz A, Li H, Malloy KJ and Lester LF, 1999 *Electron Lett*, **35**, 1163.

[3] Amtout A, Raghavan S, Rotella P, von Winckel G, Stinz A, Krishna S, 2004 *J. Appl. Phys.* **96**, 3782.

[4] Haft D, Warburton RJ, Karrai K, Huant S, Medeiros-Ribeiro G, Garsia J, Schoenfeld W, Petroff PM, 2001 *Appl. Phys. Lett.*, **78,** 2946.

[5]. Stintz A, Liu GT, Gray L, Spillers R, Delgado SM, Malloy KJ, 2000 *J. Vac. Sci. Technol.* B. **18**(3), 1496.

[6]. Torchynska TV, 2008 *J. Appl. Phys.* **104** (7), 074315.

[7]. Torchynska TV, Casas Espinola JL, Borkovska L, Ostapenko S, Dybic M, Polupan O, Korsunska N, Stintz A, Eliseev PG, Malloy KJ, 2007 *J. Appl. Phys.* **101**, 024323.

[8]. Torchynska TV, Ostapenko S, Dybic M, 2005 *Phys. Rev.B,* **72**, 195341.

[9] Dybiec M, Ostapenko S, Torchynska T, Velasquez Losada E, 2004 *Appl.Phys.Lett.* **84**, 5165.

[10] Dybiec M, Borkovska L, Ostapenko S, Torchynska T, Casas Espinola JL, Stintz A. and Malloy KJ, 2006 *Applid. Surface Science*, **252,** 5542.

[11] Dybiec M, Ostapenko S, Torchynska TV, Velasquez Lozada E, Eliseev PG, Stintz A, Malloy KJ, 2005 *phys. stat. sol. (c),* **2**, 2951.

[12] Yeh CY, Lu ZW, Froyen S and Zunger A, 1992 *Phys. Rev.* B, **46,** 10086.

Growth of low density InP/GaInP quantum dots

C J Elliott[1], E A Chekhovich[1,2], A B Krysa[3], M S Skolnick[1] and A I Tartakovskii[1]

[1]Department of Physics and Astronomy, University of Sheffield, Sheffield S3 7RH, UK
[2]Institute of Solid State Physics, 142432, Chernogolovka, Russia
[3]Department of Electronic and Electrical Engineering, University of Sheffield, Sheffield S1 3JD, UK

E-mail: c.elliott@sheffield.ac.uk, a.tartakovskii@sheffield.ac.uk

Abstract. We demonstrate low density InP quantum dot growth by metalorganic vapour phase epitaxy on (100) GaAs substrates with a 3° misorientation towards ⟨111⟩. Using micro-photoluminescence characterisation, we identify the transitional region from single to bimodal quantum dot size distributions where InP deposition thicknesses, over a range of $\approx 1.1Å$, provide a region of spatially and spectrally isolated quantum dots suitable for single dot studies.

Strong confinement effects and the resulting isolation of charge-carriers from the surrounding bulk material in individual quantum dots (QDs) have, in the last 10 years, received close attention[1]. The resulting phenomena which have been revealed, show the suitability of both InGaAs/GaAs and InP/GaInP QDs to a wide variety of interesting applications, ranging from single photon emitters [2, 3, 4] to spin qubits for quantum information processing [5, 6, 7, 8]. Recent progress in optically pumped InP/GaInP QDs, grown by metalorganic vapour phase epitaxy (MOVPE), has recorded high degrees of nuclear spin polarisation $\approx 65\%$ as well as ultra-long nuclear depolarisation times up to 5000s[9, 10, 11].

However, multi-modal distributions of QD sizes are common - with QD heights ranging from 1 to 15nm[12]. This leads to spatially non-isolated QDs and therefore additional relaxation pathways and uncontrollable interactions with highly charged neighbours[13]. Thus growth of InP/GaInP QDs with a single distribution of QD sizes is essential for single dot studies.

In this work, we demonstrate crystal growth leading to reproducible and finely controlled samples containing only small QDs with densities lower than 10^9cm^{-2}, avoiding multi-modal QD size distributions. We discuss the growth requirements for such structures and the influence of uni-directional reactant flow on the QD growth. We identify InP deposition thicknesses over a range of $\approx 1.1Å$, where the transition from single to a bimodal QD size distribution provides a region of spectrally low density QDs.

Growth was performed using low-pressure MOVPE in a horizontal flow quartz reactor, with two wafers exposed side-by-side (upstream and downstream) for each growth run. All samples were grown on (100) GaAs substrates with a 3° misorientation towards ⟨111⟩ and an InP growth rate of 1.1Å/s. The growth temperature of the GaAs buffer and bottom GaInP layer was 700°C. Before deposition of the InP (QD) and GaInP (capping) layers, the wafer was cooled to 650°C. For samples grown with InP deposition time (t_{dep}) of 10s, the temperature for the GaInP capping layer was 700°C. The grown GaInP layers were nominally lattice matched to GaAs.

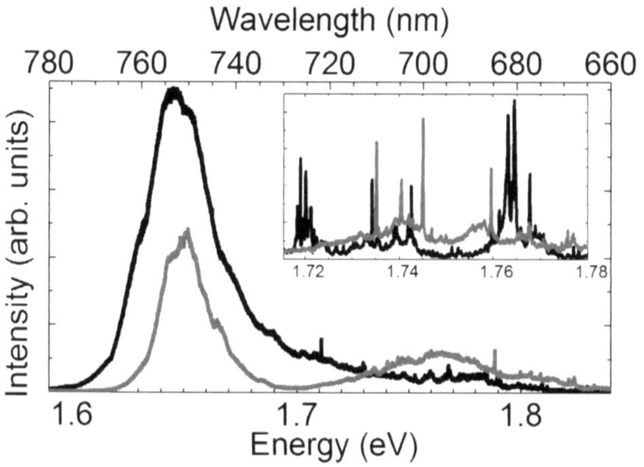

Figure 1. Low temperature μPL spectra of InP/GaInP QDs (P = 15μW) at two positions are shown. Comparing the spectra at an upstream (black) and downstream (grey) position, the change in size distribution of the QDs depending on position in the growth chamber is observed. Inset at $P = 1\mu$W.

All samples were measured at a temperature of 15K using a micro-photoluminescence (μPL) set-up. A laser with a wavelength of 543nm or 633nm, focussed onto a 2μm-diameter spot on the sample, was used to excite carriers in all spectra presented. μPL signal was measured with a high spectral resolution using a double spectrometer/CCD detection, enabling detection of photoluminescence (PL) from individual QDs.

Figure 1 shows typical μPL spectra recorded at relatively high excitation power, $P = 15\mu$W, for previously studied QD samples grown with t_{dep} = 10s. As seen for both upstream (black line) and downstream (grey line) samples, a pronounced bimodal size distribution is observed with two broad peaks centred at 1.77 and 1.65eV, corresponding to small and large QDs, respectively. Sharp peaks corresponding to PL of individual QDs are observed in the range 1.71−1.77eV at a lower excitation power, $P = 1\mu$W, for which population of excited states in QDs is suppressed [see Figure 1 inset]. However, even at such low power the peak at 1.65eV remains broad and featureless implying markedly higher densities of large QDs. Figure 1 also shows that the relative intensity of the low energy peak decreases for the downstream sample - indicating a lower density of large QDs forming at this point in the growth chamber.

The effect of reducing t_{dep} is shown in Figure 2, measured at $P = 0.15\mu$W. At the shortest growth time, 2.0s (lower grey line), we measure emission from only small QDs with a narrow size distribution peaked at 1.85eV and the wetting layer is observable at 1.89eV. As t_{dep} is increased the QD PL shifts to lower energy, becomes spectrally broader and wetting layer emission is no longer observed. At t_{dep} = 3s (upper grey line), a weak peak at 1.64eV is first observed, corresponding to initial formation of large QDs.

The optimum conditions for the growth of low density QDs with a single size distribution, are reached at t_{dep} = 3s. Between t_{dep} = 4 and 2.5s, over a range of InP deposition times, $\Delta t_{dep} \approx$ 1s, we observe a relatively small number of individual QD peaks, distributed in a wide spectral range, allowing clear observation of optical properties of individual QDs. This corresponds to a change in the InP deposition thickness of $\approx 1.1\mathring{A}$ and the QD density changes from 1.1x10^9 to 0.8x10^{9}cm^{-2}, for t_{dep}=2.5 and 3s respectively (where the density is estimated by counting the number of lines in the spectra at low powers).

We note that the observed range of deposition times which leads to growth of suitable samples, $\Delta t_{dep} \approx$ 1s, is large in comparison to the mechanical growth-control time. This results in very reproducible growth confirmed in our further growth experiments. We also note a smooth variation of QD density over the length of the reaction chamber is observed. As the measurement position is moved further downstream, the QD and bulk GaInP PL peaks shift to higher energy. This may be due to reduced concentration of In and Ga reactants on passing down the growth chamber.

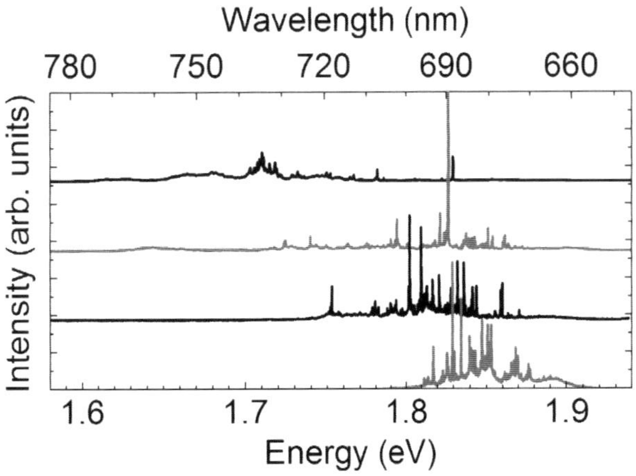

Figure 2. Low temperature μPL spectra of InP/GaInP QDs. Spectra with t_{dep} = 4.0s (upper black spectrum), 3.0s, 2.5s and 2.0s (lower grey spectrum) at P = 0.15μW are shown. The trend to increase spectral density of small QDs upon decreasing t_{dep} and the presence of large QDs at long t_{dep} can be clearly seen.

In conclusion, MOVPE growth conditions for spectrally and spatially isolated InP/GaInP QDs have been achieved. The obstacle of multi-modal size growth has been overcome and densities of less than 10^9cm^{-2} have been observed. We find a reproducible, smooth transition in QD size distribution and density for varying InP deposition times and position along the growth chamber. A wide range of InP deposition times corresponding to $\approx 1.1\AA$ change in deposition thickness, where conditions are suitable for single dot studies, has been found.

Acknowledgments
This work has been supported by the EPSRC Programme Grant EP/G601642/1 and the Royal Society. AIT is grateful for support by an EPSRC ARF.

References
[1] See a collection of reviews on QDs: 2009 *Semiconductor Quantum Bits* ed F Henneberger and O Benson (Singapore: Pan Stanford Publishing)
[2] Santori C, Pelton M, Solomon G, Dale Y and Yamamoto Y 2001 *Phys. Rev. Lett.* **86**(8), 1502
[3] Michler P, Kiraz A, Becher C, Schoenfeld W V, Petroff P M, Zhang L, Hu E and Imamoğlu A 2000 *Science* **290** 2282
[4] Schulz W -M, Roßbach R, Reischle M, Beirne G J, Bommer M, Jetter M and Michler P 2009 *Phys. Rev. B* **79** 035329
[5] Stevenson R M, Hudson A J, Bennett A J, Young R J, Nicoll C A, Ritchie D A and Shields A J 2008 Phys. Rev. Lett. **101** 170501
[6] Stievater T H, Xiaoqin L, Steel D G, Gammon D, Katzer D S, Park D, Piermarocchi C and Sham L J 2001 *Phys. Rev. Lett.* **87**(13) 133603
[7] Scholz M, Aichele T, Ramelow S and Benson O 2006 *Phys. Rev. Lett.* **96** 180501
[8] Xu K J, Huang Y P, Moore M G and Piermarocchi C 2009 *Phys. Rev. lett.* **103** 037401
[9] Skiba-Szymanska J, Chekhovich E A, Nikolaenko A E, Tartakovskii A I, Makhonin M N, Drouzas I, Skolnick M S and Krysa A B 2008 *Phys Rev B* **77** 165338
[10] Chekhovich E A, Makhonin M N, Kavokin K V, Krysa A B, Skolnick M S and Tartakovskii A I 2010 *Phys. Rev. Lett.* **104** 066804
[11] Chekhovich E A, Makhonin M N, Skiba-Szymanska J, Krysa A B, Kulakovskii V D, Skolnick M S and Tartakovskii A I 2010 *Submitted to Phys. Rev. B*
[12] Persson J, Holm M, Pryor C, Hessman D, Seifert W, Samuelson L and Pistol M -E 2003 *Phys Rev B* **67** 035320
[13] Hessman D, Persson J, Pistol M -E, Pryor C and Samuelson L 2001 *Phys. Rev. B* **64** 233308
[14] Bour D P 1993 *Quantum Well Lasers: AlGaInP Quantum Well Lasers* ed P S Zory Jr. (London: Academic Press) p415

Using the small Si cluster as a building block

A. J. Lu

Physics department of Donghua University, Shanghai, 201620, P. R. China

ajlu@dhu.edu.cn

Abstract. Assembly between Si_4 clusters is calculated to search for the Si_8 cluster with the lowest energy. Tetrahedral Si_4 cluster leads to a larger binding energy of the assembled Si_8, compared with the rhombus Si_4 cluster, although the latter is confirmed as the configuration with the lowest energy among the isomers of Si_4. So the prediction of larger silicon clusters should involve the building blocks with the lowest energy as well as those are meta-stable. The binding energy of silicon clusters versus the number of building blocks shows a unique curve. Thus, using the binding energy of the building blocks to predict the energy evolution of larger silicon clusters is suggested as a new feasible scheme for the medium-sized silicon clusters.

1. Introduction

Silicon clusters have been studied for decades as they are hopeful candidates in the application of electric and opto-electric fields [1-6]. It was demonstrated as the switch in the current circuit and reported as the possible blue-light emitter due to the large energy bands in the quasi-0D structure [7].

Unique properties are tightly related with the structure of the cluster, so the prediction of the atomic structure of the silicon clusters is hot topic in physics and material science [8]. Several attempts were made to seek for the most efficient method of finding the configuration of the cluster with the lowest energy. For example, generic scheme is one of the most popular methods in the work of searching for the medium-sized silicon clusters [9-12]. It is successful and has been used to predict many important structures of silicon clusters in theory.

The biggest problem occurred in the structure searching is the determination on the found structure whether it is stable. That was usually treated as the differentiation of the total energy of obtained structure whether the lowest one. Only when the cases with the lower total energies were found, the comparison ruled out the ones with larger energy and the others were reserved. It is natural to suspect the dependence of the searching scheme whether efficient and complete. We used our theoretical calculation to show a new way to determine the energy evolution of the silicon cluster, based on the smaller cluster regarded as the building block.

Based on the previous work, we calculate the energy of silicon clusters Si_n ($n<12$) coupled with the smaller silicon clusters used as the building blocks. These small clusters used as building blocks are so-called seeds in our work, and we find a unique curve between the number of seeds and the binding energy of the assembled result. Furthermore, the result indicates that, the larger silicon clusters showing larger binding energy which is calculated from the building blocks should be the structure with lower energy more possibly. In another word, the curve of binding energy of the building blocks could be a criterion to differentiate the clusters from the unstable candidates. Thus a new method to determine the configuration of the silicon cluster at medium-size is suggested.

2. Methods

Density functional theory has been confirmed one of the most efficient methods to calculate the energy evolution of the nano-sized materials. It is our choice to deal with the silicon clusters. A well-developed *SIESTA* package [13-15] was used and the exchange-correlation functional was selected as General Gradient Approximation (GGA) proposed by Perdew, Burke and Ernzerhof (PBE). Double-zeta basis with polarization was employed and the potential of silicon was checked in the calculation of the bulk silicon crystal. The cubic supercell was considered as $15\times15\times15$ Å^3 so that the distance between neighboring clusters is large enough to ensure the interaction between them neglectful.

During the full-relaxation, the Si_4 cluster in different configuration was optimized. The one with rhombus structure was found the most stable as the total energy of it is lower than the tetrahedral one by 0.54 eV/atom. Such a result is consistent with the early report submitted by Khakomov *et al* [16]. The assembling result of two Si_4 clusters was shown as two possible structures in different symmetry which were demonstrated in Krishnan's work [17]. We did obtain the same two structures in our calculation and the energy comparison agreed with the published result. All of this confirmed that our computational scheme is valid and acceptable at this level of precision.

3. Results and discussions

It has been predicted that the small silicon clusters would assemble when they are close to each other, due to the activity of the cluster related with the large ratio between surface area and the volume. Such a phenomenon was confirmed in our calculation. When the distance between two Si_4 clusters was fixed manually, we calculated the total energy of the Si_4-Si_4 system. Fig. 1 shows the energy evolution of the assembly, starting from (a) two rhombus Si_4 clusters and (b) two tetrahedral Si_4 clusters. Both of the procedures were found exothermal, which indicates the assembling taking place automatically.

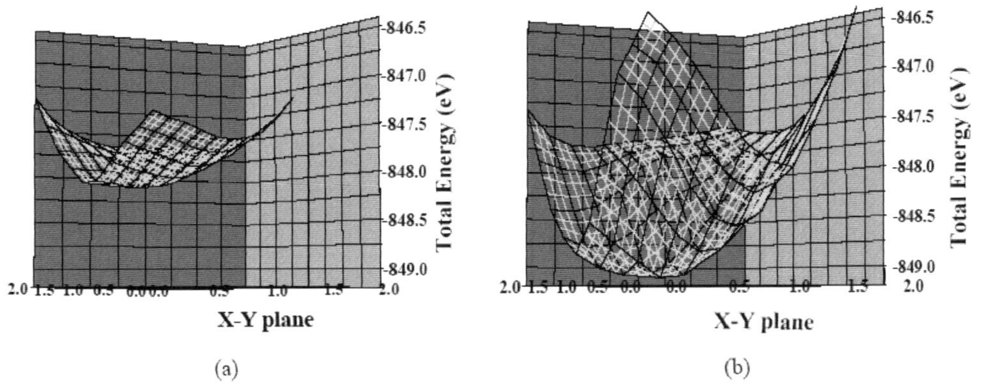

Fig. 1 Energy evolution of two Si_4 cluster assembling. X-Y plane shows the distance and orientation between the bottom centers of the two clusters. The variance of the total energy shows the energy lowering is much sharper in (b) between two tetrahedral Si_4 clusters than that in (a) between a pair of rhombus Si_4 clusters.

In Fig. 1(a), the distance between the clusters is measured between the centers of two rhombus configurations when they are approaching parallel. In Fig. 1(b), the distance is measured between the bottom centers of two neighboring tetrahedral Si_4 clusters. In fact, we found the energy evolution different when the two tetrahedral Si_4 clusters approaching in different ways. Here is the typical case that leads to the lowest energy. The orientation of the distance was indicated in Fig.1, and such a mapping of total energy revealed energy variation during the assembling in the real space totally.

As the definition of the distance varies in the figures, the quantity at each point is not of importance. It is emphasized the slope of the energy surface which shows the stabilization of the assembled structures. The slope of the energy surface is much larger in Fig. 1(b) than that in Fig. 1(a), indicating the tetrahedral Si_4 clusters are more favorable candidates in the assembling. As it has been confirmed that the tetrahedron is a meta-stable structure of Si_4 cluster, such a result shows that the assembling happens on meta-stable structures more possibly. It means that during the searching of the medium-sized silicon clusters, the seed clusters at meta-stable states must be included. The total energy is not always the best proof in the cluster searching.

In the analysis of the binding energy, defined as $E_b = E_{seed} \times n - E_{product}$, such a finding is even more remarkable. Although the total energy of the Si_4-Si_4 assembled structures varies slightly, indicated in the inset of Fig. 2, the binding energy shows a large difference between the main products. It is shown in Fig. 2 that the binding energy of the tetrahedral Si_4 clusters induced structure is larger than that coming from rhombus Si_4 clusters of 5 eV or so. Such a large difference would be hidden if we are only concerned of the total energy of the structures. A further detailed observation revealed that the final structures in these two assembles are different, as they are in C_{2h} symmetry and D_{3d} symmetry respectively. These two configurations were discussed in ref [17], where they concluded the former one as the isomer with lower energy, which coincides with our work perfectly.

Such a result notified us again that the Si_4 structure at meta-stable is important in assemble process. If only the rhombus Si_4 clusters were considered in assemble, the Si_8 in C_{2h} would be missed in the searching procedure. We could only obtain the Si_8 in D_{3d} symmetry starting from rhombus Si_4 clusters. Moreover, we were enlightened by the amplification of the binding energy difference in Fig. 2. It is hopeful using binding energy to assist the seeking of the larger sized silicon clusters as it is sensitive to the atomic structure among the isomers.

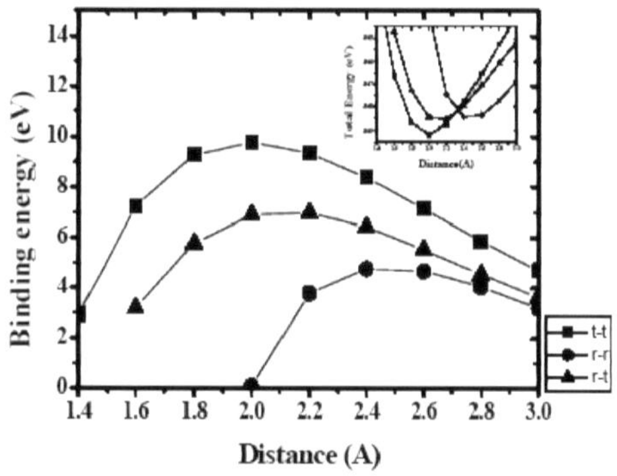

Fig. 2. Binding energy of Si_4-Si_4 assembling versus the distance between two building blocks. Inset shows the total energy of the Si_4-Si_4 system. Filled squares display the energy evolution of two tetrahedral Si_4 assembling, filled circles show the assembling between two rhombus Si_4 clusters, and filled triangles indicate the approaching between a tetrahedral Si_4 and a rhombus Si_4 cluster.

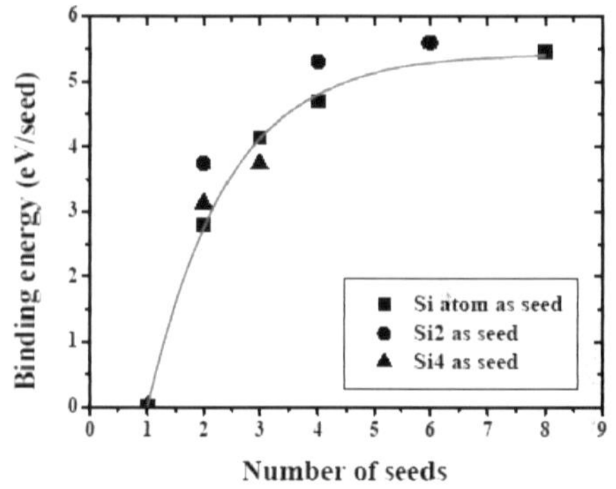

Fig. 3 Binding energy of the silicon clusters versus the number of the seeds that are building blocks in the assembling. Filled squares are results of cluster using silicon atoms as seeds, filled circles show the clusters obtained from Si-Si dimmer, and filled triangles indicate the silicon clusters made up of Si_4 clusters.

Based on our definition of the binding energy per seed, $E_{b^*} = (E_{seed} \times n - E_{product}) / n$, we calculated the binding energy of a series of silicon clusters. The data points of the binding energy of the clusters versus the number of seeds are displayed in Fig. 3. It is interesting that all of them obey the same rule as they fall in a common line no matter which seed we used. The filled squares show the binding energy of the clusters using the atomic silicon as seed; the filled circles indicate the binding energy of the cluster using the Si-Si dimmer as the seed; the filled triangles show the binding energy of the cluster using the Si_4 as the building blocks. Red line is guidance to our eyes. It is clear that the binding energy increases as the number of seeds increases and saturates. That is a reasonable result because the binding energy should converge to the binding energy of silicon bulk crystal gradually.

In fact, in Fig.3 we do not show results of the structure with the lowest energy only. That is, the data point showing two dimmers assembling is related with the rhombus Si_4 cluster, the data point showing two Si_4 coupling results in the Si_8 cluster in D_{3d} symmetry. Both of them are accepted as the structure with the lowest energy. The points according to them are higher than the red line in position, showing the stable building blocks leading to the stable assembled result. The point displaying three Si_4 clusters is just one case we test on, and its binding energy is some smaller than the comparative dot in the red line. In the further confirmation calculation we found that, such a structure is not the configuration with the lowest energy. Such a result demonstrates that, it is hopeful to judge the configuration whether the one with the lowest energy through taking the curve as a reference. If the found structure showing a binding energy smaller than the corresponding point in the line shown in Fig. 3, the structure should not be the structure with the lowest energy. It means that, we could use the binding energy of the cluster to search for the new configurations, and even estimate the minimum of the binding energy of medium-sized silicon clusters based on the energy of the building blocks. For example, the binding energy of Si_{12} cluster (made up of three Si_4 clusters) should be not smaller than 4.14 eV, so that e may filter the structures in the geometry searching of the silicon clusters firstly.

4. Conclusion

Based on DFT calculation we study the assembling of smallest silicon clusters. We confirmed the assembling an exothermal process and this assembling tightly related with the searching of larger silicon clusters in theoretical study. It is found that the binding energy of the cluster could be estimated from the energy of the smaller clusters which could be regarded as building blocks. The reference line comes from the binding energy of silicon clusters versus the number of building blocks—silicon atoms. Only the structure shows the binding energy larger than the corresponding point indicated by the curve could be the structure with the lowest energy. It is a new suggestion to filter the structure in the work of structure searching of silicon clusters.

Acknowledgement

This project was supported by the Fundamental Research Fund for the Central Universities (No.2009C03-1-01) and the Starting Funding for Youth Teacher in Donghua University.

[1] L. Mitas, J. C. Grossman, I. Stich, and J. Tobik, Phys. Rev. Lett. (2000)84, 1479.
[2] F. Baletto and R. Ferrando, Rev. Mod. Phys. (2005)77, 371.
[3] U. Landman, R. N. Barnett, A. G. Scherbakov, and P. Avouris, Phys. Rev. Lett. (2000)85,1958.
[4] K. D. Hirchman et al, Nature (London) (1996)384, 338.
[5] S. Y. Lin et al, Nature (London) (1998)394, 251.
[6] S. Yoo et al, J. Am. Chem. Soc. (2003)125, 13318.
[7] Robert Dalpé, Elaine Gauthier and Marie-Pierre Ippersiel. The State of Nanotechnology Research: Report to the National Research Council of Canada[P]. May,1997.
[8] L. S. Pedroza and A. J. R. da Silva, Phys. Rev B (2007)75, 245331.
[9] S. Yoo and X. C. Zeng, Ang. Chem. Int. Ed. (2005)44, 1491.
[10] K. –M. Ho, A. A. Shvartsburg, B. Pan, Z. –Y. Lu, C. –Z. Wang, J. G. Wacker, J. L. Fey, M. F. Jarrold, Nature.(1998)392, 582.
[11] B. Liu, Z. –Y. Lu, B. Pan, C. –Z. Wang, K. –M. Ho, A. A. Shvartsburg, M. F. Jarrold, J. Chem. Phys. (1998)109, 9401.
[12] I. Rata, A. A. Shvartsbrug, M. Horoi, T. Frauenheim, K. W. M. Siu, K. A. Jackson, Phys. Rev. Lett. (2000)85, 546.
[13] P. Ordejón, E. Artacho and J. M. Soler, Phys. Rev. B (1996)53, 10441.
[14] E. Anglada, J. M. Soler, J. Junquera and E. Artacho, Phys. Rev. B (2002)66, 205101.
[15] E. Artacho, E. Anglada, O. Dieguez, J. D. Gale, A. García, J. Junquera, R. M. Martin, P. Ordejón, J. M. Pruneda, D. Sánchez-Portal and J. M. Soler, J. Phys.: Condens. Matter (2008)20, 06420.
[16] Z. M. Khakomov, P. L. Tereshchuk, N. T. Sulaymanov, F. T. Umarova and M. T. Swihart, Phys. Rev B (2005)72, 115335.
[17] K. Raghavachari and C. Mc. Rohlfing, J. Chem. Phys (1988)89, 2219.

Synthesis of high quality monodisperse Nickel Oxide Nanocrystals

P Khagram*, F Tomson, R M D Brydson, R Crook

School of Process, Environmental and Materials Engineering, University of Leeds, Leeds, LS2 9JT

*Corresponding Author: pmpk@leeds.ac.uk

Abstract. Monodisperse nickel oxide (NiO) nanocrystals with a particle size of 2.7±0.7nm have been synthesised using low temperature reflux of hydrated nickel acetate and methanol. It has also been possible to create NiO nanocrystals without refluxing. The fabrication of NiO nanorods with size varying from 2 – 10 μm in length and about 0.25μm in width and crystalline pellets with approximate dimensions of 150nm by 300nm was also achieved.

1. Introduction

The unique quantum properties displayed by nanomaterials make them very attractive to use for a wide range of applications. There has been extensive research in the synthesis and potential applications of these materials and structures which include nanocrystals, nanorods, nanowires, nanofilms and core shell structures to name but a few. Nickel oxide has attracted particular interest due to a wide range of beneficial properties such as catalytic behaviour, anomalous electronic and magnetic properties. These properties open up numerous potential applications such as catalysis, electrochromic windows, sensors, magnetic resonance imaging, targeted drug delivery, solid oxide fuel cells as well as functional layers, dye-sensitized photocathodes and electrodes in solar cells [1-6]

Here we report the successful synthesis of Nickel oxide nanocrystals with optical and electrical properties tailored for applications in solar cells. NiO is an inherently p-type semiconductor with numerous potential applications in solar cells mentioned above. The band gap of nickel oxide is 3.55 eV which is considered to be a wide band gap and therefore it would not be possible to use it as an efficient single junction solar cell device. Nickel oxide nanocrystals with an average size of 2.7±0.7nm have already successfully been synthesised and now attempts are being made to add a Zinc oxide shell to form a p-n junction. It is hoped that successful synthesis of these Nickel oxide nanocrystals will allow them to be used as a functional layer in a tandem solar cell exploiting the higher energy end of the solar spectrum.

Nickel oxide nanocrystals are synthesized by reflux of alcohol solutions of hydrated metal acetate under ambient conditions, following a method similar to that used by Wu et al. for ZnO [7,8]. Chemical and physical properties, including the configuration of deposited nanocrystals, are measured using transmission electron microscopy (TEM), energy-dispersive x-ray spectroscopy (EDX), selected area electron diffraction (SAED) and high angle annular dark field HAADF imaging. Optical properties are measured using spectrofloremetry, photoluminescence, and a novel 'nanocrystal test bed' semiconductor device which is being developed at Leeds by Malvi et al.

TEM images, EDX spectra and SAED confirm the successful synthesis of high quality NiO nanocrystals. The highly crystalline NiO nanocrystals produced are spherical with excellent dispersivity and an average size of 2.7±0.7nm. This tight size distribution is very important for practical applications as the properties of the nanocrystals vary with size particularly the optical and electrical properties which are both of paramount importance when it comes to solar cell applications. Numerous samples have been produced and variations to the synthesis have been successfully attempted to control the size of the particles. Particles have even been produced without the application of heat which is unique in terms of producing nanocrystals of this quality. NiO nanorods have also been produced which is an exciting prospect as these structures are also showing great promise in photovoltaic applications.

2. Methodology

The synthesis of NiO nanocrystals via simple reflux methods similar to Wu et al. [7,8]. Hydrated metal acetate was added to methanol in the ratio of 0.03:4.00 and refluxed for 6 hours under ambient conditions, at a temperature of 67°C (the boiling point of the solution) and vigorous stirring. The above ratio was the starting point from which the concentration of the metal ions was varied by adding more metal acetate at various times during the reflux. In addition to this the reflux times were also varied. Cooling to room temperature (23°C) the organic capping agent 3-aminopropyl trimethoxysilane (AM), which has the chemical formula $H_2N(CH_2)_3Si(OCH_3)_3$ is added to different samples at different times. A similar reflux hydrolysis method has been reported to produce CuO nanocrystals [9].

3. Results

TEM, EDX and SAED analysis confirm the synthesis of NiO nanocrystals. It can be seen from figure 1 that the NiO nanocrystals produced have a tight size distribution with an average particle diameter of 2.5±0.5nm. Figure 2 is the SAED pattern taken from the same area as the TEM image in figure 1 and confirms that the lattice spacings are consistent with that found in NiO. The NiO nanocrystals produced here were done so after a 6 hour reflux with the above mentioned Ni : methanol ratio.

Figure 1. TEM image of NiO nanocrystals capped with AM. The nanocrystals can clearly be seen and their crystalline structure is also evident.

Figure 2. SAED pattern of the NiO nanocrystals produced. The atomic spacing from the electron diffraction confirms that the structure is NiO.

NiO nanocrystals produced here were done so after a 6 hour reflux with the above mentioned Ni : methanol ratio. In figure 3 the nanocrystals produced are also from a 6 hour reflux but are uncapped. They are the same size as the capped sample and the electron diffraction pattern clearly shows that they are crystalline.

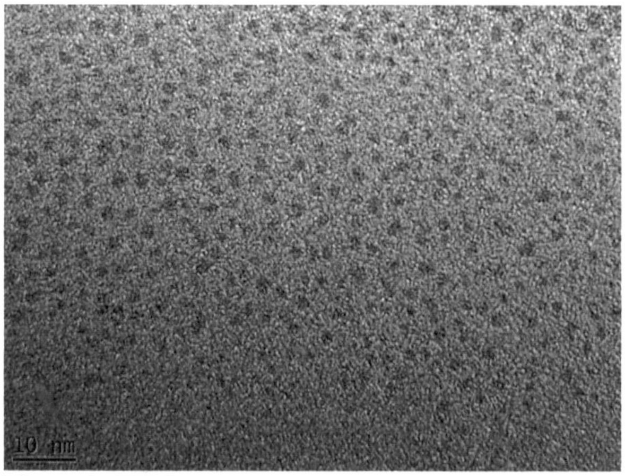

Figure 3. TEM image of NiO nanocrystals produced after a 6 hour reflux. This sample is uncapped

Figure 4. SAED pattern of the NiO nanocrystals produced. The atomic spacing from the electron diffraction confirms that the structure is NiO.

Figure 5 shows a TEM image of an eight hour reflux. The NiO particles produced here are significantly larger and elongated rod like particles. There are not any small circular nanocrystals present in this sample.

Figure 6 shows nanocrystals produced from just mixing hydrated nickel acetate with methanol and ultrasonicating the solution. This is a remarkable result as in all methods seen in the literature for producing NiO nanocrystals require significantly large temperatures. Note the nanocrystals produced are the same size as those produced in the 6 hour reflux with the same concentration. The same results are achieved for 2 hour and 4 hour refluxes respectively.

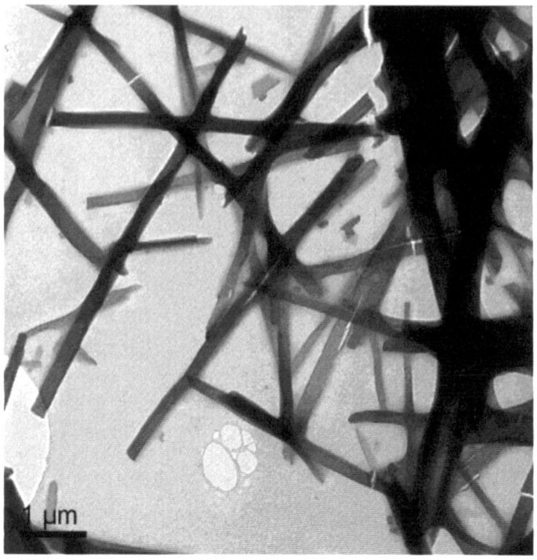

Figure 5. TEM image of NiO rods produced after an 8 hour reflux.

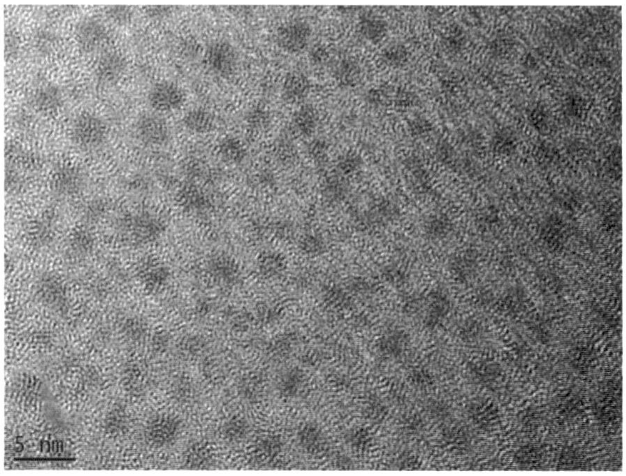

Figure 6. TEM image of NiO nanocrystals produced from no reflux.

In figure 7 the nanocrystals produced are from a 6 hour reflux which had the same initial concentration as mentioned above and in the previous refluxes. However after two and four hours the same mass of nickel acetate was added. Therefore the final concentration of nickel ions in the methanol was tripled. As can be seen the nanocrystals produced are the same size as the previous refluxes.

However as seen in figure 8 there are some larger particles produced which are micron sized. These micron sized particles show signs of agglomeration. It was possible to separate these large particles from the small nanocrystals via sedimentation as the micron sized particles tend to sink to the bottom of the container. When the sample was sonicated before being deposited on the TEM grid the nanocrystals could be seen around close to the larger particles but not dispersed all along the grid like when these larger particles were not present.

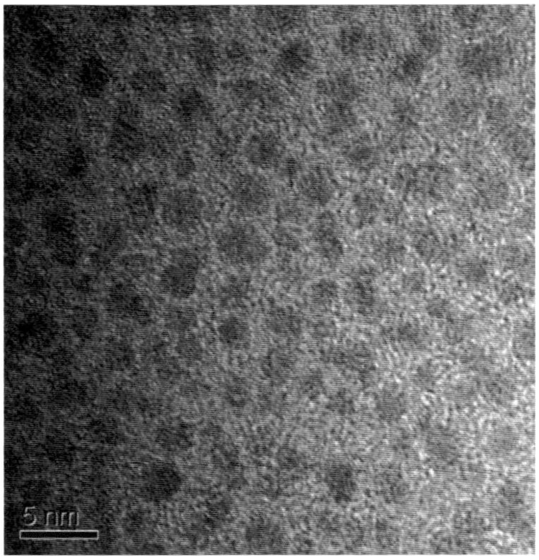

Figure 7. TEM image of NiO nanocrystals produced from a 6 hour reflux but adding nickel acetate at various times during the reflux.

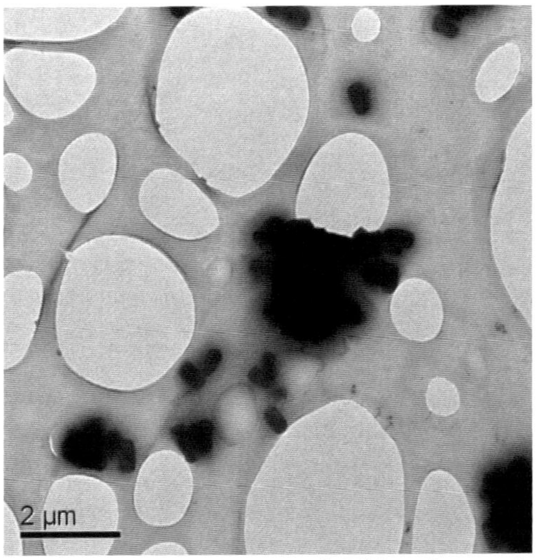

Figure 8. TEM image of NiO micron sized particles produced from a 6 hour reflux but adding nickel acetate at various times during the reflux.

4. Summary

It can be seen from the above images that the NiO particles produced have a consistent size of 2.7±0.7nm and that some of the conventional approaches to increasing the size have little effect. However it is also possible to produce very large particles with increased reflux time and concentration which can easily be separated from the smaller particles via filtration and sedimentation. Remarkably it is possible to produce nanocrystals without heating which seems to be unprecedented when producing nanocrystals.

References

[1] C An, R Wang, S Wang and Y Liu, Materials Research Bulletin **43**, 2563-2568 (2008)
[2] Y Li, M Afzaal and P O'Brien, J. Mater. Chem. **16**, 2175-2180, (2006)
[3] I Hotový et al, Journal of Electrical Engineering, **58**, 347-350, (2007)
[4] L Chen, L Li and G Li, Journal of Solid State Chemistry, **181**, 2073-2077 (2008)
[5] L Chen, L Li and G Li, Solid State Ionics **179**, 712-717 (2008)
[6] A Nattestad et al, Nanotechnology, **19**, 295304 (2008)
[7] Wu YL, Lim CS, Tok AIY, Lau HM, Boey FYC and Zeng XT *Nanotechnology* **18** 215604 (2007)
[8] Wu YL, Tok AIY, Boey FYC and Zeng XT, 2007 *IEEE Trans Nanotech* **6,** 497 (2007)
[9] Junwu Zhu et al, Materials Science and Engineering A 384 172–176(2004)

Journal of Physics: Conference Series **245** (2010) 012064 doi:10.1088/1742-6596/245/1/012064

Optical Analysis of a series of Size and Shape-controlled Type-II CdTe/CdSe/CdTe Heterostructure Nanorods

Hongming Zhao[1,3], Bonil Koo[2], J.L. Movilla[4], Dongmin Chen[3], Brian A. Korgel[2] and Kuntheak Kheng[1]

[1]Nanophysics and Semiconductor Group, CEA/CNRS/Université Joseph Fourier, Institut Néel, 38054 Grenoble Cedex 9, France

[2]Department of Chemical Engineering, Center for Nano- and Molecular Science and Technology, and Texas Materials Institute, University of Texas, Austin, TX 78712, USA

[3]Beijing National Laboratory for Condensed Matter Physics, Institute of Physics, Chinese Academy of Sciences, 100190 Beijing, P.R. China

[4]Department of Physical and Analytical Chemistry, University Jaume-I, E-12080 Castelló, Spain

Email : kuntheak.kheng@ujf-grenoble.fr

Abstract. Two series of size and shape-controlled type-II CdTe/CdSe/CdTe heterostructure nanorods with different geometrical structures were prepared by the colloidal sequential reactant injection technique. The samples, including pure CdSe seeds, were optically analysed by steady state and time resolved photoluminescence spectra. The experimental results demonstrate that the energy and lifetime of the photoluminescence from charge-separated band structure can be tuned by the nature of nanorods' band edges in both "linear" and "core-shell" heterostructure nanorods. But between different groups of nanorod samples, the spectral properties show different variation tendencies with the increasing CdTe tip. As one possible explanation, the compensation between the variations of kinetic and Coulomb contributions was discussed by numerical calculations based on the effective mass and envelope function approximations.

1. Introduction

The type-II CdTe/CdSe nanocrystals have been extensively studied for potential applications in optoelectronic devices and for their novel electronic and optical properties that come from quantum confinement in their unique structure different from type-I nanocrystals [1, 2, 3]. At the interface between CdSe and CdTe segments, the band edges have a staggered type-II alignment. The conduction and valence band energies within the CdSe segment are respectively lower than the conduction and valence band energies in CdTe, as showed in Figure1.(a). After the photo-excitation, it is inclined to give a spatial separation of confined holes and electrons in type-II band alignment, making easier to harvest the photo-generated carrier. This makes CdTe/CdSe heterostructures an attractive candidate for the fabrication of solar cells [4].

Recent series of works described the progress of synthesis and characterization of nonspherical CdTe/CdSe heterostructure [5, 6, 7]. Comparing to the spherical nanocrystals, the optical and electronic properties of size- and shape-controlled nonspherical nanorods can be manipulated more easily. Due to the tunable properties, these heterostructures nanorods provide exciting opportunities for fundamental research of nanostructures and potential optoelectronic device application. The basic optical properties of CdTe/CdSe/CdTe heterostructure nanorods have been obtained by recent studies [7]. It was found that the PL spectra at low temperature originated from two different optical transitions: the direct (type-I) radiative recombination of CdSe and the red-shifted radiation from type-II transition due to an electron in CdSe segment and a hole in CdTe. For further understanding of the characteristic of indirect type-II transition in this system, two different geometrical structure of CdTe/CdSe/CdTe heterostructures nanorods were designed and synthesized.

2. Experiment

Two series of samples (A and B) have been synthesized. From a same CdSe seed (sample N1), increasing CdTe tips at both ends are grown (samples N2 to N5). The nanorods of series A are perfect linear heterostructures, while the nanorods of series B have an additional core-shell heterostructure interface, see Figure1.(b). The CdTe/CdSe/CdTe heterostructure nanorods were synthesized by the colloidal sequential reactant injection technique [5, 6]. Firstly, a Cd-TDPAcomplex [8], which is necessary for nanorod formation, was prepared and degassing. Then, a mixture of 0.856 g of Cd-TDPA complex and 1.1436 g of TOPO was degassed in the reaction flask for one hour and heated to 320 °C under nitrogen. A selenium precursor reactant solution was prepared by dissolving 0.0316 g of Se in a mixture of 0.117 mL of TBP, 0.8705 mL of TOP, and 0.1735 mL of toluene at 120 °C. The tellurium reactant solution was made under the same condition except with an increased amount (1.25 mL) of TOP to lower the solution viscosity for easy injection through a syringe pump. The selenium precursor was rapidly injected into the mixture of Cd-TDPA and TOPO at 320 °C. The temperature was then immediately decreased to 275 °C to promote the formation of relatively high aspect ratio CdSe nanorods [8]. After heating for 15 minutes, the tellurium precursor was injected by syringe pump for 7.5 minutes at 250 °C, which yields linear CdTe/CdSe/CdTe heterojunction nanorods. In contrast, to make core-shell type heterojunction nanorods, the reaction solution was heated to 275 °C and then the tellurium precursor was injected at the same injection rate. During Te-injection, nanorods were removed from solution at different times in ~0.3 mL aliquots for characterization.

Careful size calculations based on the TEM micrographs give the mean diameter and length of both the seed CdSe (named N1) and the CdTe/CdSe/CdTe heterostructure nanorods (named N2, N3, N4 and N5). The serial number and corresponding size of samples-A and –B are presented in Table 1. The standard deviation of the size ranged between 10 and 15% for all samples.

Table1. Serial number and corresponding size of samples-A and -B.

Samples-A	N1(seed)	N2	N3	N4	N5
Diameter(nm)	3.48	3.50	3.50	3.50	3.51
Length(nm)	20.3	25.3	32.1	37.9	40.1
Samples-B	N1(seed)	N2	N3	N4	N5
Diameter(nm)	3.74	3.79	3.77	3.81	4.16
Length(nm)	25	28.1	37.4	41.7	50.7

For samples-A, the diameter from N2 to N5 keeps at about 3.5nm and they are considered as “linear” heterostructures nanorods. But the difference of diameters between different samples in samples-B is out of just error range of measurement, meaning that there is additional CdTe growth as a shell. They are “core-shell” heterostructure nanorods.

Steady-state photoluminescence (PL) spectra have been measured (using the 448nm line of an Argon ion laser) on these samples to investigate the competition between emissions from type-I and type-II band structure. To confirm the long lifetime expected for the type-II transition and characterize the relationship between the emission lifetime and wavelength in PL of- CdTe/CdSe/CdTe heterostructure nanorods, time resolved PL (TRPL) experiments were carried using a frequency doubled, pulsed Ti-sapphire laser with a pulse interval modulation system which can adjust the interval of laser pulse ("time window") in a large range

3. Results and Discussions

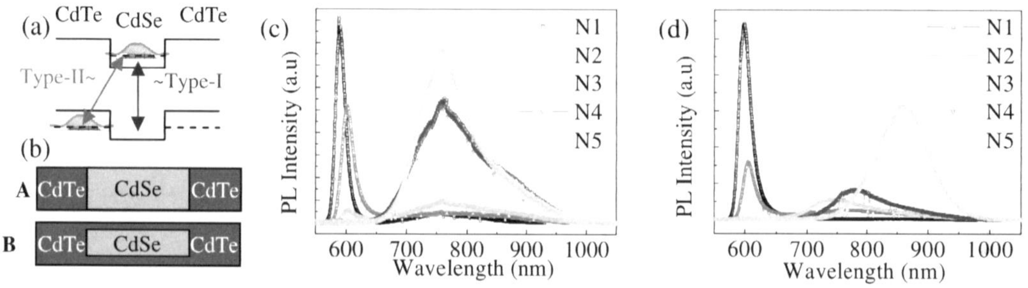

Figure1.(a) Type-II band alignment at the interface between CdSe and CdTe; (b) The geometrical structure of samples-A and –B; (c) PL of CdSe seed (N1) and samples-A (from N2 to N5); (d) PL of CdSe seed (N1) and samples-B (from N2 to N5); All the data were obtained at 5K.

Figure1.(c) and (d) shows the -PL spectra at 5K of pure CdSe seeds (N1), the CdTe/CdSe/CdTe heterostructure nanorods (N2-N5) from samples-A and –B respectively. The type-I direct transition line is observed at about 600nm of N1 samples (pure CdSe nanorod seed). The PL spectra of CdTe/CdSe/CdTe heterostructure nanorods (N2-N5) show an additional broad PL peaks (700-850nm) whose intensity increases with the length of the CdTe tips . This broad peak is attributed to type-II emission [9]. Due to the weak overlap of wavefunction within the spatially separated electrons and holes in type-II structure [3], a long lifetime of type-II emission in samples-A and -B nanorods is expected. The TRPL was carried out to confirm it. We show the results for each series A and B, in samples N3 (Fig. 2) because both the type-I (at about 600nm) and the type-II (about 770nm) transitions have substantial intensity in the same sample. We found indeed a very long lifetime which is more than 1000ns for the indirect emission at about 770nm as compared to the 4.0 ns (measured with a more accurate time scale) decay time of the CdSe direct emission at 600nm, as shown in Figure2.

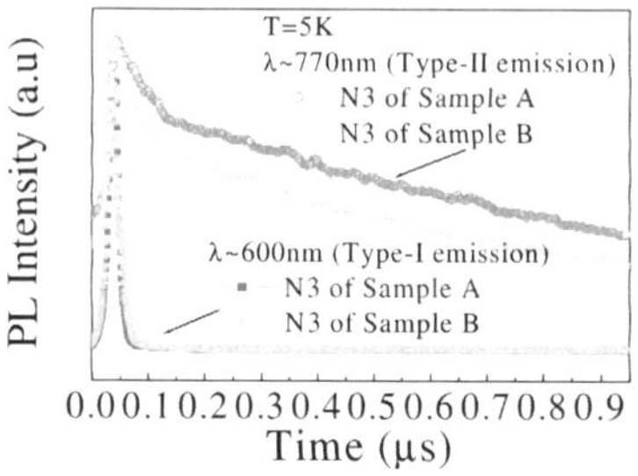

Figure2. PL decays measured at 770nm (Type-II emission) and 600nm (Typr-I emission) by TRPL. The curve in red is for the data of "linear" CdTe/CdSe/CdTe heterostructure nanorods samples-A N3; the curve in green is for the data of "core-shell" CdTe/CdSe/CdTe heterostructure nanorods samples-B N3. All the data were obtained at 5K.

As presented in Figure1.(c) and (d), for samples-B, with the increasing length of CdTe tip from nanorods N2 to N5, the intensity of additional broad PL peaks increased with a red shift of the wavelength of peak from 790nm to 850nm. That is, bigger CdTe tips (act as shells) had lower energy gap. But for A samples, the broad emission peaks (~770nm) in happened all at the same position. In order to confirm theoretically the possible interpretations, we carried out numerical calculations based on the effective mass and envelope function approximations to describe the one-band electron and hole single-particle Hamiltonians, followed by a full configuration-interaction scheme to include the electron-hole Coulomb attraction (see e.g. [12] for details).

The results of the simulations were consistent with the experimental observations. For samples-A, the "linear" heterostructure nanorods, except for very low CdTe injection, the type-II transition energy tends rapidly to a constant value as the CdTe length increases, which is due to a compensation between the variations of kinetic and Coulomb contributions. Thus, as the length of the CdTe tip increases, the hole kinetic energy is relaxed, favoring a decrease of the gap energy. However, such decrease is compensated by the effect of a less effective Coulomb stabilization of the exciton yielded by the increase of the mean distance between electron and hole. But the compensation is not present in samples-B. The red shift of the type-II emission as the CdTe injection increases is ascribed mainly to the increase of the rod width (which is produced because of the growth of the CdTe shell), and not because of the increase in length. The reason is that, due to the large anisotropy of the confining potentials, the (strong) lateral confinement is much more effective than the longitudinal (growth direction) confinement and, as a consequence, now both electron and hole undergo a large relaxation of the kinetic energy which is no longer compensated by Coulomb contributions.

4. Conclusions

The optical analyses of a series of type-II CdTe/CdSe/CdTe heterostructures nanorods with different geometrical structures by steady state PL and TRPL was carried out. The size dependent wavelength shift of PL spectra at 5K shows the transfer of charge carries recombination from type-I (CdSe) to type-II transition (CdSe/CdTe) and the competition between the two routes of recombination. The origin of the apparently different behavior of samples-A and -B is the large anisotropy of the rod shape, which makes the electron and hole energy much more sensitive to changes in the (strong) lateral confinement than in the (weak) longitudinal one. These results and tentative conclusions, which are described by the numerical calculations, should help to shed light on our understanding of the type-II band alignment in CdTe/CdSe/CdTe heterostructures nanorods. Further experiments and theoretical calculation about the optical properties in this system are in progress.

Acknowledgments

This work was support by China Scholarship Council (CSC, File No.2008491036). We would like to thank Dr. Joël BLEUSE for the critical assistance and discussions of TRPL measurements.

References

[1] Kim S, Fisher B, Eisier H J and Bawendi M 2003 *J. Am. Chem.Soc.* **125** 11466.
[2] Sandeep Kumar, Marcus Jones, Shun S. Lo, and Gregory D. Scholes 2007 *Small* 3, **9**, 1633 – 1639.
[3] Wang C H, Chen T T, Chen Y F, Ho M L, Lai C W and Chou P T 2008 *Nanotechnology* **19** 115702.
[4] Gur I, Fromer N A, Geier M L and Alivisatos A P 2005 *Science* **310** 462.
[5] Shieh F, Saunders A E and Korgel B A 2005 *J. Phys. Chem. B* **109** (18), 8538-8542.
[6] Koo B, Korgel B A 2008 *Nano Lett.* 8, **8**, 2490–2496.
[7] Saunders A E, Koo B,Wang X, Shih C K and Korgel B A 2008 *ChemPhysChem* **9**,1158-1163.
[8] Peng Z A, Peng X J 2002 *Am. Chem. Soc* **124**, 3343.
[9] Brus L 1986 *J.Phys.Chem* **90** 2555.
[10] Efros Al L and Efros A L 1982 *Soviet Phys. Semicond* **16** 772.
[11] Alivisatos A P 1996 *J. Phys. Chem* **100**, 13226-13239.
[12] J.L. Movilla, J.I. Climente, J. Planelles 2010 *Computer Physics Communications* **181** 92–98.

Control of Strain in GaSbAs/InAs/GaAs Quantum Dots

V. Haxha[1], I. Drouzas[2,3], J.M. Ulloa[3], M. Bozkurt[3], P.M. Koenraad[3], D.J. Mowbray[2], H.Y. Liu[4], M.J. Steer[4], M. Hopkinson[4], M.A. Migliorato[1]

[1]School of Electrical and Electronic Engineering, University of Manchester, Sackville Street, Manchester M13 9PL United Kingdom

[2]Department of Physics and Astronomy, University of Sheffield, Sheffield S3 7RH, United Kingdom

[3]Department of Applied Physics, COBRA Inter-University Research Institute, Eindhoven University of Technology, P.O. Box 513, 5600 MB Eindhoven, The Netherlands

[4] EPSRC National Centre for III-V Technologies, Sheffield S1 3JD, United Kingdom

max.migliorato@manchester.ac.uk

Abstract. We discuss strain simulations of quantum dot structures covered with a GaSbAs strain reducing capping layer in the presence of Sb segregation. Cross Sectional Scanning Tunneling Microscopy shows strong Sb and In segregation in the material surrounding the quantum dot. Using the three layer model originally proposed for the SiGe system by D. J. Godbey, M. G. Ancona, J. Vac. Sci. Technol. A **15**, 976 (1997) we accurately calculate the segregation profile and include a non uniform composition to our models. Using atomistic modeling, we present strain maps of the quantum dot structures that show the propagation of the strain into the GaAs region is strongly affected by the shape and composition of the strain reduction layer.

1. Introduction

The development of semiconductor self assembled Quantum Dot (QD) Lasers [1-2] follows the exploitation of the 3D confinement in self assembled InAs/GaAs Stranski-Krastanow islands and has demonstrated low threshold current and reduced temperature dependent lasing characteristics.

Telecommunication applications dictate that the emission wavelength of InAs/GaAs QDs needs to be in the range 1.31 and 1.55 μm. In order to achieve this QD layers need to be formed as substantially larger islands compared to those used in the first generation of QD laser devices. Such large QDs can be synthesized by overgrowth, epitaxial growth control or by growth on higher index surfaces. However such large islands, when capped with a GaAs layer, tend to develop very large strain fields in the surrounding matrix which result in threading dislocations. [3]

The most viable solution to this problem is to use a "strain reduction layer (SRL)" to embed the InAs (or InGaAs) islands in a material which is an intermediate lattice constant between itself and the substrate. This material, usually grown as a pseudomorphic layer, acts to reduce the band gap and increases the aspect ratio of the QD. Further more the hydrostatic strain present in the island is diffused into the 2D layer and reduces dislocations. The resulting QD/QW combination is called a "dot-in-well" structure (DWELL) [4-8]. Of the various material systems useful as SRLs, Sb alloys are of particular interest because of the potential type II band alignment which could offer more functionality to the devices. [9]

One technological issue when growing semiconductor epitaxial layers via either MBE or MOCVD is that of elemental segregation. [10-11] The occurrence of segregation has been observed for both group III and group V atoms and generally results in non abrupt interfaces and deviations from the designed properties of the heterostructures. Various theoretical models have been proposed in the years to explain experimental observations for both SiGe and III-V semiconductor alloys. [12-15]

We have recently reported [16] how the Muraki [12], Kinetic [13-15] or Thermodynamic [15] models cannot reproduce satisfactorily the experimentally measured segregation profiles in quantum dot layers covered by a GaSbAs SRL. In fact only the three layer model originally proposed by Godbey and Ancona for SiGe alloys [17] reproduces the long tail of the Sb distribution. Earlier experimental observation by Dorin [18] led to postulate that in films containing Sb, where significant roughening may open paths for Sb segregation, the exchange processes are more likely to take place between at least three layers. This explains the reason for which earlier models, all based on a two layer exchange mechanism, fail to predict the extended diffusion of Sb atoms. Furthermore In segregation appears partially suppressed compared to QD layers covered by GaAs. However it is yet sufficiently present to allow the experimental observation of a GaInAsSb alloy within the SRL.

2. Modelling Strain in the SRL and Quantum Dots

In this section we report our investigation of the influence of the presence of Sb atoms together with In, Ga and As on the strain properties of QDs capped with a SRL. Our intention was to compare our models with direct experimental observation by means of Cross Sectional Scanning Tunneling Microscopy (X-STM), as reported in our previous work [16]. The samples consisted of QD layers capped with GaAsSb, grown by solid source MBE, as follows: an n+ Si doped GaAs [001] substrate was covered by a 200 nm GaAs buffer (590°C), followed by a 2.8 ML InAs layer (500°C), after a 30s growth interrupt a 6 nm GaSbAs (475°C), 10 nm GaAs (475°C) and finally 40nm GaAs (590°C). With the exception of the buffer layer the growth sequence was repeated 3 times but each time changing the nominal Sb content to 12%, 15% and 20%. The sample was finally capped with a further 10nm of GaAs (590°C). The InAs layers were deposited at a growth rate of 0.094 ML/s.

To determine the strain tensor in these structures we have implemented a series of Molecular Statics simulations. The atomistic models are designed as follows: we initially build a GaAs (001) substrate section of dimensions 40x40x10nm followed by a thin InAs 2D layer (wetting layer) of thickness of 2 ML. The pure InAs QD island sits on top of the 2D layer and has dimensions of roughly 28nm in diameter and 10nm in height. The shape of the island is taken from that proposed by Costantini *et al* [19] where a detailed analysis of the facets observable in STM microscopy was presented.

In the six different models presented here, the region around the island and up to the height of the QD (SRL) is formed by either pure InGaAs, GaSbAs or GaInAsSb, either as a uniform alloy or incorporating a non uniform profile taken from model data of segregation. Such data was already presented in our earlier work [16]. The entire structure is then further capped with a 10nm layer of pure GaAs. Periodic boundaries are used in the [100] and [010] crystallographic directions. The structures comprise usually about 2.8 million atoms and the computational effort is minimized by using a 64-processor parallel implementation of the IMD™ software [20] and using optimally parameterized Tersoff empirical potentials [21-22]. The strain is then directly extracted from the atomic bonds, by comparing the strained lattice to an ideal unstrained one and taking the local composition into account, in a sort of Virtual Crystal Approximation.

3. Results

We present the strain maps for the biaxial and dilation strain, which are obtained from the diagonal components of the strain tensor and express the overall tetragonal distortion and compression of the unit cells, respectively. In Fig. 1 we show maps of the dilation (i.e. $e_{xx} + e_{yy} + e_{zz}$) and biaxial strain components (i.e. $e_{zz} - \frac{1}{2}(e_{xx} + e_{yy})$), for the case of three uniformly alloyed SRLs with 10% In and 20% Sb, 50% In and no Sb, 30% Sb and no In. Including an SRL during growth is a very effective way of reducing the strain in the GaAs matrix, a fact clearly supported and confirmed in these simulations when compared to results for a GaAs/InAs/GaAs island previously reported [16]. In all cases the change of the propagating strain is evident, as well as having a marked effect on the elastic distortion inside the islands.

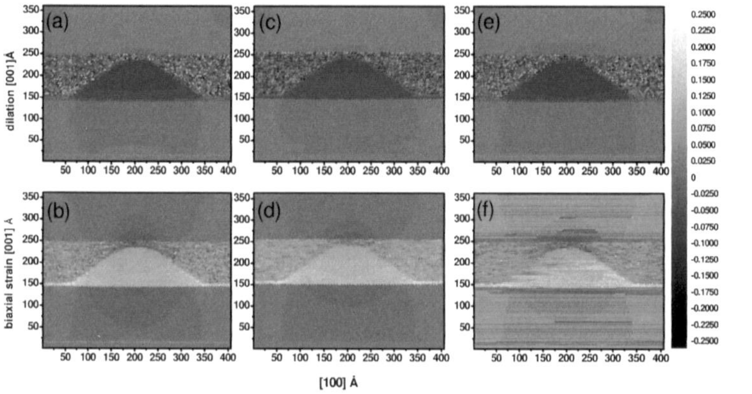

Fig. 1: Strain maps: the SRL is composed by InGaSbAs with 10% In and 20% Sb ((a) and (b)), InGaAs with 50% In ((c) and (d)) or GaSbAs with 30% Sb ((e) and (f)).

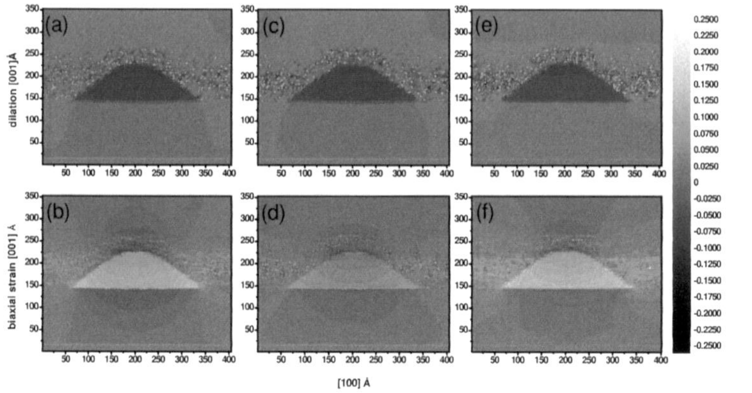

Fig. 2: Strain maps: the SRL is composed by In and Sb segregated layers with a nominal Sb composition of 12% Sb ((a) and (b)), 15% Sb ((c) and (d)) and 20% Sb ((e) and (f)).

However the SRLs of the models used in Fig. 1 is an idealized situation. As previously reported [16] the SRL tends to have very non uniform compositions that are well described by segregation profiles. We therefore also simulated in Fig. 2 the intermixing with In of a GaSbAs SRL with nominally 12%, 15% and 20% Sb, using calculated segregation profiles from our earlier work [16]. All other parameters are kept the same. The experimental X-STM topographies [16] also indicate that the material in the SRL does not form a uniform two dimensional layer. In fact Sb atoms are clearly seen to "wrap" around the island and strongly limit the uniformity of the planes above the tip of the QD. This effect was reproduced in our input models. The results of the simulations shown in Fig. 2 indicate that the strain distribution is altered by the presence of a strongly alloyed SRL compared with having a uniform composition.

Of the three compositions tested we found that the biaxial strain is more strongly reduced in the island for the intermediate Sb content of 15%. The dilation instead appears almost constant in all three islands. Upon comparing the highest Sb composition in both the uniform and non uniform case, it is manifest that the more strongly alloyed SRL is more effective in reducing the strain in the GaAs regions above and below the QD island.

4. Conclusions

We reported a study of the strain in strongly alloyed GaSbAs/InAs/GaAs QDs and showed how the presence of Sb is able to tune the strain of the structures. Such strain is likely to have a profound influence on the electronic properties as the percentage of Sb controls the amount of internal strain of the islands. Furthermore the strain is also tuned above and below the islands allowing for a more precise morphological coupling should multistacks of QD islands be prepared in such a way.

References

[1] N.N. Ledentsov, V. A. Shchukin, M. Grundmann, N. Kirstaedter, J. Bohrer, O. Schmidt, D. Bimberg, V. M. Ustinov, A. Yu. Egorov, A. E. Zhukov, P. S. Kop'ev, S. V. Zaitsev, N. Yu. Gordeev, Zh. I. Alferov, A. I. Borovkov, A. O. Kosogov, S. S. Ruvimov, P. Werner, U. Gosele, J. Heydenreich, Phys. Rev. B **54**, 8743 (1996).

[2] D. Bimberg, N.N. Ledentsov, M. Grundmann, N. Kirstaedter, O. G. Schmidt, M. H Mao, V. M. Ustinov, A. Yu. Egorov, A. E. Zhukov, P. S. Kopev, Zh. I. Alferov, S. S. Ruvimov, U. Gosele, J. Heydenreich, Phys. Status Solidi B **194**, 159 (1996).

[3] M. Gutierrez, M. Hopkinson, H. Y. Liu, M. Herrera, D. Gonzalez, R. Garcia, J. Cryst. Growth. **278**, 151 (2005).

[4] S. K. Ray, H. Y. Liu, T. L. Choi, K. M. Groom, S. L. Liew, M. Hopkinson, R. A. Hogg, Jpn. J. Appl. Phys. **46**, 2418 (2007).

[5] J. L. Casas Espinola, T. V. Torchynska, E. Velasquez Lozada, L. V. Shcherbyna, A. Stintz, R. Pena Sierra, Physica B.**401**, 584 (2007).

[6] J. S. Ng, H. Y. Liu, M. J. Steer, M. Hopkinson, J. P. R. David, Microelectron. J. **37**, 1468 (2006).

[7] C. H. Lin, W. W. Pai, F. Y. Chang, H. H. Lin, Appl. Phys. Lett. **90**, 063102 (2007).

[8] S. Krishna, J. Phys. D: Appl. Phys. **38**, 2142 (2005)

[9] J. M. Ripalda, D. Granados, Y. Gonzalez, A. M. Sanchez, S. I. Molina, J. M. Garcia, Appl. Phys. Lett. **87**, 202108 (2005).

[10] D. Mc Lean, Grain Boundaries in Metals (Oxford University Press, Oxford 1957).

[11] J. M. Moison, C. Guille, F. Houzay, F. Barthe, M. Van Rompay, Phys. Rev. B **40**, 6149 (1989).

[12] K. Muraki, S. Fukatsu, Y. Shiraki, Appl. Phys. Lett. **61**, 3 (1992).

[13] S. Fukatsu, K. Fujita, H. Yaguchi, Y. Shiraki, R. Ito, Appl. Phys. Lett. **59**, 2103 (1991).

[14] K. Fujita, S. Fukatsu, H. Yaguchi, Y. Shiraki, R. Ito, Appl. Phys. Lett. **59**, 2240 (1991).

[15] O. Dehaese, X. Wallart, and F. Mollot, Appl. Phys. Lett. **66**, 52 (1995).

[16] V. Haxha, I. Drouzas, J.M. Ulloa, M. Bozkurt, P.M. Koenraad, D.J. Mowbray, H.Y. Liu, M.J. Steer, M. Hopkinson, M.A. Migliorato, Phys Rev B **80**, 165334 (2009)

[17] D. J. Godbey, M. G. Ancona, J. Vac. Sci. Technol. A **15**, 976 (1997).

[18] C. Dorin, J. Mirecki Millunchick C. Wauchope, J. Appl. Phys. **94**, 1667 (2003).

[19] G. Costantini, C. Manzano, R. Songmuang, O.K. Schmidt, and K. Kern, Appl. Phys. Lett. 82, 3194 (2003)

[20] J. Stadler, R. Mikulla, H.-R. Trebin, Int. J. Mod. Phys. C **8**, 1131 (1997).

[21] J. Tersoff, Phys. Rev. Lett. **56**, 632 (1986); Phys. Rev.B **37**, 6991 (1988); Phys. Rev. B **39**, 5566 (1989)

[22] D. Powell, M. A. Migliorato, and A. G. Cullis, Phys. Rev. B **75**, 115202 (2007).

Journal of Physics: Conference Series **245** (2010) 012066 doi:10.1088/1742-6596/245/1/012066

Effect of the In (Ga) inter-diffusion on the optical properties in InAs/GaAs annealed quantum dots

Ouasli Maher, Khemiri Témim , Béchir Jlassi , Jalloul Balti and Sihem Jaziri
Laboratoire de physique des matériaux, Faculté des sciences de Bizerte,7021Jarzouna,Tunisie

E-mail :sihem.jaziri@fsb.rnu.tn

Abstract. We theoretically consider the effect of post growth annealing on the electron-hole interaction in the self-assembled InAs/GaAs Quantum Dots (QD). We offer a numerical model of the QD which is based on an experiments data with such heterostructure. Our model allows us to simulate the process of annealing and to calculate the distribution profiles of (In and Ga) atoms over the heterostructure. Using this model we have calculated the energy states of carriers and the optical transition in annealed QDs. We used the parameters of our model referring to the experimental data on Photoluminescence (PL) spectra for a set of heterostructure with InAs/GaAs annealed at different temperatures and we are shown the effects of the interdiffusion into QD on the optical properties which explain the tuning of PL profiles caused by the diffusion of In atoms from QDs into the barrier layers and leads to this modification in these profiles.

Till now, there is doubt that the self assembled semiconductor quantum dots have created many topics in research due to their fundamental physics and promising prospects for applications in highly competitive electronic and optoelectronic devices. According to the development of new semiconductor growth technique, it is possible to control the shapes and sizes of self assembled QDs by changing the growth mode or post growth thermal annealing. The experimental characterization of the QDs structural properties shows that they have a shape of disk-like, lens-like and truncated cone-like and leads us to conclude that it has a cylindrical symmetry. The strong confinement of charge carriers inside the QD facilitates the formation of bound states of carriers. The inter band transition between them gives a new born of quasi particles such as exciton which is play an important role in optoelectronic devices. Many experimental research existed on InAs/GaAs self assembled QDs which shown a tuning on the optical properties specially the inter band transition between carriers in an annealed QDs for this set of heterostructure. In our work we theoretically consider the effect of post growth annealing on this transition in a self assembled InAs/GaAs QD. We offer two numerical models for the shape of QD, these models allow us to simulate the annealing process and to calculate the distribution profiles of In atoms over the heterostructure which we can calculate the eigenstates and the energies of carriers, the knowledge of these energies we calculated the optical transition at different temperatures. Lattice-mismatch-induced growth of self-assembled nanoscale InAs islands on GaAs substrates has emerged as one of the most promising ways for in situ QD fabrication. When the thickness of the InAs layer is beyond the critical thickness around 1.7 monolayer (ML), the structure is usually composed of an InAs wetting layer (WL) and cone-like InAs islands deposited on it [1]. The thermal annealing tunes the energy levels in QDs and affects the QD PL intensity [2].

Our aim in this paper is how the optical transition will be modified if we theoretically consider the effect of post growth annealing QD which leads to indium and gallium interdiffusion this phenomenon is described by Fick's law that can be written as: $\frac{\partial x(\vec{r},t)}{\partial t} - D\Delta x(\vec{r},t) = 0$. We assume the diffusion coefficient to be independent of space coordinates, where $x(\vec{r},t)$ is a position dependent function of indium fraction in the $In_xGa_{1-x}As$ as solid solution forming the QD annealed during the time t, the D coefficient is assumed to be the constant and Δ is the laplacian operator. Holding account by cylindrical symmetry, the function $x(\vec{r},t)$ has got a separable form in azimuthal angle θ $x(\vec{r},t) = \chi(\rho,z,t)\varphi(\theta)$ because the cylindrical symmetry has been conserved during the annealing process. So the indium distribution is independent from azimuthal angle, i.e. $\varphi(\theta)$ =1. After having dividing on D and separating the variables, we can write the diffusion equation at the following form for the function $\chi(\rho,z,t)$: $\frac{1}{D}\frac{\partial}{\partial t}\chi - \frac{1}{\rho}\frac{\partial}{\partial \rho}\rho\frac{\partial}{\partial \rho}\chi - \frac{\partial^2}{\partial z^2}\chi = 0$. The functions χ and x have the same initial conditions. So the diffusion equation should be supplemented with initial condition on the function x which is chosen to be 1 inside QD/WL and zero outside. We choose two different shapes for the QD Fig.1a and Fig.1b.

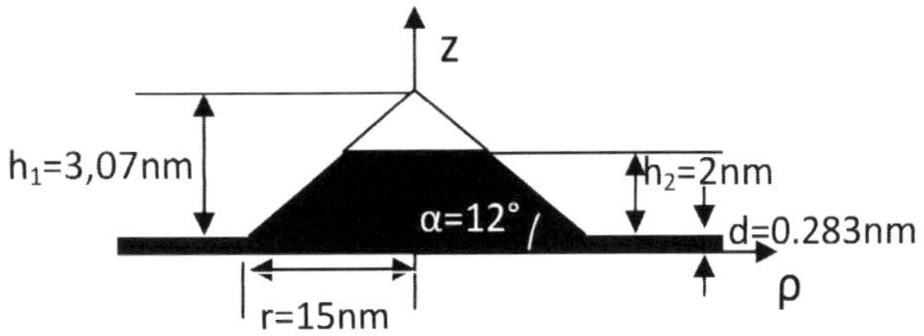

Fig.1a.the dot shape of truncated cone

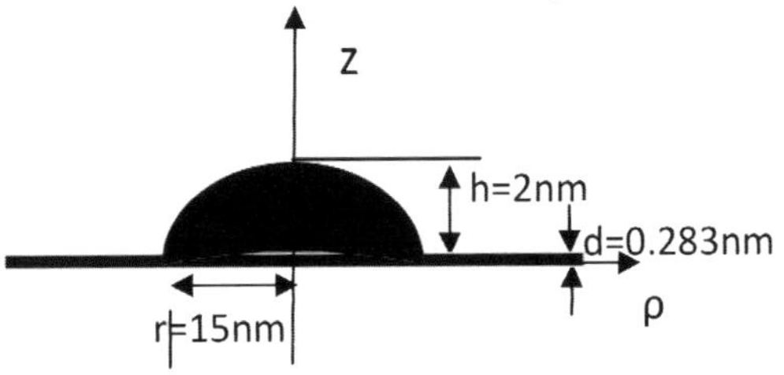

Fig.1b.the dot shape of lens

Using finite element technique, we solve the described diffusion problem with different diffusion constants and according the time. We assumed the Arrhenius equation for the temperature dependence of the diffusion constant: $D(T_a) = D_0 \exp\left(-\frac{E_a}{k_B T_a}\right)$ where T_a is the annealed temperature, is E_a the activation energy of the inter-diffusion process which is found 1.23 eV [4], k_B is the Boltzmann constant and D_0 is the pre-exponential factor (D_0 =8.5 10^{-14} m^2/s) [5].

In Fig.2 shows examples of In distribution over the hetero structure calculated for the QDs unannealed and annealed at different temperatures. As seen, annealing of these heterostructures leads for a rapid dissolution of the InAs QD into GaAs barrier layers. The average indium content does not exceed in one hand, 20% for lens QD shape annealed at temperature above 980°C.and around 8% in truncated cone annealed at the same temperature. We have presented a theoretical study of the phenomenon of inter-diffusion activated by thermal annealing in nanostructures types $In_xGa_{1-x}As$. In general, it is known that the distribution of chemical composition of such hetero-structures is capable to be strongly modified by inter-diffusion of constituents in the various processes needed for the formatting of a component. It follows a change in the potential seen by the charge carriers and therefore a variation of the electronic states energy at stake. In order to determine the true nature of focal mechanisms, it is essential to know more precisely as possible the actual shape of the potential. The diffusion of indium from the QDs into barrier layer leads to a modification in the profile of the confinement potential which is modeled using the calculated indium content $x(\vec{r},30s)$ for the annealed QD after 30s(in the rest of the text we will consider $x(\vec{r},30s) = x(\vec{r})$). For this purpose we used a linear approximation for the band gap $In_xGa_{1-x}As$ using relevant quantities for InAs and GaAs: $Eg(\vec{r}) = Eg(InAs).x(\vec{r}) + Eg(GaAs).(1 - x(\vec{r}))$, where $Eg(InAs) = 0.415$ eV [6], $Eg(GaAs) = 1.519eV$ [6], are the band gap for InAs and GaAs respectively. The annealing leads to redistribution the indium concentration to relaxation the built-in strain. We assume the strain energy linearly depends on the indium concentration in the annealed QD. Thus, the confinement potential profiles of carriers have been determinate by the following formula:

$V_{e,h}(\vec{r}) = Q_{e,h}\big(Eg(InAs) - Eg(\vec{r})\big) - \varepsilon_{ST}^{e,h}.x(\vec{r})$ where $Q_{e,h}$ the band-offset ratio which used $Q_e/Q_h = 66/34$ [7,8], and ε_{ST}^{e}, ε_{ST}^{h} are the strain energies of the conduction and valence bands which obtained 450 meV [9], and 90 meV [9], respectively. We can also determine the shape of the potential inside the dot at different temperatures. Note that the band offset of the electron decreases under the effect of temperature. In order to calculate the carriers' energies, we introduce the calculated

band gap and confinement potential over the hetero structure in the excitonic Hamiltonian which written as:
$H_{ex} = Eg(\vec{r_e}) + T(\vec{r_e}) + V_e(\vec{r_e}) + T(\vec{r_h}) + V_h(\vec{r_h}) + V_c(\vec{r_e},\vec{r_h})$where $T_{i=e,h} = -\frac{\hbar^2}{2}\vec{\nabla}\left(\frac{1}{m_i^*}\vec{\nabla}\right)$ denotes the motion of carriers,m_e^*, m_h^* are the effective mass of electron and hole respectively ,$\hbar$ is the Planck's constant divided by 2π ,$V_c(\vec{r_e},\vec{r_h}) = -\frac{e^2}{\epsilon|\vec{r_e}-\vec{r_h}|}$ is the Colombian potential. In the accepted model of cylindrical symmetry of QDs, we modelize our dot shape by truncated cone and lens shape lying on a WL and each shape are described like the following relations: $D^{truncated\ cone}(\rho,z) = \theta(z)\theta(d-z) + \theta(z-d)\theta(h_1+d-z)\theta\big(h_2+d-z-\rho tan(\alpha)\big)$ with d the WL thickness ,h_1 is the maximum height of the cone and h_2 is the height of truncated cone(see Fig.1a),$D^{lens} = \theta(z)\theta(d-z) + \theta(z-d)\theta(h+d-z)\theta(R^2-(z+R-d-h)^2-\rho^2)$ here R the base radius,d is the WL thickness and h is the maximum height over this layer(see Fig.1b).θ is the Heaviside function.
The first part of the excitonic Hamiltonian, $H_{e=}Eg(\vec{r_e}) + T(\vec{r_e}) + V_e(\vec{r_e})$ is the motion of the electron which its engine energies and associated wave function are calculated by numerical diagonalization on a Fourier-Bessel basis over a large cylinder domain so we can write: $|\psi_n\rangle = \sum_{i>0,j>0} C_{i,j}\,|\phi_{i,j}^n >$ here $\phi_{i,j}^n(\vec{r}) = \alpha_i^n\,exp(in\theta)J_n\left(\frac{\lambda_i^n\rho}{R}\right)sin\left(\frac{\pi jz}{Z}\right)$, $C_{i,j}$ are the basis coefficients, λ_i^n is the ith root of the nth order Bessel function, α_i^n is the normalization constant. The introduction of the quantum number n=0, ±1, ±2… gives a natural way to label the wave functions.
The states will be called following the usual nomenclature in atomic physics by the letters (iS, iP, iD…) we denote n=0 to 1S, 2S, 3S states, n=±1 to 1P, states, n=±2 to D states….
The same method is applied to the hole Hamiltonian $H_h = T(\vec{r_h}) + V_h(\vec{r_h})$ to determine their energy states.
Under the regime of strong confinement, the Coulomb energy is usually neglected in quantum dots as a first approximation. We assume the energy confinement of carriers is much higher than the energy corrections made by the Coulomb interaction. In order to further consider the situation of strong confinement, the dot would have a radius much smaller than the Bohr radius of exciton. As a second approximation, we will consider the effective masses as constants due to their small changes during the process of inter-diffusion. After these approximations the Hamiltonian of the exciton is reduced as a sum of energy of electron and hole. Our approximation to calculate the excitonic energies is to consider that the annealing process contribute more in confinement potential than in the effective mass.
The figures Fig.4 and Fig.5 show the evolution of the energy of the exciton depending on the radius of the dot and we note that the energy decreases with increasing the size of the dot for the three spectra at different temperatures. As described before, the annealing process affects the energy levels of carriers which causes a change in the excitonic energy levels, since we have neglect the Coulomb interaction the excitonic energy is written as a sum of carriers'energy. This remarkable increase can be explained by the fact that the diffusion causes the penetration of Ga has the effect of increasing the optical gap and a subsequent increase in the optical transition

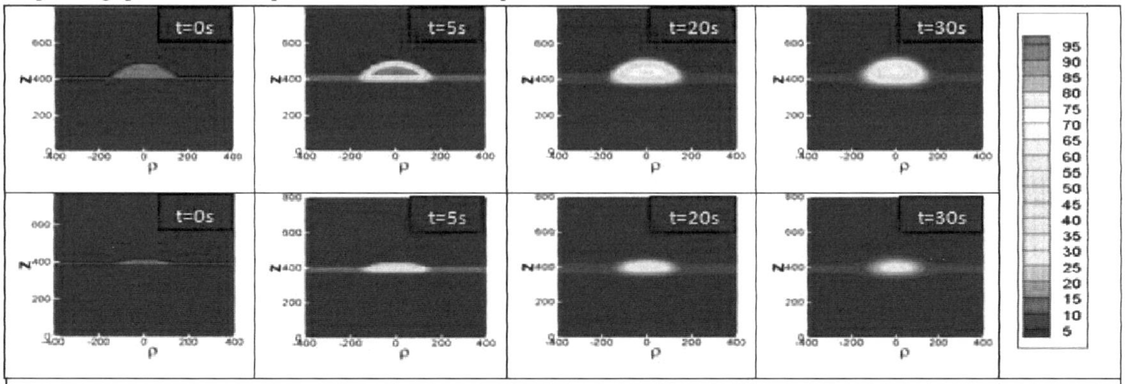

Fig.2. cross section of calculated distribution of indium content x by % over the hetero structure annealed at T=820°C

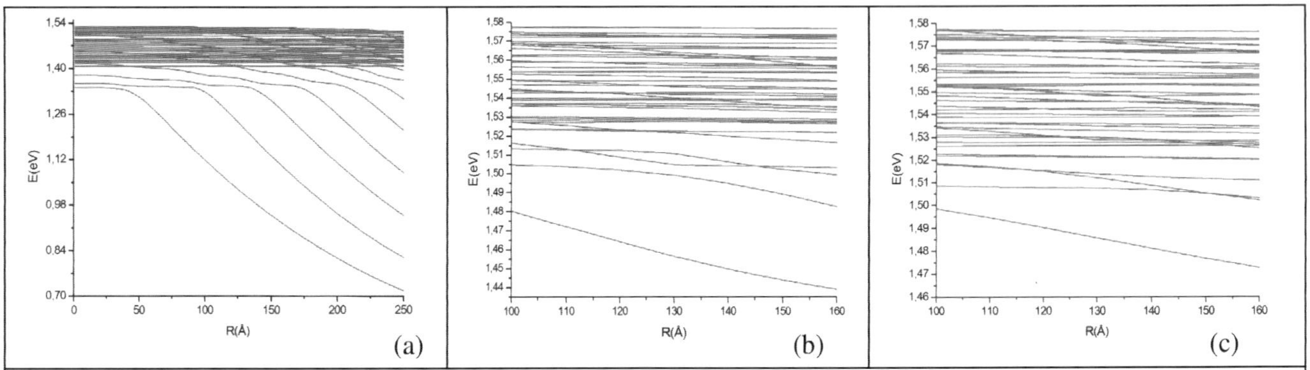

Fig.4: Evolution of S exciton energy levels versus radius for truncated cone shape: (a)without annealing, (b)annealed at 820°C, (c) annealed at 900°C

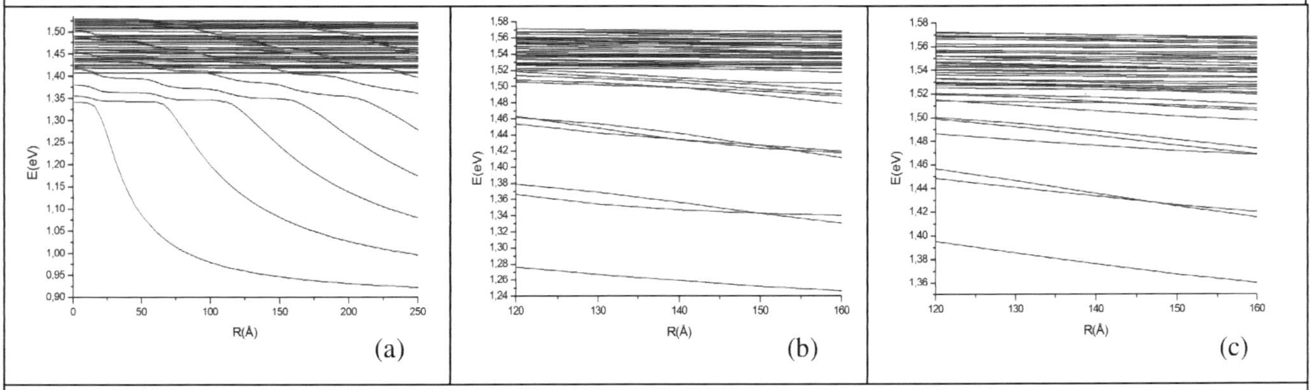

Fig.5: Evolution of S exciton energy levels versus radius for lens shape: (a)without annealing, (b)annealed at 820°C, (c)annealed at 900°C

In summary, we believe that the results presented in this paper constitute direct evidence that a post-growth annealing process can change the electronic properties of the self-assembled quantum dots system. Homogeneous samples were shown to occur due to the atomic inter-diffusion process between the barriers and the islands (dots) which changes the sizes and the shapes of the dots. Thus, the controlled heat treatment presents a way to achieve a quantum dot device with tailored features, since a controlled annealing process can tune the electronic properties. After the temperature annealing, the inter-diffusion process of Indium (out the dot) / Gallium (in to the dot) was clearly seen. The annealing at 600°C-900°C induced a diffusion process forming an $In_xGa_{1-x}As$ alloy self-assembled quantum dot. A similar temperature diffusion process inducing a post-growth structural modification
of the typically the self-assembled InAs/GaAs quantum dots have the truncated cone or lens shape was also detected This implies in a decrease of the carrier coupling between the dots caused by the corresponding increase of the strength of the electron localization. The calculated excitonic transitions 1Se → 1Sh well reproduce the experiment results.

REFERENCES

[1] D. Leonard, M. Krishnamurthy, C.M. Reaves, S.P. Den Baars and P.M. Petroff, *Appl. Phys. Lett.* 63, 3203 (1993)
[2] A.O. Kosogov, P. Werner, U. Gosele, N.N. Ledentsov, D. Bimberg, V.M. Ustinov, A.Yu. Egorov, A.E. Shukov, P.S. Kopev, N.A. Bert and Zh.I. Alferov, *Appl. Phys. Lett.* 69,3072 (1996).
[3] J. Y. Marzin and G. Bastard *Solid state commun*,92:437,1994
[4] S.Malik, C. Roberts, R. Murray, and M. Pate, *Appl. Phys. Lett.* 71,1987 (1997).
[5] M. Yu. Petrov, I. V. Ignatiev, S. V. Poltavtsev, A. Greilich, A. Bauschulte, D. R. Yakovlev, and M. Bayer, *physical review B* 78, 045315 (2008).
[6] H. Landoldt and R. Börnstein, Numerical Data and Functional Relationships in Science and Technology, Landolt-Börnstein, *New Series, Group III*, Vol. 17, Pt. A (Springer, Berlin, 1987).
[7] C. G. Van de Walle, *Phys. Rev. B* 39, 1871 (1989).
[8] O. Stier, M. Grundmann, and D. Bimberg, *Phys. Rev. B* 59, 5688 (1999).
[9] M. Califano and P. Harrison, *Phys. Rev. B* 61, 10959 (2000).

One pot synthesis of bi-linker stabilised CdSe quantum dots

I Concina[1,§], M M Natile[2], A Braga[1,3], A Vomiero[1], V Morandi[4], L Ortolani[4], M Ferroni[1,3], G Sberveglieri[1,3]

[1] CNR-IDASC, SENSOR Laboratory, Via Valotti, 9 – 25133 Brescia, Italy
[2] ISTM-CNR, INSTM - Padova, Dept. of Chemical Sciences, University of Padova, via Marzolo, 1 –35131 Padova, Italy
[3] Dept. of Chemistry and Physics, University of Brescia, via Valotti, 9 – 25133 Brescia, Italy
[4] CNR-IMM Sezione di Bologna, via Gobetti 101 - 40129 Bologna, Italy

E-mail: isabella.concina@ing.unibs.it

Abstract. In this study we exploited the classic Murray's synthesis for generating a hydrophilic CdSe quantum dot system in a single step procedure, with the aim of directly obtaining a material responding to the characteristic of polarity required in many end applications. 6-phosphonohexanoic acid was used as both ligand for generating the active monomer during the synthesis of the quantum dots and final stabiliser. Diffraction measurements identified the cubic phase of cadmium selenide. Energy dispersive spectroscopy analysis revealed non-stoichiometric quantum dots, being the Cd/Se ratio 60/40. This feature suggests a configuration in which Cd^{2+} ions are present on the nanocrystal surface. Diffuse reflectance infrared Fourier transform analysis was applied in order to investigate the structure of the quantum dot system: the results indicate a configuration in which the carboxylic function of 6-phosphonohexanoic acid establishes only a partial interaction with the quantum dot surface, being set in a pseudo-ester configuration.

1. Introduction

Semiconductor nanocrystals are attracting a relevant interest because of their potential applications in many domains, spanning from medicine to photovoltaics [1]. The most common synthetic approach relies on the fundamental works by Murray [2] and Peng [3], both based on the La Mer's theory [4] and exploiting a burst reaction in an oversaturated mixture of a cadmium precursor. This approach allows to obtain crystalline quantum dots (QDs) surrounded by an aliphatic shell. In view of end applications, often the native shell has to be substituted by a more suitable stabiliser by means of subsequent ligand exchange reactions. Since semiconductor QDs appear as very promising materials to be used as light harvesters in third generation solar cells [5], a bi-linker ligand ended by a carboxylic group is almost essential because of the capability of this group in interacting with metal oxide surfaces.

In this study we adapted the Murray's synthesis to perform a single step procedure to directly obtain a CdSe QD system stabilised by a bi-linker molecule, namely 6-phosphonohexanoic acid (HPA). The obtained CdSe nanocrystals were tested as light harvester in a solar cell device.

§ To whom any correspondence should be addressed.

2. Experimental

2.1. CdSe Quantum Dot synthesis

Standard airless techniques (Schlenk line) were used in the preparation of CdSe quantum dots. In a typical synthesis the Cd precursor is generated in a three-neck 25 ml flask by heating at 360 °C suitable amounts of CdO (4.10 x 10^{-4} mol) as cadmium source, 6-phosphonohexanoic acid (HPA) (8.20 x 10^{-4} mol) as ligand and trioctylphosphine oxide (TOPO) (9.80 x 10^{-3} mol) as coordinating solvent. The selenium precursor is generated under nitrogen atmosphere by stirring Se powder (8.20 x 10^{-4} mol) with trioctylphosphine (4.03 x 10^{-3} mol) and then injected into the reactor at a temperature of 280 °C. The raw material is purified by washing with anhydrous methanol and n-hexane. HPA-stabilised CdSe quantum dots were finally dried under vacuum at room temperature.

2.2. CdSe Quantum Dot characterisation

Diffuse reflectance infrared Fourier tranform (DRIFT) spectra were collected in a Bruker TENSOR 27 spectrometer (32 scans at a resolution of 2 cm^{-1}) and displayed in Kubelka-Munk units [6], [7]. Few milligrams of the sample were mixed with KBr powder and loaded in the accessory for diffuse reflectance (Harrick Scientific). KBr was used as a background. Transmission electron microscopy (TEM) observations were carried out with a Fei Tecnai F20 TEM with a Schottky emitter and operating at 200 keV. The instrument is equipped with an Edax EDS PV9761N SUTW Energy Dispersive X-ray spectrometer (EDX) for X-ray microanalysis. The QDs were dissolved in methanol and the solution dropped and evaporated at room temperature and atmospheric pressure on a copper grid with a holey carbon film for the analyses. The glancing incidence X-ray diffraction (GIXRD) pattern was collected by a Bruker D8 Advance diffractometer equipped by a Göbel mirror at a fixed incidence angle $\theta_i = 0.5°$. The Cu Kα line of a conventional X-ray source powered at 40 kV and 40 mA was used for the experiments. The average crystallite dimensions were evaluated by means of the Scherrer equation. Ultraviolet-visible (UV-Vis) measurements were carried out in a T80 PG instrument (quartz cells, optical path 1 cm).

2.3. Solar cell fabrication and device characterization

A film (3 μm thickness) of TiO_2 nanoparticles (Solaronix) of nominal diameter of 13 nm was deposited by tape casting onto a SnO_2-F glass and fired at 450 °C for 30 minutes in atmospheric ambient. HPA-stabilised CdSe QDs were dissolved in methanol and the solution dropped and evaporated on the photoanode. QD-sensitised cells were prepared using the I^-/I_3^- redox couple and a Pt/ITO coated glass counterelectrode. The functional properties of the cells were investigated under simulated sun light irradiation using an ABET 2000 solar simulator at AM 1.5G irradiation with intensity 100 mW/cm^2, tunable with mechanical filters of different transmittance.

3. Results and discussion

The DRIFT spectrum of the as prepared CdSe QDs is perfectly superimposable to that of the tri-n-octylphosphine oxide used as reaction medium (Figure 1 (a)). A procedure of purification was then carried out in order to eliminate the residual solvent. After the purification the IR spectrum completely changes (Figure 1 (b)) and the profile is more similar to that of HPA. Phosphonic acids are characterised by three adjacent bands in the region from about 1070 to 933 cm^{-1} ascribable to several vibration modes of the -PO_3H_2 group, in particular the asymmetric and symmetric stretching modes of -PO_3^{2-} are visible at 1035 and 1005 cm^{-1}, respectively, while the contribution at 968 cm^{-1} is due to P-OH stretching (Figure 1 (b)). The characteristic P=O stretching mode in HPA is visible as a rather, though well distinguished, weak peak at 1211 cm^{-1}. The very intense and sharp peak centred at 1710 cm^{-1} is, instead, characteristic of the C=O stretching mode of the carboxylic group of HPA (Figure 1 (b)). The analysis of the DRIFT spectrum recorded on the purified HPA-stabilised CdSe QDs reveals that the peak due to P=O stretching is disappeared while the whole group of P-O stretching modes results broadened and more defined. Moreover, a shift of ν_{asym} and ν_{sym} of P-O modes is well evident:

the vibrational modes are now centred at 1170 and 1088 cm^{-1}, respectively. These changes suggest a multidentate coordination of the phosphonate group to the CdSe QDs. Concerning the carboxylic function in the spectrum of the purified HPA-stabilised CdSe QDs, besides the band centred at 1710 cm^{-1} due to the stretching of C=O group which is perfectly superimposable to the analogous in HPA, an other contribution at 1548 cm^{-1} is noteworthy. This last signal can be ascribed to the asymmetric stretching vibration of the -C(O)O group. No peaks appear in the quantum dot spectrum due to symmetric mode of -C(O)O group.

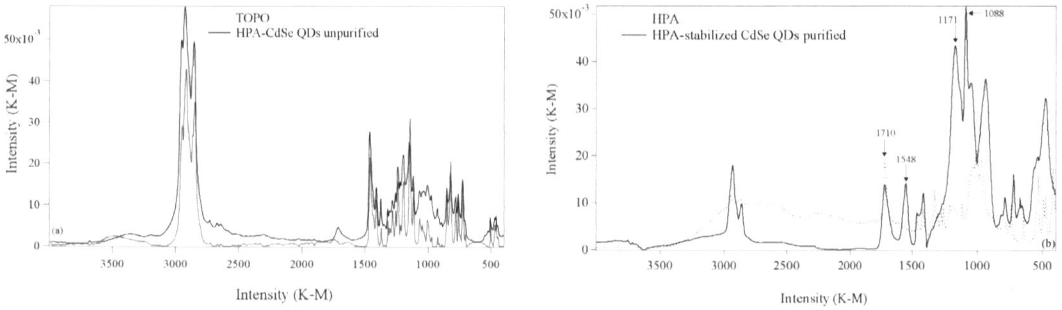

Figure 1: comparison between DRIFT spectra of (a) TOPO (dashed line) and unpurified HPA-stabilised CdSe QDs (solid line) and (b) HPA (dashed line) and purified HPA-stabilised CdSe QDs (solid line).

The obtained results suggest that HPA binds the surface Cd^{2+} ions by means of the phosphonate group: this hypothesis is supported by the re-organization and the global red shift of the vibrational modes of phosphonate group (Figure 1 (b)). On the other hand, the carboxylic function shows the stretching mode of carbonyl unaltered with respect to peak shape and position after coordination and the appearance of the asymmetric vibration of the –C(O)O group, while the O-H stretching mode is almost completely lost, only a shoulder being visible as a minor contribution in the region above 3000 cm^{-1}. When acting as a monodentate ligand, carboxylate group takes an ester-like configuration. In such a kind of configuration, the symmetry of carboxylic function RC(O)O is broken and the infrared spectrum would show the contribution from the stretching of C=O and an asymmetric vibration of the whole group [12]. We then suggest that the HPA molecules allocate all around the nanocrystal in a configuration like that shown in the inset of Figure 2 (c).
Evaluation of particles sizes was performed by means of TEM measurements (Figure 2 (a)), by the full width of half maximum (FWHM) of the XRD peak (Figure 2 (b)) and by the position of the first excitonic peak [8] (Figure 2 (c)). An optimum agreement was found between XRD (2.4 nm) and UV-Vis (2.7 nm) evaluation, whereas the presence of the stabiliser, visible in the micrograph as an amorphous material, could induce the TEM overestimating (5.4 nm) the nanocrystal sizes.
Selected area electron diffraction (SAED) analysis reveals the cubic structure of CdSe (a= 0.6077 nm, S.G: F-43m), also confirmed by XRD measurements. EDX analysis indicates that the quantum dots are non-stoichiometric, showing a molar ratio Cd/Se of 60/40. This finding is coherent with data previously reported by other authors: Moreels [9] and Taylor [10] found indeed similar ratios between the metal M and the counterion E in analogous systems. Recently, Owen [11] proposed a sphere-like structure for these materials, in which Cd^{2+} ions are present on the QD's surface, thus justifying the non stoichiometry. In such a configuration, metal ions can easily interact with the molecules of ligand shell by coordination, thus contributing in stabilising the system.

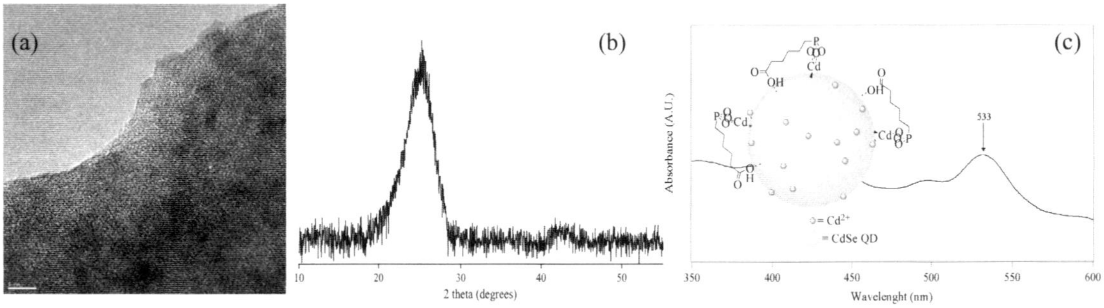

Figure 2: (a) TEM image (scale bar = 5 nm) and (b) XRD and (c) UV-Vis spectrum of HPA-stabilised CdSe quantum dots. (c) Proposed structure of HPA- stabilised CdSe quantum dots. HPA coordinates as a multidentate ligand by employing both the phosphonate and the carboxylate groups.

Preliminary measurements on a QD-sensitised solar cell showed short-circuit current density of 0.75 mA/cm^2 and open-circuit voltage of 0.43 V when illuminated with 100 mW/cm^2 simulated AM 1.5 spectrum, while an optimum Fill Factor was determined (0.64).

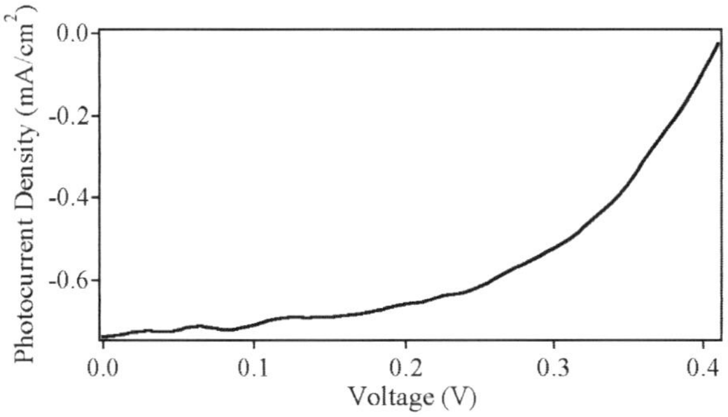

Figure 3: Current-voltage characteristics of QD-sensitised solar cell measured under simulated air mass (AM) 1.5 sun light (100 mW cm^{-2}).

4. Conclusions

The classic Murray's approach to generate crystalline CdSe quantum dots can be easily adapted to the one pot preparation of bi-linker stabilised system, which resulted soluble in polar solvents. Preliminary measurements on QD solar cells showed promising results.

5. References

[1] Gill R, Zayats M and Willner I 2008 *Angew. Chem.* **47** 7625
[2] Murray C B, Norris D J and Bawendi M G 1993 *J. Am. Chem. Soc.* **115** 8706
[3] Peng Z A and Peng X 2001 *J. Am. Chem. Soc.* **123** 183
[4] La Mer V K and Dinegar R H 1950 *J. Am. Chem. Soc.* **72** 4847
[5] Bang J H and Kamat P V 2009 *ACS Nano* **3** 1467
[6] Kubelka P and Munk F 1931 *Z. Technol. Phys.* **12** 593
[7] Kortum G 1969 *Reflectance Spectroscopy* (New York: Springer)
[8] Yu W W, Qu L, Guo W and Peng X 2003 *Chem. Mat* **15** 2854
[9] Moreels I, Fritzinger B, Martins J C and Hens Z 2008 *J. Am. Chem. Soc.* **130** 15081
[10] Taylor J, Kippeny T and Rosenthal S J 2001 *J. Cluster Sci.* **12** 571
[11] J.S. Owen, J. Park, P-E Trudeau, A. P. Alivisatos, 2008 *J. Am. Chem. Soc.*, **130**, 12279
[12] Deacon G B and Phillips R J 1980 *Coord. Chem. Rev.* **33** 227

Photocurrent characterization of intraband transition in GaN\AlN quantum dots

A Vardi[1], G Bahir[1], S E Schacham[2], P K Kandaswamy[3] and E Monroy[3]
[1]Dept. of Electrical Engineering, Technion-Israel Institute of Technology, Haifa 32000, Israel
[2]Dept. of Electrical and Electronic Engineering, Ariel University Center, Ariel 40700, Israel
[3] Equipe Mixte CEA-CNRS « Nanophysique et Semiconducteurs », INAC/SP2M/NPSC, CEA-Grenoble, 17 rue des Martyrs, 38054 Grenoble cedex 9, France

E-mail: valon@tx.technion.ac.il

Abstract. Photoconductivity (PC) in GaN/AlN quantum dot (QD) matrix due to in-plane transport is characterized. When exposed to near infrared (NIR) radiation a peak is observed at 1.4 μm. Based on its energy and the polarization dependence, it is associated with the polarized S to P_z intra-band transition in the QDs. This PC signal turns from positive to negative as temperature is raised, increasing exponentially from 50K to 300K. UV excitation at 2.8 eV renders a positive PC at all temperatures, and when combined with NIR radiation, negative PC is observed even at 12 K. We claim that following NIR excitation the electrons get trapped in deep levels in the AlN barrier, from which the UV radiation re-excite them into the QDs.

1. Introduction

Epitaxially grown GaN QDs embedded in AlN form an intriguing quantum structure due to huge internal electric field and the large band offset of up to 1.8 eV. The large band discontinuities facilitate intersubband (ISB) / intraband (IB) transitions in the near-infrared (NIR) spectral range (1-3 μm) [1]. Recently devices based on ISB transitions in GaN\AlN structures were implemented, such as all-optical switches [2], electro-optical modulators [3], lateral Quantum Dot IR Photodetectors (QDIPs) [4] and Quantum Cascade Detectors (QCD) [5].

Photoconductivity in ISB quantum structures is usually based on excitation of carriers from bound states in the QDs to extended states in the wetting layer (WL) or in the continuum. Most ISB devices relay on vertical carrier transport, in which photoexcited carriers in the continuum are drifted by the electric field along the growth direction. Photocurrent related to Bound-to-Bound (BtB) electronic transition is not likely, since the excited-state energies are below the effective potential barrier and excited carriers have a low escape probability [6].

In this work we investigate the photo response of lateral QDIPs. Exposing the device to NIR radiation, result in a reduction of the device conductivity with respect to its dark value. This phenomenon, known as Negative Photo Conductivity (NPC), was observed in several material systems [7-10] and is mainly attributed to either a reduction in the number of carriers available for transport [7-9] or to a reduction of carriers mobility [10], induced by the radiation. In our samples, the NPC intensity decreases exponentially as the temperature is reduced, and eventually the PC signal turns positive at low enough temperature (12K). The response time of the NPC is extremely slow, up to hundreds of seconds. Further insight to the origin of the NPC was obtained by employing ultra-violet (UV) excitation along with the NIR one. The physical processes involved in these photo responses are investigated in the following.

2. Experimental

The measurements were performed on a sample consisting of a single GaN QD layer deposited on a 1 μm thick AlN-on-sapphire substrate using plasma-assisted molecular-beam epitaxy. QD growth starts by deposition of 2 monolayers of GaN as wetting layer (WL) followed by 2 additional GaN monolayers from which QDs are self-assembled by the Stranski-Krastanov growth mode. The structure was covered by a 3 nm AlN cap. AFM measurements on a similar uncapped sample were used to determine the QD density (10^{12} cm^{-2}), average height (1.2±0.5 nm), and diameter (15±3 nm) [11]. A lateral photodetector was formed by implementing 2 contacts, 150μm wide and 6μm apart. Rapid- thermal annealing (RTA) was used for the ohmic contacts which consist of a multi-layer metallization of Ti:Al:Ni:Au (15 nm: 220 nm: 40 nm: 300 nm). The substrate was polished at 45° to enable irradiation at both S and P configurations, allowing for analysis of polarization effects on intersubband transitions.

The spectral response at the NIR range was investigated. FTIR measurements, using step-scan at fixed frequency, show photocurrent spectra which peaks at 1.4 μm. These spectra are completely P polarized. The peak intensity increases exponentially with temperature, as seen in figure 1.

Additional photocurrent experiments were performed by exposing the sample to a 20 mW laser radiation at 1.55 μm. The measurements show negative photoconductivity, i.e. the current decreases following irradiation, as shown in the inset of figure 1. As the temperature is lowered the negative effect decreases, and eventually the PC signal turns positive, as described later-on. The recovery of the current to its dark value is orders of magnitude faster at 300 K than at 50 K.

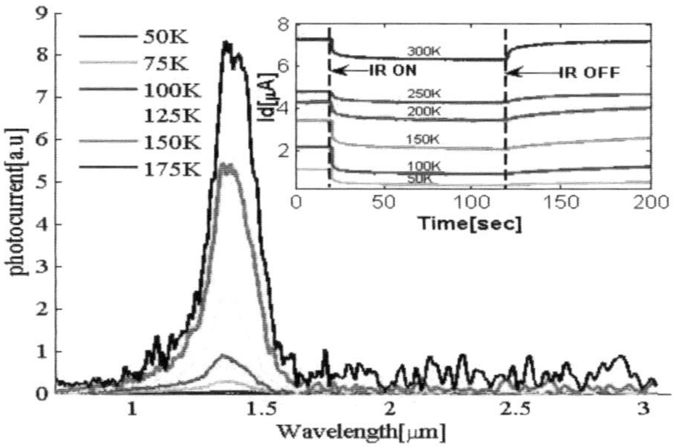

Figure 1. Intraband PC spectra, under 1.55 μm laser radiation at increasing temperatures for single QDs layer.

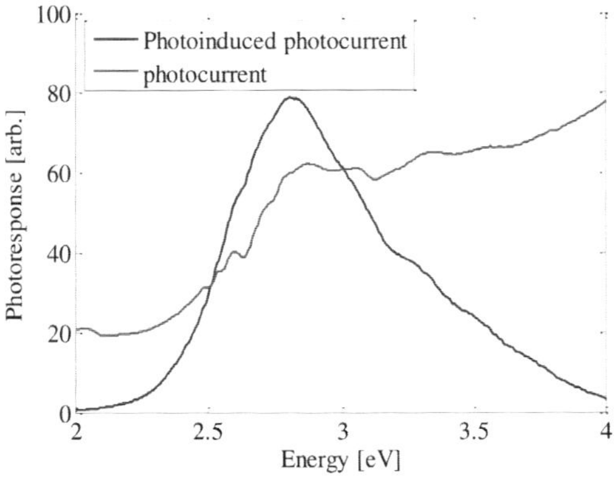

Figure 2. Photocurrent (green) and photo-induced NPC (blue) for UV excitation

Further insight into the processes involved was obtained by combining the NIR PC measurements with UV radiation at various temperatures. First, we examined the photo-response of the structure to UV excitation. The room temperature photocurrent spectrum, shown in figure 2 (in green), shows a peak at 2.8 eV. Also shown is the photo-induced experimental results, in which the photocurrent generated by the 1.55 μm is modulated by UV radiation from a Xenon source, showing a clear peak at 2.8 eV as well.

When the sample is exposed to both NIR and UV radiation, different processes are observed. The NIR light reduces the photocurrent generated by UV excitation even at a low temperature (12 K) at which the pure NIR PC signal is positive. The experimental results are shown in figure 3, presenting the effects of temperature and UV illumination on the sign and time response of the intraband NIR PC. At room temperature the NIR PC is always negative while the UV illumination results in a positive PC (figure 3a). At low temperature (12 K) the response to NIR radiation becomes positive with very short rise time as shown in figure 3b. However, in the presence of UV illumination, the NIR signal turns from positive to negative as shown in this figure.

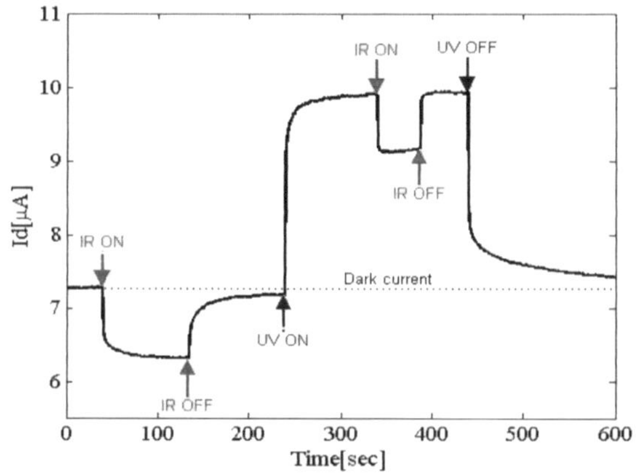

Figure 3a. Current vs. time under IR (1.55 μm) and UV (300 nm) radiation at room temperature.

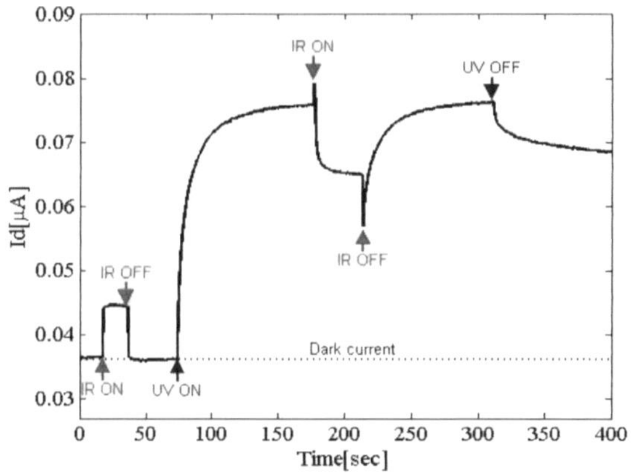

Figure 3b. Current vs. time under IR and UV radiation at low temp. (12 K).

3. Discussion

Lateral transport processes in the 2D QD layer is governed by tunnelling. Recently we have identified the origin of dark current as hopping conductivity between the QDs, facilitated by their high density, typical for this material system [12]. The negative response to the NIR radiation is attributed to trapping of conductive electrons in deep levels located in the AlN barriers, resulting in a reduction of available carriers for hopping.

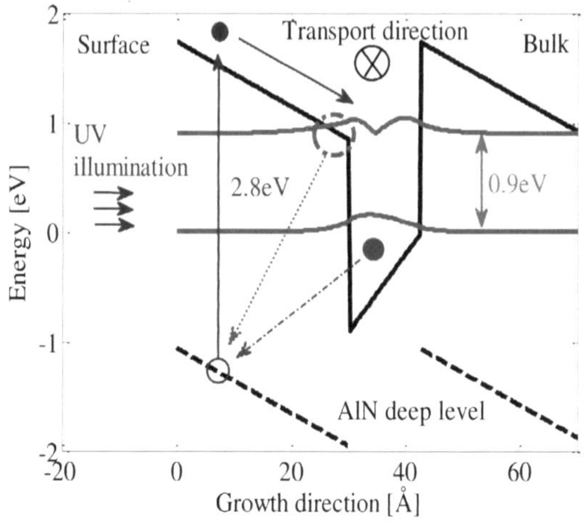

Figure 4. Band diagram of QD along with the S and Pz wavefunctions. Arrows indicate different capture and emission processes of carriers.

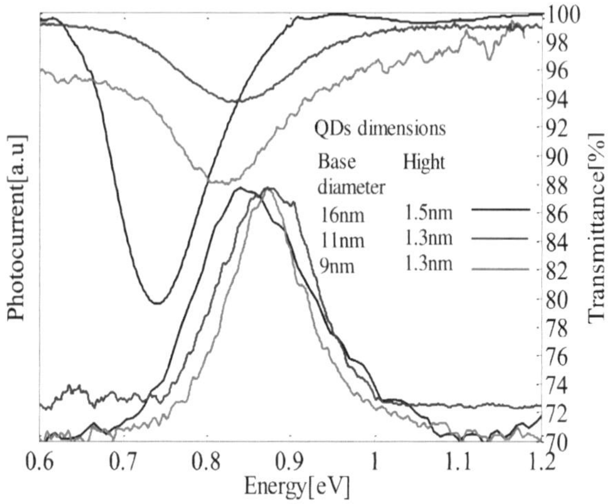

Figure 5. Photocurrent Vs. transmittance spectra, obtained on lateral QDIPs containing 20 layers of QDs of different dimensions.

The energy and the polarization of the radiation associated with NIR PC peak correspond to the S to P_z intra-band transition in the QDs [12]. Due to the polarization field in nitrides, the wavefunction of carriers in the Pz state extends into the AlN conduction band, as described in figure 4 (dashed purple circle). Thus, the probability for electrons in these states to be trapped in the AlN barriers deep levels increases substantially, with respect to trapping of carriers from the QDs ground state (brown dotted-dashed arrow in figure 4). This probability is determined mostly by the cross-section for trapping, which is proportional to the thermal velocity and to the square root of temperature [13]. Therefore, the trapping becomes less effective as the temperature is lowered. The extension of the Pz wavefunction into the AlN barrier is larger for smaller QDs, therefore, the negative PC favours small QDs. Indeed, a blue shift between the absorption spectrum to the negative PC at room temperature was observed in samples containing 20 layers of QDs (figure 5). The energy of the AlN deep level can be estimated by the UV photocurrent spectrum (figure 2) which is peaked at 2.8 eV and marked by the dashed line in the band diagram (Figure 4). These results are consistent with recent results of cathodoluminescence obtained on pure AlN [14].

As the AlN deep levels are lower in energy than the QDs ground state, the reverse process of re-emission of the trapped carriers into the QDs requires thermal energy and thus takes seconds at room temperature but minutes at 50 K (inset of figure 1). The extremely long time constant results in two opposing findings. On one hand the relative decrease in current is much larger at low temperatures. On the other hand, since the PC spectrum is recorded at a sampling rate of 130 Hz, as the temperature is lowered, the very slow response results in an ever decreasing measured peak amplitude (figure 1).

Figure 2 shows the effects of temperature and UV illumination on the sign and time response of the intraband NIR PC. At room temperature the NIR PC is always negative. UV illumination results in a positive PC due to release of trapped carriers from the AlN deep levels into the QDs (figure 3a). As the temperature is lowered, the traps become populated and the cross-section for trapping decreases. Hence, at low enough temperatures, carriers which are excited into the Pz level of the QDs remains in the GaN channel, thus, contributing to the hopping conductivity. Since the hopping probability of excited Pz carriers is higher than that of carriers in the S states [12], the response to NIR radiation becomes positive with very short rise time as shown in figure 3b. UV illumination, even at low temperatures, releases trapped carriers and results in a positive PC (blue arrows in figure 4). Finally, the NIR signal in the presence of UV illumination evolves with time from positive to negative as shown in figure 3b. As the UV excitation is turned off, the dark current original value is not restored, suggesting that carriers which originates from the AlN deep levels (by the UV illumination) stay within the QDs. Carriers which are excited by the UV light into the AlN continuum are drifted by the polarization field into the QDs, while the transfer of carriers from the QDs to the AlN deep levels relays on tunnelling. This irreversible process is demonstrated schematically in figure 4. Further modelling of trapping processes is under study.

References

[1] Tchernycheva M, Nevou L, Doyennette L, Helman A, Colombelli R, Julien F H, Guillot F, Monroy E, Shibata T, and Tanaka M 2005 *Appl. Phys. Lett.* **87** 101912
[2] Iizuka N, Kaneko K, and Suzuki N 2006 *IEEE Journal of Quantum Electronics* **42** 765
[3] Kheirodin N et al. 2008 *IEEE Photonics Technology Letters* **20** 724
[4] Vardi A, Akopian N, Bahir G, Doyennette L, Tchernycheva M, Nevou L, Julien F H, Guillot F and Monroy E 2006 *Appl. Phys. Lett.* **88** 143101
[5] Vardi A, Bahir G, Bougerol C, Monroy E, Schacham S E, Tchernycheva M, Julien F H 2008 *Appl. Phys. Lett.* **92** 11112
[6] Fu L, Tan H H, McKerracher I, Wong-Leung J, Jagadish C, Vukmirovic N, and Harrison P 2006 *J. Appl. Phys.* **99** 114517
[7] Kukushkin I V, von Klitzing K, Ploog K, Kirpichev V E and Sheped B N 1989 *Phys. Rev. B* **40** 4179
[8] Hopkins P F, Rimberg A J, Westervelt R M, Tuttle G, and Kroemer H 1991 *Appl. Phys. Lett.* **58** 1428
[9] Lo I, Mitchel W C, Monasreh M O, Stutz C E, and Evans K R 1992 *Appl. Phys. Lett.* **60** 751
[10] Yakimov A I, Dvurechenskii A V, Nikiforov A I and Pchelyakov O P 2000 *Phys. Rev.* B **62** 16283
[11] Guillot F et al. 2006 *J. Appl. Phys.* **100** 044326
[12] Vardi A, Bahir G, Schacham S E, Kandaswamy P K and Monroy E 2009 *Phys. Rev. B* **80** 155439
[13] Miller G L, Lang D V and Kimerling L C 1977 *Ann. Rev. Mater. Sci.* 377
[14] Sarusi G, Moshe O, Khatsevich S, Rich D H, Salzman J, Meyler B, Shandalov M and Golan Y 2006 *J. Electron. Mater.* **35** (12) L15

Self-assembled colloidal PbS quantum dots on GaAs substrates

Wei Lü[1], Fumihiko Yamada and Itaru Kamiya

Toyota Technological Institute, Nagoya 468-8511, Japan

E-mail: luwei@toyota-ti.ac.jp, kamiya@toyota-ti.ac.jp

Abstract. We report the fabrication and analysis of self-assembled monolayer and bilayer films of colloidal PbS quantum dots (QDs) on GaAs (001) substrates. 1,6-hexanedithiol is used as link molecule between QDs and GaAs substrates. Atomic force microscopy (AFM) and photoluminescence (PL) measurements confirm the formation of PbS QD film on GaAs. For the monolayer PbS QD film, the temperature-dependent PL shows a feature typical of close-packed film. For the bilayer PbS QD film fabricated from two different mean-sized PbS QDs, we find that the stacking sequence of QDs with different size affects the quantum yield and emission wavelength of the film.

1. Introduction

Colloidal quantum dots (QDs) are characterized by their high fluorescence efficiency, narrow spectral emission and broadly tunable color-emission. These properties, combined with the high photochemical stability, are expected to increase the device lifetime at the high luminescence and current regimes necessary for lighting applications.[1] Among such QDs, PbS QDs are receiving great attention due to the narrow energy gap relevant for optical applications in the near-infrared region such as long wavelength telecommunications, photovoltaics, and bioimaging in low absorption of biological systems.[2-4] In particular, the integration of colloidal semiconductor QDs on substrates to form two-dimensional arrays is becoming a promising field of scientific research due to their potential applications in microelectronics.[5]

In this work, we describe the assembly and photoluminescence (PL) characterization of monolayer and bilayer of PbS QDs on GaAs substrates. The PL intensities of monolayer PbS QDs show increase followed by decrease with rise in temperature, which is a feature typical for close-packed PbS QD films in agreement with our previous report.[6] For the bilayer PbS QD films fabricated from two PbS QD ensembles with different mean sizes, we find that different stacking sequence of QDs with different size affects the quantum yield and emission wavelength of the bilayer film.

2. Experimental details

PbS QDs were synthesized using an organometallic route previously reported.[7] The resulting PbS QDs were precipitated from the reaction mixture with methanol and redispersed in tetrachloroethylene (TCE). PbS QDs with two different mean sizes (3.6 and 4.1 nm dia.) were used in this work.

The GaAs wafers were cleaned in an ultrasonic bath with acetone and 2-propanol, and dried in a stream of nitrogen. Surface oxide was then removed by a 80 s etch in concentrated (35%) HCl followed by rinse with deionized water and 2-propanol. Surface modification of the GaAs wafers was performed in a solution containing 90% 2-propanol and 10% 1,6-hexanedithiol (volume ratio) for 8 h in dark at room temperature. The surface modification was followed by several thorough rinses with 2-propanol. The self-assembly of monolayer PbS QDs was performed in a TCE solution containing PbS QDs for 12 h in dark at room temperature. After assembly the samples were thoroughly rinsed with TCE and dried with a stream of nitrogen. The bilayer samples was fabricated by repeating dipping in 1,6-hexanedithiol solution and PbS QD solution. Figure 1 shows the schematic description of the monolayer and bilayer films of PbS QDs on GaAs.

The 860 nm beam from a continuous-wave Ti:sapphire laser was used as an excitation source for the PL measurements. The samples were mounted in a compressed helium cryostat under vacuum to avoid air-exposure.

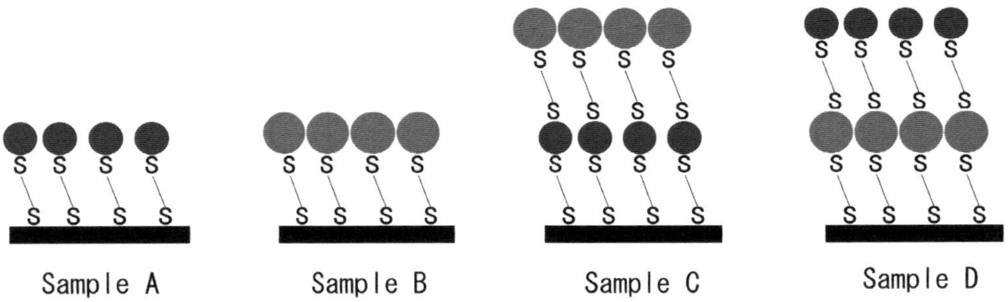

Figure 1. Schematic description of monolayer PbS QDs of two different sizes on GaAs (samples A and B) and bilayer PbS QDs on GaAs with different stacking sequences (samples C and D). The black lines represent 1,6-hexanedithiol link molecules.

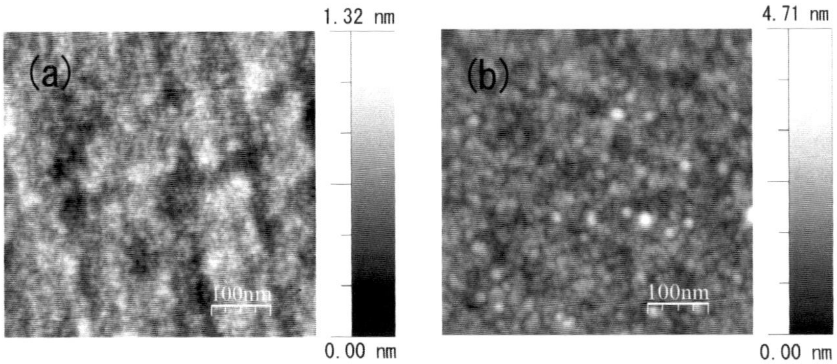

Figure 2. AFM images of (a) the untreated (001) GaAs substrate and (b) monolayer PbS QDs assembled on GaAs (Sample A).

3. Results and Discussions

To assess the coverage of the PbS QDs on the GaAs substrate, atomic force microscopy (AFM) was performed using SII Nanotechnology SPA300/E-SWEEP with Nanonis QC4-S. The topography of the samples was measured immediately after assembly, under vacuum with a base pressure below 10^{-5} Pa at room temperature. Figure 2 shows the AFM images of the untreated GaAs substrate and the PbS QDs (~3.6 nm dia.) assembled on GaAs substrate. The images shown are typical of images measured in several different areas of the substrate. The untreated GaAs shows a peak-to-valley roughness of less than 1.32 nm. The PbS assembled GaAs shows dot-like features with heights of up to 4.71 nm and typical lateral sizes of 10~15 nm. The lateral sizes of the AFM images are several times larger than the actual size of the QDs used (~3.6 nm), which is due to the AFM tip diameter (~20 nm). Considering the AFM tip diameter, this image is consistent with coverage of disordered QD monolayer on the GaAs surface.

Figure 3 shows PL spectra of samples A and B measured at 4.2 K. The emission of GaAs is observed at higher energy edge, which is not shown here. The emission peaks of samples A and B are located at 1100 and 1340 nm due to the different sizes. The emission peak is not Gaussian unlike the solution spectra, which is probably caused by energy transfer between QDs in the close-packed film.[8] Figure 4 shows the integrated PL intensities of sample A as a function of measurement temperatures. Sample B shows the similar behaviour with that of sample A. The PL intensity shows increase followed by decrease with rise in temperature. According to our previous report[6], the increase followed by decrease in PL intensity with rise in temperature is a typical feature of close-packed PbS QD films and can be explained by thermally activated electronic transfer as shown in inset of Figure 4.

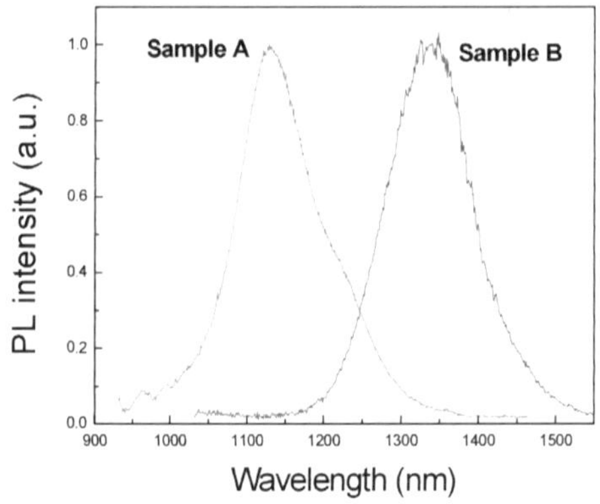

Figure 3. Normalized PL spectra of samples A and B measured at 4.2 K.

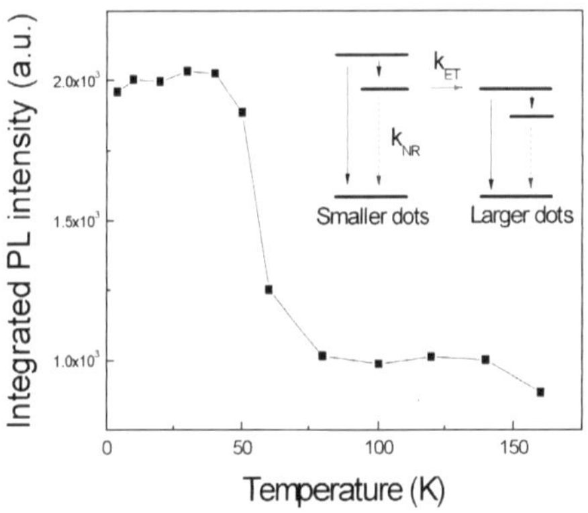

Figure 4. The integrated PL intensities of sample A as a function of measurement temperatures. The inset describes the thermally activated electronic transfer between PbS QDs in close-packed film.

Figure 5 shows PL spectra of samples A, B, C and D measured at 4.2 K. We try our best to insure that the measurements are fulfilled under the same condition so that we can compare the four samples quantitatively. Compared with sample B, both samples C and D exhibit stronger emission due to energy transfer from smaller to larger QDs. Klimov *et al.* showed the evidence of rapid directed energy transfer in a bilayer structure comprising differently sized core-shell CdSe/ZnS QDs bound by dithiol linkers.[9] Further, funnel-like band gap profiles are realized applying layer-by-layer assembly to CdTe nanocrystals of distinct sizes, which greatly improves the PL quantum yield of the QD films.[10] However, to our best knowledge, there is no report on the effect of stacking sequence of the bilayer structures on PL behavior of films. Our results indicate that the stacking sequence of bilayer structure comprising different sized QDs affects the quantum yield and emission wavelength of the films thus prepared. Compared with emission peak of sample B, the PL peak of sample C exhibits blueshift and that of sample D exhibits redshift. However, the emission intensity of sample C is stronger than that of sample D. These abnormal phenomena are explained as followed.

For the surface modified GaAs wafers, many thiol molecules are distributed on the surface of GaAs, which serve as anchor to catch QDs in solution to form single monolayer close-packed QD film. To stack the second QD monolayer on top, the first QD monolayer has to be modified by another thiol molecule layer on the surface of the first QD monolayer. However, in this case, the quantity of available thiols serve as anchor to catch the next QD monolayer would be less than that on the GaAs wafer since it is unlikely that thiol molecules form a self-assembled layer as it did on the GaAs substrate. Also considering the probability that a thiol molecule catches a QD, there would be less QDs on the second monolayer than on the first monolayer. As a result, the mean distance between adjacent QDs is larger than that of the first monolayer.

According to the discussion above, the mean distance between larger QDs (the second monolayer) in sample C should be larger than that of sample B. The larger distance between QDs induces smaller energy transfer rate. As we discussed previously[6], the energy transfer occurs in close-packed QD films due to inhomogeneous size distribution, which induce the redshift of emission line of QDs in close-packed film. Therefore, the blueshift of emission line of sample C compared with sample B is due to larger distance between the QDs in the second monolayer.

The smaller QDs serve as donors in bilayer films. Hence, there are more donors in sample C than in sample D according to our discussion. The energy transfer from smaller to larger QDs increases with the number of donors. Therefore, the stronger emission from sample C can be attributed to the existence of more donors.

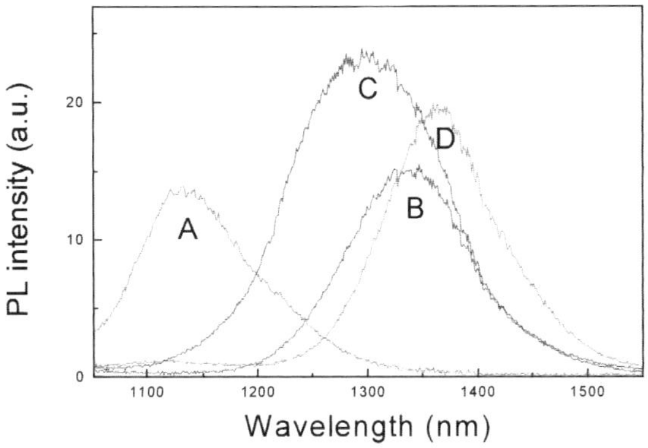

Figure 5. PL spectra of samples A, B, C and D measured at 4.2 K under the same conditions

4. Conclusion

In summary, monolayer and bilayer structures of PbS QDs on GaAs have been fabricated. AFM and PL measurement confirm the formation of PbS QD film on GaAs. The monolayer structure exhibits typical PL feature of close-packed film. For the bilayer structure fabricated from two different mean-sized PbS QDs, the stacking sequence of smaller and larger QDs affects the quantum yield and emission wavelength of the films thus prepared.

Acknowledgement

This work was supported by the High-Tech Research Center Project and Strategic Research Infrastructure Project, the Ministry of Education, Sports, Science and Technology, Japan.

References

[1] Rizzo A, Mazzeo M, Biasiucci M, Cingolani R and Gigli G 2008 *Small* **4** 2143
[2] Rogach A L, Eychmüller A, Hickey S G and Kershaw S V 2007 *Small* **3** 536
[3] McDonald S A, Konstantatos G, Zhang S, Cyr P W, Klem E J, Levina L and Sargent E H. 2005 *Nat. Mater.* **4** 138
[4] Hyun B R, Chen H, Rey D A, Wise F W and Batt C A 2007 *J. Phys. Chem. B* **111** 5726
[5] Tseng J. Y, Lin M. H and Chau L K 2001 *Colloids Surf A* **182** 239
[6] Lü W, Kamiya I, Ichida M and Ando H 2009 *Appl. Phys. Lett.* **95** 083102
[7] Hines M A and Scholes G D 2003 *Adv. Mater.* **15** 1884
[8] Stouwdam J, Shan J, Veggel F, Pattantyus-Abraham A and Raudsepp J 2007 *J. Phys. Chem. C* **111** 1086
[9] Crooker S A, Hollingsworth J A, Tretiak S and Klimov V I 2002 *Phys. ReV. Lett.* **89** 186802
[10] Franzl T, Klar T A, Schietinger S, Rogach A L and Feldmann J 2004 *Nano Lett.* **4** 1599

Optical response of quantum dot multilayer structures

L. M. Thu and O. Voskoboynikov

Department of Electronics Engineering, National Chiao Tung University, 1001 Ta Hsueh Rd., Hsinchu 300, Taiwan, R.O.C.

Email: leminhthu.ee95g@nctu.edu.tw

Abstract. We theoretically study optical response of *InAs/GaAs* quantum dot multilayered structures. Based on the multi-scale hybrid discrete-continuum description the optical response of an isolated layer of embedded quantum dots has been determined. Using the propagation-matrix approach we express the amplitudes of incident, reflected and transmitted electromagnetic waves in a multilayered structure by the reflection and transmission coefficients of the isolated layers. We study the overall reflectance and transmittance dependencies on the number of layers and distance between consecutive layers. The increase of the number of the layers considerably enhances the overall reflectance of the structures. Interference effects become significant for certain distances between layers in the structures.

1. Introduction

In the past years progress in modern technology makes it possible to fabricate semiconductor nano objects such as quantum dots within wide ranges of shapes and sizes. Those nano objects become very promising elements in constructing small building blocks for integrated photonics and nano-structured metamaterials [1,2,3]. Generally the metamaterials are designed as artificial multilayered structures. Optics of isolated layers of semiconductor quantum dots, quantum rings and quantum dot molecules was studied in details recently (see [4,5,6] and references therein). Therefore a proper modeling of the optical response from multilayered structures is on demand. In this study we performed simulations of multilayered structures made from *InAs* lens-shaped quantum dots embedded in infinite *GaAs* matrix. In our model each layer is characterized by the reflection and transmission coefficients obtained in [5, 6]. Based on the propagation-matrix approach [7] the overall reflection and transmission coefficients of multilayered structures are derived. We study the overall optical coefficients as functions on the optical transition's energy and distance between consecutive layers.

2. Theoretical approach

The structure to be investigated consists of consecutive layers of quantum dots composed of *InAs/GaAs* with interlayer distance d, (as it is shown in figure 1). In our consideration the electromagnetic plane wave is incident to the first layer, propagates through the structure and emerges at the last layer. We assume that the consecutive incident angles of the wave on each layer are identical. Therefore all layers in the multilayered structure are described by the same reflection and transmission coefficients. Using the propagation-matrix approach [7] for the following amplitudes of the incident (A_i, A_{i+1}) and reflected (B_i, B_{i+1}) waves at ith and (i +1)th layers we can write

$$\begin{pmatrix} A_i \\ B_i \end{pmatrix} = P_i \begin{pmatrix} A_{i+1} \\ B_{i+1} \end{pmatrix} \quad (1)$$

The propagation matrix P_i is commonly written as [7]

$$P_i = \frac{1}{t_i}\begin{pmatrix} e^{-i\phi_i} & r_i e^{i\phi_i} \\ r_i e^{-i\phi_i} & e^{i\phi_i} \end{pmatrix}$$

where r_i and t_i present the reflection and transmission coefficients of the ith layer respectively, $\phi_i = ik_z d$, k_z is the perpendicular to the layer component of the wave vector $k = n\omega/c$ (n is the semiconductor matrix refractive index, ω is the wave frequency, and c is the light speed in the vacuum).

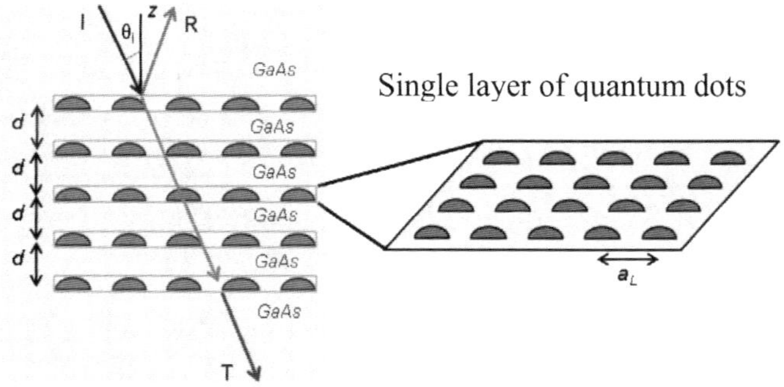

Figure 1. Scheme of the quantum dot multilayer structure.

By iteratively carrying out equation (1) we can obtain the overall propagation matrix P which connects the amplitudes of the incident, reflected and transmitted waves for the structure of finite number of layers N:

$$\begin{pmatrix} E_0 \\ rE_0 \end{pmatrix} = \begin{pmatrix} P_{11} & P_{12} \\ P_{21} & P_{22} \end{pmatrix} \cdot \begin{pmatrix} tE_0 \\ 0 \end{pmatrix} \quad (2)$$

where E_0 is amplitude of incident wave at the first layer, r and t are the total reflection and transmission coefficients, respectively. From (2) we obtain the reflectance and transmittance of the multilayered structure:

$$R = |r|^2 = \left|\frac{P_{21}}{P_{11}}\right|^2, T = |t|^2 = \left|\frac{1}{P_{11}}\right|^2 \quad (3)$$

To simulate the overall optical response (reflectance and transmittance) of multilayered structures the reflection and transmission coefficients of each ith layer are required. We use the Vlieger's expressions from [8] to obtain them:

$$r_i^{(ss)} = f(A_y \cos\theta_i - f)^{-1}, t_i^{(ss)} = 1 + r_i^{(ss)},$$

$$r_i^{(pp)} = \frac{f\cos\theta_i}{A_x - f\cos\theta_i} - \frac{f\sin^2\theta_i}{A_z\cos\theta_i - f\sin^2\theta_i},$$

$$t_i^{(pp)} = \frac{f\cos\theta_i}{A_x - f\cos\theta_i} - \frac{A_z\cos\theta_i}{A_z\cos\theta_i - f\sin^2\theta_i}, \tag{4}$$

where subscripts "*ss*" and "*pp*" refer to the light polarization, θ_i is the incidence angle,

$$A_u = \alpha_0\alpha_{BE,uu}^{-1}(\omega) - \varepsilon_m^{-1}(f + \alpha_0 t_u), f = 2\pi i a_L k,$$

$$t_u = -\frac{N_u}{\varepsilon_0 V} + \frac{ik^3}{6\pi\varepsilon_0}, \alpha_0 = 4\pi\varepsilon_0 a_L^2, \varepsilon_m = n^2.$$

In the last equations: N_u $\{u = x, y, z\}$ and V denote the depolarization factor and the volume of the quantum dot, respectively. The inter-planar transfer tensor **f** is defined for two dimensional lattice of the quantum dots as: $f_x = f_y = - f_z/2$, fz = 9.03362 [4], $\alpha_{EB,uu}(\omega)$ is the polarizability of an isolated quantum dot defined in [4, 5]. The polarizability includes the static and dynamic parts and the last one depends on the optical transition energies ($E = \hbar\omega$) [4, 5, 6].

3. Simulation results and discussion

For all numerical simulations to determine the polarizability of isolated quantum dots we use the COMSOL MultiPhysics package (www.comsol.com). We accept realistic semiconductor material parameters from [6, 9, 10]. The dimensions of a single *InAs/GaAs* lens-shaped quantum dot are taken as 2 nm in height and 15 nm in radius. The lattice constant in a layer is a_L = 100 nm. The reflection and transmission coefficients of a layer of the dots are calculated by using equation (4). Finally we used computed coefficients to simulate the overall reflectance and transmittance of multilayered structures. The reflectance of the structures including several layers for *s*-polarized light is shown in figure 2. The peaks in the reflectance relate to the allowed optical transitions in the quantum dots. Clearly, for a single layer the reflectance is weak. However the reflectance still reproduces important information on quantum mechanics of individual quantum dots.

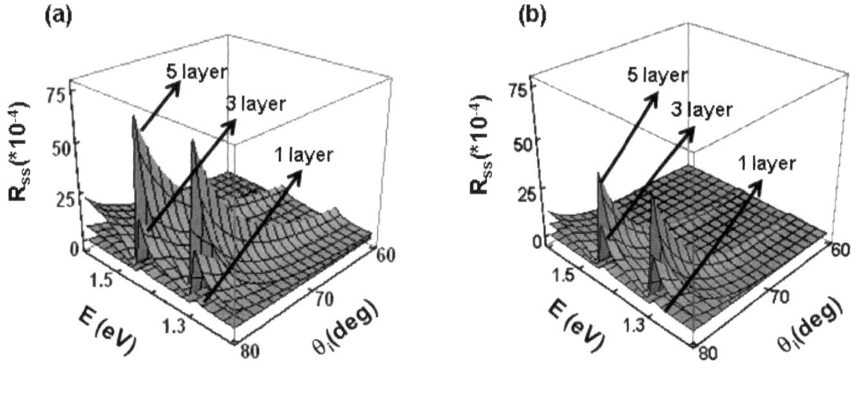

Figure 2. Reflectance of the quantum dot multilayered structures as a function on the transition energy and angle of incidence: (a) d = 10nm and (b) d = 50nm.

For structures with small *kd*, the electromagnetic wave phase change formed between consecutive layers can be neglected and no interference effects can be seen in the total reflectance. When the number of the layers in the structure increases the overall reflectance of the structure considerably enhances (see figure 2). For large enough *d* the interference becomes significant and this leads to the appearance of the periodical peaks in the reflectance. The dependence of the overall reflectance on the transition energy and distance *d* (incident angle θ_i =60^0) for the structure consisting of five layers is presented in figure 3. Clearly at some specific distances the peaks related to the allowed optical transitions in the quantum dots can be intensified or vanished because of the interference.

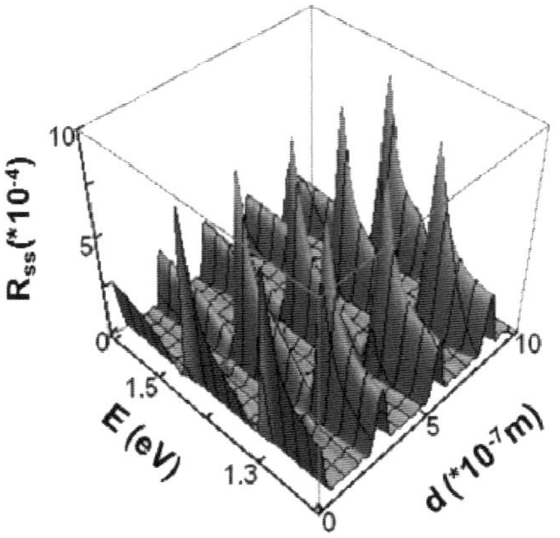

Figure 3. Reflectance of the quantum dot multilayered structure consisting of five layers as a function on the transition energy and distance between the consecutive layers.

In short conclusion, using the propagation matrix approach we theoretically study magneto-optical response of the quantum dot multilayered structures. Our simulation results show that when the distance between layers is comprehensively small the overall reflectance of the structures increases with increasing number of layers. The interference effects appear for the appropriate distances between the layers. Our approach to model semiconductor multilayered nano structures is useful for simulations optical properties of semiconductor metamaterials.

Acknowledgements
This work is supported by the National Science Council of the Republic of China under Contracts no. 97-2112-M-009-012-MY3 and NSC 97-2120-M-009-004 and by the Aim for the Top University Plan of the National Chiao Tung University and Ministry of Education of Taiwan, ROC.

References

[1] Shalaev M 2007 *Nature Photonics* **1** 41
[2] Smith D R, Pendry J B, Wiltshire M C K 2004 *Science* **305** 788
[3] Ramakrishna S A 2005 *Rep. Prog. Phys.* **68** 449
[4] Voskoboynikov O, Wijers C M J, Liu J L, and Lee C P 2005 *Phy. Rev.* B **71** 245332
[5] Wijers C M J, Chu J-H, Liu J L, and Voskoboynikov O 2006 *Phys. Rev.* B **74** 035323
[6] Thu L M and Voskoboynikov O 2009 *Phys. Rev.* B **80** 155442
[7] Chuang S L 1995 *Physics of Optoelectronics Devices* (Wiley, New York) chapter 5.
[8] Vlieger J 1973 *Physica* **64** 63
[9] Vurgaftman I, Meyer J R, and Ram-Mohan L R (2001) *J. Appl. Phys.* **89** 5815
[10] Pryor C E and Pistol M E (2005) *Phys. Rev.* B **72** 205311

Journal of Physics: Conference Series **245** (2010) 012071 doi:10.1088/1742-6596/245/1/012071

Ge/Si nanostructures with quantum dots grown by ion-beam assisted heteroepitaxy

J.V. Smagina, V.A. Zinoviev, P.L. Novikov, V.A. Armbrister, Koptev E.S., A.V. Dvurechenskii

Institute of semiconductor physics SB RAS, av Lavrentjeva, 630090, Novosibirsk, Russia

E-mail: smagina@isp.nsc.ru

Abstract. Effect of low-energy ion-beam irradiation during Ge on Si(100) growth on characteristics of Ge nanoislands array is studied. It is shown by STM that pulsed ion-beam action leads to an increase in Ge nanoislands density. The dependence of Ge nanoislands density upon ion energy is non-monotonous with maximum at 150 eV. RBS data indicate a perfect crystal structure of Ge nanoislands. By Raman spectroscopy the Ge content in SiGe quantum dots is estimated to be higher than 75%.

1. Introduction

Self-assembled Ge quantum dots (QD) on Si(100) have been intensively investigated as the basis of novel electronic and optical devices. The well-defined sizes with little dispersion and high density of QD are generally required for any practical applications [1, 2]. It is a challenge to meet all these requirements simultaneously. A promising way to achieve the homogeneous size distribution and high density of QD is using pulsed irradiation by low-energy (~ 100 eV) Ge^+ ions during the heteroepitaxy. We have found that the pulse low-energy ion irradiation during Ge/Si heteroepitaxy stimulates nucleation and growth of three-dimensional Ge islands characterized by high density of islands and small average size and low size dispersion [3]. The observed phenomena were studied by molecular dynamics (MD) and Monte-Carlo (MC) simulation. It was shown that about 5 interstitials are produced in Si bulk by an ion impact [4]. Areas of local tensile strain above the interstitial atoms are responsible for the change in the binding energy of the Ge adatoms. MC modeling has shown that the places where the ions impact the surface become the centers of preferential nucleation of three-dimensional islands. The results of the MC simulations have demonstrated a good agreement with the experimental data [5].

In the present work we study the low-energy ion-beam-assisted growth of tree-dimensional Ge nanoislands on the Si(100) as dependent on the growth temperature, energy of ions and degree of ionization of molecular beam. The effect of ion irradiation on the defect generation and the composition of Ge/Si(100) nanostructures were studied by the Raman spectroscopy, Rutherford backscattering and transmission electron microscope techniques.

2. Experimental

The samples were grown by molecular-beam epitaxy (MBE) on p-Si(100) wafers. Two growth modes were implemented: 1) conventional MBE of Ge on Si (sample M); 2) MBE with pulsed irradiation by a beam of low-energy Ge^+ ions (MI). The ion energy could be controllably varied from 50 to 250 eV. A system of ionization and acceleration of Ge^+ ions provided the degree of ionization of Ge molecular beam from 0.02% to 0.5%. The pulsed irradiation (with the pulse duration τ~0.5 s) was applied after each completed germanium monolayer (ML) was deposited. The incident angle of the molecular and ion beams was 54°. The deposition rate was approximately 0.1 ML/s and the substrate temperature was varied within interval from 250 °C to 450 °C. In all experiments the total amount of Ge deposited was 5 ML. In addition, multilayer Si/Ge structures were grown. 15-nm-thick Si spacer layers and a 50-nm-thick cap Si layer were deposited at 500°C by the conventional MBE (with no irradiation).

The morphology of the surface was *in situ* examined by scanning tunneling microscopy (STM). Crystal perfection of Ge/Si structures were analyzed by RBS and TEM. The elemental composition of the multilayer structures was studied by Raman spectroscopy. The spectra were recorded at room temperature using an automated setup based on a DFS-52 spectrometer. An Ar^+ laser (λ=514/5 nm) was used as the source of excitation.

3. Experimental results

Figure 1 presents STM images of surface with three-dimensional (3D) islands formed after the deposition of 5 Ge ML. In the M type sample (Figure 1, a) the density of 3D islands is 2.1×10^{11} cm^{-2} with the mean size of the islands 22 nm and the size dispersion 3 nm. In the MI-type sample (Figure 1, b) the density of Ge islands is 9×10^{11} cm^{-2}, the mean size of the islands is 6.5±0.7 nm.

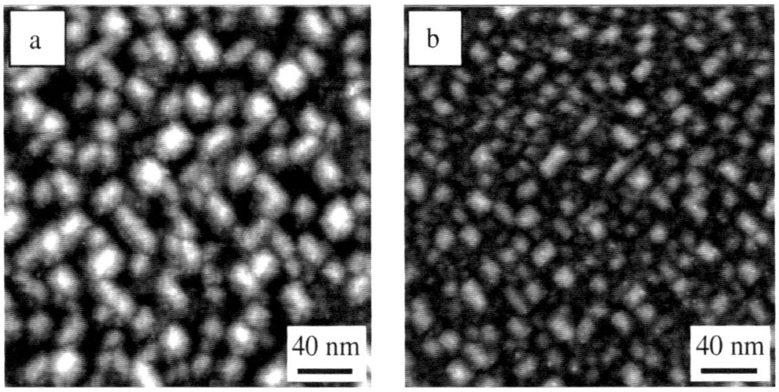

Figure 1. STM images (200 × 200 nm) of surface with 3D islands: (a) conventional MBE and (b) MBE with pulsed irradiation by Ge^+ ions. Conditions: the molecular Ge flux density is 7×10^{13} cm^{-2}s^{-1}, the temperatures of substrate is 350°C, the ion beam energy is 140 eV, the pulse duration is 0.5 s, and ion flux density is 3.1×10^{11} cm^{-2}s^{-1}.

Using STM data the temperature dependences of Ge islands density were obtained. In the M-type experiment (Figure 2, curve 1) the density goes down at temperature varying from 250 °C to 350 °C, whereas at the further increase of temperature (from 350 °C to 450 °C) no significant change in the density is observed. The temperature dependence of Ge islands density for MI-type samples (Figure 2, curve 2) exhibits a similar behavior. However, the density of Ge islands observed in MI-type samples was higher as compared with that in M-type ones.

The effect of ion energy and integrated flux of Ge ions on the density of 3D islands is presented in Figure 3 and Figure 4, respectively. The density of islands increases as the ion energy is varying from 50 eV to 150 eV. At higher energies the density goes down. The dependence of islands density on the integrated flux of Ge ions also has a maximum at the flux value of 10^{12} cm^{-2}.

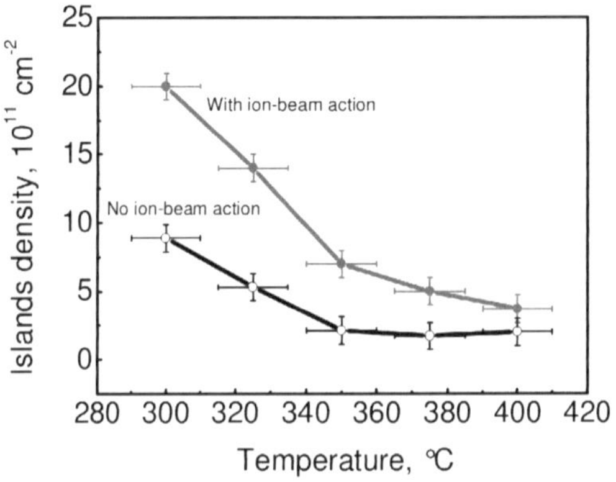

Figure 2. Temperature dependence of Ge nanoislands density built from STM data. Black line corresponds to conventional MBE regime, grey line corresponds to ion-beam-assisted growth.

RBS spectra from SiGe nanostructures are shown in Figure 5. The backscattering yield from Ge layers turns out to be sensitive to the growth conditions [3]. Raman spectra of multilayer samples M and MI with $Ge_xSi_{(1-x)}$ QD are presented in fig.5. Raman peaks at 307-315 cm^{-1} and 370-430 cm^{-1} correspond to Ge-Ge and Ge-Si optical phonons, respectively. Using the standard analysis procedure [6, 7, 8] we estimated that the Ge content x in $Ge_xSi_{(1-x)}$ QDs is higher than 0.75.

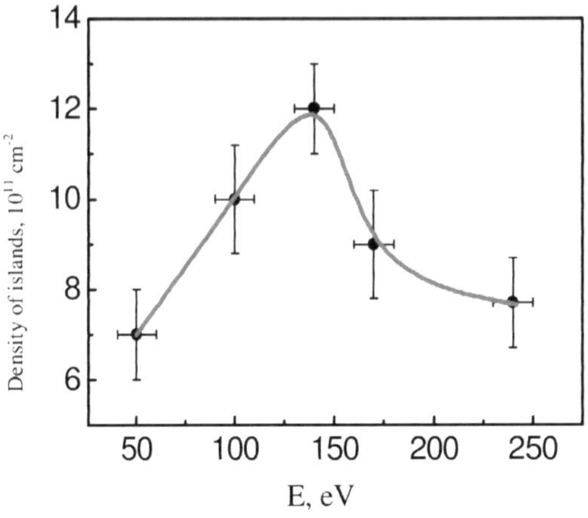

Figure 3. Density of Ge nanoislands grown by ion-beam-assisted heteroepitaxy as dependent on ion energy.

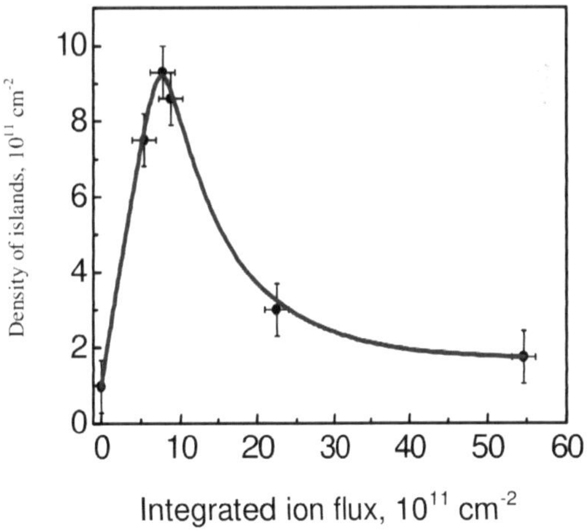

Figure 4. Density of Ge nanoislands grown by ion-beam-assisted heteroepitaxy as dependent on integrated ion flux. Ion energy E=140 eV.

4. Conclusions

It was demonstrated that, at specific values of the integrated ion flux (less than 10^{12} cm^{-2}) and ion energy (100-140 eV), the mean size of 3D Ge islands decreases, their density increases, and the island size dispersion is reduced as compared to those observed for the conventional MBE. The experimentally revealed size ordering of islands is associated with the synchronization of island nucleation under pulsed ion irradiation. Our experimental results demonstrated that Ge/Si(100) heteroepitaxy with pulsed low-energy ion beam action enables creation of defect-free 3D Ge islands. The pulsed ion irradiation during the growth of Ge/Si structures does not lead to significant mixing of the Ge and Si layers.

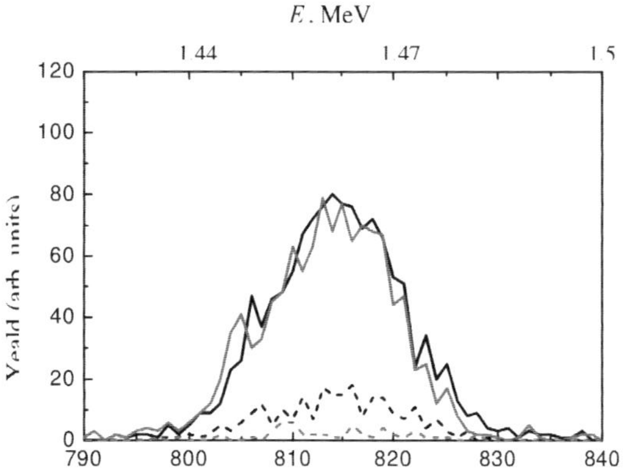

Figure 5. RBS spectra from Ge layers grown by pulsed ion-beam-assisted MBE at 350 °C. Oriented (dashed line) and non-oriented (solid line) He+ ion beams were used. Ge ion energy was 140 eV (grey line) and 240 eV (black line), respectively. Ge layer thickness is 5 monolayers.

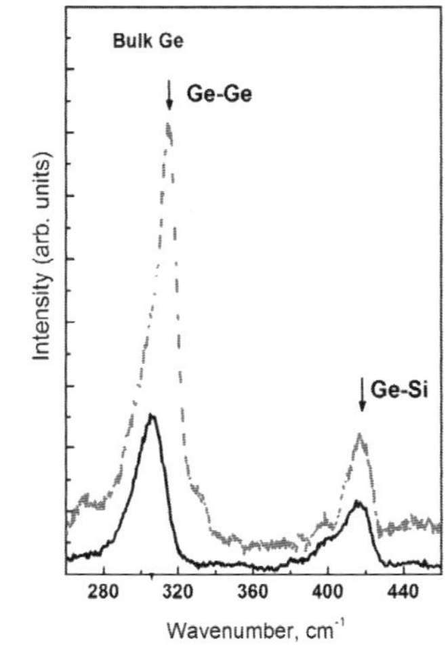

Figure 6. Raman spectra from Si/Ge multilayer structures grown by conventional (black line) and ion-beam-assisted (grey line) MBE.

Acknowledgements
This work was supported by RFBR (Grants 09-02-00882, 10-02-01181), by Program of SB RAS and Program of Nanotechnology Department of RAS.

References

[1] Ledentsov N N, Ustinov V M, Shchukin V A, et al. 1998 *Semicinductors* **32**, 343.
[2] Alferov Zh I, 1998 *Semiconductors* **32**, 1.
[3] Dvurechenskii A V, Smagina J V, Zinoviev V A, Teys S A, Gutakovskii A K, Groetzschel R 2005 *Surf. Coat.Technol.* **196**, 25.
[4] Smagina J V, Novikov P L, Armbrister V A, Zinoviev V A, Nenashev A V, Dvurechenskii A V 2009 *Phys. B* **404**, 4712.
[5] Smagina J V, Zinovyev V A, Neshev A V, Dvurechenskii A V, Armbrister A V, Teys S A 2008 *JETP* **106**, 517.
[6] Tan P H, Brunner K, Bougeard D and Abstreiter G 2003 *Phys. Rev. B* **68**, 125302.
[7] Mooney P M, Dacol F, Tsang J C, and Chu J O 1993 *Appl. Phys. Lett.* **62**, 2069.
[8] Dvurechenskii AV, Smagina Zh V, Zinovyev V A, et al. 2004 *JETP Let.* **79**, 333.

Optical and electrical properties of stacked binary InAs-GaAs quantum dot structures prepared under Surfactant-mediated growth conditions

M. Alduraibi[1], M. Missous[1], P. Luke Sam[2], A. Tierno[2], S. Keatings[2] and T. Ackemann[2]

[1]School of Electrical and Electronic Engineering, The University of Manchester, Manchester, M60 1QD, UK.
[2]SUPA and Department of Physics, University of Strathclyde, Glasgow, Scotland, UK

E-mail: mduraibi@postgrad.manchester.ac.uk

Abstract. The structural, optical and electrical properties of a 10-layer InAs/GaAs quantum dots (QDs) system having InAs layers (2.9ML) grown under surfactant growth conditions, using only an impinging In beam, were investigated. This growth mode still resulted in the formation of quantum dots, but with dot sizes smaller and sample quality better than those for normal growth (NG) of ~3ML InAs-GaAs QD structures. Room temperature photoluminescence measurements showed PL emission from this sample at 1200-1300 nm, i.e. reaching the telecom O-band. At low substrate growth temperatures (LT), 250°C, and under the same "Arsenic free" growth condition an InAs/GaAs superlattice structure without the formation of QD was successfully grown with up to 2.9MLs of InAs, which was not achievable under NG conditions. Both samples showed a noticeable photocurrent when illuminated with 1.2 - 1.3 μm lasers. Thus they can be used as photoconductive materials that can be excited with wavelength longer than that used for the well-known LT-GaAs ultrafast material and close to the telecom wavelengths of 1.3 μm for Terahertz imaging or other optoelectronic applications.

1. Introduction

Semiconductor quantum dots (QDs) have been extensively investigated in recent years due to their unique optical properties arising from their discrete energy levels, i.e., delta function-like density of states, which make them useful for optoelectronic applications. InAs QDs have attracted much attention due to the effective bandgap energy which allows emission and absorption at one of the telecom wavelengths (~1.3 μm), where silica fibers have one of their minimum losses windows and minimum dispersion. For the Stranski-Krastanow growth mode the density of the dots is between 10^9-10^{11} cm^{-2} depending on the growth parameters, and this density corresponds to a small fractional area occupancy ~ 10% [1]. This low surface coverage of QDs reduces some of the physical values that are important in optoelectronic applications, such as the low normal incident absorption and low model gain for a single dot plane lasers [2]. Therefore, multiple QD layers, stacked QD structures, are normally grown in order to enhance the performance of QD-based devices. [3-5]

It has been found that reducing the thickness allows the strain fields to act as nucleation centres for the self-assembled QDs in the subsequent grown layer. Such strain leads to a correlation between QDs to form stacks of vertically aligned QDs [1, 6] which alter the optical and electrical properties of the multilayered QD materials[1]. Moreover, Ng *et al.* [7] studied the effect of increasing the InAs' nominal thickness on the properties of the same QDs system, i.e. InAs/GaAs multiple QD. The study showed that, in addition to the formation of InAs QDs, there was a formation of volcano-like defects after increasing the thickness to 2.73 MLs. These defects, extending from the initial InAs layers to the surface of the sample, degraded the structural and optical quality of the structure. Although such extended defects can be reduced by increasing the thickness of the spacer layer [8], this study will show another method by which these defects can be eliminated without changing the thickness of the spacer layers.

In this work, two stacked binary InAs/GaAs QD structures were grown under surfactant-mediated growth conditions at two different temperatures. The growth technique will be illustrated in the following section followed by a discussion of some of the structural, optical and electrical properties of these samples and some applications they can be used to.

2. Growth technique

The two samples analysed in this study were grown using a RIBER VG V90H MBE system. Both samples consist of a 170 nm buffer layer grown at 580°C, followed by ten periods of 25 nm thickness GaAs spacers and InAs with a nominal thickness of 2.9 monolayers (MLs). The growth rate of GaAs and InAs was 1 $ML.sec^{-1}$ and 0.033 $ML.sec^{-1}$ respectively. The only difference between the samples was the growth temperature of the InAs/GaAs layers which was 450°C for sample no. 1949 and low temperature LT, i.e. 250°C, for sample no. 1753. The InAs layers of both samples were grown using only the In beam while the As source was closed. The group V elements were obtained under such a "surfactant growth technique" from the residual As in the growth chamber.

3. Results

In comparison to the normal growth condition (NGC) technique, the surfactant growth condition (SGC) mode resulted in an improvement in the quality of the sample structure. This can be seen from the double crystal x-ray diffraction (DCXRD) curves presented in Figure 1. Such quality improvement is apparent from the sharpness of the satellite peaks with FWHM of 69.5 arcsec for the zeroth order peak as can be seen from the rocking curve of the VMBE#1949 sample in comparison with the broader peaks with FWHM of around 105 arcsec for the zeroth order DCXRD peak of the structure grown under NGC [7]. In addition, there was no presence of peaks associated with the satellite peaks in the DCXRD spectrum of the SGC sample unlike the very weak second periodic peaks in the spectrum of the NGC sample which is largely correlated with the formation of volcano-like defects in such NGC structures, as reported previously [7].

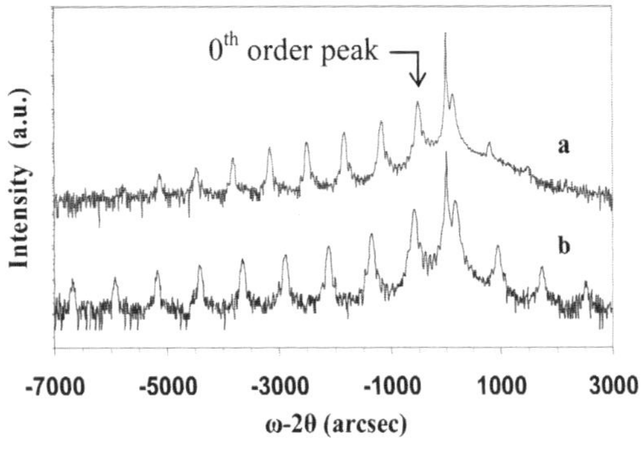

Figure 1. DCXRD rocking curves of the samples grown at: (a) 450°C and (b) 250°C.

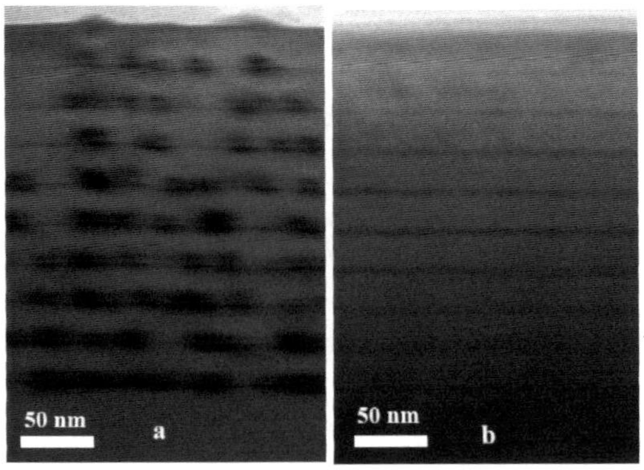

Figure 2. TEM Images of the samples grown at: (a) 450°C and (b) 250°C.

In addition to the sharp satellite peaks, the excellent quality of the samples can be seen from the bright field TEM images which show the formation of InAs QDs in sample no. 1949, Figure 2a. As can be noticed from the strained regions around the InAs QDs, the average width of the QDs is less than 20nm which is smaller than that of the 2.73MLs-InAs/GaAs QDs grown under NGC by Ng et al. where the width of QDs appeared to be above 50nm. This result shows the strong effect of the As pressure reduction during the growth of the QDs and such size reduction can be the reason for the absence of the volcano-like defects in the SGC sample as this reduces the extended strain between the adjacent InAs layers which are thought to be the reason for the defect formation.

As shown in Figure 2b, there was no formation of QDs in the LT grown sample, no. 1753, which can be due to the lower migration length of the adatoms at LT which leads to the formation of a superlattice structure instead of QDs. Thus, by controlling the amount of arsenic at low temperatures as occurs in the SGC mode, the critical thickness of InAs can be extended well beyond the 1.7 ML limit seen at higher growth temperatures. It should be mentioned here that attempts to achieve such a thickness of InAs layers, 2.9 MLs, under normal growth conditions were not successful as the InAs layers turned into a halo–like structure within ~1.5 ML at low substrate growth temperature.

Room temperature photoluminescence (PL) measurements carried out using 532 nm excitation wavelength showed clear PL emission from the QD sample, no. 1949, at around 1246 nm, i.e. close to the 1.3 um telecom O-band, with FWHM of around 50 nm, indicating a reasonable uniformity in the size of the QDs (see Figure 3). In addition, there was another peak at about 1180 nm having an intensity almost 10 times less than the intensity of the main PL peak and with about 90 nm FWHM. This peak can be related to a transition between exited states, whereas the main peak is a result of ground state e-h recombination in the QDs. On the other hand and as expected, there was no noticeable emission from the LT grown sample, no. 1753, which can be due to the point defects created during LT growth acting as nonradiative recombination centres.

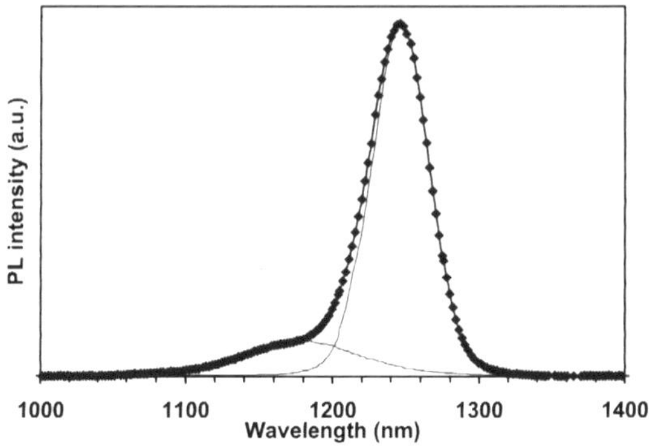

Figure 3. PL spectrum of sample no 1949 showing the Gaussian fits for the two PL peaks at 1180 nm and 1246 nm.

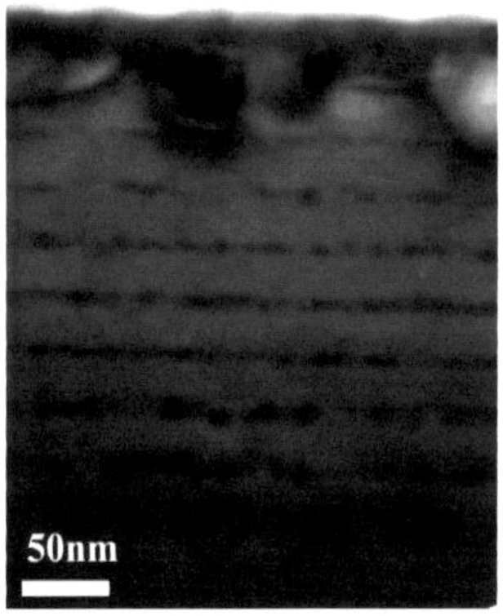

Figure 4. The formation of As nano-particles in sample no. 1753 after post-growth annealing at 550°C for 10 minutes.

Post-growth annealing on the LT grown sample showed no improvement in its optical quality. In addition the sample maintained its superlattice structure after annealing with no changes towards the QD morphology. However, the excess arsenic incorporated in the material due to the LT growth started to interact with the InAs layers at annealing temperature above 400°C as such a process enhances the point defects to form As nanoparticles very close to the InAs layers which consequently alter the morphology of these layers as shown in Figure 4. The size of these As clusters increased from an average diameter of 3 nm after 450°C to 13 nm after 550°C annealing temperatures; such interaction between the As clusters and the InAs layers were studied in more details elsewhere. [9].

Due to the presence of the QDs in the higher temperature grown sample and the As defects in the LT sample which can reduce the carrier lifetime, these samples can be used as ultrafast photoconductive PC materials for generating and detecting terahertz radiation using photomixing or broadband pulsed techniques. Both techniques required PC materials with an ultrashort carrier lifetime for fast response and a very high resistance to minimize the dark current in the THz emitters and detectors. LT GaAs is one of the well known materials that are used for this purpose; however, its band gap prevents the use of telecom lasers for excitation. The two samples studied in this work can be used as alternatives due to the narrower bandgap material imbedded in the structure, i.e. InAs. In order to measure the resistance and the responsivity of the samples to the 1.3μm excitation, PC devices have been fabricated by depositing bow-tie shape metal contacts with a gap length of 5 μm. Figure 5 shows the dark resistance of the sample as a function of the annealing temperature. Sample no 1949 shows the high resistance caused by the QDs acting as traps for the carriers, but annealing showed no significant effect which was expected as the sample was grown at a relatively high temperature. In contrast, the resistance of sample no. 1753 was very low for an unannealed sample due to the hopping conduction between midgap states created by As point defects. The reduction in the point defects' density after annealing, in addition to the formation of As clusters which form depletion regions around themselves acting as buried Schottky barriers that lead to a dramatic increase in the resistance. Both samples showed a resistance as high as that reported for LT-GaAs material, that is, the well-known material used for THz devices [10].

Photocurrent measurements were performed with a tunable laser (spot radius 6 μm) and fairly high photocurrent peaks around the ground state transition peak were observed. Figure 6 shows the current responsivity as a function of the excitation wavelength at 20 V bias. The photocurrent measured from the unannealed LT grown sample was smaller than the one from unannealed QDs sample which is likely due to the shorter carrier lifetime in the LT material. The photocurrent peaks were more pronounced at high bias voltages.

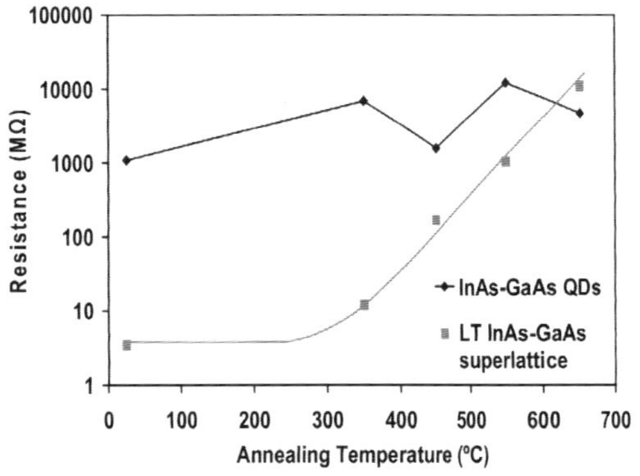

Figure 5. Dark resistance as a function of the annealing temperature. The gap length of the metal contacts was 5 μm.

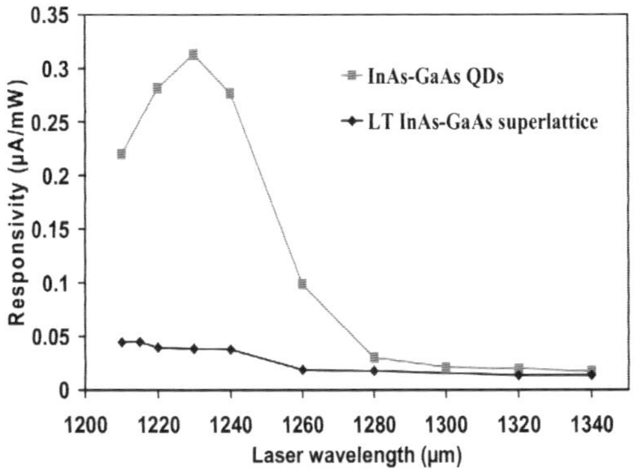

Figure 6. Current responsivity as a function of the excitation wavelength, both samples are unannealed, gap length 5 μm, bias 20 V.

4. Conclusion

Structural, optical and electrical properties of novel normal and LT 10-layer InAs/GaAs surfactant grown materials were investigated. The results presented here showed that the use of SGC technique to grow stacked 2.9ML InAs-GaAs quantum dot structures at normal growth temperature eliminated the formation of extended (volcano-like) defects that were present in NGC samples while using the same technique for LT growth resulted in the formation of up to 2.9ML-thick InAs layers without quantum dot formation thus extending greatly the critical thickness of InAs at these very low growth temperatures.

The dependences on exciting wavelength, power and bias voltage were studied and the data obtained suggests that they can be used as photoconductive materials for Terahertz imaging or other optoelectronic applications. Unlike LT-GaAs, these samples can be excited at wavelength above 870nm and close to the telecom wavelengths of 1.3 μm and compared to other lower bandgap materials used for THz applications, e.g. LT InGaAs [11,12], the InAs/GaAs superlattice and QDs structures have the additional advantage of a high resistivity of the material. Further investigations of these samples such as lifetime measurements are on-going

The authors from The University of Strathclyde acknowledge the financial support from EPSRC, Project No. EP/E025021.

References

[1] Migliorato M, Wilson L, Mowbray D, Skolnick M, Al-Khafaji M, Cullis A and Hopkinson M 2001 *J. Appl. Phys.* **90** 6374-8
[2] Fry P W, Harris L, Parnell S R, Finley J, Ashmore A D, Mowbray D J, Skolnick M S, Hopkinson M, Hill G, and Clark J C 2000 *J. Appl. Phys.* **87** 615-7
[3] Schmidt O G, Kirstaedter N, Ledentsov N N, Mao M H, Bimberg D, Ustinov V M, Egorov A Y, Zhukov A E, Maximov M V, Kopev P S, and Alferov Z I 1996 *Electron Lett.* **32** 1302-4
[4] Heinrichsdorff F, Mao M, Kirstaedter N, Krost A, Bimberg D, Kosogov AO and Werner P 1997 *Appl. Phys. Lett.* **71** 22-4
[5] Tatebayashi J, Hatori N, Ishida M, Ebe H, Sugawara M, Arakawa Y, Sudo H and Kuramata A 2005 *Appl. Phys. Lett.* **86** 053107-9
[6] Henini M *Microelectr. J.* 2003 **34** 333-6
[7] Ng J, Bangert U and Missous M *Semicond. Sci. Tech.* 2007 **22** 80-5
[8] Ng J and Missous M *Microelectr. J.* 2006 **37** 1446-50
[9] Alduraibi M, Mitchell C, Chakraborty S and Missous M *Microelectr. J.* 2009 **40** 550-3
[10] Gregory I S, Baker C, Tribe W R, Evans M J, H. Beere E, Linfield E H, Davies A G and Missous M 2003 *App. Phys. Lett.* **83** 4199-201
[11] Baker C, Gregory I, Evans M, Tribe R, Linfield EH and Missous M *Opt. Express* 2005 **13** 9639-44
[12] Takazato A, Kamakura M, Matsui T, Kitagawa J and Kadoya Y Appl. Phys. Lett. 2007 **91** 011102

A Novel Route to Zeolite-Templated ZnO Quantum Wires

G.J. Baker, F.F. Ouali and G.B. Hix

School of Science and Technology, Nottingham Trent University, Clifton Lane, Nottingham, NG11 8NS

fouzia.ouali@ntu.ac.uk

gary.hix@ntu.ac.uk

Abstract. ZnO nanoparticles were synthesised within the pores of the zeolite ferrierite (FER) by a novel molten salt impregnation method and compared to a traditional wet impregnation method. The structure and optical properties were characterised by powder XRD, photoluminescence and UV-visible diffuse reflectance spectroscopy. The XRD results reveal a distinct difference in the effect of the impregnation method on the zeolite framework. The optical data show blue-shifts in the absorption and emission profiles of encapsulated ZnO with respect to the bulk material.

1. Introduction

Wide band gap semiconducting materials are of great interest to photonic materials research because of their potential applications as blue/UV emitters, solar cells, etc. Over the past decade there has been intensive research into the growth and synthesis of nanostructured, semiconducting materials, such as quantum dots, wires and nanorods, with the promise of altered and improved optical properties as a result of quantum confinement effects. [1]

ZnO; a wide direct band-gap II-VI semiconductor with a bulk band gap of 3.37 eV at 298 K and a relatively high exciton binding energy of 60 meV[2] that allows significant exciton emission at room temperature, has been studied in this work. Previous attempts at nanoscale production of ZnO particles have utilized methods such as colloidal precipitation[3] and metal organic chemical vapour deposition (MOCVD) [4], which can lead to aggregation of ZnO particles. From 2006 Su *et al.*[5] investigated a method of nanoscale ZnO production that aimed to resolve this problem by wet impregnation of nano- and mesoporous silica and zeolite templates, which has been followed by several other groups. [6, 7]

The advantages of using aluminosilicate zeolites as templates for optoelectronic materials are the relative rigidity of the frameworks, their chemical and thermal stability, pore size, shape and distribution regularity and the vast choice of available framework geometries. The typical pore dimensions of zeolites range from ~0.2 nm to ~0.8 nm and they are generally stable to temperatures of approx. 800 °C. Su *et al* studied the use of both ZSM-5 and faujasite (FAU) to encapsulate ZnO. ZSM-5 consists of long channels with diameters of 0.55 nm that run the length of the zeolite particles and FAU contains well defined, 0-dimensional cages with diameters of 1.3 nm, i.e.; quantum wire and quantum dot templates respectively. The zeolite investigated in this work is ferrierite (FER); an aluminosilicate zeolite with a 2D pore system of intersecting perpendicular channels with diameters of 0.54 nm and 0.48 nm that run the length of the particle.

This work investigates a novel method of zeolite impregnation and the effect of impregnation method on the optical properties of ZnO templated by FER with the aim of altering the band structure of ZnO as a result of quantum confinement.

2. Experimental

The FER used in this work was supplied by Zeolyst International in the NH_4^+ form with Si:Al ratio of 20. The ammonium zeolite was converted to the protonated form (HFER) by calcining at 400 °C in air overnight. This was then dried prior to being subjected to the two impregnation treatments.

Incorporation of ZnO via the molten salt method was achieved by grinding the dried FER together with $Zn(NO_3)_2.6H_2O$ (melting point = 36 °C) in a 1:10 molar ratio and heating to 110 °C under nitrogen for several hours. The cooled sample was then washed with a small amount of distilled water (~20 cm^3) and filtered under vacuum to remove any surface $Zn(NO_3)_2.6H_2O$ not absorbed by the zeolite. Oxidised samples are hereafter denoted FER/ZnO-M

Wet impregnation was carried out by adding a known amount of FER to an aqueous solution of $Zn(NO_3)_2$ and stirring for several hours with gentle heating until the solvent evaporated. In using this method a precise Zn loading of the zeolite was achieved. Zn loadings of 5 – 50 wt% were used. Oxidised samples are hereafter denoted FER/ZnO-Wx%, where x is the Zn wt% loading.

The dried FER/$Zn(NO_3)_2$ samples were then heated to 800 °C in air at a ramp rate of 1 °C min^{-1}, held at 800 °C for 5 hours and cooled at 1 °C min^{-1} to produce ZnO nanoparticles encapsulated within the pores of the host zeolite.

3. Characterisation

Powder X-ray diffraction (XRD) patterns were recorded at room temperature using a Philips X'Pert Pro Diffractometer with Ge-monochromated Cu-Kα_1 radiation with a wavelength of 1.54056 Å. Data were collected over the range 3 – 60 ° in 2θ with a step size of 0.02 ° and were analysed using Rietveld refinement with Topas Academic. [8] UV-visible diffuse-reflectance spectra were recorded using a Jasco V-630 equipped with a 60 mm integrating sphere. Data were collected over a range of 200-1000 nm at a scan rate of 400 nm min^{-1}. Photoluminescence (PL) data were recorded at room temperature using a Nd:YAG laser at 255 nm as the excitation source with an optical fibre to transmit the luminescence signal to a spectrometer (S2000, Ocean Optics Inc (OOI)) and the acquired spectrum was interpreted *via* OOI software.

4. Results and Discussion

4.1. Powder XRD

Figure 1 shows powder XRD of ZnO (Sigma Aldrich, ≥ 99.0%), HFER, FER/ZnO-M and FER/ZnO-W10%.

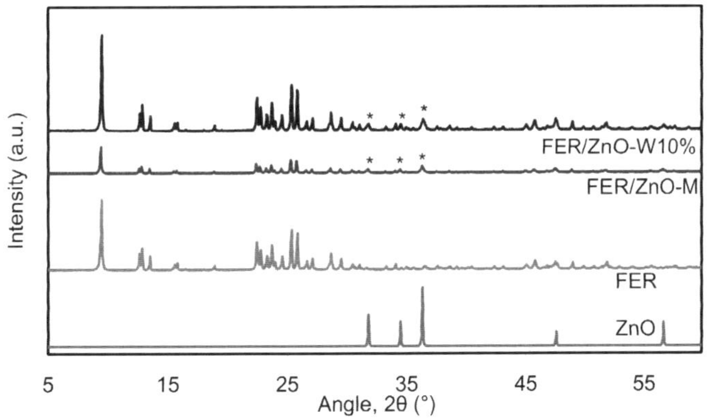

Figure 1: XRD patterns with ZnO reflections marked with *.

It is evident from Figure 1 that during both the molten salt and solution impregnation and subsequent oxidation processes the zeolite retains its structure. A good Rietveld refinement profile fit (R < 20) is achieved with only 1.57 % ZnO content in FER/ZnO-M, suggesting the absence of large, extra-zeolite ZnO crystallites. The profile of FER/ZnO-W10% shows 3 % ZnO, which is more prevalent than in FER/ZnO-M, at 2θ = 31.8, 34.5 and 36 ° (*). This is attributed to surface ZnO as the wet impregnation process relies on diffusion in order to distribute $Zn(NO_3)_2$ throughout the pore system, whereas the molten salt process relies on capillary forces.

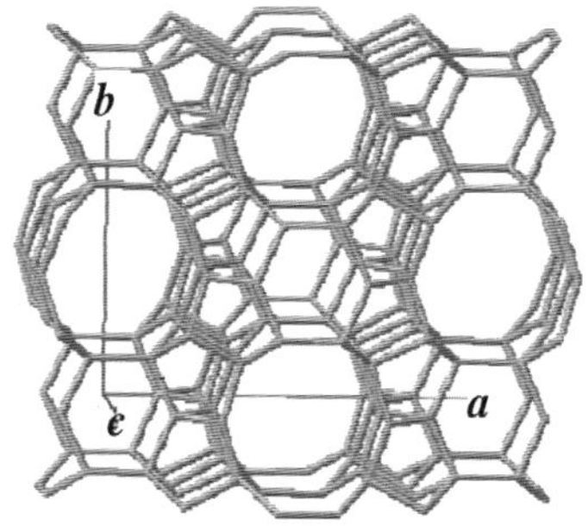

Figure 2: Zeolite FER viewed along [001]. [9]

What is less noticeable from Figure 1 is that the unit cell parameters for each FER sample are slightly different. By performing Rietveld refinement and including both FER and ZnO phases in the model it is found that although the zeolite structure is retained, the unit cell for FER is slightly larger along all 3 axes for FER/ZnO-M than for FER/ZnO-W5% to -W50% by 0.004 Å along *a* to 0.02 Å along *c*. When compared with FER/ZnO-W0%, which has smaller *a* and *c* cell parameters (see Figure 2), one can see that the presence of $Zn(NO_3)_2$ within the pores has the effect of supporting the structure during the high temperature oxidation, which causes contraction of the framework. The b parameter is larger when x < 10 % than when x > 10 %, but is larger still in FER/ZnO-M. This phenomenon has also been consistently recorded in work with other zeolites, such as ZSM-5 and faujasite, and other guest species, such as GaN and Zn_3N_2 and will be published elsewhere at a later date.

4.2. Optical Properties

The PL emission and UV-visible diffuse-reflectance absorption spectra are presented in Figure 3.

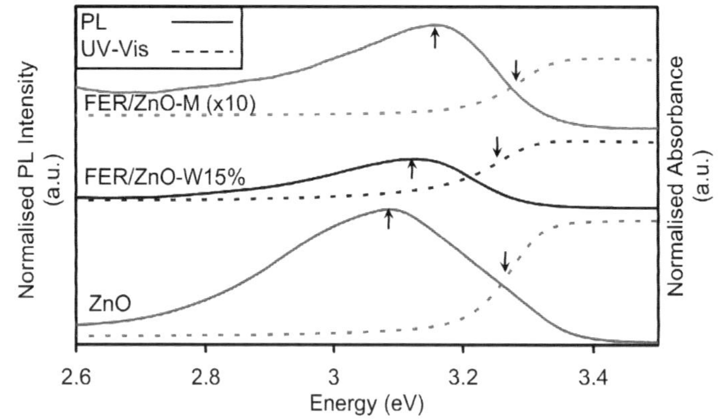

Figure 3: PL and diffuse-reflectance spectra. Arrows mark E_{PL} and E_{Abs}.

As is clear, the emission wavelength of ZnO (Sigma Aldrich, ≥ 99%) is blue shifted upon zeolite inclusion and is dependant on the impregnation process. There is a shift in PL energy from 3.084 eV for bulk ZnO to 3.125 eV for FER/ZnO-W15% and to 3.158 eV for FER/ZnO-M.

By applying the Brus equation[10], the measured shift in the band gap can be compared to the expected sizes of the zeolite-templated nanoparticles (i.e. comparable to the pore sizes):

$$\Delta E = \frac{h^2}{8\mu R^2} - \frac{1.8e^2}{4\pi\varepsilon_0\varepsilon_\infty R}$$

where μ is the exciton effective mass (0.217 m_e for ZnO[11]), R is the particle diameter and ε_∞ is the high-frequency dielectric constant (3.7). By applying this equation to the data presented in this work we obtain estimated ZnO particle sizes of 2.2 nm for FER/ZnO-W15% and 2.05 nm for FER/ZnO-M. At first glance these figures suggest the formation of particles larger that the pore size within which ZnO is intended to be confined, but this is not the case as the Brus equation assumes spherical particles. The intention of this work is to create confined particles; resulting in the formation of nanorods with the long axis several times greater than the two short axes; therefore the Brus equation doesn't fully apply, but can be used as a fair estimation of confinement effects.

The absorption data show a slight deviation from the emission spectra, in accordance with a Stokes shift of 181 meV for bulk ZnO, 131 meV for FER/ZnO-W15% and 125 meV for FER/ZnO-M. The band gaps according to absorption data were calculated using a 1st derivative approach.

5. Conclusions

To conclude, this work presents a novel method of producing nanoscale optical materials using nanoporous zeolites as templates and a solvent-free route to impregnation of zeolites with semiconductor precursor materials. Powder XRD studies suggest the inclusion of dense guest phases within the pores that support the pore structure of the host zeolite during the subsequent relatively harsh reaction conditions. Room temperature optical measurements confirm that the band gaps of semiconducting materials are blue-shifted upon zeolite templated quantum confinement and the magnitude of Stokes shift is similar to those observed in bulk ZnO. Although the Brus equation cannot be fully relied on, it has been used to show that a degree of quantum confinement has been achieved with approx. particle sizes (averaged over all 3 axes) of 2.05 nm for FER/ZnO-M and 2.2 nm for FER/ZnO-W15%. HRTEM measurements are planned to confirm these findings.

6. Acknowledgements

Thanks are due to EPSRC for funding and Dr. Robert Ranson for assistance with PL measurements.

References

[1] M. Henini, M. Bugajski, *Microelectron. J.* , *36,* **2005**, 950-956.
[2] C. Klingshirn, R. Hauschild, H. Priller, M. Decker, J. Zeller, H. Kalt, *Superlattices and Microstructures, 38,* **2005**, 209-222.
[3] M. Wang, E. K. Na, J. S. Kim, E. J. Kim, S. H. Hahn, C. Park, K. Koo, *Mater. Lett., 61,* **2007**, 4094-4096.
[4] B. Hahn, G. Heindel, E. Pschorr-Schoberer, W. Gebhardt, *Semicond. Sci. Technol.,* **1998**, 788-791.
[5] C. Bouvy, W. Marine, R. Sporken, B. L. Su, *Chem. Phys. Lett., 428,* **2006**, 312-316.
[6] C. S. Lim, S. J. Lee, W. Oh, I. J. Kim, *Journal of Ceramic Processing Research , 10,* **2009**, 382.
[7] J. Shi, J. Chen, Z. Feng, T. Chen, X. Wang, P. Ying, C. Li , *J. Phys. Chem. B, 110,* **2006**, 25618.
[8] A. Coelho, *Topas Academic,* **2007**.
[9] Ch. Baerlocher and L. B. McCusker, *Database of Zeolite Structures: www.iza-structure.org/databases.*
[10] L. E. Brus, *J. Chem. Phys , 80,* **1984**, 4403-4409.
[11] W. R. L. Lambrecht, A. V. Rodina, S. Limpijumnong, B. Segall, B. K. Meyer, *Phys. Rev. B , 65,* **2002**, 075207.

Journal of Physics: Conference Series **245** (2010) 012074 doi:10.1088/1742-6596/245/1/012074

Low Density Metamorphic Quantum Dot structures with emission in the 1.3 - 1.55 μm window

L Seravalli, G Trevisi, P Frigeri and C Bocchi

IMEM-CNR Institute, Parco Area delle Scienze, 37/a, 43100 Parma, Italy

e-mail:seravall@imem.cnr.it

Abstract. We report on the molecular beam epitaxy growth of InAs/InGaAs metamorphic quantum dot structures for single-photon operation at long wavelengths. Low density of quantum dots has been achieved by depositing sub-critical coverages of InAs, while the redshift of emission was obtained by growing the nanostructures on relaxed InGaAs buffers. By optimizing the design and growth parameters, such as the InAs coverage, the post growth annealing time and the compositions of InGaAs confining layers, we were able to obtain structures with quantum dot densities of the order of the 10^8 cm^{-2} and emission in the whole range 1.3 - 1.55 μm at low temperature.

1. Introduction

Future applications in the fields of quantum cryptography and quantum communication will need efficient single photon sources emitting in the telecom windows (1.3 - 1.55 μm). Self-assembled InAs/GaAs quantum dots (QDs) [1-2] have been recognized as ideal nanostructures for such innovative devices [3-4]. However, particular growth procedures are necessary to have a QD surface density low enough to allow study of single photons (in the order of a few dots/μm^2) [5-6]. The deposition of an InAs layer with a coverage smaller (sub-critical) than the critical thickness for the two- to three- dimensional (2D–3D) growth mode followed by a growth interruption was demonstrated as being particularly useful for different systems [7-8]. Notwithstanding, low density structures with emission in the telecom windows for structures grown on GaAs have been rarely reported: while 1.3 μm emission at low temperature has been reached [5-6], 1.55 μm emission from low density QDs is still a difficult task. For high density structures, however, this objective has been reached by depositing QDs on metamorphic buffers [9-10]: thanks to the decrease of the QD strain and of band discontinuities the emission wavelength may be consistently redshifted [11]. Very recently, low density QDs on metamorphic buffers have been obtained by stopping the substrate rotation during InAs deposition: single dot emission at 1.55 μm was reported, but only in a narrow region of the sample [12].

Here we report on a new approach to obtain low density InAs QD structures grown on GaAs substrate suitable as single photon sources at 1.55 μm, relying on the deposition of sub-critical InAs coverages on relaxed InGaAs layers. The experimental work was based on the growth by Molecular Beam Epitaxy (MBE) and the characterization by Photoluminescence (PL) and Atomic Force Microscopy (AFM) of different structures with different compositions of relaxed InGaAs lower confining layers (LCLs) and upper confining layers (UCLs). The aim was not limited to the achievement of 1.55 μm emission from low density QD structures, but it included also the study of properties of the metamorphic system.

2. Experimental details

QD structures consist of: i) 100 nm of GaAs grown by MBE at 600 °C, ii) an $In_xGa_{1-x}As$ LCL of 500 nm deposited by MBE at 400 °C, iii) 5 nm of GaAs grown by ALMBE at 400 °C to smooth the surface, iv) an InAs layer of coverage θ deposited by MBE at 490 °C followed by a Post Growth Annealing (PGA) of time τ under As flux at 490 °C and v) an $In_yGa_{1-y}As$ UCL of 20 nm grown by ALMBE at 360 °C. Compositions of LCLs were x = 0.15 or x = 0.30, resulting in a mismatch between QDs and LCLs of 6.34 % and 5.19%, respectively: in the InAs/GaAs system the QD-CL mismatch is 7.16 %. These mismatch values were calculated on the basis of the Mareè theory of strain relaxation [13], experimentally confirmed by different techniques [14-15]. AFM measurements were performed on uncapped QD structures to study surface morphology and evaluate the QD density. PL characterization was performed by a 532 nm excitation at 10 K with a Fourier transform spectrometer with 1 meV resolution. Spectra were corrected for the response of the set-up.

3. Results and discussion

As reported in some works on QD growth on metamorphic buffers [16-18], the differences in mismatch, composition and surface morphology affect the evolution from the 2D growth regime to the nucleation of 3D nanosized islands: in particular the value of the critical coverage θ_c for 2D-3D transition depends on the composition x of the InGaAs buffer layer. Henceforth, it was necessary to measure the value of θ_c, to derive the sub-critical coverages to be used for low density structures. θ_c was found to be 1.8 ML and 2.2 ML for x = 0.15 and x = 0.30, respectively, by observing the transition from a RHEED streaky diffraction pattern to a spotty one. These values are larger as compared to critical coverage θ_c = 1.6 ML for the deposition of InAs on GaAs.

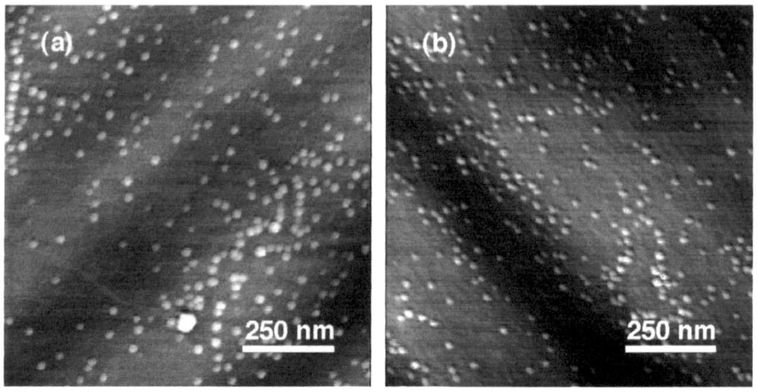

Figure 1. AFM images of the $InAs/In_{0.15}Ga_{0.85}As$ QD structures. (a) θ = 1.8 ML and τ = 0 s and (b) θ = 1.65 ML and τ = 15 s.

Figure 1 shows the AFM images of uncapped QDs grown on $In_{0.15}Ga_{0.85}As$ LCLs with (a) 1.8 ML InAs coverage and τ = 0 s and (b) 1.65 ML InAs sub-critical coverage and PGA time τ = 15 s. An important feature is evident: the undulation of the surface underlying InAs islands, that is known to affect QD nucleation and positioning [17,19]. As can be seen from Figure 1 the growth of a sub-critical InAs coverage slightly affects the QD density that decreases from 4-5 x 10^{10} cm^{-2} for the θ = 1.8 ML sample to 1-2 10^{10} cm^{-2} for the θ = 1.65 ML one. Ultra-low QD density lower than 2-3 10^8 cm^{-2} was achieved by further reducing the sub-critical coverage: AFM micrograph in Figure 2(a) shows the morphology of InAs QDs grown on $In_{0.15}Ga_{0.85}As$ LCL with θ = 1.5 ML and τ = 22 s. Similar low densities were also obtained for InAs QDs grown on $In_{0.30}Ga_{0.70}As$ LCL (Figure 2(b)). In this case the optmized values to obtain densities in the order of few 10^8 cm^{-2} were θ = 1.8 ML and τ = 300 s. It is important to note that all the density data were deduced from the statistical analysis of larger scan area AFM images not reported here.

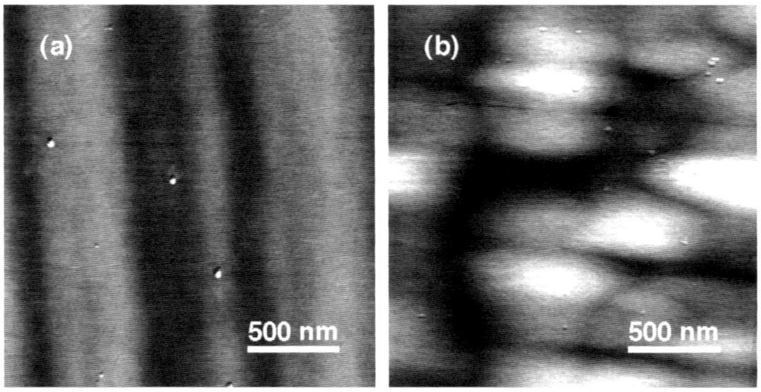

Figure 2. AFM images of (a) InAs/$In_{0.15}Ga_{0.85}As$ QDs with θ = 1.5 ML and τ = 22 s and (b) InAs/$In_{0.30}Ga_{0.70}As$ QDs with θ = 1.8 ML and τ = 300 s.

PL emission spectra of structures capped with $In_{0.15}Ga_{0.85}As$, shown in Figure 3, can be deconvolved in three components that, on the basis of the AFM studies, can be attributed to different families of QDs. The effects of different PGA times (that will not be discussed here) have been considered, resulting in a change of the density of QDs belonging to different families and in a broadening of their size distribution that, while little affecting emission energies, cause an increase of PL linewidth.

In Figure 3 we report also the PL emission spectra of structures with x = 0.30 capped with $In_{0.30}Ga_{0.70}As$: clear contributions of different QD families are not evident as in the x = 0.15 case, possibly due to increase of dishomogeneities in the dimensions of island families, as seen in AFM images, and the emission wavelength is redshifted to 1.3 μm. As widely discussed in [11], when the QD-CL mismatch is reduced, the QD strain is decreased, causing a lowering of the QD energy gap. This effect adds up to the energy decrease due to the reduction of QD-CL band discontinuities and results in a redshift of the emission energy. To extend the emission wavelength towards 1.55 μm, instead of increasing x and incur in worsening of the surface morphology, we opted to increase the UCL composition y beyond the value of 0.30, as it was done in pseudomorphic structures [5-6].

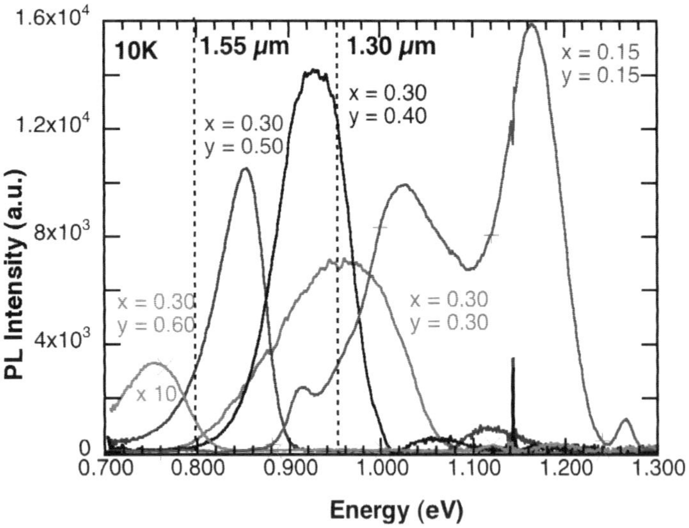

Figure 3. PL spectra of metamorphic low density QD structures at 10K for different x values of the $In_xGa_{1-x}As$ LCL and y values of the $In_yGa_{1-y}As$ UCL.

In this way the QD emission was further redshifted, as shown in Figure 3, up to and even beyond 1.55 μm, at the expenses of a reduction of the emission intensity, in particular when y = 0.60 is used. The wavelength redshift may be due to: i) the lowering of UCL-QD band discontinuities, ii) the change in QD size and/or composition due to QD-cap intermixing effects and iii) a reduction of the

QD strain. It is interesting to note how the width of the PL emission reduces when values of y larger than x are used, an observation that could hint to interaction processes between QDs and UCLs that affect the homogeneity of the QD ensemble. Finally, it can be appreciated how by changing the composition y of the UCLs the emission wavelength can be tuned in the whole 1.3 - 1.6 μm range, while keeping the same QD growth scheme and the same metamorphic LCL parameters.

4. Conclusions

In conclusions, we grew by MBE low density QD structures on GaAs with emission tunable at 1.3 and 1.55 μm, by depositing sub-critical InAs coverages on $In_xGa_{1-x}As$ metamorphic LCLs. The relevant parameters for this design have been studied, namely the sub-critical coverage and the PGA time: the peculiarities of the metamorphic system were put in evidence, in particular the different 2D-3D critical coverages and the undulated surface morphology. Values of relevant parameters to obtain QD density in the order of few QDs / μm^2 were found for x = 0.15 and 0.30. We showed how, by choosing suitable values of x and y, the PL emission wavelength can be tuned in the whole range 1.3 - 1.6 μm.

In particular, for x = 0.30 and y = 0.60 an emission at 1.64 μm was reached, a very high value for nanostructures grown on GaAs. We believe these results demonstrate that the metamorphic approach can be very valuable to obtain nanostructures for single photon sources. Moreover, this design has the potentiality for tuning emission and other single dot properties, by applying the so-called Quantum Dot Strain Engineering approach to low density structures.

References

[1] Bimberg D 2008 *Semiconductor Nanostructures* (Berlin: Springer)
[2] Henini M 2008 *Handbook of self assembled semiconductor nanostructures for novel devices in photonics and electronics* (Amsterdam;London: Elsevier)
[3] Yuan Z, Kardynal B, Stevenson R, Shields A, Lobo C, Cooper K, Beattie N, Ritchie D and Pepper M 2002 *Science* **295** 102-105
[4] Stevenson RM, Young RJ, Atkinson P, Cooper K, Ritchie DA and Shields AJ 2006 *Nature* **439** 179–182
[5] Alloing B, Zinoni C, Li L, Fiore A and Patriarche G 2007 *J. Appl. Phys.* **101** 024918
[6] Trevisi G, Seravalli L, Frigeri P and Franchi S 2009 *Nanotechnology* **20** 415607
[7] Song H, Usuki T, Nakata Y, Yokoyama N, Sasakura H and Muto S 2006 *Phys. Rev. B* **73** 115327
[8] Li L, Chauvin N, Patriarche G, Alloing B and Fiore A 2008 *J. Appl. Phys.* **104** 083508
[9] Balakrishnan G, Huang S, Rotter T, Stintz A, Dawson L, Malloy K, Xu H and Huffaker D 2004 *Appl. Phys. Lett.* **84** 2058
[10] Seravalli L, Frigeri P, Trevisi G and Franchi S 2008 *Appl Phys Lett* **92** 213104
[11] Seravalli L, Minelli M, Frigeri P, Franchi S, Guizzetti G, Patrini M, Ciabattoni T and Geddo M 2007 *J Appl Phys* **101** 024313
[12] Semenova ES, Hostein R, Patriarche G, Mauguin O, Largeau L, Robert-Philip I, Beveratos A and Lemaıtre A 2008 *J Appl Phys* **103** 103533
[13] Maree P, Barbour J, Vanderveen J, Kavanagh K, Bullelieuwma C And Viegers M 1987 *J. Appl. Phys.* **62** 4413-4420
[14] Bellani V et al. 2007 *Eur. Phys. J. B* **56** 217-222
[15] Geddo M, Guizzetti G, Patrini M, Ciabattoni T, Frigeri P, Franchi S and Seravalli L 2005 *Appl Phys Lett* **87** 263120
[16] Kryzhanovskaya N et al. 2004 *Semiconductors* **38** 833-836
[17] Kanjanachuchai S, Maitreeboriraks M, Thet C, Limwongse T and Panyakeow S 2009 *Microelectron. Eng.* **86** 844-849
[18] Seravalli L, Frigeri P, Nasi L, Trevisi G and Bocchi C *submitted to Nanotechnology*
[19] Mano T, Notzel R, Hamhuis G, Eijkemans T and Wolter J 2002 *Appl. Phys. Lett.* **81** 1705-1707

Formation of highly anisotropic GaAs quantum dots on GaAs(001) substrates

M Jo[1], T Mano[1] and K Sakoda[1,2]

[1]Quantum Dot Research Center, National Institute for Materials Science, 1-2-1, Sengen, Tsukuba, Ibaraki 305-0047, Japan
[2]Graduate School of Pure and Applied Sciences, University of Tsukuba, 1-1-1 Tennodai, Tsukuba, Ibaraki 305-8577, Japan

E-mail: JO.Masafumi@nims.go.jp

Abstract. We report on the formation of highly anisotropic GaAs quantum dots on GaAs (001) substrates using droplet epitaxy. Uncapped annealing of the dots brings significant elongation along the $[\bar{1}10]$ azimuth, resulting in highly anisotropic GaAs quantum dots. Effects of structural anisotropy of the quantum dots are studied in terms of polarized emission.

1. Introduction

Nanostructure self-assembly has been extensively studied from fundamental as well as technical points of view [1]. In order to derive the desired physical properties it is important to control the morphology of the nanostructure. For example, highly symmetric quantum dots (QDs) are believed to play a crucial role in suppressing the exciton fine-structure splitting that is a major obstacle to producing entangled photon pairs from the QDs [2, 3]. In contrast elongated QDs or quantum dash is another interesting nanostructure which, due to the preferential elongation, allows the realization of polarization-sensitive functionalities such as vertical-cavity surface-emitting lasers with distinct polarization states [4, 5].

So far, growth of quantum dashes has been demonstrated in lattice-mismatched systems such as In(Ga)As/InP [6-8] and InGaAs/GaAs [9], in which the dots spontaneously elongate in order to minimize the total surface/interface energy and elastic strain energy of the epitaxial islands [10]. In such strained systems, however, pure geometric confinement is greatly modified by strain-induced potential. This complicates a straightforward clarification of the effects of structural anisotropy on the physical properties of the nanostructures. As an alternative self-assembling method, droplet epitaxy (DE) has been developed in lattice-matched systems to produce various nanostructures [11-15]. In self-assembled GaAs/AlGaAs nanostructure grown by DE, size and shape of the nanocrystals are dependent on the As pressure as well as on the temperature in crystallizing Ga droplets. For example, dots are formed under intense As flux while ring-like shapes are obtained at relatively low As supply [14]. In both cases, however, the shape of the nanocrystals grown at low temperature has so far been isotropic in plane.

Here, we report on the formation of highly anisotropic GaAs QDs grown on GaAs (001) substrates using DE. Uncapped annealing induces the QDs elongation along the $[\bar{1}10]$ azimuth, which is ruled by the anisotropy in surface step energy.

2. Experimental

Samples were grown on semi-insulating GaAs (001) substrates by molecular-beam epitaxy. After the growth of a 100-nm $Al_{0.35}Ga_{0.65}As$ buffer layer at 580°C, 1.5 monolayers (MLs) of Ga were supplied to form Ga droplets with a flux of 0.1 ML/s at 300°C. Next, the droplets were crystallized into GaAs under As_4 flux irradiation of 2.5×10^{-4} Torr at 200°C. Then, the samples were annealed for 10 min under As_4 supply (1.0×10^{-5} Torr) without capping, at temperatures ranging from 350°C to 500°C. During the annealing stage, the surface states were monitored by reflection high-energy electron diffraction (RHEED). Atomic force microscopy (AFM) was used to characterize the surface morphology of the open dots. Optical measurement was performed on the samples capped with a 50-nm $Al_{0.35}Ga_{0.65}As$ layer followed by a 10-nm GaAs protective coat. The $Al_{0.35}Ga_{0.65}As$ capping layer was grown in two stages. First, a 40-nm-thick $Al_{0.35}Ga_{0.65}As$ layer was grown at the same annealing temperature of the GaAs QDs, and then the rest of the $Al_{0.35}Ga_{0.65}As$ layer was deposited at 580°C. A rapid thermal annealing process (800°C, 4 min.) was maintained for all capped samples to improve the optical quality. Photoluminescence (PL) spectra of the samples were taken at 6 K, using the 532-nm line of a frequency-doubled Nd:YAG laser. The PL signals were dispersed by a monochromator and detected by a cooled Si charge-coupled device array.

3. Results and Discussion

Figure 1 shows AFM images of the uncovered sample surfaces annealed at various temperatures: (a) 350, (b) 400, (c) 450, (d) 475, and (e) 500°C. The base length along [110] and $[\bar{1}10]$ of the dots obtained from the AFM scans are plotted in Fig. 2.

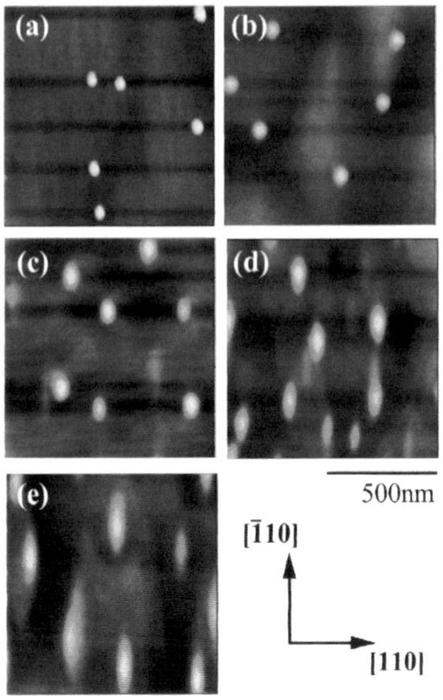

Figure 1. 1×1 μm^2 AFM images of GaAs QDs annealed at different temperatures: (a) 350, (b) 400, (c) 450, (d) 475, and (e) 500°C. Colour scale (*z* height) differs for each scan: 20 nm for (a) and (b), 8 nm for (c) and (d), and 3 nm for (e).

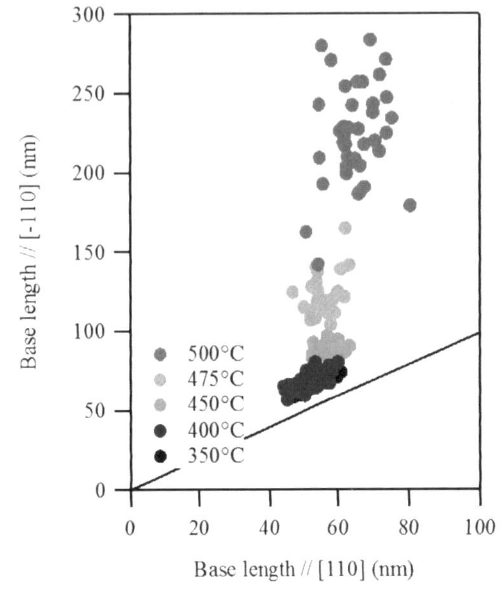

Figure 2. QD base length along $[\bar{1}10]$ direction plotted against the base along [110] direction for different annealing temperatures. The line denotes an aspect ratio of 1.0. Note that data at 400°C almost overlap those at 350°C.

As seen in figure 1, the dots become laterally anisotropic with increasing the annealing temperature, while the dot height decreases monotonically. At temperatures lower than 400°C, the dot shape is almost isotropic in plane, reflecting the isotropy of seeding Ga droplets. Then, the base length parallel to $[\bar{1}10]$ starts to increase at temperatures higher than 450°C, whereas the other base length along [110] remains nearly constant over the entire temperature range, which is clearly seen in figure 2. Finally, at 500°C annealing, the mean aspect ratio of the long axis to the short axis reaches ~ 3.4.

Highly anisotropic QDs are formed as a result of the rearrangement of constituent atoms on an anisotropic GaAs(001)-(2×4) surface at high-temperature annealing [16,17]. Figure 3 compares RHEED patterns of the surface after annealing at (a) 400 and (b) 450°C. At 400°C, the surface shows a clear *c*(4×4) reconstruction with a spotty feature due to the QDs. Then the reconstruction changes to a complex structure at 450°C, which is observed at the transition regime from *c*(4×4) to (2×4). This is consistent with the temperature at which the elongation of the QDs starts to occur, as confirmed by AFM observations.

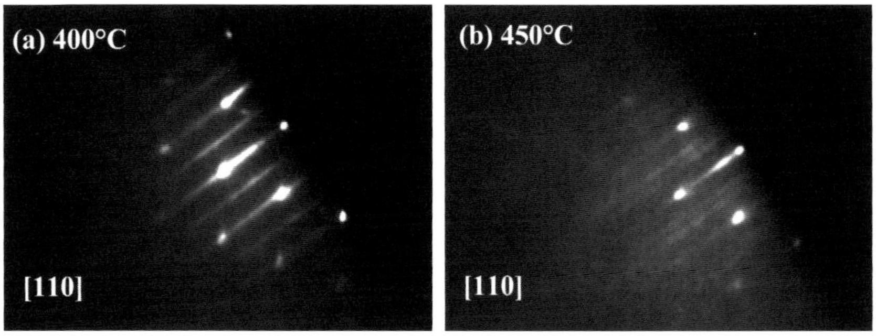

Figure 3. RHEED patterns of the annealed surface (a) 400 and (b) 450°C, taken along the [110] azimuth.

The shape of the GaAs QDs is governed both by kinetic and thermodynamic aspects. Kinetically, different surface diffusion constants and sticking coefficients will produce the anisotropic islands. Thermodynamically, the islands become elongated along low-energy steps to minimize the total free energy. In the GaAs(001)-(2×4) surface, the energy of the A-step, which is parallel to the As-As dimers, is lower than that of the B-step, which is perpendicular to the dimers [18]. In addition, Ga atoms are incorporated more easily at the B-steps than at the A-steps [19]. Thus, the GaAs islands on the GaAs(001)-(2×4) surface elongate along $[\bar{1}10]$.

As a result of the shape transition, emission from the QDs becomes polarized in plane. Figure 4 depicts PL spectra of a 500°C-annealed sample with the polarization field along [110] and $[\bar{1}10]$. The PL intensity is larger in $[\bar{1}10]$ than in [110], illustrating the lateral confinement anisotropy. In the inset, the degree of polarization (DOP), defined as $\mathrm{DOP} = (I_{[\bar{1}10]} - I_{[110]})/(I_{[\bar{1}10]} + I_{[110]})$, where $I_{[\bar{1}10]([110])}$ is the PL intensity parallel to the $[\bar{1}10]$ ([110]) direction, is plotted as a function of the annealing temperature. As can be seen, the DOP increases in general with increasing shape anisotropy due to heavy- and light-hole mixing [20]. Considering the relatively large size of the QDs grown in this study, DOP of 10% in the 500°C-annealed sample is within a reasonable range compared to that of GaAs/AlAs quantum wires on (775)B [21]. By narrowing the width of the QDs, further enhancement of DOP can be expected.

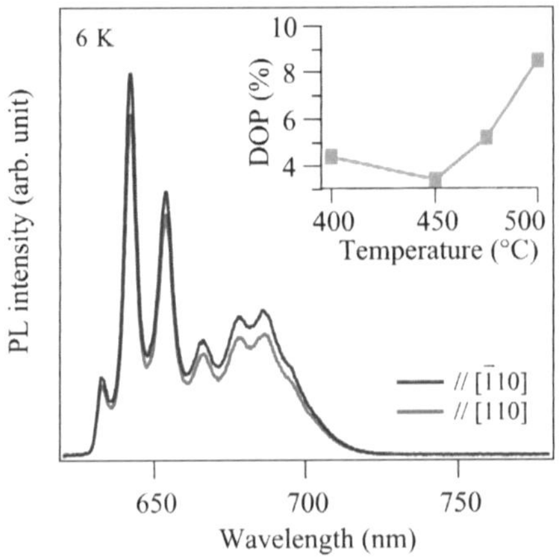

Figure 4. Polarization-resolved PL at 6 K from the sample annealed at 500°C. Emission along the long axis of the QD ([$\bar{1}$10]) is stronger than that along the short axis ([110]). Inset: $DOP = (I_{[\bar{1}10]} - I_{[110]})/(I_{[\bar{1}10]} + I_{[110]})$ is plotted as a function of the annealing temperature.

4. Conclusion

We have reported the self-assembly of highly anisotropic GaAs QDs on GaAs (001) substrates using droplet epitaxy. Thermal annealing at higher temperature than 450°C causes significant dot elongation along [$\bar{1}$10], accompanied by a decrease in dot height. Change in the surface reconstruction from *c*(4×4) to (2×4) is considered to be responsible for this shape transition. Emission from the QDs is [1 10]-polarized as a result of structural anisotropy.

References

[1] Kiravittaya S, Rastelli A and Schmidt O G 2009 *Rep. Prog. Phys.* **72** 046502
[2] Bayer M *et al.* 2002 *Phys. Rev. B* **65** 195315
[3] Stevenson R M *et al.* 2006 *Nature* **439** 179
[4] Ohno Y *et al.* 2004 *Vac. Sci. Technol. B* **22** 1526
[5] Lamy J M *et al.* 2009 *Appl. Phys. Lett.* **95** 011117
[6] Sauerwald A *et al.* 2005 *Appl. Phys. Lett.* **86** 253112
[7] Balakrishnan G *et al.* 2004 *Appl. Phys. Lett.* **84** 2058
[8] Tzeng T E, Chen C Y, Feng D J and Lay T S 2009 *J. Cryst. Growth* **311** 1851
[9] Ma W, Nötze, R, Schönherr H -P and Ploog K H 2001 *Appl. Phys. Lett.* **79** 4219
[10] Tersoff J and Tromp R M 1993 *Phys. Rev. Lett.* **70** 2782
[11] Koguchi N, Takahashi and Chikyow T 1991 *J. Cryst. Growth* **111** 688
[12] Koguchi N and Ishige K 1993 *Jpn. J. Appl. Phys.* Part 1 **32** 2052
[13] Watanabe K, Koguchi N and Gotoh Y 2000 *Jpn J. Appl. Phys.* Part 2 **39** L79
[14] Mano T *et al.* 2005 *Nano Lett.* **5** 425
[15] Lee J H *et al.* 2008 *Cryst. Growth Des.* **8** 1945
[16] LaBella V P *et al.* 2001 *J. Vac. Sci. Technol. B* **19** 1640
[17] Johnson M D, Leung K T, Birch A, Orr B G and Tersoff J 1996 *Surf. Sci.* **350** 254
[18] Zhang S B and Zunger A 1996 *Phys. Rev. B* **53** 1343
[19] Horikoshi Y, Yamaguchi H, Briones F and Kawashima M 1990 *J. Cryst. Growth* **105**, 326
[20] Léger Y, Besombes L, Maingault L and Mariette H 2007 *Phys. Rev. B* **76**, 045331
[21] Higashiwaki M, Yamamoto M, Shimomura S and Hiyamizu S 1997 *Appl. Phys. Lett.* **71**, 2005

Vertical stacking of InAs quantum dots for polarization-insensitive semiconductor optical amplifiers

T Inoue[1], M Asada[1], N Yasuoka[2], T Kita[1] and O Wada[1]

[1]Department of Electrical and Electronic Engineering, Kobe University, 1-1 Rokkodai, Nada, Kobe 657-8501, Japan

[2]Fujitsu Laboratories Ltd., 10-1 Morinosato-Wakamiya, Atsugi, Kanagawa 243-0197, Japan

t_inoue@person.kobe-u.ac.jp

Abstract. In this paper, we report a control of the polarization property in quantum dot semiconductor optical amplifiers (QD-SOAs) using vertically-stacked, electronically-coupled InAs/GaAs QDs grown by molecular beam epitaxy. By optimizing the number of stacked layers and intermediate GaAs thickness, the 9-stacked QDs demonstrated the polarization-insensitive operation within 1.2 dB in the 1.3-μm optical communication band. Our results demonstrate that the electronically-coupled QDs are useful to realize the polarization-insensitive QD-SOAs.

1. Introduction

The atom-like density of state in semiconductor quantum dots (QDs) is very promising for semiconductor optical amplifiers (SOAs) realizing high saturation power, multi-channel signal processing, and high-speed response based on nonlinear effects [1]. Self-assembled InAs QDs on GaAs or InP substrates have been attracting much interest because their operating wavelengths cover the 1.3- and the 1.55-μm optical communication bands. However, for the practical application of such devices, controlling the polarization is a key issue. In conventional SOAs, there has been a problem that the transverse-electric (TE) mode gain exceeds the transverse-magnetic (TM) mode gain, even when the QDs are used as the active region in SOAs. The reason for the high TE mode gain is the flattened QD structure and the strong biaxial compressive strain applied to the QD. In our previous works, we proposed artificial control of the photoluminescence (PL) polarization [2] and optical gain [3] by using closely stacked QDs called columnar QDs. Furthermore, recently, polarization-insensitive optical gain in the 1.55-μm band has been demonstrated using InP-based columnar QD-SOA [4].

On the other hand, closely stacked QDs with intermediate layers have been considered as a candidate for realizing polarization-insensitive SOA [5,6]. In these QDs, each QD layer is geometrically well separated, which results in the formation of electronically coupled states in the growth direction [7]. On the basis of this concept, we have demonstrated the TM-mode dominant emission at the 1.3 μm optical communication band by measuring edge-emitted photoluminescence (PL) [8]. In this work, we fabricated a SOA structure containing a series of vertically stacked self-assembled InAs/GaAs QDs with GaAs intermediate and examined effects of the number of the stacking layers on the polarization characteristics of electroluminescence (EL).

2. Quantum dot growth and device fabrication

Samples were grown on n-type GaAs (001) substrates by solid-source molecular-beam epitaxy. After thermally cleaning the substrates, a 150-nm n-GaAs buffer layer and 1.0-µm-thick n-$Al_{0.3}Ga_{0.7}As$ cladding layer were grown at 550°C under the As flux of 1.0 × 10^{-5} Torr. Then, 0.3-µm un-doped GaAs waveguide layer including the stacked QDs layer was grown. In the centre of the waveguide layer, vertically stacked 3-, 6- and 9- InAs QD layers with GaAs intermediate layers were embedded. The nominal thickness of deposited InAs was 2.6 MLs for the first QD layer and 1.9 MLs for the subsequent layers. The GaAs intermediate layer thickness was 4.5 nm. To stabilize the growth surface, growth interruptions of 10 s and 60 s were introduced after growing the GaAs and InAs layers, respectively. During QD growth, the substrate temperature and the As flux were set at 450°C and 5 × 10^{-6} Torr, respectively. Nominal growth rate of InAs and GaAs were 0.012 and 0.5 ML/s, respectively. Then, the substrate temperature and the As flux were changed to 500°C and 1 × 10^{-5} Torr, respectively, and a 1.0-µm-thick p-$Al_{0.3}Ga_{0.7}As$ cladding layer and a 0.15-µm-thick p-GaAs contact layer were grown.

A mesa-type SOA structures having 13-µm-wide and 1.9-mm-lengh stripe were fabricated by standard lithography and wet chemical etching. Polarized EL spectra from fabricated SOAs were measured by under dc current injection at 18°C. EL from the SOAs was collimated using an objective lens and was separated into the TE and TM components using a linear polarizer. The polarized EL signal was refocused on an optical fiber using an objective lens. The EL signal was dispersed by a monochromator and was detected by a liquid nitrogen cooled InGaAs diode array.

3. Results and discussion

Figure 1 shows a typical (-110)-cross-sectional transmission electron microscope (TEM) image of the 9-layer stacked QDs grown at the same condition except $Al_{0.3}Ga_{0.7}As$ cladding layers. Vertically-aligned QDs separated by GaAs intermediate layers are clearly seen. This feature is obviously different from columnar QDs in which stacked InAs layers form large QDs as a whole [2,3,4].

The wavelength-integrated EL intensities are shown in figure 2. The EL intensities of the TE- and TM-mode components are plotted as a function of the injected current densities. As the number of the stacked layers increases, the TM-mode EL intensity against to the TE mode one increases. This result can be explained by the electronic coupling of the stacked QDs. As the result, almost polarization-independent future is achieved in the EL of the 9-layer stacked QDs.

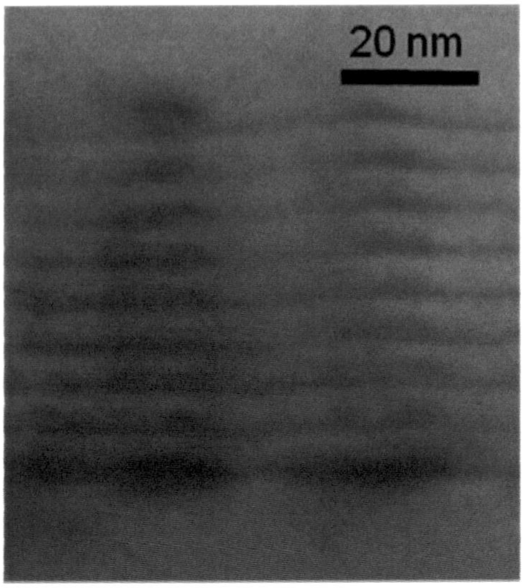

Figure 1. (-110)-cross-sectional TEM image of 9-layer stacked QDs.

Next, EL spectra from the SOAs containing the (a) 3-, (b) 6-, and (c) 9-layer stacked QDs were characterized. The upper panels in Figure 3 show linearly polarized EL spectra represented on a logarithmic scale. The injected current density was 40.5 A/cm^2. The lower panels show the polarization difference spectra in dB which means the ratio of the TE mode EL intensity to the TM mode one. Along with the polarized PL results reported in reference 8, the polarized EL spectra show enhancement of the TM-polarization component with the increase in the number of the stacking layers. The polarization differences acquired at the wavelengths of each ground-state transition indicated by arrows are 9.6, 5.8, and 1.2 dB for the 3-, 6- and 9-layer

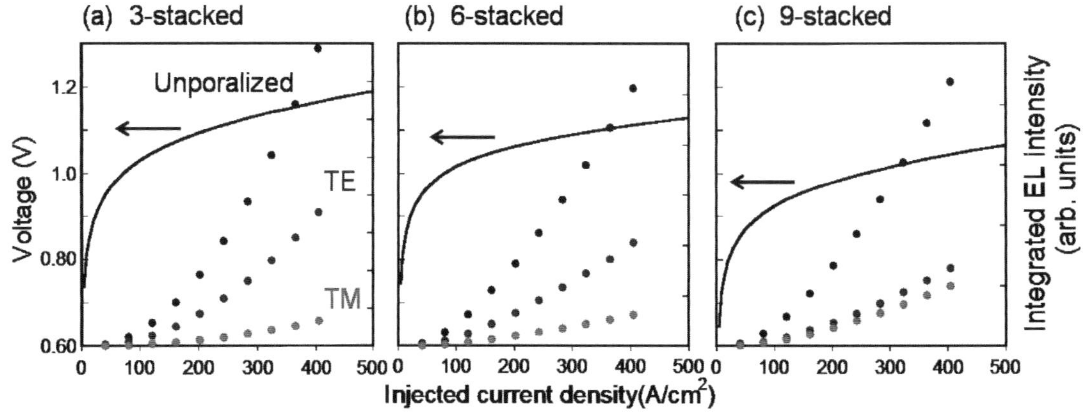

Figure 2. I-V curves (black curves) and integrated EL intensities of unpolarized (black dots) and TE-(blue dots) and TM-(red dots) mode polarized components from SOAs including (a) 3-, (b) 6- and (c) 9-stacked QDs.

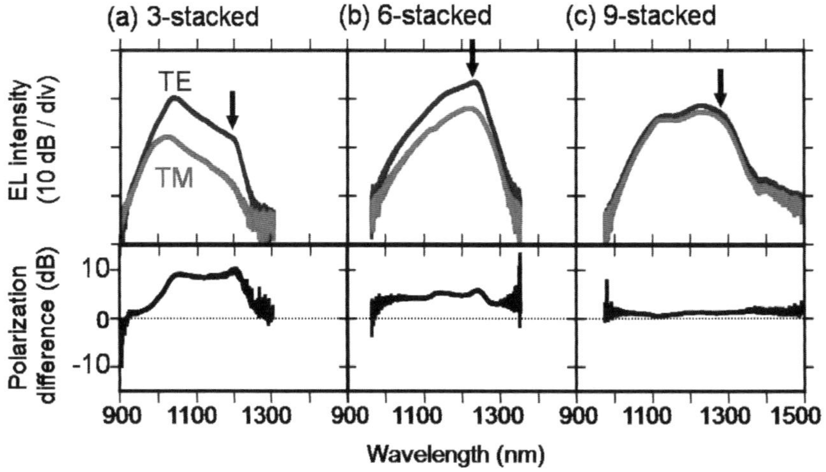

Figure 3. (Upper panels) Polarized EL spectra of the (a) 3-, (b) 6- and (c) 9-layer stacked QDs. (Lower panels) Polarization difference spectra derived from EL spectra. Polarization differences at black arrows are 9.6, 5.8 and 1.2 dB for the (a) 3-, (b) 6-, and (c) 9-layer stacked QDs.

stacked QDs, respectively. For the 9-layer stacked QDs, it is noted that broad-band and almost isotropic EL with a polarization difference of less than 1.2 dB covers the 1.3-μm optical communication band.

4. Conclusion

In conclusion, we have demonstrated the control of the polarization dependence in QD-SOAs using a series of vertically stacked self-assembled InAs/GaAs QDs with GaAs intermediate layers. The broad-band and almost isotropic EL with a polarization difference of less than 1.2 dB has been successfully observed for the fabricated SOA containing 9-layer stacked QDs.

Acknowledgement

We thank Prof. H. Yasuda of Kobe University for helping in TEM observations. We also thank Fujitsu Laboratories Ltd.'s engineers for process of SOA devices. One of the authors (T.I) is supported by Research Fellowships of the Japan Society for the Promotion of Science (JSPS) for young scientists. This work is partially supported by Grant-in-Aid for Scientific Research from Ministry of Education, Culture, Sports, Science, and Technology (MEXT) of Japan (Contracts No. 21360151 and 22241035).

References

[1] Sugawara M, Hatori N, Akiyama T, Nakata Y and Ishikawa H 2001 *Jpn J Appl Phys* **40** L488
[2] Kita T, Wada O, Ebe H, Nakata Y and Sugawara M 2002 *Jpn J Appl Phys* **41** L1143
[3] Kita T, Tamura N, Wada O, Sugawara M, Nakata Y, Ebe H and Arakawa Y 2006 *Appl Phys Let.* **88** 211106
[4] Yasuoka N, Kawaguchi K, Ebe H, Akiyama T, Ekawa M, Tanaka S, Morito K, Uetake A, Sugawara M and Arakawa Y 2008 *Appl Phys Lett* **92** 101108
[5] Anantathanasarn S, Nötzel R, van Veldhoven P J, van Otten F W M, Eijkemans T J and Wolter J H 2006 *Appl Phys Lett* **88** 063105
[6] Anantathanasarn S, van Veldhoven P J, Eijkemans T J, de Vries T, Smalbrugge E, Geluk E J, E Bente A J M, Oei Y S, Smit M K and Nötzel R 2008 *App. Phys Lett* **92** 123113
[7] Kojima O, Nakatani H, Kita T, Wada O, Akahane K and Tsuchiya M 2008 *J Appl Phys* **103** 113504
[8] Inoue T, Asada M, Kojima O, Kita T and. Wada O 2010 *Proc. SPIE* **7597** 75971J

Journal of Physics: Conference Series **245** (2010) 012077 doi:10.1088/1742-6596/245/1/012077

InP quantum dots for applications in laser devices and future solid-state quantum gates

E Koroknay, W-M Schulz, M Eichfelder, R Roßbach, M Jetter and P Michler

Institut für Halbleiteroptik und Funktionelle Grenzflächen, Universität Stuttgart, Allmandring 3, 70569 Stuttgart, Germany

E-mail: e.koroknay@ihfg.uni-stuttgart.de

Abstract. Single and stacked layers of InP quantum dots in $(Al_xGa_{1-x})_{0.51}In_{0.49}P$ barriers were grown by metal-organic vapor-phase epitaxy for applications in semiconductor laser devices and optical gate structures. An in-plane laser structure with a single layer of InP QDs is presented, emitting at 1.942 eV and exhibiting low threshold current densities of 780 A/cm^2 in electrically pulsed laser operation at 288 K for a 2000 μm long device with uncoated facets. In order to realize an externally driven optical gate structure the stacking behavior of InP in $(Al_xGa_{1-x})_{0.51}In_{0.49}P$ barriers was investigated. To ensure the optical addressability of each quantum dot layer, a special double dot structure where the high energetic smaller sized quantum dot is situated above the low energetic larger sized dot, was produced. The coupling between these quantum dots can be adjusted by the thickness of the spacer layer. The structures are embedded in a ni-Schottky structure and the influence of an external electric field on the emission of the quantum dot ensemble is investigated.

Due to their discrete density of states, systems of single and coupled quantum dots (QDs) are of high interest for applications in quantum information processing. Also the implementation of QDs as active medium in semiconductor lasers is a promising concept [1]. For applications in the visible spectral range the material system of InP QDs embedded in $(Al_xGa_{1-x})_{0.51}In_{0.49}P$ barriers is an interesting candidate. The authors showed that in this case the emission wavelength can be tuned over a wide spectral range from red to green by changing the charge carrier confinement via the aluminum content in the barrier material and also the QD bandgap energy itself by aluminum interdiffusion into the QD for elevated growth temperatures [2]. Due to the improved charge carrier confinement for aluminum containing barriers the QDs show enhanced temperature stability and single-photon emission could be proved up to 80 K [2, 3], which makes them interesting candidates for applications in quantum information processing. The main task for future device application is to develop an externally driven optical gate. This can be realized for example by a controllable interaction (e.g. charge carrier tunneling) between vertical aligned QDs. An external electric field can influence this tunnel coupling via the Quantum Confined Stark Effekt (QCSE) [4, 5]. As our InP/AlGaInP QDs, grown by metal-organic vapor-phase epitaxy (MOVPE), exhibit high QD densities with values between $1.3{\cdot}10^{10}/cm^2$ and $8{\cdot}10^{10}/cm^2$ [2] additional structuring of the sample surface is necessary for measurements on single QDs and vertically coupled quantum dot molecules (QDMs). In contrast, for applications of InP QDs as active medium in laser structures, a high QD density is preferable.
All sample structures were fabricated by MOVPE with standard precursors at low pressure

(100 mbar) on (100) GaAs:Si substrates oriented 6° toward the $[111]_A$ direction. For elementary studies on single and stacked layers of InP QDs, a 100 nm thick GaAs buffer layer followed by a 100 nm thick $(Al_xGa_{1-x})_{0.51}In_{0.49}P$ layer is deposited on which the first layer of self-assembled QDs is grown in Stranski-Krastanow growth mode at 650 °C or 710 °C by depositing 2.1 monolayers (ML) of InP with a growth rate of 1.05 ML/s. A growth interruption of 20 s is included to ripen the QDs. For photoluminescence (PL) measurements the samples were capped by a 30 nm thick layer of the corresponding barrier material. Fig. 1(a) and (b) show temperature

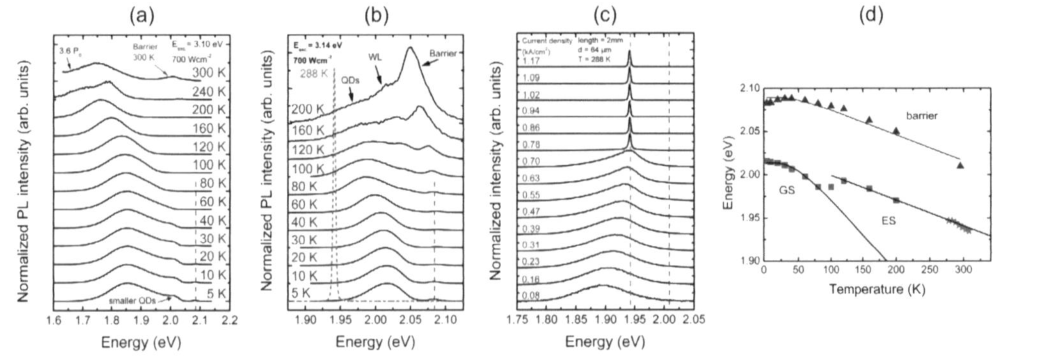

Figure 1. (a) Ensemble PL spectra of InP QDs in $Al_{0.20}Ga_{0.80}InP$ grown at 650 °C and at (b) 710 °C. The red dashed line shows the lasing spectra at 288 K. (c) EL spectra of QD laser at different drive current densities. The dashed lines at 2.01 eV and 1.942 eV mark the spectral position of the $Al_{0.20}Ga_{0.80}InP$ barrier and the lasing wavelength respectively. (d) Temperature dependent transition wavelength of the $Al_{0.20}Ga_{0.80}InP$ barrier (triangles), QD ensemble (boxes) and lasing wavelength (stars) for a QD growth temperature of 710°C. The solid curves are fits according to [8].

dependent PL spectra for QDs grown on $Al_{0.20}Ga_{0.80}InP$ at 650 °C and 710 °C, respectively. To perform PL measurements the samples were placed in a He-flow coldfinger cryostat and excited optically with a pulsed frequency doubled Ti:Al_2O_3 laser with excitation energies of 3.10 eV and 3.14 eV respectively. The PL was dispersed by a 0.32 nm Jobin Yvon monochromator and detected using a multichannel plate photomultiplier tube. For low temperatures in both cases the emission of the barrier at 2.083 eV is clearly visible. The QD ensemble of the sample with 650 °C QD growth temperature shows emission at around 1.85 eV at 5 K. Additionally, a second emission peak at around 2.01 eV is visible. As these QDs usually show a bimodal height distribution, this shoulder can be attributed to the emission of smaller sized QDs in the ensemble. By increasing the temperature, the emission shifts to lower energies, due to bandgap shrinkage. Previous studies show, that these QDs show a high luminescence efficiency per QD with single photon emission up to 80 K [2]. In the case of higher growth temperatures (710 °C) the ensemble PL of the QDs is centered at around 2.02 eV at 4.2 K with a FWHM of 46 meV. At elevateted temperatures the spectra are dominated by the emission of the barrier. The additional shoulder at around 2.04 eV can be attributed to the emission of the wetting layer (WL). By fitting gaussian distributions on the data in Fig. 1(b) the transition energy of the barrier (triangles) and the QD ensemble (boxes) can be evaluated and are displayed in Fig. 1(d). The QD emission exhibits a kink at around 100 K due to thermal excitation of charge carriers to higher energy states in the QD. The data are modelled using simplified Bose functions [8] (solid curves) with two branches for the QD emission, one for the ground state (GS) and for the excited state (ES), respectively. With values of the energetical distance between the excited states and the barrier layer of only 66 meV, the carrier confinement in the QDs is quite weak in this case.

The presented laser structure consists of a silicon doped superlattice as buffer, followed by a seperate confinement heterostructure as reported in reference [6]. The single layer of self-assembled InP QDs grown at 710 °C was placed in the center of a 20 nm thick undoped $(Al_{0.20}Ga_{0.80})_{0.51}In_{0.49}P$ barrier surrounded by 2×150 nm thick undoped $(Al_{0.55}Ga_{0.45})_{0.51}In_{0.49}P$ waveguide layers. The resonator structure was made by cleaving and the facets were left uncoated. The device was mounted junction-side up on a brass heatsink and contacted with a Au coated tungsten needle. Self-heating of the devices was reduced by testing them in pulsed mode with a pulsewidth of 300 ns and a repetition frequency of 10 kHz. Fig. 1(c) shows spectra of the QD laser at different drive current densities as reported in reference [6], measured with a fibre coupled spectrometer at room temperature. The current densities are calculated using the broad area stripe geometry of the top contact with a length of 2000 μm and a width of 64 μm. Current spreading is neglected, therefore the effective current density is expected to be smaller. At low current densities electroluminescence of the QD ensemble can be observed. The QD emission shifts from 1.895 eV at 0.08 kA/cm^2 to 1.943 eV at 0.70 kA/cm^2, which indicates additional recombination of excitons from higher energetic states as well as the occupation of higher energetic smaller sized QDs in the ensemble. Lasing was observed at 1.942 eV for values above the threshold current density of j_{th}=0.78 kA/cm^2. The shift of the lasing wavelength to higher energies (ΔE=48 meV) indicates lasing from excited states, which can be confirmed due to the good agreement of the spectral position of the lasing wavelength and the emission from excited states of the QD ensemble for elevated temperatures (shown in Fig. 1(b)).

For samples with stacked layers of QDs, a (Al)GaInP spacer layer with variable thickness ranging from 2.5 nm to 20 nm is deposited after the growth of the first QD layer and the second layer of InP QDs was grown at the same temperature as the bottom layer (650 °C). It could be demonstrated in several studies, that due to the additional strain field the QDs in the second layer tend to be larger if the same amount of QD material is deposited [5, 7]. In this case the amount of deposited InP is reduced for the growth of the upper QDs to realize low energetic larger sized QDs in the bottom layer and high energetic smaller sized QDs in the upper layer. This structure was used to achieve high spectral separation of the luminescence of both QD layers in order to be able to distinguish between both QD layers even in ensemble PL measurements. To study the influence of an electric field, the QDs were embedded in the intrinsic region of a ni-Schottky structure with an ohmic back contact. The additional 30 nm GaAs cap layer on top of this sample structure is removed in the region where photoluminescence measurements are performed. The top contact is processed with apertures with diameters of several μm to provide both, electric field and optical access. To achieve a homogeneous electric field the whole sample was covered with a 5 nm thick semitransparent Au layer. Previously we have proven the vertical odering of InP QDs in GaInP barriers, due to a strain driven self-alignment, as well as the formation of quantum dot molecules (QDMs) with decreasing spacer layer thickness with an electronic coupling by a charge carrier tunneling process [9, 10]. In order to achieve enhanced temperature stability the growth of stacked layers of InP QDs in $Al_xGa_{1-x}InP$ barriers with 20 % aluminum content was investigated. The influence of the barrier thickness on the tunneling probability can already be seen in ensemble PL spectra. For ensemble PL measurements the samples were mounted in a He-flow cold finger cryostat and excited with a cw HeCd-Laser at 325 nm from the top. Therefore, electron hole pairs are mainly created in the $Al_{0.20}Ga_{0.80}InP$ cap layer and relax first into the high energetic QDs of the upper layer. If the tunnel barrier thickness is small enough, the charge carriers are able to tunnel phonon assisted into the QDs of the bottom layer, where they recombine. Fig. 2(a) shows the ensemble PL spectra of the QD stacks with different spacer thicknesses (d_{sp}). For small d_{sp} the luminescence spectra are dominated by the emission of the bottom layer. By increasing the spacer thickness from 2.5 nm to 20 nm the coupling strength gets weaker. The luminescence intensity of the top layer is increasing as the probability of carriers tunnelling from the top to the bottom QD layer is reduced. By incorporating the coupled QDs with 10 nm spacing in an electrical contactable

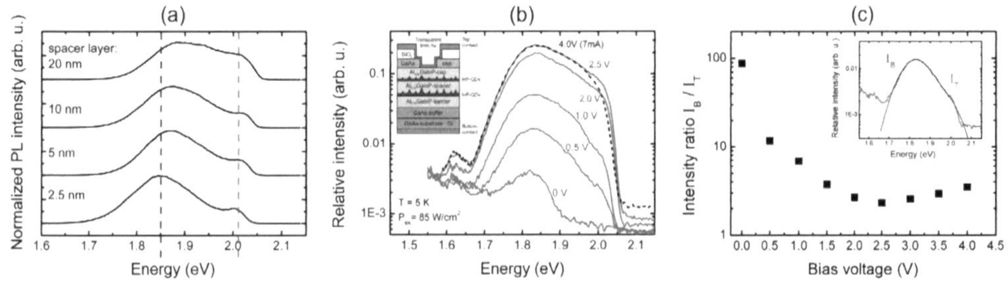

Figure 2. Ensemble PL spectra of stacked InP QD layers in a $Al_{0.20}Ga_{0.80}InP$ barrier (a) with variable spacer thickness at 5 K and (b) with applied electric field and 10 nm spacer. (c) Intensity ratio of the bottom (I_B) and top (I_T) QD layer for applied electric field. The inset shows exemplarily the fitted Gaussian curves on the spectra.

structure as described above, the changes of the PL under applied electric field can be observed (Fig. 2(b)). For 0 V bias the PL spectra shows a maximum at 1.83 eV. This emission can be attributed to the larger QDs of the bottom layer. The emission of the smaller QDs of the bottom and top layer is totally suppressed. Due to the sample structure, the band edge is tilted and the luminescence intensity is quenched. For the smaller QDs the charge carriers are able to escape into the barrier. For the larger QDs of the bottom layer the electron and hole wavefunction are strongly delocalized and the recombination rate is therefore reduced. With increasing bias voltage the PL intensity increases as the built in electric field is reduced and an additional shoulder appears on the high energetic side. This shoulder first shows an increasing intensity relative to the low energetic peak. For a bias voltages higher than 2.5 V the intensity decreases again and an increasing current can be measured. Fig. 2(c) displays the ratio I_B/I_T of the integrated luminescence intensity of the bottom and the top layer. For 2.5 eV flat band condition is assumed.

We have shown, that the InP/AlGaInP QDs have the potential for either laser devices or quantum optics applications. The single layer QD broad area laser shows emission at 1.942 eV with a threshold density of 780 A/cm^2. Furthermore we have shown the influence of an electric field on the PL of asymmetrically QD layers. By precisely adjusting the QD growth, the spacer layer thickness, the doping and the processing of the structures, future quantum gates should be realisable in this material system.

References

[1] Arakawa Y and Sakaki H 1982 *Appl. Phys. Lett.* **40** 939

[2] Schulz W M, Roßbach R, Reischle M, Beirne G J, Bommer M, Jetter M and Michler P 2009 *Phys. Rev. B* **79** 35329

[3] Roßbach R, Schulz W M, Reischle M, Beirne G J, Jetter M and Michler P 2007 *J. Cryst. Growth* **298** 595

[4] Ortner G, Bayer M, Lyanda-Geller Y, Reinecke T L, Kress A, Reithmaier J P and Forchel A 2005 *Phys. Rev. Lett.* **94** 157401

[5] Krenner H J, Sabathil M, Clark E C,Kress A, Schuh D, Bichler M, Abstreiter G and Finley J J 2005 *Phys. Rev. Lett.* **94** 57402

[6] Schulz W M, Eichfelder M, Roßbach R, Jetter M and Michler P 2009 *Appl. Phys. Express* **2** 112501

[7] Schmidt O G, Kienzle O, Hao Y and Eberl K 1999 *Appl. Phys. Lett.* **74** 1272

[8] O'Donnell P and Chen X 1991 *Appl. Phys. Lett.* **58** 2924

[9] Roßbach R, Reischle M, Beirne G J, Jetter M and Michler P 2007 *J. Cryst. Growth* **298** 603

[10] Reischle M, Beirne G J, Roßbach R, Jetter M, Schweizer H and Michler P 2007 *Phys. Rev. B* **76** 85338

Low Thermal Budget Fabrication of III-V Quantum Nanostructures on Si Substrates

S Bietti[1], C Somaschini[1], S Sanguinetti[1], N Koguchi[1], G Isella[2], D Chrastina[2] and A Fedorov[2]

[1] L-NESS and Dipartimento di Scienza dei Materiali, via Cozzi 53, I-20125 Milano, Italy

[2] CNISM, L-NESS and Dipartimento di Fisica, Politecnico di Milano, via Anzani 42, I-22100 Como, Italy

E-mail: stefano.sanguinetti@mater.unimib.it

Abstract. We show the possibility to integrate high quality III-V quantum nanostructures tunable in shape and emission energy on Si-Ge Virtual Substrate. Strong photoemission is observed, also at room temperature, from two different kind of GaAs quantum nanostructures fabricated on Silicon substrate. Due to the low thermal budget of the procedure used for the fabrication of the active layer, Droplet Epitaxy is to be considered an excellent candidate for implementation of optoelectronic devices on CMOS circuits.

The achievement of III-V self-assembled quantum nanostructures (QNs) integration on silicon substrates could allow the use of silicon infrastructure, based on complementary metal oxide semiconductor (CMOS) technology, as base for the fabrication of optoelectronic devices like nano emitters and intersubband detector realized with III-V semiconductor materials [1].
The integration of III-V optoelectronic devices and structures on Si is far from optimized, notwithstanding its great technological relevance. Some important issues have to be overcame in order to obtain high quality III-V materials direcly on Si. The first is the large lattice mismatch (about 4.1%) between Si and GaAs, which introduces misfit dislocation nucleation as soon as the GaAs epilayer thickness exceeds a critical value. A second issue is the antiphase disorder, that arises from different symmetry of the zincblend and diamond lattice. A last major issue is the strict thermal budget required during the growth of an epilayer to mantain the compatibility with the underlying CMOS circuit.
To avoid these issues we used a Ge virtual substrates (GeVS), that is a fully relaxed, micron thick, Ge epitaxial buffer. GeVS can be used to accommodate the lattice mismatch between GaAs-based III-Vs and the Si substrate. GeVS growth technique is also compliant with front-end CMOS integration. Anti Phase Domains (APDs) can be effectively eliminated by using intentionally misoriented substrates [2,3]. For the growth of III-V nanostructures we selected an intrinsecally low thermal budget (LTB) technique, the droplet epitaxy (DE) [4,5]. Being performed at 200-350° C, DE is perfectly suited for the realization of growth procedures compatible with back-end integration of III-V nanostructures on CMOS that is, after the CMOS circuitry has been already realized. The fabrication of quantum nanostructures by DE is consistent with the strict constraints on thermal budget imposed by the compatibility with the underlying CMOS circuit. Moreover, the use of DE technique opens the possibility to fabricate a large variety of three dimensional nanostructures tunable in density, size and shape [6-8].

1. Experimental setup

For our experiments we used a silicon (001) substrate, 6° misoriented towards [110]. A fully-relaxed 2 μm layer of Ge buffer layer was deposited on the silicon substrate by Low-Energy Plasma-Enhanced Chemical Vapour Deposition (LEPECVD) [9,10] in order to form a Ge virtual substrate (GeVS). Just after deposition, the GeVS was cyclically annealed six times at 600 - 780° C to reduce the threading dislocation density to 2×10^{7} cm^{-2} [11]. GeVS technology we adopted is reported to be used for the fabrication of ultra-low dark current Ge photodiodes and to be fully compliant with CMOS frond-end technology [12].
The GeVS was then transferred to a Gen II MBE machine. In the growth chamber we started an MBE growth of 1 μm GaAs buffer keeping the temperature at 580° C. Confirmation of APD-free sample came from the clear (2x4) surface recontruction of the RHEED pattern during GaAs growth. Our first task was then to test the self aggregation of QNs by DE on a GaAs/Ge/Si substrate. So for the first sample we

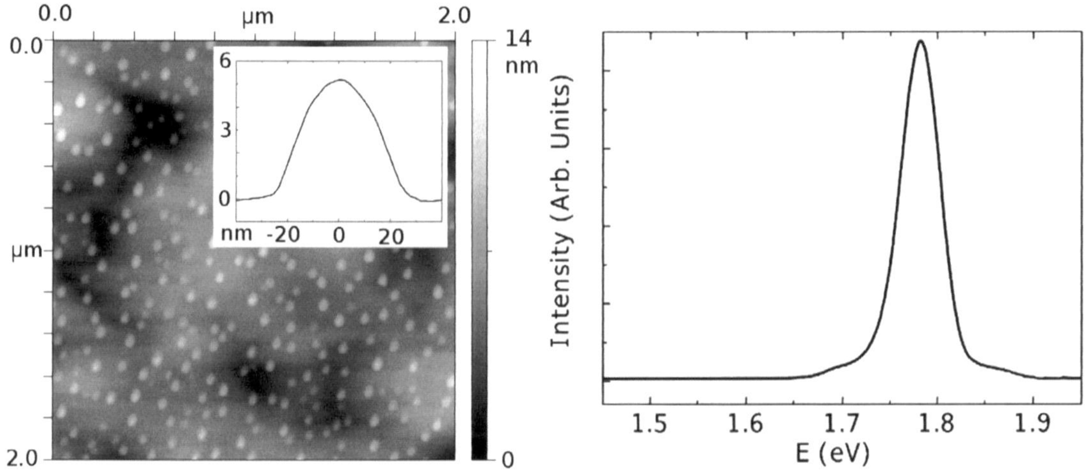

Figure 1. Left panel: 2x2 μm AFM image for uncapped QD sample. Inset shows profile of a typical nanostructure. Right panel: photoluminescence spectrum for QD sample.

used standard high temperature MBE growth for the deposition of an high crystal quality $Al_{0.30}Ga_{0.70}As$ buffer. A temperature range between 150 and 200° C compatible with CMOS technology was mantained only for the fabrication of the quantum dots (QDs) by DE. Details for the growth procedure are reported in ref [13]. The sample was finally capped with 150 nm of $Al_{0.30}Ga_{0.70}As$ by conventional MBE at 580° C for PL measurements and with a GaAs capping layer. This sample was then submitted to an annealing step at 700° C in As atmosphere for 30 min to improve barrier crystal quality [14].

We then fabricated a second sample with an active layer fully compliant with LTB requirements. For this purpose we reduced the thermal budget for the annealing procedure and for the creation of the $Al_{0.30}Ga_{0.70}As$ barrier. Using migration enhanced epitaxy [15], temperature was kept constant at 350° C during the creation of the whole active layer. Details for the growth procedure are reported in ref [16]. For this sample another 80 nm $Al_{0.30}$ $Ga_{0.70}$ As layer was deposited by MEE at 350° C for photoluminescence measurements. The sample was finally submitted to a 4 minutes Rapid Thermal Annealing (RTA) procedure at 600° C in nitrogen atmosphere to improve barrier crystal quality [17]. Also this step is compliant with LTB requirements for $>0.25\mu$m CMOS back-end technology [18].
Surface morphology was analyzed on uncapped samples by a Veeco-Innova Atomic Force Microscope (AFM). The photoluminescence (PL) spectra of the capped samples were measured at 14 K using a closed-cycle cold-finger cryostat. PL was excited with a neodymium doped yttrium aluminum garnet laser with λ_{exc} = 532 nm and an excitation power density of P_{exc} = 6 W/cm^2. The spectrum was measured by a grating monochromator operating with a Peltier-cooled charge-coupled device detector.

2. Results and discussion

Left panel of figure 1 shows the AFM image of the first uncapped sample. It demonstrates the QD formation on the surface of the first sample. The density is about 1×10^{10} cm^{-2}, in good agreement with data reported for DE experiments on GaAs substrates [6]. An interesting point is to observe that the size and density of QDs is not constant across the surface of the sample, being smaller and with higher density in the substrate valleys. This behaviour could be associated qualitatively with the interaction of the gallium droplets with surface steps. Is known that surface steps are preferential nucleation sites for gallium droplets, as shown on (110) substrates [19]. The QD ensemble shows a mean height of ≈ 5 nm and a base radius of ≈ 20 nm with a size dispersion for both values around 25%. PL spectrum of QD capped sample (figure 1, right panel) shows an intense emission peak at 1.78 eV, in good agreement with the calculated value of E_{gs}=1.75 eV. This emission energy was predicted for a single randomly choosen single quantum dot, applying the method of effective mass approximation outlined in [21] and using material parameters reported in [22] .
Left panel of figure 2 shows AFM image of the second uncapped sample. It demonstrates the formation of quantum nanostructures with a density of 6×10^8 cm^{-2} characterized by a regular, nanometers high, flat disk with a diameter of hundreds of nanometers and a hole at the center of ≈ 80 nm. The rim of the

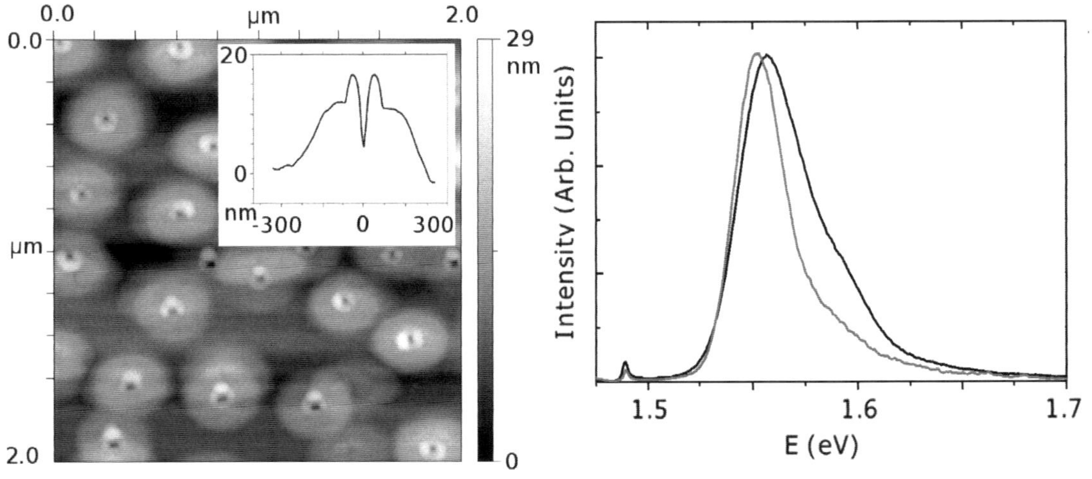

Figure 2. Left panel: 2x2 μm AFM image for uncapped CRD sample. Inset shows profile of a typical nanostructure. Right panel: normalized photoluminescence spectra for CRD sample with excitation power density of 6 W/cm^2 (black line) and 0.6 W/cm^2 (red line).

inner hole is protruded over the disk surface by some nanometers. Because of their shape we call these nanostructures Coupled Ring-Disks (CRDs) [20].

PL measurements of the CRD sample show (figure 2, right panel) an intense emission peak at 1.55 eV, and a shoulder at 1.59 eV. The first value is in good agreement with the energy of the ground state E_{gs}=1.56 eV obtained using effective mass approximation. The shoulder should be attributed to the excited state. This calculation shows also that electron and hole are confinated in the inner ring, and that the excited state is extended along the external disk.

Comparing PL data makes clear that is possible to obtain efficient and high quality III-V QNs on Si substrate using a LTB procedure not only for the fabrication of the QNs, but also for the whole active layer. Measurements performed on the CRD sample shows a quantum efficiency of η=3$\times$10^{-3} similar to the one obtained by DE on standard GaAs substrate [16]. Calculations were performed evaluating the number of incoming and emitted photons.

The PL emission is clearly detectable at room temperature, where PL intensity is reduced by a factor $\approx$ 400 respect to the low temperature case.

3. Conclusions

We demonstrated the growth of high quality and high efficiency III-V QNs on Silicon substrate, using a LTB procedure. In particular, the fabrication procedure for the GeVS we proposed is compliant with front-end CMOS integration, while the growth procedure for the AlGaAs/GaAs active region is fully compliant with CMOS back-end integration. With small changes for the GaAs layer growth, the whole structure would be easily made fully compliant with the requirements of CMOS technology.

References

[1] Wu J, Li Z, Shao D, Manasreh M, Kunets V P, Wang Z M, Salamo G J and Weaver B D 2009 *Appl. Phys. Lett.* **94** 171102
[2] Bringans R D, Biegelsen D K and Swartz L E 1991 *Phys. Rev. B* **44** 3054
[3] Ting S M and Fitzgerald E A 2000 *J. Appl. Phys.* **87** 2618
[4] Koguchi N, Takahashi S and Chiykow T 1991 *J. Cryst. Growth* **111** 688
[5] Koguchi N and Ishige K 1993 *Jpn. J. Appl. Phys. Part 1* **32** 2052
[6] Watanabe K, Koguchi N and Gotoh Y 2000 *Jpn. J. Appl. Phys. Part 1* **39** 79
[7] Mano T and Koguchi N 2005 *J. Cryst. Growth* **278** 2005
[8] Somaschini C, Bietti S, Koguchi N and Sanguinetti S 2009 *Nano Letters* **9** 3419
[9] Rosenblad C, von Kanel H, Kummer M, Dommann A and Muller E 2000 *Appl. Phys. Lett.* **4** 427
[10] Isella G, Chrastina D, Rossner B, Hackbarth T, Herzog H-J, Konig U, von Kanel H 2004 *Solid-State Electron.* **48** 1317
[11] Isella G, Osmond J, Kummer M, Kaufmann R and von Kanel H 2007 *Semicond. Sci. Technol.* **22** S26
[12] Osmond J, Isella G, Chrastina D, Kaufmann R, Acciarri M and von Kanel H 2009 *Appl. Phys. Lett.* **94** 201106
[13] Bietti S, Sanguinetti S, Somaschini C, Koguchi N, Isella G, Chrastina D and Fedorov A 2009 *IOP Conf. Ser.: Mater. Sci. Eng.* **6** 012009
[14] Mantovani V, Sanguinetti S, Guzzi M, Grilli E, Gurioli M, Watanabe K and Koguchi N *Physica E* **23** 377

[15] Horikoshi Y, Kawashima M and Yamaguchi H 1987 *Appl. Phys. Lett.* **50** 1686
[16] Bietti S, Somaschini C, Sanguinetti S, Koguchi N, Isella G and Chrastina D 2009 *Appl. Phys. Lett.* **95**
[17] Sanguinetti S, Mano T, Gerosa A, Somaschini C, Bietti S, Koguchi N, Grilli E, Guzzi M, Gurioli M and Abbarchi M. 2008 *J. Appl. Phys.* **104** 113519
[18] Takehuci H, Wung A, Sun X, Howe R T and King T-J 2005 *IEEE Trans. Electron. Devices* **52** 2081
[19] Mano T, Kuroda T, Noda T and Sakoda K 2009 *Phys. Stat. Sol. (b)* **246** 729
[20] Somaschini C, Bietti S, Sanguinetti S, Koguchi N and Fedorov A 2010 *Nanotechnology* **21** 125601
[21] Marzin J-Y and Bastard G 1994 *Solid State Commun.* **92** 437
[22] Kuroda T, Mano T, Ochiai T, Sanguinetti S, Sakoda K, Kido G and Koguchi N 2005 *Phys. Rev. B* **72** 205301

Quantum confinement in MOVPE-grown structures with self-assembled InAs/GaAs quantum dots

K Kuldová[1,2,4], M Molas[2], J Borysiuk[2,3], A Babiński[2], Z Výborný[1], J Pangrác[1], J Oswald[1]

[1] Institute of Physics of the AS CR, v. v. i., Cukrovarnická 10, CZ-162 00 Praha 6, Czech Republic

[2] Institute of Experimental Physics, University of Warsaw, Hoża 69, PL-00-681 Warszawa, Poland

[3] Institute of Physics PAS, Al. Lotników 32/46, PL-02-668 Warszawa, Poland

E-mail: kuldova@fzu.cz

Abstract. In this communication we report on low-temperature, micro-photoluminescence study of quantum confinement in MOVPE-grown structures with InAs/GaAs quantum dots (QDs) with GaAs and/or strain reducing InGaAs/GaAs capping. We focus our attention on sharp emission lines, which appear in both structures at energies up to 80 meV below the wetting line emission. Power-dependent measurements confirmed their attribution to single excitons as well as biexcitons. Negative binding energy of biexcitons with systematic dependence on their energy was observed. It has been proposed that the investigated emission lines result from radiative recombination in flat non-fully developed QDs in the investigated structure. The attribution is confirmed by transmission electron microscopic analysis of investigated structures.

1. Introduction

Scientific interest in self-assembled InAs/GaAs quantum dots (QDs) is driven by interesting fundamental physics and numerous potential applications. One of those applications is the area of inexpensive, low threshold and high output lasers emitting in the 1.55 μm communication band. Tuning the emission energy within the infrared band is a challenge for technology. Standard approach, namely covering the QDs with a GaAs capping layer (CL), leads to a rapid shrinkage of the QD heights, which results in a blue shift of the QDs luminescence at least to ~ 1.2 μm. A promising way to avoid the blue-shift of the QD emission is the application of a thin (5 nm) InGaAs strain reducing layer (SRL), which covers QDs [1, 2]. The understanding of processes, which are at play during formation of such QDs with SRL is of prime importance from technological point of view. Also important is the knowledge of possible confinement potentials, which can trap photoexcited carriers, potentially affecting the performance of optoelectronic devices. The study of carrier localizations in the energy region of the wetting layer (WL) is the first step of such study [3, 4].

---

[4] To whom any correspondence should be addressed.

2. Experimental

Structures with InAs/GaAs QDs and ternary SRLs (23% of In) were prepared by low pressure metalorganic vapor phase epitaxy (LP MOVPE) in AIXTRON200 on SI(100) GaAs substrates using Stranski–Krastanow growth mode. The InAs QDs were covered by 30 nm of GaAs CL or by 4 nm of $In_{0.23}Ga_{0.77}As$ SRL and 30 nm of GaAs CL. Reflectance anisotropy spectroscopy in-situ measurement (EpiRAS 200TT–LayTec) was used to monitor the formation and development of InAs QDs, as well as the growth of InGaAs and GaAs layers [5]. Mesa-patterning on the samples limits the number of photoexcited dots and allows to reproducibly locate investigated dots. Samples for optical measurements were mounted in a He-flow cryostat and measured at temperature 4.2 K. The 532 nm laser line was used for non-resonant optical excitation. The excitation light was delivered and the photoluminescence (PL) collected via a microscope objective (spot size ~1 μm), dispersed using a 0.5 m monochromator and detected using a liquid-nitrogen-cooled Si charged coupled device camera. The macro PL was detected by a Ge detector using standard lock-in technique.

3. Results and discussion

The PL spectrum from the investigated structure is composed of the emission from the GaAs barrier, self-assembled QDs, and the emission from the WL (figure 1). Note the red shift of PL from the QDs and the WL due to the SRL capping. Micro-PL spectra from both samples in the WL energy region measured on a 5×5 μm mesa are presented in figure 2. The PL band at the 1.36 eV is attributed to radiative recombination to the doubly ionized Cu_{Ga} acceptor, which corresponds to the 0.156 eV acceptor level [6]. In the case of the SRL sample the Cu_{Ga} – related band co-exists with the WL emission. The Cu_{Ga} – related band originates from the GaAs substrate and can be bound up with Cu contamination of polished and etched surface of SI GaAs wafers [7].

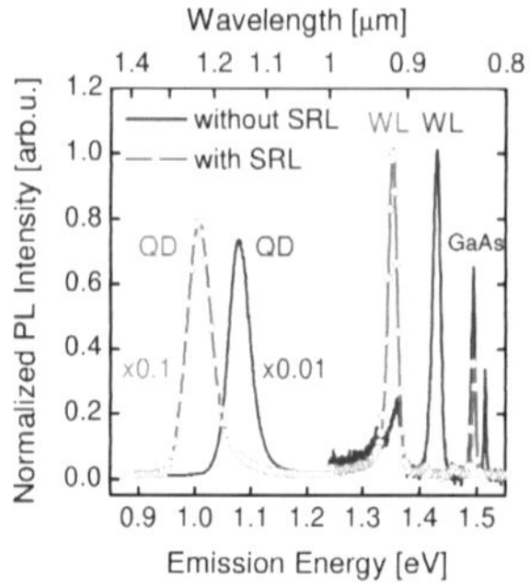

Figure 1. Low temperature PL spectra of the sample without (blue line) and with the SRL (red line) taken from unmasked regions.

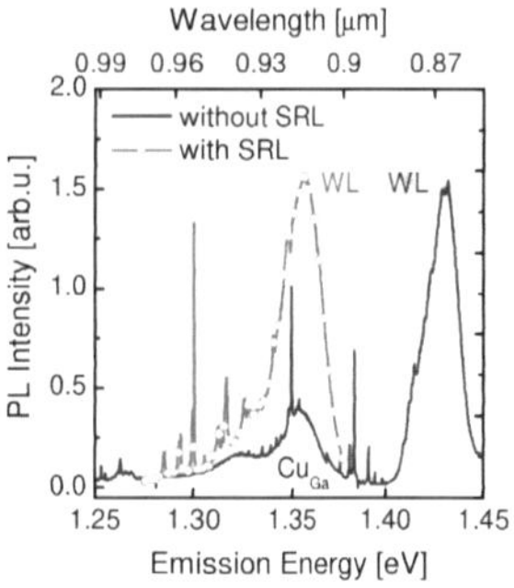

Figure 2. Low temperature μ-PL spectra of the sample without (blue line) and with the SRL (red line) taken from a 5×5 μm mesa.

We focus our attention on sharp emission lines, which appear in both structures at energies up to 80 meV below the WL emission.

To classify these emission lines, power-dependent PL spectra were measured. In figures 3 and 4 are represented μ-PL spectra of excitonic lines for increasing excitation power density P_{excit} of both samples respectively. The power dependence of each line for low excitation is characterized by $(P_{excit})^n$, the exponents n are presented for each line in figures. In insets are shown examples of corresponding fits.

In both structures pairs of sharp lines (B, C and D, E for the sample without the SRL and J, K for the sample with the SRL) can be distinguished.

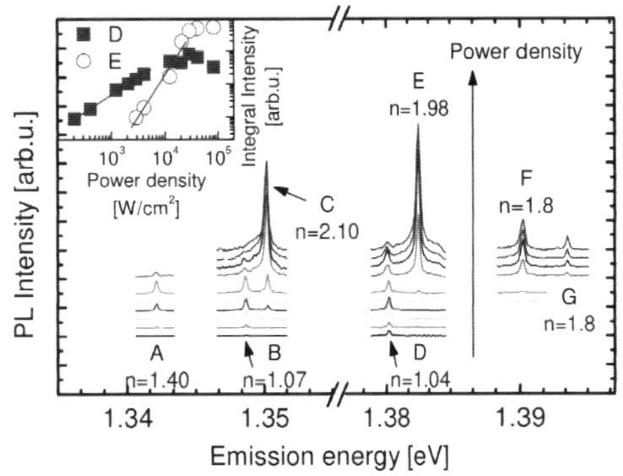

Figure 3. Power-dependent μ-PL spectra of the sample without the SRL. Inset: The integrated PL intensity of D and E emission lines as a function of power density.

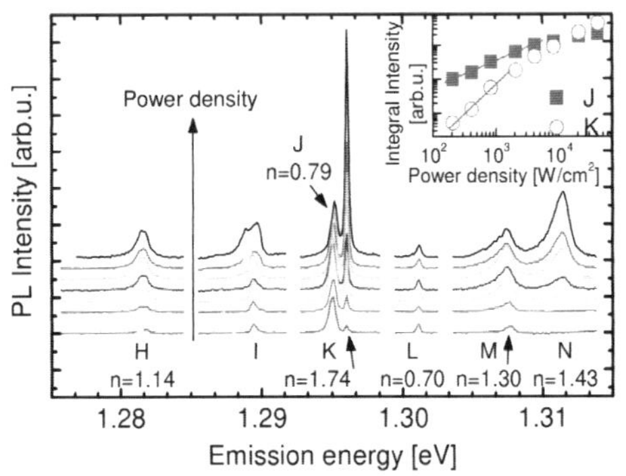

Figure 4. Power-dependent μ-PL spectra of the sample with the SRL. Inset: The integrated PL intensity of J and K emission lines as a function of power density.

The emission lines have similar power-dependence: lower (higher) energy line from each pair emerges with linear (quadratic) excitation power density dependence. Limited number of lines observed in the energy region 1.3 - 1.4 eV for the sample without the SRL allows us to attribute lines B and D to two different excitons and lines C and D to their respective biexcitons in QDs. Binding energies of those biexcitons are equal to $E_{B\text{-}C} = -1.59$ meV and $E_{D\text{-}E} = -2.51$ meV. There are two types of features in the spectra of the SRL sample: lines J, K and L are sharp with the FHWM comparable to that of figure 3; the linewidth of H, M and N lines is at least 5 times larger. The J, K and L lines have linear, quadratic and sublinear excitation power density dependence. We attribute these lines to the exciton, biexciton and a charged trion respectively. Binding energies of the biexciton is equal to $E_{J\text{-}K} = -1$ meV. The line I has a complex structure composed of at least three closely spaced peaks. In our opinion this line is composed of the neutral exciton, biexciton and charged exciton from a single QD with a very low binding energy. In agreement with [8], the biexciton binding energies in both samples decrease with increasing transition energy and their values follow a linear dependence (see figure 5).

The order of appearance of few-particle states (exciton X, biexciton XX and charged excitons, i.e., positive and negative trions X^+, X^-) are discussed in [9] for a large number of QD structures of varying size, shape, and composition. In [8] the binding energy for flat InAs/GaAs MOVCD QDs decreases with increasing transition energy changing its sign at about 1.24 eV. The binding to antibinding transition is attributed to three-dimensional confinement, quenching correlation, and exchange and causing local charge separation.

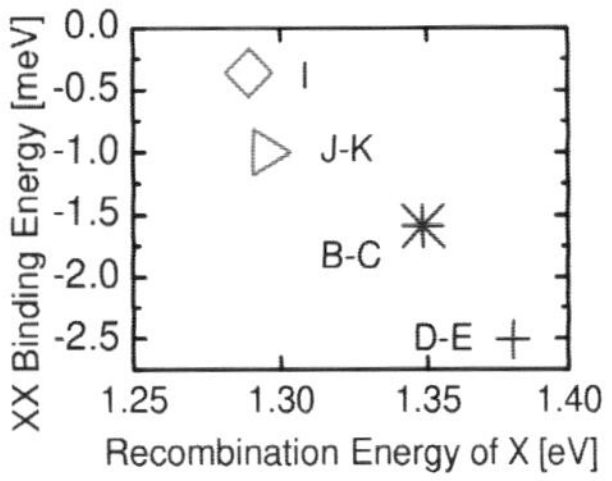

Figure 5. Biexciton binding energy as a function of exciton energy for the sample without (blue) and with the SRL (red symbols). Letters refer to excitonic lines denoted in figures 3 and 4 respectively.

The results of optical measurements can be confronted with an analysis of the sample morphology. Cross-sectional transmission electron microscopy (TEM) reveals that self-assembled InAs QDs in sample without the SRL are flat, with the height of 2.5 nm and base diameter ~25 nm (figure 6). This confirms results of atomic force microscopy of an uncapped sample QDs of height 2.5 - 3.5 nm and diameter 15 - 25 nm (not shown here). The QDs in SRL samples (figure 6, bottom) are higher (5 nm) and thinner with diameter ~20 nm as predicted in [1]. Beyond fully developed self-assembled InAs QDs the presence of lower objects (~2 × WL height) have been observed on both samples with significantly lower density. Their diameter is comparable with this of fully developed QDs. In our opinion these not-fully developed QDs are the features confining excitons investigated in our work. Although their density seems to be much lower than the density of self-assembled QDs, they may influence the performance of possible devices relaying on transport of excitons in the WL. They also seem to be interesting from fundamental point of view, as the biexciton binding energy may vanish in those objects.

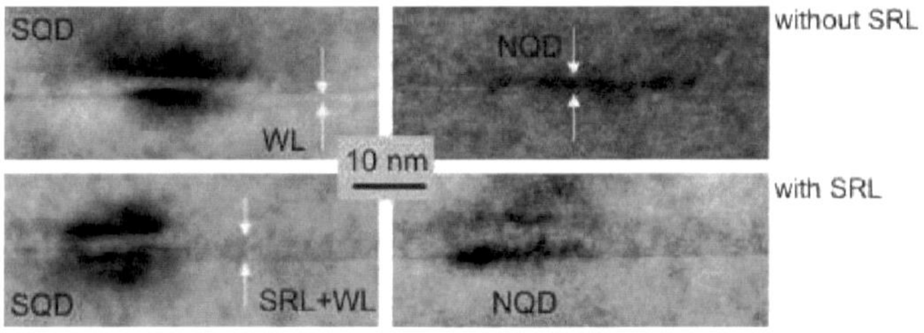

Figure 6. The TEM images of the sample without (top) and with the SRL (bottom). The InAs WL for the GaAs CL sample, $In_{0.23}Ga_{0.77}As$ SRL in contact with WL for the SRL sample, self-assembled InAs QDs (SQD), and non-fully developed InAs QDs (NQD) are identified.

4. Conclusions

In conclusion, we have performed microluminescence studies of MOVPE-grown structures with InAs/GaAs QDs with GaAs capping layer and InGaAs strain reducing layer. We have observed excitonic and biexcitonic emission lines, which have been attributed to the not-fully developed QDs in those structures. The biexciton binding energy was observed to linearly depend on the energy of exciton, vanishing below 1.30 eV. The spectroscopic results were confirmed by transmission electron microscopy, which revealed the presence of those objects.

Acknowledgments

Financial support from the EU (MTKD-CT-2005-029671), Polish funds for science, Czech Science Foundation projects 202/09/0676 and P102/10/1201, and the research program of Institute of Physics, Prague AV0Z 10100521 is acknowledged. We would like to thank Karel Melichar for technical assistance with the structure growth and the FMSE WUT for using TEM microscope JEOL JEM 3010.

References

[1] Hazdra P, Oswald J, Komarnitskyy V, Kuldová K, Hospodková A, Hulicius E and Pangrác J 2009 *Superlattices and Microstructures* **46** 324
[2] Hospodková A, Hulicius E, Pangrác J, Oswald J, Vyskočil J, Kuldová K, Šimeček T, Hazdra P and Caha O 2010 *J. Crystal Growth* **312** 1383
[3] Babinski A, Jasinski J 2002 *Thin Solid Films* **412** 84
[4] Babiński A, Borysiuk J, Kret S, Czyż M, Golnik A, Raymond S and Wasilewski Z R 2008 *Appl. Phys. Lett.* **92** 171104
[5] Hulicius E, Oswald E, Pangrác J, Vyskočil J, Hospodková A, Kuldová K, Melichar K and Šimeček T 2008 *J. Crystal Growth* **310** 2229
[6] Wang Z G, Gislason H P, and Monemar B 1985 *J. Appl. Phys.* **58** 230
[7] Fang Z Q, Look D C, and Jones R L 1997 *J. Electronic Materials* **26** L29
[8] Rodt S, Heitz R, Schliwa A, Sellin R L, Guffarth F, and Bimberg D 2003 *Phys. Rev.* B 68 035331
[9] Schliwa A, Winkelnkemper M, and Bimberg D 2009 *Phys. Rev.* B 79 075443

Electro- and photoluminescence of InAs/GaAs quantum dot structures

J Oswald[1,3], P Hazdra[2], K Kuldová[1], A Hospodková[1], E Hulicius[1], J Pangrác[1], and J Vyskočil[1]

[1] Institute of Physics of the AS CR, v. v. i., Cukrovarnická 10, 162 00 Prague 6, Czech Republic

[2] Department of Microelectronics, Faculty of Electrical Engineering, Czech Technical University in Prague, Technická 2, 166 27, Czech Republic

E-mail: oswald@fzu.cz

Abstract. In this communication, we study the effect of composition of the low-temperature grown quantum dot (QD) capping layer on electro- (EL) and photoluminescence (PL) from self assembled InAs QDs grown by metal organic vapour phase epitaxy (MOVPE). QD structures were prepared by low-pressure MOVPE in a reflectance anisotropy spectroscopy LayTec equipped AIXTRON 200 machine, using Stranski–Krastanow growth mode on n-type Si doped and on semi-insulating GaAs substrates. To enable EL characterization, QD structures grown on the conducting substrate were embedded into a p-i-n structure. We focused on the shift of the emission maximum towards longer wavelengths by using InGaAs strain reducing layer (SRL). Three concentrations of In were used 0%, 13% and 23%. The shift of the emission maximum towards the low energy side of spectrum was observed only for the p-i-n sample with 13% In in SRL, while further increasing of In content to 23% caused the blue shift of the position of both PL and EL maxima.

1. Introduction

Self-assembled InAs/GaAs quantum dot (QD) lasers operating at 1.3 μm or 1.55 μm represent possible efficient light sources for high-speed communication systems. Although marked progress has been achieved [1], still many questions are opened. While molecular beam epitaxy prepared QD structures are already used in the laser structures, QD structures grown by metalorganic vapour phase epitaxy (MOVPE) exhibit a considerable blue shift of electroluminescence (EL) maximum or lasing wavelength compared to the maximum position of their photoluminescence (PL). The preparation of InAs/GaAs QD structures with a strong room temperature PL in the range of 1.3 – 1.55 μm can be realized by using a thin strain reducing InGaAs layer (SRL) [2]. However, preparation of EL structures emitting in this range is more complicated due to the blue shift of emission with increasing thickness of the capping layer [3] and emission from the higher excited states. In this contribution, we

[3] To whom any correspondence should be addressed.

describe the effect of composition and thickness of the low-temperature grown QD capping layer on EL and PL from self assembled InAs QDs grown by MOVPE.

2. Experimental

QD structures were prepared by low-pressure MOVPE in a reflectance anisotropy spectroscopy (RAS) LayTec equipped AIXTRON 200 machine, using Stranski–Krastanow growth mode on n-type Si doped (carrier concentration 2 10^{18} cm^{-3}) and semiinsulating (SI) GaAs substrates. To enable EL characterization, QD structures grown on the conducting substrate were embedded into a p-i-n structure. TMGa, TMIn, and AsH_3 were used as precursors for the growth of GaAs, InAs, and $In_xGa_{1-x}As$ layers. The first GaAs buffer layer was grown at 650°C, then the temperature was decreased to 490°C for the growth of the second buffer layer; undoped GaAs layer; InAs QD layer; undoped 5 nm $In_xGa_{1-x}As$ SRL and 10 nm GaAs QD capping layers; then the temperature was increased to 650°C and the rest of the structure undoped GaAs, p-type C doped GaAs (4 10^{17} cm^{-3}) and p^+-type C doped GaAs (1 10^{18} cm^{-3}) layers was prepared at this temperature. A set of the p-i-n structures with different concentration of In (x=0; 0.13 and 0.23) in QD capping SRL was prepared. Growth of the structure was checked by the time resolved RAS at 2.6 eV for buffer, capping and conducting layers and at 4.2 eV for QDs formation [4]. The reason for the simultanously preparation of the p-i-n structures on n-type Si doped and SI GaAs substrates was the possibility to compare EL and PL spectra since the PL spectrum obtained on n-type Si doped substrate shows only a strong broad luminescence band from the substrate. PL and EL spectra were detected by a Ge detector using standard lock-in technique. PL was excited by a semiconductor laser with an emission wavelength of 808 nm and an excitation density ~10 Wcm^{-2}. The quality of the p-n junction was check by the measurement of current-voltage characteristics, which did not depend on the composition of the SRL. Three reference QD samples with the same active QD layer without p-n junction only with 50 nm thick GaAs capping layer were prepared. The set of prepared samples is shown in table 1.

Structure/ SRL	without SRL	13% In	23% In
PIN on n-doped substrate	A1	A2	A3
PIN on SI substrate	B1	B2	B3
Single QD layer on SI substrate	R1	R2	R3

Table 1. Summary of measured samples

3. Results and discussion

In situ time resolved RAS measurements showed that QDs were created in all samples. The example

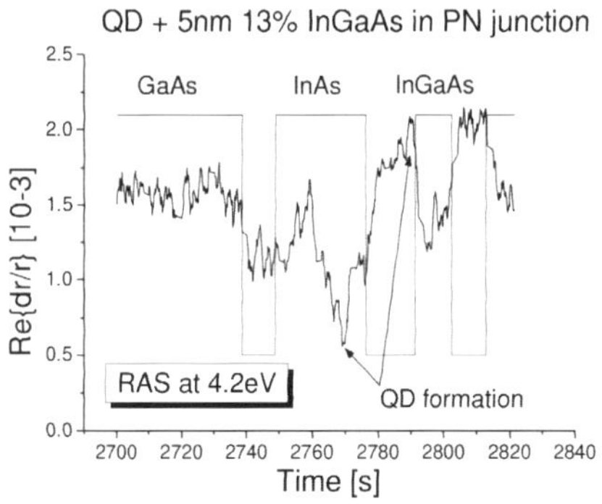

Figure 1. Time resolved RAS signal of sample A2, measured at 4.2 eV.

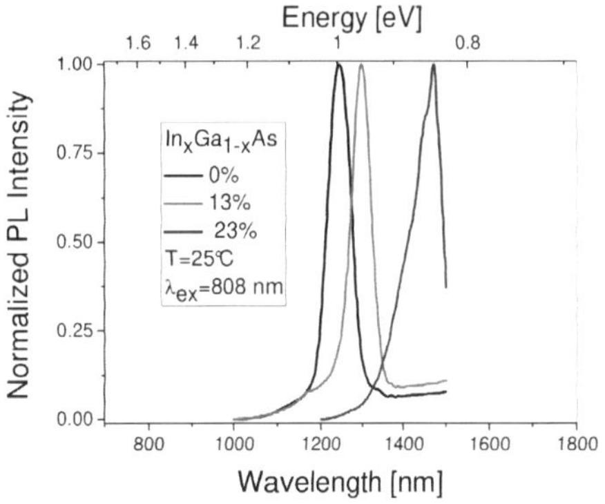

Figure 2. Comparison of normalized PL spectra of reference samples R1, R2 and R3.

of the results for the sample B2 during InAs and InGaAs deposition is shown in figure 1. PL spectra of reference QD samples show the shift of emission maximum to longer wavelengths with increasing In contend in SRL (figure 2). The full width at half maximum was about 50 meV, no clear dependence on In content was observed. The comparison of PL and EL spectra for samples B1 and A1 shows the same position of maxima at 1230 nm (figure 3). This maximum was ascribed to the ground state transition in QDs. The position of this maximum is in agreement with the result on a reference QD sample, where the PL maximum was at 1245 nm. The shoulders on the high energy side of spectra at 1120 nm are interpreted as the first excited transition in QDs. It was proved by measurements of the dependence of the PL and EL spectra on excitation intensity. The dependence of EL spectra on the excitation current is shown in figure 4.

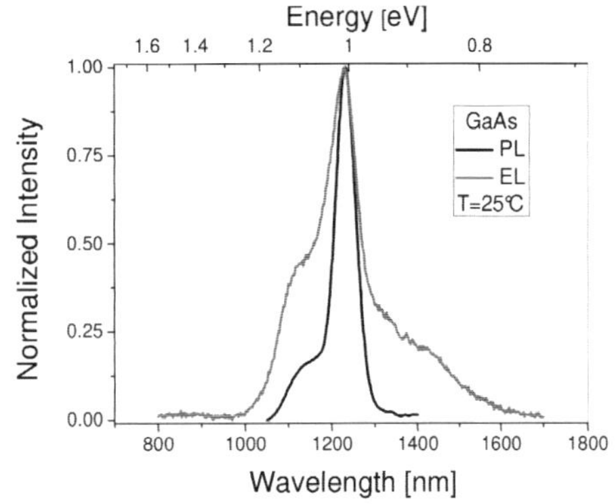

Figure 3. Comparison of PL and EL spectra of samples B1 and A1, at room temperature.

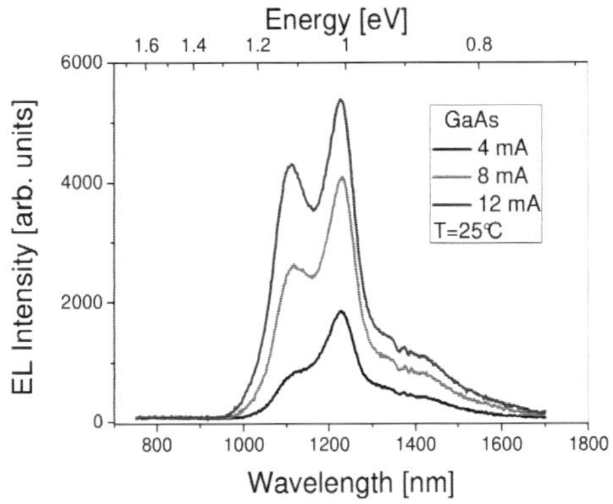

Figure 4. Dependence of EL spectra on the excitation current for sample A1

For samples B2 and A2 the wavelengths of PL and EL emission maxima were shifted to ~1289 nm and 1270 nm, respectively (figure 5). This agrees with the shift of the sample R1 (~1297 nm) and it can be interpreted as the ground state transition in QDs. The most intensive EL maximum at 1170 nm can be ascribed to the first excited transition in QDs. This agrees with PL results, where the first excited transition in QDs was found at ~1180 nm. For samples B3 and A3 surprisingly the blue shift of the PL and EL maxima were observed (figure 6). The position of PL maximum was ~1150 nm. This is much less than for the reference sample R3 where the maximum was at 1470 nm. The EL maximum is at 1120 nm. Another maximum at 970 nm was observed. Similar maximum was observed for sample A2 at 942 nm (figure 5).

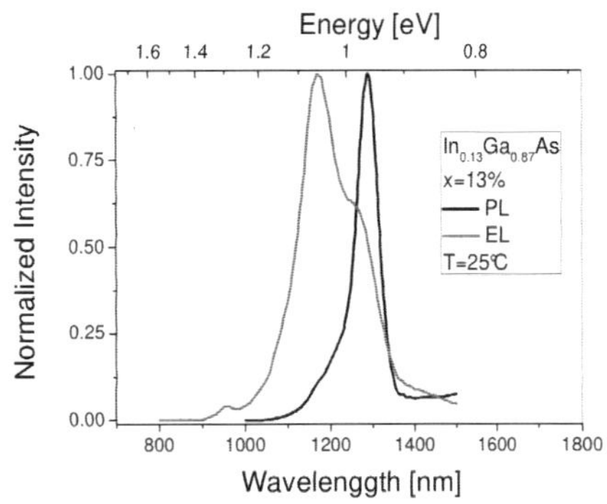

Figure 5. Comparison of PL and EL spectra of samples B2 and A2, at room temperature.

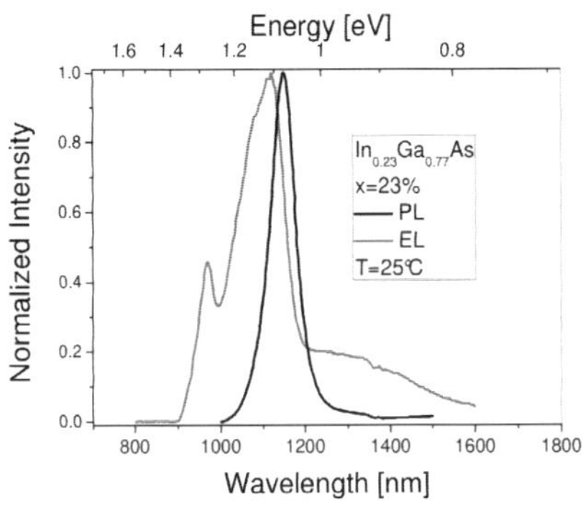

Figure 6. Comparison of PL and EL spectra of samples B3 and A3, at room temperature.

Blue shifted transitions cannot be interpreted as the transitions in QDs. The band at 1120 nm may be ascribed to a transition in InGaAs quantum well covering QDs as InGaAs SRL. This explanation is partially supported by our recent results on embedded thin InGaAs layers in GaAs [5]. The origin of narrow bands at 942 nm for sample A2 and at 970 nm for A3 is not clear yet.

4. Conclusions

We prepared a set of QD structures with different content of In in SRL which were embedded into a p-i-n structure. We observed room temperature electro- and photoluminescence. The supposed shift of the emission maxima towards low energy side of spectrum with increasing concentration of In in SRL was observed only for the p-i-n sample with 13% In in SRL (figure 7). For the p-i-n sample with 23% In in SRL the blue shift of the position of PL and EL maxima was observed (figure 8). We propose the interpretation of the blue shifted band by emission from the InGaAs quantum well covering QDs as InGaAs SRL. This interpretation has to be verified by next studies.

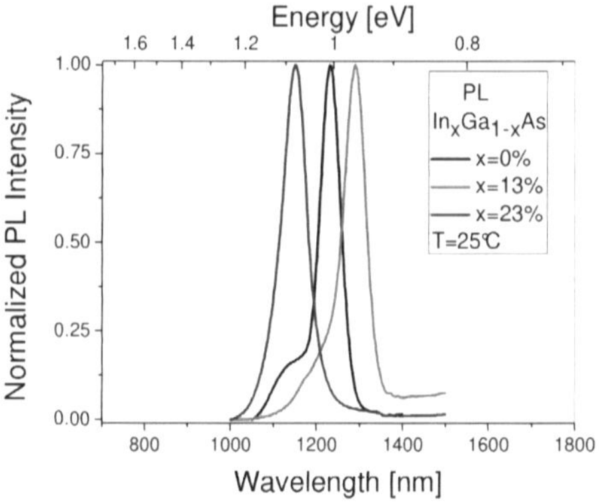

Figure 7. Comparison of normalized PL spectra of samples B1, B2 and B3, at room temperature.

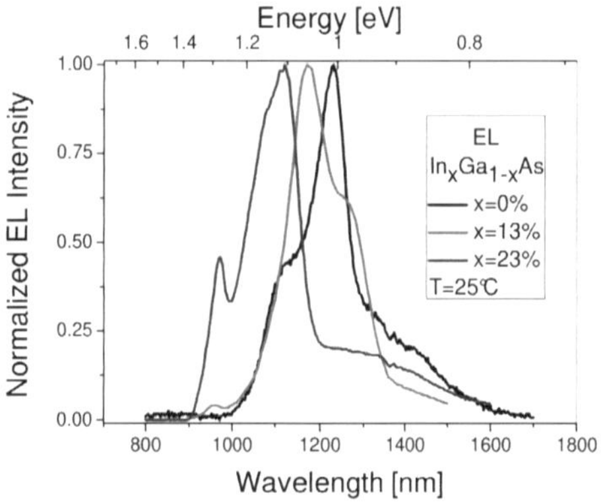

Figure 8. Comparison of normalized EL spectra of samples A1, A2 and A3, at room temperature.

Acknowledgments

This work was supported by Grant Agency of the Czech Science Foundation projects 202/09/0676 and P102/10/1201, the research program of the Institute of Physics AS CR AVOZ 10100521, and the program MSM 6840770014.

References

[1] Maximov M V, Ustinov V M, Zhukov A E, Kryzhanovskaya N V, Payusov A S, Novikov I I, Gordeev N Yu, Shernyakov Yu M, Krestnikov I, Livshits D, Mikhrin S and Kovsh A 2008 *Semicond. Sci. Technol.* **23** 105004

[2] Hazdra P, Oswald J, Komarnitskyy V, Kuldová K, Hospodková A, Vyskočil J, Hulicius E and Pangrác J 2009 *IOP Conf. Series: Materials Science and Engineering* **6** 012007

[3] Hazdra P, Oswald J, Komarnitskyy V, Kuldová K, Hospodková A, Hulicius E and Pangrác J 2009*Superlattices and Microstuctures* **46** 324-327

[4] Hospodková A, Vyskočil J, Pangrác J, Oswald J, Hulicius J, Kuldová K 2010 *Surface Science* **604** 318–321.

[5] Hazdra P, Voves J, Oswald J, Hulicius E, Pangrác J and Šimeček T 2003 *Journal of Crystal Growth* **248** 328-332

Theoretical modelling of quaternary GaInAsSb/GaAs self-assembled quantum dots

J. M. Llorens[1], A. G. Taboada[1], J. M. Ripalda[1], D. Alonso-Álvarez[1], B. Alén[1], J. Martín-Sánchez[1], J. M. García[1], Y. González[1], A. M. Sánchez[3], A. M. Beltrán[2], P. L. Galindo[2] and S. I. Molina[2]

[1] IMM-Instituto de Microelectrónica de Madrid, (CNM-CSIC), Isaac Newton 8, PTM, E-28760 Tres Cantos, Madrid, Spain.
[2] Universidad de Cádiz, Puerto Real, 11510 Cádiz, Spain
[3] Department of Physics, University of Warwick, Coventry CV4 7AL, United Kingdom

E-mail: jose.llorens@imm.cnm.csic.es

Abstract.
InAs/GaAs quantum dots exposed to Sb after growth exhibit spectral changes. We study in the present paper an idealized nanostructure consisting of a homogeneous distribution of the quaternary GaInAsSb surrounded by a barrier of GaAs. We find that the valence band offset is a critical parameter in modelling its electronic structure. Depending on this value, we predict a transition from type-I to type-II band alignment at a different Sb concentration. The addition of Sb to reduce the transition energy while keeping a type-I alignment is only of benefit at low Sb concentration.

1. Introduction

Light sources operating efficiently at 1.55 μm are highly desired by the telecommunication industry. Laser diodes based on semiconductor quantum wells are able to work at high rates, with low power consumption and high temporal and spatial coherence. An excellent candidate technology for lasing at the telecommunication window of interest relies upon InGaAsP/InP multi-quantum-wells. It would be however preferable to replace the InP substrate by an inexpensive one. The most extended approach to achieve this goal consists on growing InAs self-assembled quantum dots (QDs) on top of GaAs substrates. Even though the bandgap of InAs at low temperature lays at $\sim$ 3.0 μm, the emission of InAs/GaAs QDs occurs at $\sim$ 1.3 μm or shorter wavelengths. This blue shift results mainly of confinement and strain effects. The effect of the strain can be partially alleviated using a metamorphic layer [1] or a strain-reducing layer. Alternatively, it is possible to increase the emission wavelength by replacing As atoms by Sb ones, given that InSb presents an even narrower bandgap than InAs. This approach has been very promising, giving rise to room temperature emission at 1.6 μm from InAs/GaAs QDs covered with GaAsSb[2].

We constraint the current study to analyse the impact of Sb on the electronic structure of an InAs/GaAs QD. We leave size effects and the impact of metamorphic and strain-reducing layers for future work. We propose a model for describing such a QD exposed to Sb: a lens shaped volume of 10 nm radius and 8 nm height sited on top of a 0.5 nm wetting layer. Its composition consists of a homogeneous distribution of the quaternary $Ga_xIn_{1-x}As_ySb_{1-y}$. The amount of Sb in the alloy is limited to 0.5, since higher values are difficult to achieve experimentally. The surrounding barrier material is pure GaAs. Further details on actual samples of InAs/GaAs containing Sb can be found in [3, 4].

2. Method

The whole theoretical study has been performed with the Nextnano++ software package.[5] The strain distribution is computed by minimising the strain energy of the whole structure given the lattice mismatch between the QD material and the barrier. The QD electronic structure is then obtained by means of the 8×8 Kane's Hamiltonian in the framework of the effective mass approximation. Strain effects on the electronic structure are accounted for by adding the Bir-Pikus Hamiltonian to the Schrödinger equation. Further details can be found in Ref. [5].

We have followed the recommendations of Vurgaftman *et al.* [6] to set all material parameters, except the valence band offset (VBO). During the study, VBO has been shown to be critical in determining a type-I or type-II heterostructure band-alignment, which has a major impact on the electronic confinement energies. This has motivated to perform the calculation of the electronic structure for three different sets of VBO values: i) the recommended values of Vurgaftman *et al.* [6], ii) the theoretical values reported by Wei and Zunger [7] and iii) the recently revised VBO of Li *et al.* [8]. In the following, we will refer to these parameter sets as set A, B and C, respectively. The conduction and valence band alignment with respect to GaAs is shown in Figure 1 panels a) and b), respectively. Only GaSb presents a type-II conduction band alignment with respect to GaAs. However, if we introduce the strain due to the lattice mismatch, the band edges shift and the layout changes notably. In Fig. 1 c) can be seen that InSb/GaAs also exhibits a type-II band alignment.

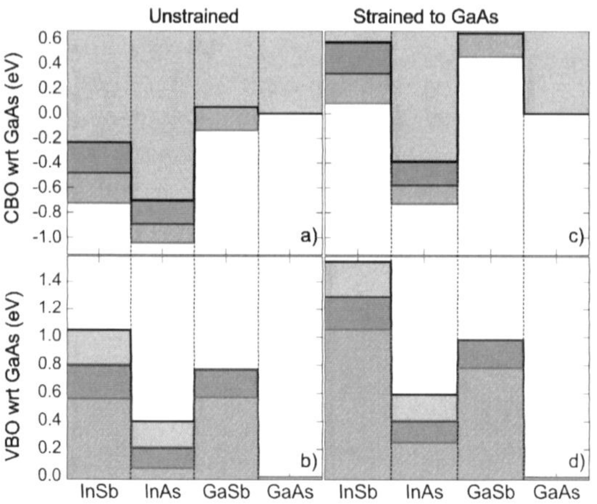

Figure 1. Band edges alignment for the conduction a) and valence b) bands. Three different set values for VBO are plotted: Vurgaftman *et al.* [6] (blue), Wei and Zunger [7] (red) and Li *et al.* [8] (black). The band edges line up under lattice mismatch strain is represented for conduction c) and valence d) bands.

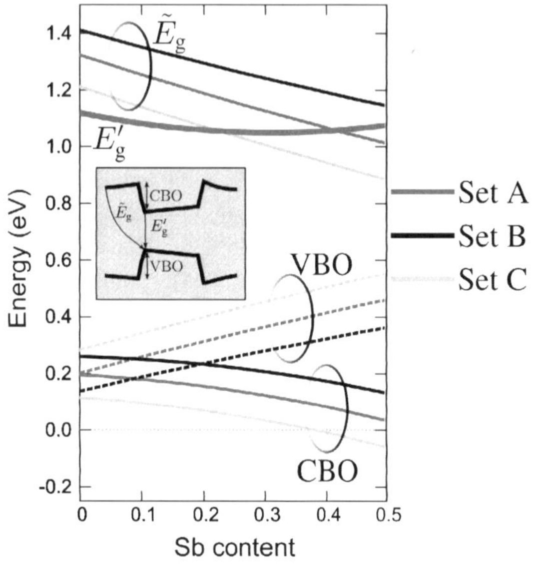

Figure 2. Spatially direct (thick red line) and indirect transitions (upper-half solid lines), and valence and conduction band-offsets (lower-half solid and dashed lines) calculated at the bottom surface of the quantum dot (inset).

The values of the quaternary parameters have been derived by interpolation of the four ternaries resulting of combining the four binary compounds involved in the alloy. We have taken the bowing parameters from Ref. [6], even for VBO, when available.

3. Results and discussion

To get a first insight into the confinement exerted by the QD on the electron and hole states, it is helpful to analyse the confinement potential at the points where the particle wavefunction tends to localise. In general, the electron wavefunction *fills* most of the QD's volume, while the hole often resides in the lower half of it. Thus, the e-h transition energy can be roughly approximated by the energy difference between the valence and conduction band edges at the centre of the bottom surface of the QD. These values differ from those of Fig. 1 c) and d) in the strain values, which now are the QD local strain ones. The effective energy gap (E'_g) as a function of the Sb concentration is shown in Fig. 2. The minimum at $y \sim 0.75$ is not related to the minimum of the ternary InAsSb bandgap, since the latter takes place at $y \sim 0.4$ [9]. It results from the increase of the strain net value due to the bigger lattice mismatch associated with the incorporation of Sb. In addition, the spatially indirect transition from the GaAs conduction band to the QD's top valence band is also of relevance here. We have labeled such energy as $\tilde{E}_g$ in Fig. 2 [1] The indirect transition crosses the direct one when sets A and C are considered. The crossing indicates at which Sb concentration the nanostructure could exhibit a type-II conduction band alignment. Finally, the effective band offsets are also included in Fig. 2. Within this qualitative picture, the holes are significantly more confined, the more Sb and conversely, the electrons.

The actual values of the electron-hole (e-h) transition energy are shown in Figure 3. There, we can see that the naive description based upon the confinement potential describes qualitatively well the dependence of the e-h energy on the Sb content. At low Sb content, the transition energy exhibits a red shift. This is followed by either a blue shift (for parameter set B) or a steeper red shift (for parameter set A and C). The latter is associated with the type-I to type-II band alignment transition. This transition takes place at y=0.43 and y=0.4 (y=0.22 and y=15.0) for x=0 and x=0.25, respectively, when set A (C) is considered. The values corresponding to set B have not been computed, being greater than y=0.5. The model predicts an emission at wavelengths greater than 1.2 μm for type-II QDs. For practical applications on light emitters, a type-II heterostructure is undesirable. The e-h overlap gets reduced in spatially indirect transitions and hence the oscillator strength and device performance. One could think that removing the Ga atoms from the QD could help in shifting the e-h transition energy towards red. This is not the case, as it is shown by the dashed lines in Fig. 3. At low y, there is a red shift of $\sim$ 100 meV. However, as more Sb is incorporated in the dot such shift becomes smaller. Although Ga atoms shift the band edges towards blue, the lattice mismatch in the quaternary with x=0.25 is smaller than in InAsSb. Thus, at $y \sim 0.5$ the transition energy becomes almost equal, for both Ga concentrations of 0% and 25%.

Ripalda *et al.* reported in Ref. [3] a photoluminescence spectrum peaked at 1.25 μm for InAs/GaAs QDs exposed to Sb and covered with GaAs. Our numerical results are consistent with such experimental observations for QDs with a composition of $Ga_xIn_{1-x}As_ySb_{1-y}$ with $0< x < 0.25$ and $0.1 \lesssim y \lesssim 0.25$. This conclusion should be taken cautiously. We have not computed the excitonic transition energy, which should appear red-shifted with respect to the energies of Fig. 3 as a result of the electrostatic Coulomb interaction. Typically, the excitonic correction is $\sim$ 20 meV in InAs/GaAs QDs[10]. Thus, to our understanding, such corrections would not have a great impact on the above estimation of the quaternary composition. However, an exact calculation should confirm our conjecture.

In conclusion, the optimal Sb concentration to get the narrowest transition energy in a $Ga_xIn_{1-x}As_ySb_{1-y}$/GaAs QD while preserving a type-I band alignment, results from the balance

[1] The conduction band energy in GaAs is taken at the centre of the top surface of the simulation box.

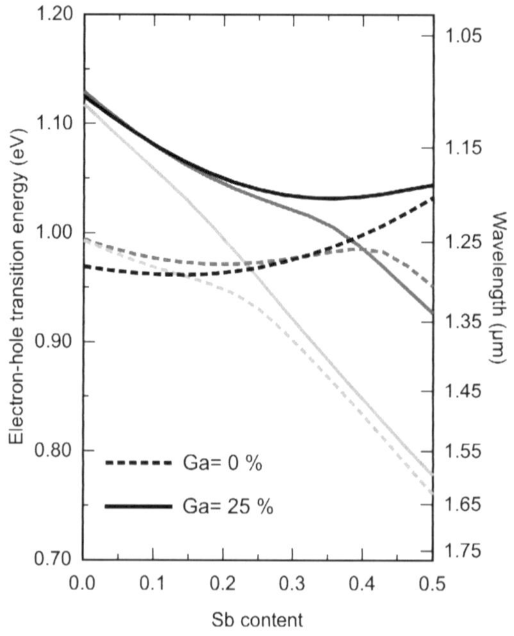

Figure 3. Electron-hole transition energy for a $Ga_{0.25}In_{0.75}As_ySb_{1-y}$/GaAs QD (solid lines) and an $InAs_ySb_{1-y}$/GaAs QD (dashed lines). The grey colour of the lines has the same meaning as in Fig 2.

of two effects: i) the red shift due to the narrower band-gap of InSb and ii) the blue shift associated to the strain induced by the higher lattice mismatch. The VBO parameter plays a critical role in the estimation of the conduction type-I to type-II transition. The results for set C show a type-II band alignment even for small concentrations of Sb. Conversely, the same calculation with set B does not show any type-I to type-II transition for $y \leq 0.5$. Finally, the results for set B show an in-between behaviour.

Acknowledgments

This work was supported by the Spanish MICINN (projects TEC2008-06756-C03-01/02/TEC, CONSOLIDER INGENIO 2010 CSD2006-0019 and CSD2009-00013), the Junta de Andalucía (PAI research groups TEP-120 and TIC-145; project P08-TEP-03516) and Comunidad Autónoma de Madrid S2009ESP-150.

4. References

[1] Ledentsov N, Kovsh A, Zhukov A, Maleev N, Mikhrin S, Vasil'ev A, Semenova E, Maximov M, Shernyakov Y, Kryzhanovskaya N, Ustinov V and Bimberg D 2003 *Electronics Letters* **39** 1126–1128
[2] Ripalda J M, Granados D, González Y, Sánchez A M, Molina S I and García J M 2005 *Appl. Phys. Lett.* **87** 202108–3
[3] Ripalda J M, Alonso-Álvarez D, Alen B, Taboada A G, García J M, González Y and González L 2007 *Appl. Phys. Lett.* **91** 012111
[4] Molina S I, Sánchez A M, Beltrán A M, Sales D L, Ben T, Chisholm M F, Varela M, Pennycook S J, Galindo P L, Papworth A J, Goodhew P J and Ripalda J M 2007 *Appl. Phys. Lett.* **91** 263105–3
[5] Birner S, Zibold T, Andlauer T, Kubis T, Sabathil M, Trellakis A and Vogl P 2007 *IEEE Transactions on Electron Devices* **54** 2137–2142
[6] Vurgaftman I, Meyer J R and Ram-Mohan L R 2001 *J. Appl. Phys.* **89** 5815–5875
[7] Wei S H and Zunger A 1998 *Appl. Phys. Lett.* **72** 2011
[8] Li Y H, Walsh A, Chen S, Yin W J, Yang J H, Li J, Da Silva J L F, Gong X G and Wei S H 2009 *Appl. Phys. Lett.* **94** 212109
[9] Fang Z M, Ma K Y, Jaw D H, Cohen R M and Stringfellow G B 1990 *J. Appl. Phys.* **67** 7034
[10] Grundmann M, Stier O and Bimberg D 1995 *Physical Review B* **52** 11969–11981

Control of the lateral growth morphology in GaAs Droplet Epitaxy

C Somaschini[1], S Bietti[1], S Sanguinetti[1], N Koguchi[1] and A Fedorov[2]

[1] LNESS and Dipartimento di Scienza dei Materiali dell'Università di Milano-Bicocca, via Cozzi 53, 20125 Milano, Italy
[2] CNISM, LNESS and Dipartimento di Fisica del Politecnico di Milano, via Anzani 42, 22100 Como, Italy

Email: stefano.sanguinetti@mater.unimib.it

Abstract. We present the detailed fabrication method of two different GaAs nanostructures with cylindrical symmetry by the Droplet Epitaxy technique. Concentric Multiple Rings or Coupled Rings/Disks are successfully obtained, exploiting the lateral growth around the Ga droplets, through a fine control of the crystallization dynamics.

1. Introduction

During the past two decades, semiconductor quantum nanostructures (NS) have been extensively investigated because of their potential application in the field of optoelectronics as active elements for lasers [1, 2], light-emitting diodes [3], detectors [4] and quantum information technology, as sources of entangled states [5, 6]. In particular, self-assembled Stranski-Krastanov (S-K) quantum dots (QDs) [7], formed by growing a lattice mismatched structure, as a result of the balance between strain relaxation and surface energy, have widely been investigated. However, these kind of systems mainly show a pyramidal-like shape and therefore the tuning of the NS optoelectronic properties through the careful design of their shape is very difficult.

Amongst the Molecular Beam Epitaxy-based techniques, the Droplet Epitaxy [8, 9] (DE) has shown an unmatched potential for the fabrication of complex III-V semiconductor NS. In DE, unlike the more conventional S-K growth, the III- and V-column elements are alternatively supplied to the substrate surface, for the formation of quantum NS. In the case of GaAs, after the irradiation of Ga molecular beam flux, nanometer-sized droplets with a low size dispersion spontaneously form at the surface. If an As flux is then supplied, new GaAs nanocrystals develop from the original droplets. Depending on the growth conditions, several kind of different morphologies have been obtained ranging from QDs [10], quantum rings [11], concentric multiple rings [12] and coupled ring/disks [13, 14].

We demonstrate the fabrication method of complex GaAs NS with cylindrical symmetry, based on the fine control of the lateral growth around the Ga droplets during the As supply. With a proper choice of the growth conditions, the morphology of the crystal can be tuned from a clear peak-and-valley profile for Concentric Multiple Rings (CMRs) to a flat disk-like structure grown in a layer-by-layer fashion, for Coupled Rings/Disks (CRDs). These structures were characterized by means of Atomic Force Microscopy (AFM) and Reflection High Energy Electron Diffraction (RHEED) analysis. The interplay between the Ga diffusion from the initially formed Ga droplets and the As adsorption onto the substrate surface in the different experimental conditions is thought to be crucial for the shape engineering of the NS.

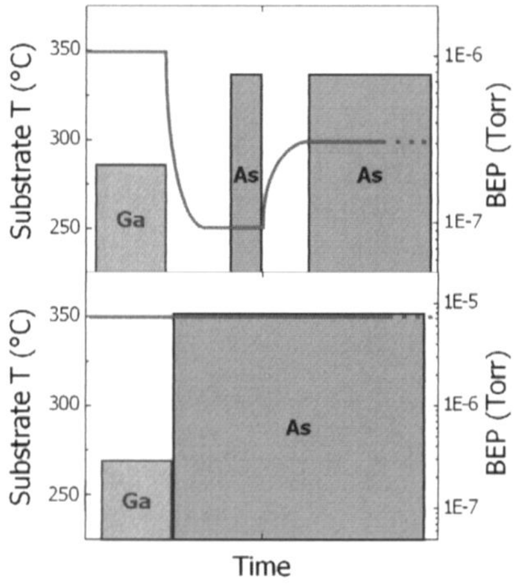

Figure 1. *Sketch of the experimental procedure for Sample A (upper panel) and B (lower panel). Red lines show the substrate temperature evolution during the growth. Green and blue areas are depicted in correspondence to the gallium and arsenic supply, respectively.*

2. Experimental

The growth experiments were performed on GaAs (100) wafers in a standard GEN II MBE system equipped with an As valved cell, which provided As_4 molecules. After the native oxide desorption at 580°C, a buffer layer of GaAs and a barrier layer of $Al_{0.3}Ga_{0.7}As$ were grown at the same temperature, taking care of the atomically flatness of the surface. After that, the substrate temperature was decreased to 350°C, the As flux cut and the background pressure in the chamber reduced to around $5x10^{-10}$ Torr. The substrate surface was then exposed to a Ga molecular beam equivalent to 10 MLs, with a deposition rate of 0.08 ML/s, to form droplets. Two samples, named A and B were prepared, following different experimental procedures, constantly monitored by means of RHEED. In the case of Sample A, after the Ga droplets formation, the substrate temperature was decreased to 250°C and a short-time (20 sec) As flux of $8x10^{-7}$ Torr was supplied. Finally the substrate temperature was raised again to 300°C and a second, long-time (20 min) arsenic exposure of the same intensity was used to completely ensure the crystallization. On the contrary, for Sample B, after the Ga deposition at 350°C, the As shutter was immediately opened at the same temperature, providing an $8x10^{-6}$ Torr arsenic partial pressure. In Fig. 1 the detailed experimental procedure for the fabrication of Sample A (upper panel) and Sample B (lower panel) is shown. After the growth, the samples were taken out from the chamber for the morphological analysis, carried out by Atomic Force Microscopy (AFM).

3. Results and discussion

As already described above, the first step for the fabrication of GaAs NS by DE is the formation of Ga droplets at the substrate surface. In our case, 10 MLs of Ga were supplied at 350°C for both Sample A and Sample B. In order to confirm the presence and the morphology of the Ga droplets, another sample was prepared, by stopping the growth just after the Ga deposition. The substrate temperature was rapidly cooled down (40 °C/min) to around 200°C and the sample taken out from the growth chamber. AFM measurements, which confirmed the existence of Ga droplets on the $Al_{0.3}Ga_{0.7}As$ barrier layer are shown in Fig. 2. The average droplets diameter and height are around 90 nm and 35 nm respectively and the surface density of the Ga droplets is $8x10^{8}$ cm^{-2}. Taking into account that 1.7 MLs of Ga are consumed to create a Ga-rich surface reconstruction [15] before the Ga droplets formation, a perfect correspondence is found when comparing the total amount of Ga atoms into the droplets and the initially supplied 10 MLs.

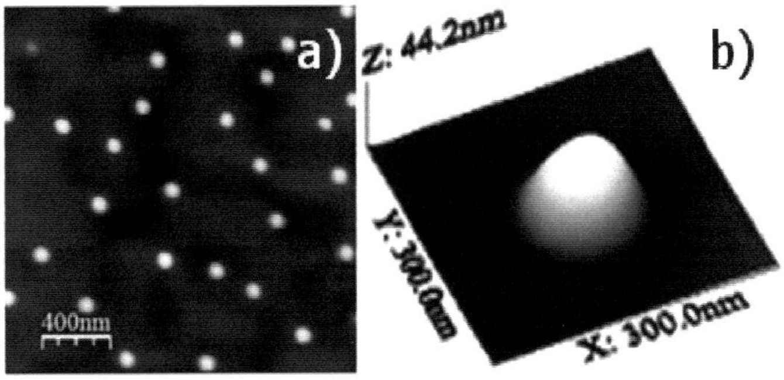

Figure 2. *Two dimensional 2 micron x 2 micron AFM image of Ga droplets just after the deposition a). Three dimensional magnification of a single droplet b).*

In Fig. 3 the AFM images (a, c) as well as the cross sectional height profiles (b, d) for Samples A (upper panels) and B (lower panels) are presented. In the case of Sample A, the final morphology consists of a concentric triple ring with inner, middle, and outer ring diameters of around 80, 140, and 210 nm, respectively and with heights around 8 nm for the inner rings, 4 nm for middle rings, and 3 nm for the outer rings. This triple ring structure present a good rotational symmetry, with a slight elongation of around 11% along the [0-11] direction, which can be caused by the anisotropic surface migration of Ga on the (100) GaAs surface [16]. The shape of Sample B is instead characterized by an inner ring with a diameter and height of around 80 nm and 11 nm respectively, surrounded by a disk with a diameter of around 200 nm and a nearly constant height of around 7 nm. The NS density at the surface is nearly equivalent to the one of the initial Ga droplets, suggesting that every droplet is transformed into the above described GaAs NS at the end of the procedures. The difference in the volume of the final GaAs structures (lower for Sample A) has to be addressed to the different experimental procedures, involving changes in the substrate temperature. During the waiting time, necessary for the substrate temperature stabilization, a partial Ga crystallization, forming a 2D thin layer all over the substrate, is observed even in absence of intentional As supply, leading to a decrease in the effective number of Ga atoms available for the formation of the final nanocrystal.

A nearly perfect correspondence is found when the diameter of the inner ring in both Sample A and B is compared to the original droplets one (80-90 nm in every case). Selective etching of pure Ga showed that, underneath the droplets, a tiny GaAs ring is already formed, just after the Ga deposition [11]. These observations imply that the inner ring formed at the end of the growth is the result of the accretion of the initially formed ring, at the droplets edge. This is the reason why, for both CMRs and CRDs, regardless of the difference in the experimental procedure, the inner ring diameter is nearly constant.

On the contrary, the lateral growth around the inner rings seems to be strongly influenced by the different procedures. Indeed the cross sectional morphology of Sample A is very different from the one of Sample B, since, in the latter, no clear peak-and-valley features are observed in the height profile. As a matter of fact, the development of GaAs NS laterally grown around the Ga reservoir depends on the mobility of Ga atoms which migrate from the droplet and are crystallized, due to the impinging As atoms. However in the first sample, where the arsenization steps are performed at 250°C and 300°C, we observed a peak-and-valley line profile, while in the second sample, where As was supplied at 350°C, a flat disk was grown. Indeed, as already discussed in [12] the first and second arsenisation steps lead to the formation of the middle and outer rings, respectively. This observation suggests that a key role in determining the final shape of the NS might be played by the substrate temperature during the arsenic irradiation to the droplets. The appearance of outer rings, with a well defined peak-and-valley profile can be addressed to the formation of As-rich c(4x4) regions around the droplets, as discovered by RHEED investigations [12]. The establishment of this As-rich regions, which are responsible for the formation of the rings, might be less favorable at $T \geq 350$°C, caused by a higher desorption of As_4 from the substrate in the high temperature regime [17]. In absence of these zones for the preferential GaAs nucleation, the lateral growth proceeds in a layer-by-layer fashion, giving rise to the disk-like structure around the original Ga droplet. The dissimilar growth dynamics thus gave rise to the distinct lateral morphologies found in Sample A and Sample B. Therefore the fine control of the interplay between the migration of Ga atoms from the droplets and the As adsorption onto the substrate surface gives the possibility of a shape engineering of the GaAs NS grown by DE. Since electronic properties depend on the geometrical features of the NS, this capability of tuning the shape of the MBE grown GaAs NS is of the outmost relevance in view of the establishment of a future generation of devices.

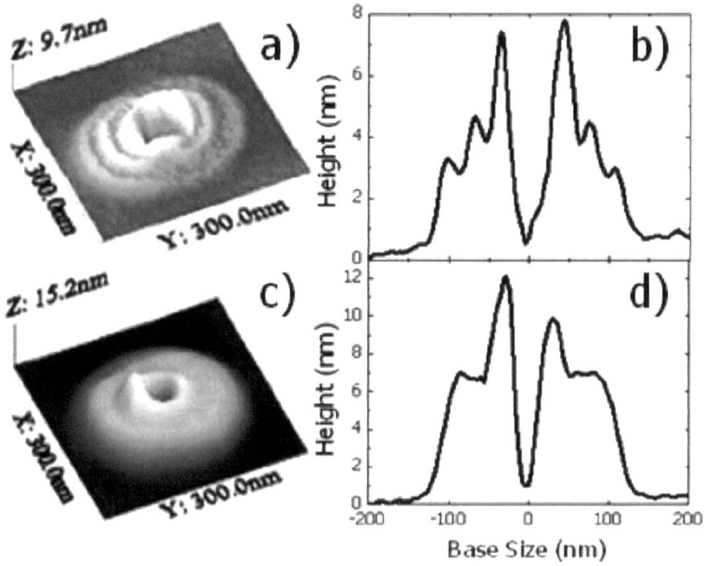

Figure 3. *AFM images of a single CMR a) and CRD c) and corresponding cross sectional height profiles taken along the (0-11) direction b), d). For each sample, every droplet is transformed into the described structure, with a size dispersion comparable to the one of the original Ga droplets.*

4. Conclusion

In conclusion, we have presented the detailed fabrication method of two different classes of GaAs NS grown by DE: CMRs and CRDs. In both cases, the lateral growth around the initially formed Ga droplets is exploited for the realization of the structure. However, while for the CMRs a clear peak-and-valley cross sectional height profile is obtained, for the CRDs a nearly flat structure is developed. We demonstrated that the morphology of the NS can be finely controlled by a suitable choice of the growth parameters. We believe these innovative GaAs NS to be promising for the future application in optoelectronics and quantum information technology.

References

[1] Arakawa Y, Sakaki H 1982 *Appl. Phys. Lett.* **40** 939.
[2] Klimov V I, Mikhailovsky A A, Xu S, Malko A, Hollingsworth J A, Leatherdale C A, Eisler H J, Bawendi M G 2000 *Science* **290** 314.
[3] Tessler N, Medvedev V, Kazes M, Kan S, Banin U 2002 *Science* **295** 1506.
[4] Vandervelde T E, Lenz M C, Varley E, Barve A, Shao J, Shenoi R V, Ramirez D A, Jang W, Sharma Y D, Krishna S 2008 *IEEE J. Quantum Electron.* **14** 1150
[5] Burkard G, Loss D, DiVincenzo D P 1999 *Phys. Rev. B* **59** 2070.
[6] Bayer M, Hawrylak P, Hinzer K, Fafard S, Korkosinski M, Wasilewski Z R, Stern O, Forchel A 2001 *Science* **291** 451.
[7] Grundmann M, Stier O, Bimberg D 1995 *Phys. Rev. B* **52** 11969.
[8] Koguchi N, Takahashi S, Chikyow T 1991 *J. Cryst. Growth* **111** 688.
[9] Koguchi N, Ishige K 1993 *Jpn. J. Appl. Phys.* **32** 2052.

[10] Watanabe K, Koguchi N, Gotoh Y, 2000 *Jpn. J. Appl. Phys.* **39** L79
[11] Mano T, Koguchi N, 2005 *J. Cryst Growth* **278** 108.
[12] Somaschini C, Bietti S, Koguchi N and Sanguinetti S 2009 *Nano Lett.* **9** 3419.
[13] Somaschini C, Bietti S, Sanguinetti S, Koguchi N, Fedorov A 2010 *Nanotechnology* **21** 125601.
[14] Bietti S, Somaschini C, Sanguinetti S, Koguchi N, Isella G, Chrastina D 2009 *Appl. Phys. Lett.* **95** 241102.
[15] Ohtake A, Kocan P, Seino K, Schmidt W G, Koguchi N 2004 *Phys. Rev. Lett.* **93** 266101.
[16] Ohta K, Kojima T, Nakagawa T 1989 *J. Cryst. Growth* **95** 71.
[17] Garcia J C, Neri C, Massies J 1989 *J. Cryst. Growth* **98** 511.

Bilayer for extending the wavelength of QD lasers

M A Majid[1], D T D Childs[1], R Airey[1], K Kennedy[1], R A Hogg[1], E Clarke[2], P Spencer[2], and R Murray[2]

[1]EPSRC National Centre for III-V Technologies, University of Sheffield, S3 7HQ, UK
[2]Physics Department, Imperial College London, SW7, UK

E-mail: - **a.m.mohammed@sheffield.ac.uk**

Abstract. This paper reports on the analysis of MBE grown bilayer quantum dot (QD) laser material. Specifically, gain and absorption characteristics of 5x 'single' QD layers with InGaAs caps, 5x 'bilayer' QDs with GaAs caps and 5x 'bilayer' QDs with InGaAs cap layers are compared. For an $In_{0.18}Ga_{0.82}As$ capped bilayer sample at room temperature, net modal gain is demonstrated beyond 1500 nm.

1. Introduction

The InAs/GaAs QD system has received strong interest due to the evidence of true zero dimensional behaviour at room temperature and above. Whilst typical emission wavelengths are found in the 1000-1300 nm range, longer wavelength QD lasers on GaAs are attractive for datacoms and telecoms applications. Many of the attractive attributes of the quantum dot (e.g. zero and tunable chirp, broad-band emission, temperature insensitivity, saturable absorption, single photon emission, etc.,) are particularly attractive at the fibre optic transmission wavelengths of ~1310 and ~1550nm. However, emission beyond 1300nm is difficult to achieve, when using independent single layers of InAs/GaAs QDs [1]. The use of bilayers of QDs has been demonstrated as a promising route to obtaining long wavelength emission and lasing [2, 3]. In the bilayer structure, each of the two QD layers is grown at different temperature, with strain interaction from the smaller QDs in the seed layer (first layer) fixing the QD density in the larger QDs of the second layer [2, 3]. In this paper we compare experimentally the gain characteristics of laser devices utilising a fivefold repeat of single QD layers with $In_{0.26}Ga_{0.74}As$ cap layers and a fivefold repeat of bilayer QDs with GaAs cap layers which are designed to operate at the same wavelength and have nominally identical QD densities (~2×10^{10} cm^{-2}). We also compare net modal gain and modal absorption spectra using the modified variable stripe length method in multi-section devices [4].

2. Experiment

The laser structures described here are shown schematically in figure 1. Growth was carried out on n^+ GaAs (100) substrates by molecular-beam epitaxy (MBE). The active region for both structures was located in a 500 nm undoped GaAs layer sandwiched between 1500 nm $Al_{0.33}Ga_{0.67}As$ cladding layers. A 400 nm p-type GaAs:Be contact layer was grown to complete the laser structure. The active region of the 'standard' structure consists of five independent QD layers which we refer to as a "single layer" structure. Before each QD layer was grown, the surface was annealed under an As_2 flux at 580 ^oC for 10 minutes to smooth the growth surface [5]. The QD layer was then subsequently grown at 485 ^oC by the deposition of 2.4 ML of InAs at a growth rate of 0.014 MLs^{-1}, giving a QD density of 2.4 x 10^{10} cm^{-2} as measured by AFM on test structures. The QDs were capped by 4 nm of $In_{0.26}Ga_{0.74}As$ then 11 nm of GaAs, also grown at 485 °C to reduce Indium segregation. After this the growth temperature was ramped during the next 45 nm of GaAs deposition to the optimum GaAs growth temperature of 580 °C [6]. The same sequence was repeated for the rest of the active region and the remainder of the structure was grown at 580 °C to reduce In-Ga inter-diffusion.

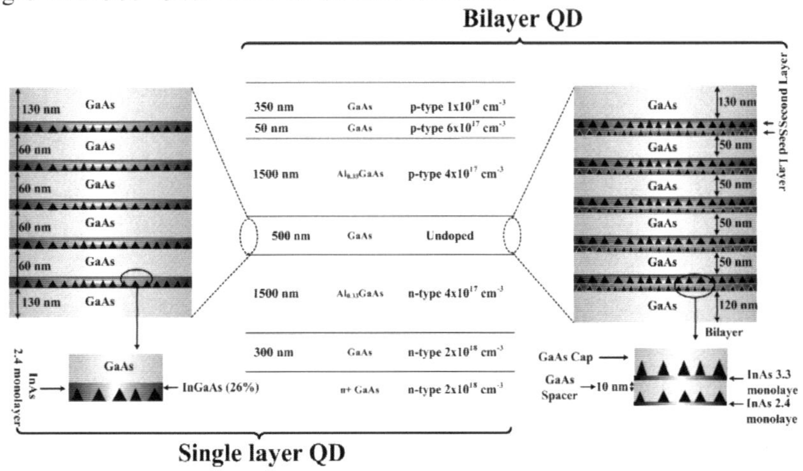

Figure 3. Single (left) and Bilayer (right) QD laser structures.

The second sample had a nominally identical waveguide structure, differing only by the replacement of the five single QD layers with 5 QD bilayers. We refer to this sample as the "bilayer QD" sample in the following. Each bilayer consists of a seed layer of QDs and an emission layer. As before, the GaAs surface was annealed at 580 °C for 10 minutes thereafter the seed layer was grown at a temperature of 480 °C by the deposition of 2.4 ML of InAs at a growth rate of 0.014 MLs^{-1}, giving a QD density of 2.7 x 10^{10} cm^{-2} (slightly higher than for the single layer structure due to the lower growth temperature). These seed QDs were capped by 10nm of GaAs spacer layer, also grown at 480 °C, before the temperature was raised again to 580 °C for 10 minutes to smooth the growth surface and desorb excess segregated indium. The emission layer was then grown at a reduced temperature of 467 °C by the deposition of 3.3 ML InAs at the same growth rate as for the seed layer. These QDs were then capped with 15 nm of GaAs also at 467°C, after which the temperature was ramped to 580 °C for subsequent GaAs growth. The same sequence was repeated for the rest of the active region with the rest of the structure grown at 580 °C. Wafers were processed into broad area lasers for the multilength analysis and segmented contact devices for the variable stripe analysis. Characteristics were measured in pulsed regime (5μs pulsed duration, 1% duty cycle) to minimise thermal effects. All measurements were performed at a tile temperature of 300K.

3. Experimental dependence of the net modal gain on the pump current density

Length dependant characterization of Fabry-Perot lasers allows a simple determination of the gain characteristics of the laser material [7, 8]. Figure 2 shows the experimental dependence of net modal gain on the drive current density for the single layer QD laser. The curves are obtained by fitting the empirical gain-current relationship given by Zhukov *et al.* [8]. As cavity length is decreased, there is an increase in the mirror loss (α_m) to a point where the total cavity loss exceeds the ground state (GS) saturated gain. At this point first excited state (ES1) lasing may occur, accompanied by increased threshold current. As the length is further decreased, so the ES1 state may also saturate, leading to lasing by the second excited (ES2) state etc. From figure 2, we observe ground state (1360 nm) saturation at a net modal gain of ~2 cm^{-1}. The first excited state (1275 nm) saturates at a net modal gain of ~8 cm^{-1}. The saturation of the second excited state is not explored in this case. Due to the in-plane symmetry of these self-assembled QDs, the first excited state has twice the degeneracy of the ground state. Consequently, the saturated modal gain for the first excited state is assumed to be twice that of the ground state [9]. Considering this, an internal loss (α_i) of ~4 cm^{-1} may be deduced, with a modal gain of ~6 cm^{-1} for the ground state and ~12 cm^{-1} for the first excited state.

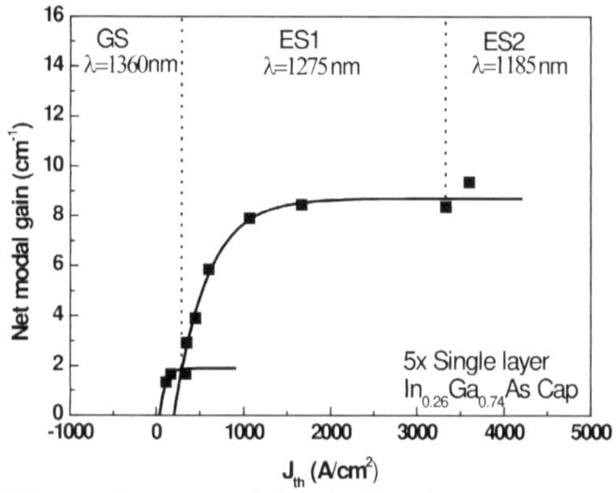

Figure 2. Net modal gain for single layer laser as function of current density. Squares represent experimental data. Curves correspond to a fitting by empirical equation from ref [8].

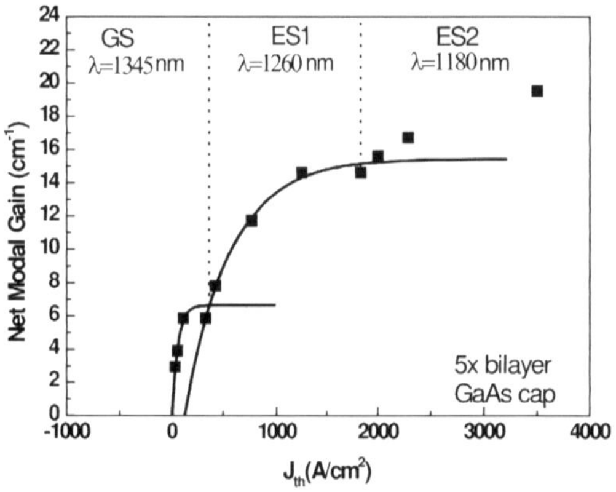

Figure 3. Net modal gain for bilayer laser as function of current density. Squares represent experimental data. Curves correspond to a fitting by empirical equation from ref [8].

Figure 3 shows the experimental dependence of net modal gain on the drive current density for the GaAs capped bilayer QD laser. We observe ground state (1345 nm) saturation at a net modal gain of ~6cm^{-1}. First excited state (1275 nm) saturates at a net modal gain of ~15 cm^{-1}. The saturation of the second excited state is not explored. If we assume a two-fold increase in degeneracy and saturated modal gain for GS and ES1 of the top QD layer, an internal loss of ~4 cm^{-1} is again deduced, with a modal gain of 10 cm^{-1} for the GS and ~20 cm^{-1} for the ES1. Considering the lower QD areal density and reduced layer number, these figures for saturated gain compare quite favourably with commercial QD material at 1310nm.

4. Measurement of Gain and Absorption spectra

Ref. [10] details the design of the multi-section device, which uses a stripe length dependent measurement to obtain gain and absorption spectra as described by Blood *et al.* [4].

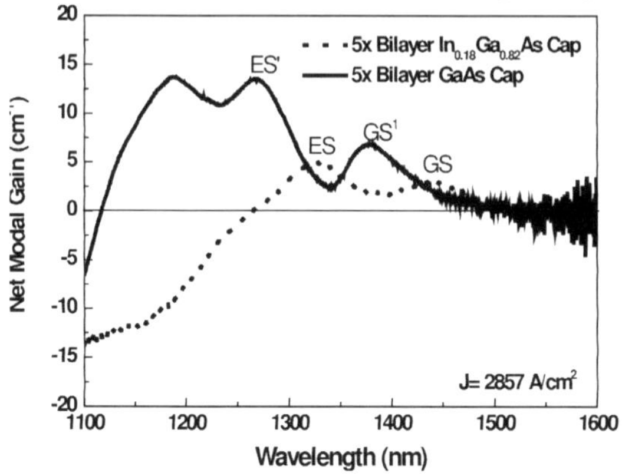

Figure 4. Net modal gain spectra of InGaAs and GaAs cap bilayer device as measured using the multi-section technique.

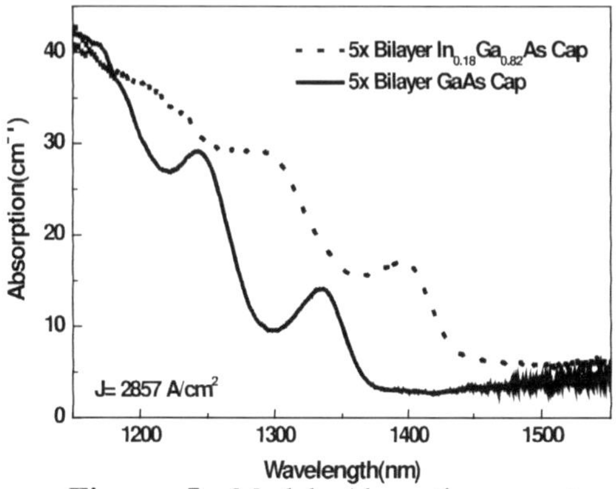

Figure 5. Modal Absorption spectra of InGaAs and GaAs cap bilayer device as measured the using multi-section technique.

The net modal gain spectra of two different bilayer devices with QDs capped by either GaAs or $In_{0.18}Ga_{0.82}As$ are shown in figure 4. In this technique, by changing the pumping a series of net modal gain spectra can be obtained, but for comparison we have selected current density of 2857 A/cm^2, where the ground and first excited state is clearly resolved for both the devices .At this current density for the GaAs cap bilayer sample we observe the GS (1370 nm) and ES1 (1270 nm) net modal gain of 6 cm^{-1} and 13 cm^{-1} respectively. Similarly, for the InGaAs cap GS^1 (1450 nm) and ES^1 (1330 nm) net modal gain of 3 and 5 cm^{-1} are observed. The gain peak wavelength is in good agreement with emission spectra obtained from the laser devices (not shown). Significantly, we observe a positive net modal gain above 1500 nm for the InGaAs cap bilayer device. This is a step towards achieving telecoms devices on GaAs substrates. Figure 5 shows the net modal absorption spectra of GaAs and InGaAs cap bilayer devices, plotted for the same current density (2857 A/cm^2) also using the multi-section devices. We note that the absorption value at a given wavelength is significantly larger than the corresponding gain [10]. We have also observed that absorption does not vary with the wavelength below the band edge of the material indicating an internal loss value of $3\pm0.5cm^{-1}$ and $6\pm0.5cm^{-1}$ for GaAs and InGaAs cap bilayer devices.

Finally, it can be seen that the absorption spectra for both the devices is similar in shape, similar in number of features and the peak modal absorption is the same. This indicates that in the bilayer, the QDs of the seed and emission layers have not just strong correlation structurally but also electrically [2].

5. Conclusion

We have shown a method of extending the emission wavelength of QDs on GaAs by using a bilayer growth technique. The GaAs capped bilayer laser demonstrated twice the modal gain for ground and excited states as compared to an InGaAs capped 'single' layer QD laser operating at similar wavelength (~1350 nm). InGaAs capping the bilayer of QDs further extends the emission wavelength to 1450 nm. Significantly, by optimization of the QD capping material it is possible to achieve long wavelength emission without sacrificing modal gain. Indeed, net modal gain above 1500 nm is noted, presenting an exciting opportunity towards fabricating optoelectronic devices with all the advantages of QDs in the telecoms windows of 1.31 and 1.55μm.

Acknowledgements

This work is supported by EPSRC (UK) grant EP/FO3427X/1.

References

[1] Zhukov A E, Kovsh A R, Maleev N A, Mikhrin S S, Ustinov V M, Tsatsul'nikov A F, Maximov M V, Volovik B V, Bedarev D A, Shernyakov Y M, Kop'ev P S and Alferov Zh I 1999 *Appl. Phys. Lett.* **75**, 1926

[2] Le Ru E C, Howe P, Jones T S and Murray R 2003 *Phys. Rev. B* **67** 165303

[3] Clarke E, Spencer P, Harbord E, Howe P and Murray R 2008 *Journal of Phys.: Conference Series* **107** 012003

[4] Blood P, Lewis G M, Smowton P M, Summers H, Thomson J and Lutti J 2003 *IEEE J. Sel. Top. Quantum Electron* **9** 1275

[5] Le Ru E C, Bennett A J, Roberts C and Murray R 2002 *J. Appl. Phys.* **91** 1365

[6] Liu H Y, Sellers I R, Gutiérrez M, Groom K M, Soong W M, Hopkinson M and David J P R 2004 *J. Appl. Phys.* **96**, 1988

[7] Kirstaedter N, Schmidt O G, Ledentsov N N, Bimberg D,Ustinov V M, Egorov A Yu, Zhukov A E, Maximov M V, Kop'ev P S and Alferov Zh I 1996 *Appl. Phys. Lett.* **69** 1226–1228

[8] Zhukov A E, Kovsh A R, Ustinov V M , Egorov A Yu, Ledentsov N N , Tsatsul'nikov A F , Maximov M V , Shernyakov Yu M, Kopchatov V I, Lunev A V, Kop'ev P S, Bimberg D and Alferov Zh I 1999 *Semicond. Sci. Technol.* **14**, 118

[9] Stevens B J, Childs D T D, Shahid H and Hogg R A 2009 *Appl. Phys. Lett.* **95**, 061101

[10] Judson P D L, Groom K M, Childs D T D, Hopkinson M and Hogg R A 2009 *IET Optoelectron.* **3**, 100-104

Magnetic and magneto-optical properties of CdS:Mn quantum dots in PVA matrix

V I Fediv[a], A I Savchuk[b], V M Frasunyak[b], V V Makoviy[b] and O A Savchuk[b]

[a]Bukovinian State Medical University, 2 Teatralna Square, Chernivtsi 58002, Ukraine
[b]Chernivtsi National University, 2 Kotsyubynsky Street, Chernivtsi 58012, Ukraine

E-mail: vfediv@mail.ru

Abstract. We have studied the magnetic and magneto-optical properties of CdS:Mn quantum dots in polyvinyl alcohol matrix synthesized by co-precipitation method. The size of quantum dots was estimated by means of absorption spectroscopy. The results of measurements of magnetic susceptibility as a function of temperature and spectral dependence of the Faraday rotation of CdS:Mn quantum dots / polyvinyl alcohol composites are presented. In this work magnetic susceptibility was investigated by Faraday's method at the temperatures of (78-300) K in magnetic fields of (0.05-0.8) T. The inverse magnetic susceptibility as a function of temperature follows a Curie Weiss law. Formation of ferromagnetic coupling between magnetic ions is supposed. Magneto-optical Faraday rotation has been investigated in the wavelength region (400-700) nm at temperature 300 K in a magnetic field up to 5 T. Sign of the Verdet constant is found to be negative.

1. Introduction

At present many studies in the field of nanotechnology are dedicated to search of bottom-up approaches for the preparation of device structures analogous to those grown by molecular beam epitaxy or related vacuum deposition techniques. The preparation and characterization of polymer/inorganic nanoparticles composites have received increasing attention in recent years because materials in the nanometer scale have provided not only new physics in reduced dimensions, but also the possibility of fabricating novel materials. Recently the attention to inorganic nanoparticles has shifted toward doping of colloidal semiconductor quantum dots (QD) by direct chemical methods because of the vast untapped potential to control their physical properties through the introduction of impurities. Diluted magnetic semiconductor (DMS) nanostructures are the basic working components of many existing and proposed devices in the emerging field of semiconductor spin-based electronics. These materials attract an increasing interest because of unusual combination of optical, magnetic and semiconducting properties. Magnetic and magneto-optical properties of DMS materials are governed by the exchange interactions between the local magnetic moments introduced by magnetic ions and band carriers, or so called s p- d exchange interactions [1-3].

In this work, to probe s p-d exchange interaction static magnetic susceptibility and Faraday rotation (FR) measurements have been carried out. The CdS:Mn quantum dots / polyvinyl alcohol (PVA) nanocomposites were synthesized by wet chemical method. The size of QD was estimated by means of absorption spectroscopy. The temperature dependences of the static magnetic susceptibility and spectral dependence of FR angle are presented. The results are analyzed from the viewpoint of contribution of Mn^{2+} ions into the observed properties.

2. Experimental

2.1. Synthesis
A CdS:Mn quantum dots / PVA composite films have been prepared by means of in situ synthesis method via the reaction of Cd^{2+}, Mn^{2+} - dispersed PVA with sodium sulfide at room temperature. The concentrations of salt solutions of $CdCl_2$, $MnCl_2$, Na_2S and the pH value were selected to optimize the QD yield, i.e. the molar concentrations of precursors and the pH value were within the range of (10^{-4} – 10^{-2}) mol/l and (3-6), respectively. The films of the CdS:Mn quantum dots / PVA composite were formed by adsorptive desiccation method.

2.2. Optical, magnetic and magneto-optical measurements
Measurements of the optical absorption spectra have served for revealing of quantum-size effect and estimation of the average size of CdS:Mn quantum dots.

Static magnetic susceptibility of CdS:Mn quantum dots / PVA composite films was measured by Faraday's method. The experiment was made at the temperature range of (78 – 300) K in magnetic fields (0,05 – 0,8) T. The force on a sample in an inhomogeneous magnetic field was measured with an electronic balance device (resolution of the order of 10^{-3} milligram) at which the sample is suspended. The measurements were calibrated using a NaCl crystal standard. The measurements of external magnetic fields applied to a sample done by a Hall probe. The chamber was firstly degazed to 10^{-3} Pa and then filled in with helium of high purity up to $1{,}3 \cdot 10^3$ Pa. Cooling of samples is accomplished by a regulated flow of nitrogen gas. The temperature is measured with copper-constantan thermocouples placed close to the sample. The temperature was changed in steps of 10 to 15 degrees and with an isothermal exposition during 5-10 min at each temperature.

Magneto-optical Faraday rotation was measured at wavelength of (400-700) nm in magnetic field up to 5 T. These experiments were performed at room temperatures. The FR was studied with a measuring setup which included diffraction monochromator, Roshon prisms, and water-cooled electromagnet.

3. Results and discussion
Figure 1 shows the optical density spectra of CdS:Mn quantum dots / PVA composites. The position of exciton peaks remained almost unchanged at 2.75 eV regardless of the precursors concentrations in the range $5 \cdot 10^{-4}$ – $5 \cdot 10^{-2}$ mol/l, therefore the size of the CdS:Mn quantum dots was also invariable. The average size of QD was calculated [4,5] to be 18 nm in diameter. However, slight increase of the amplitude of exciton peak with an increase of the amount of Na_2S precursor can be attributed to increase of the concentration of QD.

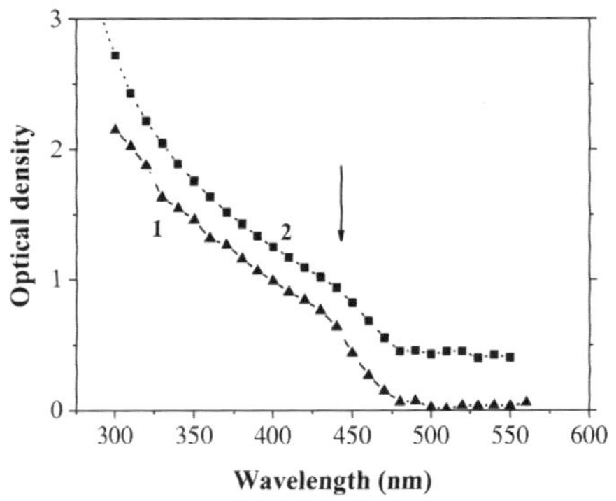

Figure 1. Optical density as a function of wavelength for CdS:Mn quantum dots / PVA composites at $[MnCl_2]$:$[CdCl_2]$ ratio is 1:9. Curves 1 and 2 correspond to $[CdCl_2]$:$[Na_2S]$ is 9:1 and $[CdCl_2]$/$[Na_2S]$ is 4:1, respectively.

3.1. Magnetic properties

In II-VI compounds Mn ions are divalent and assume the high spin d^5 configuration characterized by S = 5/2 and g = 2. The Mn ions neither introduce nor bind carriers, but give rise to the presence of the localized spin in II-VI based DMS. The spin dependent hybridization between anion p and Mn d states leads to the superexchange a short-range antiferromagnetic coupling among the Mn moments. However, the antiferromagnetic superexchange can be overcompensated by ferromagnetic interactions mediated by band holes [6].

The temperature dependence of the magnetic susceptibility is shown in Figure 2. The plot clearly shows a paramagnetic system with a temperature dependent susceptibility. Undoped CdS quantum dots /PVA composites show diamagnetic properties, which can be seen from the Figure 2.

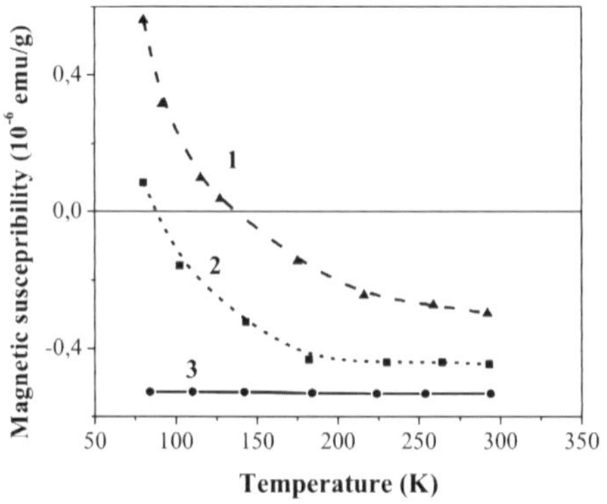

Figure 2. Temperature dependence of magnetic susceptibility of the composites. Curves 1 and 2 correspond to the samples mentioned in Figure 1, curve 3 correspont to CdS quantum dots/PVA composite.

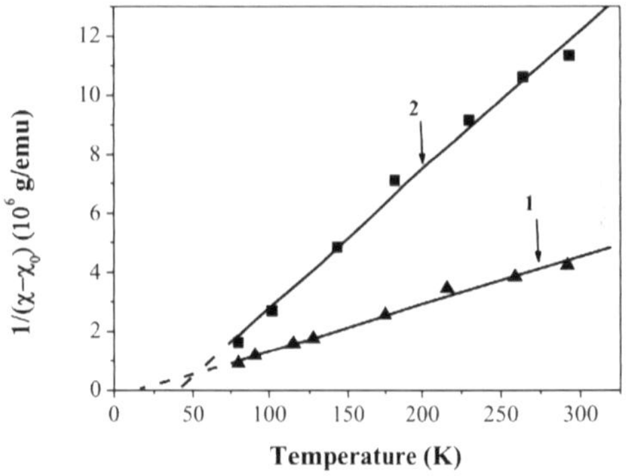

Figure 3. Inverse susceptibility as a function of temperature plotted on the base of data from Figure 2.

In the temperature interval of (75-300) K magnetic susceptibility for both samples decreases following Curie-Weiss law. The inverse susceptibility is plotted as a function of temperature in Figure 3. From the linearity of this graph it is apparent that CdS:Mn quantum dots are paramagnetic materials above 78 K. The result is consistent with the Curie–Weiss equation

$$\chi = \chi_0 + \frac{C}{T - \theta} \quad , \tag{1}$$

where C is Curie constant, Θ is paramagnetic Curie temperature and χ_0 is the temperature-independent susceptibility (diamagnetic contribution).

Analysis of experimental data was performed by a liner Curie-Weiss model. The Curie constants were obtained from the slopes of the lines and the effective Curie–Weiss temperatures by extrapolation to the temperature axis. From this fit the temperature axis intercept is found to be +17,6 K and +40,3 K for 1 and 2 samples, respectively. This indicates the ferromagnetic interactions between Mn magnetic moments in these composites below Curie–Weiss temperatures.

The features observed in the magnetic measurements of two samples are very similar, for exception of the ferromagnetic coupling between Mn^{2+} ions, which is stronger in sample 2 likely due to the fact that the Mn-Mn distance is shorter in sample 1. The increase of the ferromagnetic coupling value between Mn^{2+}ions was accompanied with an increase of the CdS phase amount. The more the amount of the phase the stronger an interaction. We admitted that these changes are caused by the embedded Mn^{2+}ions into CdS QD, because the QD size to a great extent exceeds the distance between magnetic ions which influences on a type of interaction.

3.2. Magneto-optical properties

Figure 4 shows the spectral dependence of the Faraday rotation angle of the nanocomposites after subtraction of the rotation angle of the PVA matrix.

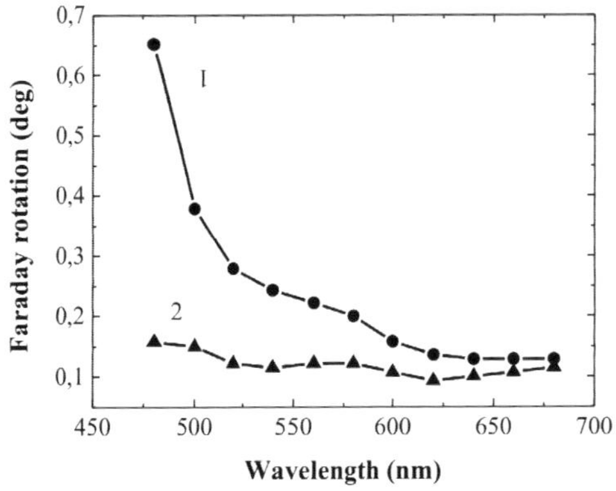

Figure 4. Spectral dependence of the Faraday rotation for PVA film samples containing CdS:Mn (1) and CdS (2) QD at 300K.

It was observed that Faraday rotation angle of semiconductor quantum dot phase has positive sign for undoped CdS quantum dots, but negative sign in case of CdS:Mn quantum dots. The revealed reversal of the direction of the Faraday rotation in its spectral dependence is associated with the competition between the paramagnetic term due to the magnetization of the Mn^{2+} ions and the diamagnetic term due to the interband transition in the CdS crystal.

Nevertheless, the calculated value of the Verdet constant in case of CdS:Mn quantum dots is - 0.16 deg/mT×cm at 475 nm at room temperature. This value is the same order as for bulk non-magnetic CdS crystals [2] because of low magnetization at room temperature and small content of Mn in the quantum dots.

4. Conclusions

We have studied magnetic and magnetooptical properties of CdS:Mn quantum dots / PVA composites. From the magnetic susceptibility data presented it is evident that nanocomposites behave as a typical Curie–Weiss paramagnet over the temperature range studied with ferromagnetic interactions between the Mn^{2+} moments. Effect of the CdS phase amount in composite on magnitude of ferromagnetic interactions can be argument for the stage of embedded magnetic ions. In the Faraday rotation spectra negative sign of the angle of rotation has been revealed in case of CdS:Mn quantum dots / PVA composites like to bulk DMS crystals.

References

[1] Furdyna J K 1988 *J.Appl.Phys.* B **64** R29.
[2] Nikitin P I and Savchuk A I 1990 *Sov.Phys.Usp.* **33** 974.
[3] Norberg N S and Gamelin D R 2006 *J. Appl. Phys.* **99** 08M104.
[4] Lippens P E and Lannoo M 1989 *Phys. Rev. B* 39, N15.
[5] Pang Q, Guo B C, Yang C L, Yang S H, Gong M L, Ge W K, Wang J N 2004 *J. Crystal Growth* **269,** 213.
[6] Dietl T 2002 *Cond-mat.* **16** 0201282.

μ-PL characterization of CdSe Quantum Dots grown on a $Zn_{0.2}Mg_{0.8}S_{0.64}Se_{0.36}$ barrier

I A Davidson[1], R T Moug[1], P A Dalgarno[1], C Bradford[1], R J Warburton[1,2] and K A Prior[1]
[1]David Brewster Building, School of Engineering and Physical Sciences, Heriot-Watt University, Edinburgh, EH14 4AS, UK
[2] Department of Physics, University of Basel, Klingelbergstrasse 82, 4056 Basel, Switzerland

E-mail: id22@hw.ac.uk

Abstract. In this paper we report on the growth and characterisation of CdSe quantum dots grown by MBE on a MgS-rich ZnMgSSe barrier and characterized by 77K PL and 4K μ-PL. The dot density was estimated to be approximately 10^{10} cm^{-2} however some emission lines were sufficiently well resolved to investigate their behaviour individually; their emission was observed to change both in intensity and energy, showing a high frequency random jittering. Correlation of the dot intensity over much longer time periods (~200s) was also observed. These results will be compared with previous results from CdSe dots grown on ZnSe barriers.

1. Introduction

Wide bandgap II-VI semiconductor devices are typically grown on commercially available III-V substrates, which have bandgaps far smaller than the epitaxial II-VI layers. This is inconvenient for many experiments and applications; including any where absorption by the substrate prohibits the transmission of light through the epitaxial layers.

For ZnSe based structures grown on GaAs, we have recently developed an epitaxial lift-off process utilizing MgS as a sacrificial layer [1] which allows the epitaxial layers to be removed easily and without the use of any mechanical or chemical processes that might damage them. The lift-off process also allows for the use of a range of new functional substrates; this is particularly useful for experiments that require a micro-cavity [2].

MgS is also an excellent barrier material with a bandgap of ~5eV [3, 4], but cannot be used as such in any structures where lift off is also required. Consequently, we have developed a wide bandgap quaternary alloy $Zn_{0.2}Mg_{0.8}S_{0.64}Se_{0.36}$ which provides both the required confinement and resistance to the etching solution [5]. Our aim is to use this material as a barrier for quantum dot containing layers and as a first step toward this we report in this paper what we believe to be the first μ-PL study of CdSe dots grown on a MgS-rich ZnMgSSe quaternary alloy.

2. Growth

All of the samples examined in this paper were grown in a VG V80H MBE system using 6N elemental sources of Zn, Cd, Mg and Se, as well as a 6N ZnS source as the only source of sulphur. All layers, except the QDs, were grown by conventional MBE [2] at 300°C. The QDs were grown by atomic layer epitaxy (ALE) using alternate deposition cycles of Cd and Se at 300°C, and were then annealed for 4 minutes at 350ºC to thermally activate the dot formation process [7]. The structures were monitored with RHEED throughout.

A series of samples were grown with the same basic structure – GaAs/34nm ZnSe (buffer)/10-15nm ZnMgSSe (barrier) /CdSe QD layer/10-15nm ZnMgSSe (barrier). All of the samples received no additional ZnSe capping layer, as the ZnMgSSe alloy has been found to resist oxidation sufficiently to protect the samples [4].

Two samples were selected for μ-PL study with 9 and 11 ALE cycles giving 4.5 and 5.5 ML of CdSe respectively. During the growth of both samples RHEED showed that they did not relax and are fully pseudomorphic.

3. Ensemble PL Characterisation

Ensemble PL measurements were made at 77K. These show (see Fig.1) a broad peak centred around ~2.5eV (496nm), from the CdSe QDs similar to those grown on ZnSe or MgS barriers [6, 9] and a smaller peak at ~2.79eV (444nm) representing emission from the ZnSe buffer layer.

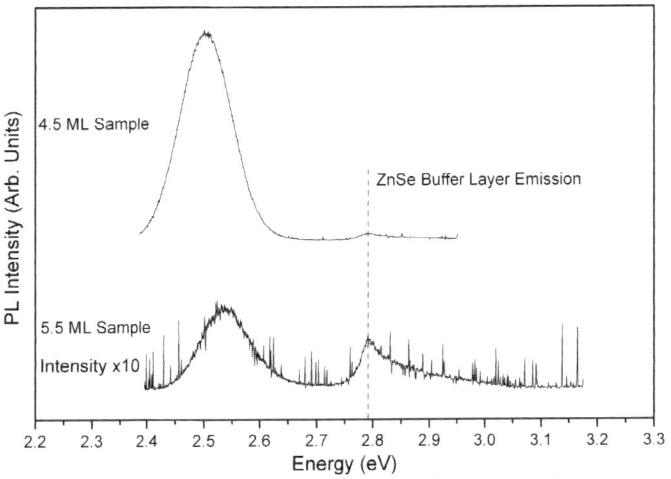

Fig. 1 – Ensemble 77K PL measurements of CdSe/ZnMgSSe QD samples

The intensity of the PL emission from these samples was quite weak, ~500 times less intense than samples grown on ZnSe, as the ZnMgSSe barrier has a bandgap far larger than the incident photon energy and hence only the CdSe QD layer absorbs incident photons. This results in a small absorption cross section and relatively low emission intensity.

4. μ-PL Characterisation

The μ-PL measurements were made at 4K using a diffraction limited confocal microscope system utilizing either a 0.85 NA microscope objective or a combination aspheric and solid-immersion lens, with an effective NA of 1.3. The samples were excited using a 60mW 405nm laser diode mounted to a TEC cooler and driven at roughly 90% of its maximum drive current, so as to ensure it produce a highly stable output. The beam was then collimated and fibre coupled to the microscope head; mounted directly about the liquid helium cryostat in which the samples were placed [8].The excitation density for all our measurements was kept constant at ~ 220 $kWcm^{-2}$. The emitted light was dispersed using either 300 or 1800 line gratings onto a LN2 cooled CCD giving energy resolutions of 0.95meV and 90μeV respectively.

Within the microscopes ~0.2 μm diameter resolution, ~15 dots were visible giving an estimated density of ~10^{10} cm^{-2} similar to the densities seen with CdSe dots grown on both ZnSe and MgS [6, 9].

Figure 2 shows a μ-PL spectrum for both samples which was obtained by averaging the emission of the samples over 30s, due to the low emission intensity. Both samples show a series of closely spaced, relatively narrow features at 4K centered around 2.5 eV representing emission from individual dots. An additional smaller feature is seen at ~2.8eV again represents the emission from the ZnSe buffer layer, as can be seen in Fig. 2A. By translating the sample using sub-nm positioning stages, it is possible to find spatially isolated individual dots that are sufficiently separated from the main ensemble in emission energy to investigate their behavior. An example of this can be seen in fig. 2B. The sharpest features observed with these samples have FWHM in the region of 3±0.95 meV.

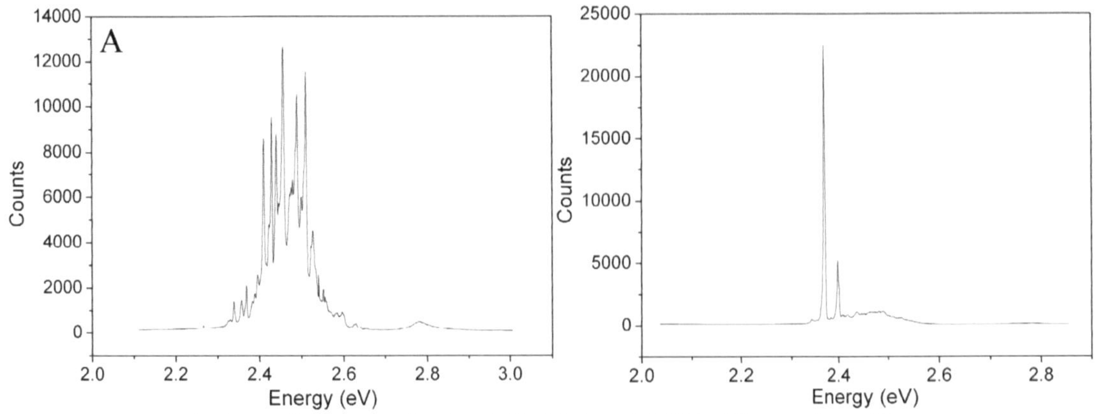

Fig. 2 – μ-PL spectra for the 5.5ML(A) and 4.5ML(B) sample at 4K

Figure 3 shows the time evolution of the spectra observed from the dot shown in fig. 2B. Over a period of time both the intensity and peak emission energy of the dot fluctuate. It should be noted that although it appears in the figure that the emission intensity periodically falls to zero (for example at times 500-550s and around 750s) this is not actually the case; in reality the intensity has just dropped to a level where it is not well resolved from the background in this colour map. It is also apparent that the large FWHM of the lines obtained in the integrated traces arises from the fluctuations in the peak position.

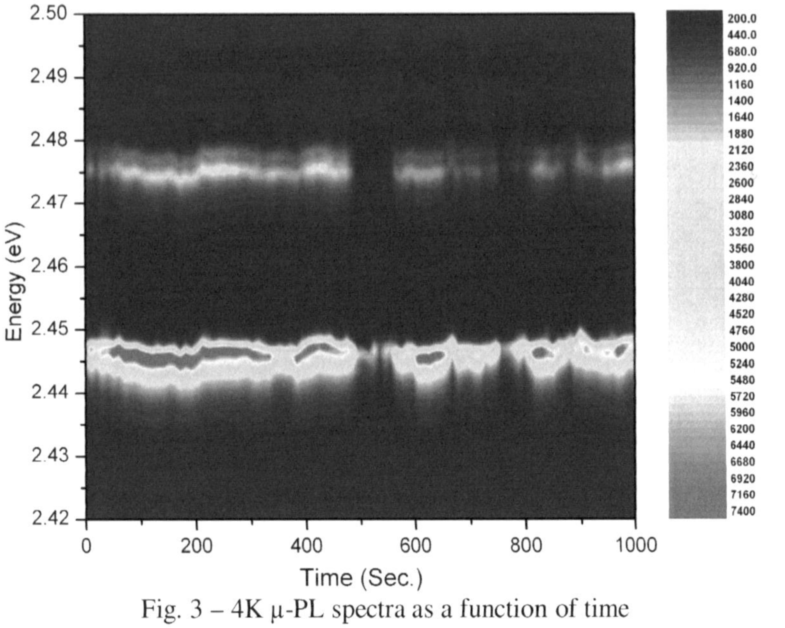

Fig. 3 – 4K μ-PL spectra as a function of time

The graph also shows a second feature whose emission energy is roughly centered on 2.475 eV. The fluctuations in intensity and energy of this feature are highly correlated to that of the lower energy line, indicating that they arise from the same dot, as has previously been observed in ZnSe/CdSe QD emission [10].

Fluctuations in both the energy and intensity of QD emission have been observed for CdSe dots grown on ZnSe buffer layers [11] and it is believed that these arise from the Quantum Confined Stark Effect produced by fluctuating charge close to the dots; either located at the sample surface [10] or at dislocations originating at stacking faults [11]. We observe a typical fluctuation (spectral diffusion) in the dot energy of 2-3meV. Assuming our dots are roughly similar in size and polarizability to CdSe dots on ZnSe [12], this could arise from single charges located approximately 10nm from the dot, which is compatible with the charge being located at the surface of the sample.

QCSD produced by a fluctuating electric field will cause the energy of the emission line to be reduced together with the emission intensity. Consequently, there should be a strong correlation between these two effects produced by the fluctuating electric field, as has been previously observed for ZnSe/CdSe dots [11-13]. However, the weaker emission from the ZnMgSSe/CdSe samples means that although we are able to identify both energy and intensity fluctuations from all dots we have examined so far, we have thus far not been able to find any statistical correlation between the two quantities.

If the quantities are indeed correlated, then our inability to resolve them signifies that the fluctuations are occurring on timescales shorter than our smallest integration time, < 25ms.

For the brightest isolated dots observed, as per fig. 3, the emission intensity shows a characteristic long period oscillation, with a strong correlation in the intensity occurring on a timescale of 200s. This effect has also been observed on ZnSe/CdSe dots by other groups where correlation times of 160s [11] and 220s [13] have been obtained.

5. Conclusion

We have investigated the behavior of CdSe QDs grown on a MgS-rich ZnMgSSe alloy barrier and found the alloy to provide good carrier confinement, however many of the individual lines observed showed significant spectral diffusion, giving peak FWHM of ~3meV. The energy and intensity of all the QDs within the field of view was also observed to fluctuate on a time scale of <25ms which is compatible with QCSD arising from isolated charges located close to individual dots.

A characteristic timescale of ~200s has been determined by statistical analysis for the variation in emission intensity and energy for some of the dots examined.

References

[1] Balocchi A, Curran A, Graham T C M, Bradford C, Prior K A and Warburton R J 2005 *Appl. Phys. Lett.* **86** 011915

[2] Curran A, Morrod J K, Prior K A, Kar A K and Warburton R J 2007 *Semicond. Sci. Technol.* **22** 1189

[3] Bradford C, O'Donnell C B, Urbaszek B, Balocchi A, Morhain C, Prior K A and Cavenett B C 2000 *Appl. Phys. Lett.* **76** 3929

[4] Davidson I A, Moug R T, Izdebski F, Bradford C and Prior K A 2010 *Phys. Stats. Sol.* In press

[5] Moug R, Bradford C, Curran A, Izdebski F, Davidson I, Prior K A and Warburton R J 2009 *Microelectronics Journal* **40** 530

[6] Urbaszek B, Bradford C, Funato M, Graham T C M, McGhee E J, Warburton R J, Prior K A and Cavenett B C 2002 *IOP Conference Series* **171** H154

[7] Funato M, Balocchi A, Bradford C, Prior K A and Cavenett B C 2002 *Appl. Phys. Lett.* **80** 443

[8] Warburton R J, Schäflein C, Haft D, Bickel F, Lorke A, Karrai K, Garcia J M, Schoenfeld W and Petroff P M 2000 *Nature* **405** 926

[9] Makino T, Andre R, Gerard J M, Romestain R, Dang Le Si, Bartels M, Lischka K and Schirkora D 2003 *Appl. Phys. Lett.* **82** 2227

[10] Turck V, Rodt S, Heitz R, Stier O, Straβburg M, Pohl U W and Bimberg D 2002 *Physica E* **13** 269

[11] Seufert J, Weigand R, Bacher G, Kummell T, Forchel A, Leonardi K and Hommel D 2000 *Appl. Phys. Lett.* **76** 1872

[12] Seufert J, Obert W, Scheibner M, Gippius N A, Bacher G, Forchel A, Passow T, Leonardi K and Hommel D 2001 *Appl. Phys. Lett.* **79** 1033

[13] Makino T, Andre R, Gerard J M, Romestain R, Dang Le Si, Bartels M, Lischka K and Schirkora D 2004 *Solid State Comms.* **130** 63

Quantum Dots 2010 IOP Publishing
Journal of Physics: Conference Series **245** (2010) 012086 doi:10.1088/1742-6596/245/1/012086

Modelling of electronic states in InAs/GaAs quantum dots with GaAsSb strain reducing overlayer

Petr Klenovský[1], **Vlastimil Křápek**[1], **Dominik Munzar**[1] **and Josef Humlíček**[1]

[1]Department of Condensed Matter Physics, Masaryk University, Kotlářská 2, 61137 Brno, Czech Republic

E-mail: klenovsky@physics.muni.cz

Abstract. We present results of our 8-band $\vec{k} \cdot \vec{p}$ calculations of the emission energy of InAs/GaAs quantum dots (QDs) covered with $GaAs_{1-x}Sb_x$ strain reducing overlayer (SRO). In agreement with previous experimental observations we find a strong red shift of the emission with increasing Sb content. We explain this effect by: (1) The lowering of the valence band offset between the QD and the SRO with increasing Sb content resulting in the type-II QDs with holes confined in the SRO for Sb concentration above 14 %. (2) The reduction of compressive strain inside the QDs. The contributions of these mechanisms to the total red shift are estimated and compared. For realistic shape and size of the QD and a realistic value of the SRO thickness the previously measured photoluminescence data are reproduced with fairly good accuracy.

1. Introduction

Self-assembled InAs/GaAs quantum dots (QDs) have been investigated for their interesting physical properties and potential applications; one of the latter is an efficient light emission on the communication wavelengths of 1.3/1.55 μm. To tune the emission wavelength towards this range, various strategies can be used. One possibility is to use QDs stacked in vertical multilayers [1]. Another option relies on capping QDs with strain reducing overlayers (SRO); here, ternary materials such as $In_xGa_{1-x}As$ [2] and $GaAs_{1-x}Sb_x$ [3] can be used. The resulting red shift with increasing In content in $In_xGa_{1-x}As$ SROs has been explained as an effect of three mechanisms: the reduction of the hydrostatic strain in the InAs QDs, lowering of the barrier height for both electrons and holes at the side of the SRO, and a change in the dot size during the growth process; the latter two were identified as the decisive contributions [2].

We present here a similar analysis for the $GaAs_{1-x}Sb_x$ SROs. The situation is slightly different since the lowering of the valence band offset with increasing Sb content results in a transition to the type-II QDs occurring at the Sb concentration of approximately 14%. Above this value holes are localized in the SRO instead of the QDs. In previous photoluminescence studies, this transition to the type-II QDs has been attributed to be the dominant cause of the observed red shift [4]. However, a detailed theoretical analysis is still missing.

2. Method

QDs capped by $GaAs_{1-x}Sb_x$ SROs were simulated using nextnano++ software [5]. The structure consisted, from bottom to top, of a GaAs substrate, InAs QD, $GaAs_{1-x}Sb_x$ SRO and a GaAs

Journal of Physics: Conference Series **245** (2010) 012086 doi:10.1088/1742-6596/245/1/012086

capping layer. Our calculations started by finding the strain inside the structure, continued by a calculation of the piezoelectric field, and ended with solving the Schrödinger equation using the 8-band $\vec{k} \cdot \vec{p}$ approximation. All these steps were performed using the nextnano++ built-in solvers. The values of the material parameters have been taken from [6], with the exception of the deformation potentials and the InAs/GaAs valence band offset, taken from [7].

To distinguish between the effects of the reduced strain in the QDs and those of the lowered valence band offset, the following calculations were performed: (1) We have calculated the strain field for the $GaAs_{1-x}Sb_x$ SRO and used it subsequently as the input for the piezoelectricity and quantum calculations. Apart from the elastic constants and the lattice parameter, we considered the SRO to correspond to a pure GaAs. In this way we have specified the contribution of the reduced strain within the QDs. (2) We have used the strain field calculated for a pure GaAs overlayer and values of the parameters of the $GaAs_{1-x}Sb_x$ SRO (except for the lattice parameter and the elastic constants) in subsequent calculations to obtain the effect of the modified valence band offset. In this case, also other parameters were changed, such as the conduction band offset and effective masses. However, the most pronounced effect is indeed connected with a modified valence band offset.

3. Results and discussion

In our calculations we have varied the dimensions of the QD and the thickness of the SRO, in order to fit the transition energy and its red shift observed in [3]. We have chosen the semiellipsoidal shape of the QD as it is common for InAs dots large in volume. We have obtained the best agreement for the QD height of 5 nm, the QD diameter of 14 nm and the SRO thickness of 3 nm.

The calculated values of the lowest transition energy are shown in Fig. 1. The experimental data measured at 15 K, taken from [3], are shown for comparison. To facilitate the comparison, the calculated data are blueshifted by 20 meV. The shift could be due, e.g., to the fact that the real QDs consist of InGaAs rather than of pure InAs.

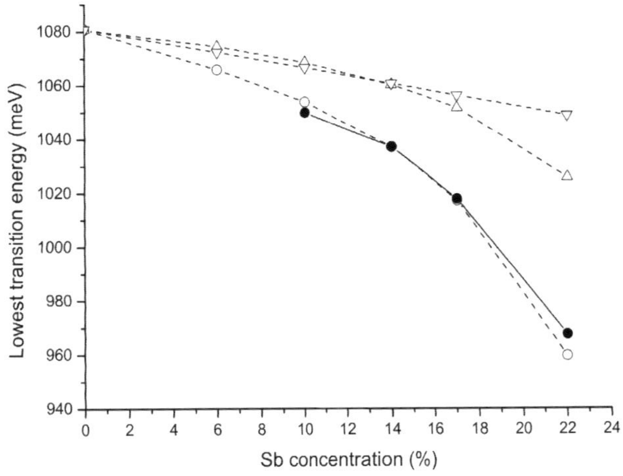

Figure 1. ● Peak emission energy measured at 15 K, taken from [3]; ○ Calculated values of the lowest transition energy; △ The same obtained considering the Sb concentration dependence of the electronic structure parameters only; ▽ The same obtained considering the strain effect only.

The calculated red shift of the lowest transition energy when going from 10 % to 22 % of Sb of 95 meV is in reasonable agreement with the experimental value of 82 meV. We decompose this shift into the contribution of the reduced strain in the QD, and that of the lowered valence band offset between the QD and the SRO. Both contributions are shown separately in Fig. 1. The modified valence band offset accounts for $\sim 45\,\mathrm{meV}$ of the red shift, while the reduced strain accounts for $\sim 20\,\mathrm{meV}$. The difference of $\sim 30\,\mathrm{meV}$ between the sum of the values and the total red shift is a result of the combination of the effects.

Below the Sb concentration of 14 %, both effects contribute nearly equally to the total shift, with a linear dependence on the Sb concentration. Around 14 %, the transition to the type-II QD occurs and increasing the Sb content further enhances the effect of the lowered valence band offset (which becomes negative). Accordingly, we identify the onset of the type-II QDs as the dominant mechanism causing the red-shift of the emission energy, confirming the assumption of Ref. [4]. However, the influence of the $GaAs_{1-x}Sb_x$ capping layer on the strain in the structure is responsible for a surprisingly large portion of the red shift as compared to the $In_xGa_{1-x}As$ SRO [2].

Fig. 2 shows the hydrostatic component of the strain ($\varepsilon_{xx} + \varepsilon_{yy} + \varepsilon_{zz}$) along the symmetry axis of the dot. The compressive strain inside the dot decreases by 0.3 % when increasing the Sb content from 10 % to 22 %. This corresponds to the red shift of the conduction band edge by $\sim 20\,\mathrm{meV}$, explaining fully the simulated contribution of the reduced strain to the red shift, shown in Fig. 1.

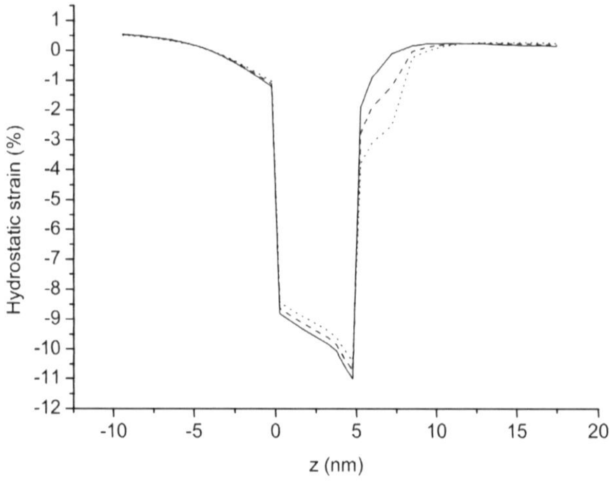

Figure 2. Hydrostatic strain ($\varepsilon_{xx} + \varepsilon_{yy} + \varepsilon_{zz}$) along the symmetry axis of the QD for —— 0 %, - - - - 10 %, ······ 22 % of Sb content in the SRO.

The reduction of the compressive strain inside the QD can be explained as follows. The lattice parameter of $GaAs_{1-x}Sb_x$ is larger than that of GaAs. The presence of the SRO therefore increases the total strain in the structure. However, the contrast in lattice parameters between the QD and the SRO is reduced as compared to the GaAs overlayer, which results in a reduced strain inside the QD. In particular, the material around the sides of the dot has a lower compressive impact. On the contrary, matching the lattice parameter to the GaAs overlayer enhances the compressive strain in the SRO, as seen clearly in Fig. 2.

There is a pronounced difference in the magnitude of the relaxation of the compressive strain inside the QDs with $GaAs_{1-x}Sb_x$ SRO considered here and inside the QDs with $In_xGa_{1-x}As$ SRO studied in Ref. [2]. The lattice parameters of the SROs are nevertheless comparable. The difference likely originates in the different aspect ratios of the QDs in both cases. The $GaAs_{1-x}Sb_x$ covered dots are rather high and the elastic interaction between them and the SRO is more intense than in the flat QDs covered with $In_xGa_{1-x}As$.

4. Conclusion

In conclusion, we have calculated the emission energy of QDs covered with the $GaAs_{1-x}Sb_x$ SRO with various values of the Sb content. We were able to reproduce the observed transition energies, and their red shift with increasing Sb content. In agreement with previous studies, it has been found that the red shift is caused mainly by the transition to the type-II QDs above the Sb concentration of 14 %. Almost one fourth of the red shift is caused by the reduction of the compressive strain inside the QDs; this is much more than in the case of $In_xGa_{1-x}As$ SROs.

Acknowledgments

The work was supported by the Institutional research program MSM0021622410 and the GACR grant GA202/09/0676.

References

[1] Shchukin V A, Ledentsov N N and Bimberg D 2004 *Epitaxy of Nanostructures* (Nanoscience and Nanotechnology) (Berlin: Springer-Verlag)
[2] Kuldová K, Křápek V, Hospodková A, Oswald J, Pangrác J, Melichar K, Hulicius E, Potemski M and Humlíček J 2006 *Phys. Stat. Sol. (c)* **3** 3811
[3] Liu H Y, Steer M J, Badcock T J, Mowbray D J, Skolnick M S, Navaretti P, Groom K M, Hopkins M and Hogg R A 2005 *Appl. Phys. Lett.* **86** 143108
[4] Liu H Y, Steer M J, Badcock T J, Mowbray D J, Skolnick M S, Suarez F, Ng J S, Hopkinson M and David J P R 2006 *J. Appl. Phys.* **99** 046104
[5] Trellakis A, Zibold T, Andlauer T, Birner S, Smith R K, Morschl R, Vogl P 2006 *J. Comp. Elec.* **5** 285
[6] Vurgaftman I, Meyer J R and Ram-Mohan L R 2001 *J. Appl. Phys.* **89** 5815
[7] Wei S and Zunger A 1998 *Appl. Phys. Lett.* **72** 2011

STEM imaging of InP/AlGaInP quantum dots

Y Qiu, AB Krysa and T Walther

Department of Electronic & Electrical Engineering, University of Sheffield, Mappin Street, Sheffield S1 3JD, UK

E-mail: y.qiu@sheffield.ac.uk

Abstract. InP quantum dot structures embedded between ternary AlInP, ternary GaInP or quaternary AlGaInP have been grown by metal organic vapour phase epitaxy on GaAs(001) at temperatures of 640 °C or 730 °C under different growth conditions and were investigated by scanning transmission electron microscopy (STEM). We found quantum dots higher than 2nm only for samples grown at 640 °C and the sample grown at 730°C with the lowest growth rate and additional growth interrupts. For the higher growth temperature and high growth rates quantum dot sizes evolve differently: the dots either stay much smaller and remain embedded into the wetting layer, essentially forming quantum wells with lateral compositional variations, or occasionally they can form agglomerates of super-size islands that protrude to the surface. This demonstrates that quantum dot formation is favoured by slow growth, while fast kinetics promotes flat layer growth.

1. Introduction

InP quantum dots (QDs) embedded between (Al)GaInP lattice matched to GaAs substrates emit around 700-800nm wavelength. This is potentially useful for applications as laser diodes (LD) [1] or semiconductor saturable absorber mirrors (SESAMs) [2]. Therefore, it is very important to control both the microstructure and the chemical composition of such QDs. In this study four samples with different growth temperatures, different growth rates and growth interrupts have been compared.

2. Experimental

InP QDs were grown by metal organic vapour phase epitaxy (MOVPE) on GaAs (001) substrates. Table 1 presents the nominal structures of the QD layers and growth conditions. Electron transparent <110> cross-sectional specimens were prepared by standard methods, gluing together stacks of material, then cutting, grinding, polishing and finally Ar^+ ion milling them until perforation.

sample number MR	2629	2620	2725	2727
growth temperature (°C)	640	730	730	730
growth rate (nm/sec)	0.38	0.76	0.76	0.38
growth interrupt	20sec	none	none	10sec
upper barrier	GaInP	GaInP	AlInP	AlInP
InP QD thickness (nm)	1.50	0.70	0.76	0.76
lower barrier	GaInP	AlGaInP	AlInP	AlInP
repeat sequence	×6	×5	×5	×5

Table 1. Structural details and growth conditions

We applied bright-field (BF) and annular dark-field (ADF) STEM in a JEOL 2010F field-emission transmission electron microscope equipped with a scan unit, providing an electron beam ~0.25nm in size at ~10mrad semi-angle of beam convergence and 0.5mm spherical aberration constant. The intensity in ADF imaging is approximately proportional to the square of the average atomic number ('Z-contrast') if the inner collection angle is very large and the specimen uniformly flat.

3. Results

BF and ADF STEM imaging have been used to study the density and the size of the QDs. Example images are shown below, with the growth direction pointing from bottom left to top right.

Figures 1a and 1b show stacks of quantum dots in sample MR2629 where the QDs in the lower two or three layers have a well-defined dot geometry while the upper ones appear flatter and laterally diffuse. The GaInP barriers are relatively narrow, so QDs in successive layers are pushed upwards. Their In content appears maximum in the upper halves of the nominally pure InP QDs, indicating some intermixing at least with the lower barriers. Also, in some areas, such as in Figure 1c, cation ordering on alternating {111} lattice planes is visible, pointing to the onset of spinodal decomposition in the ternary GaInP alloy [3] of the barrier.

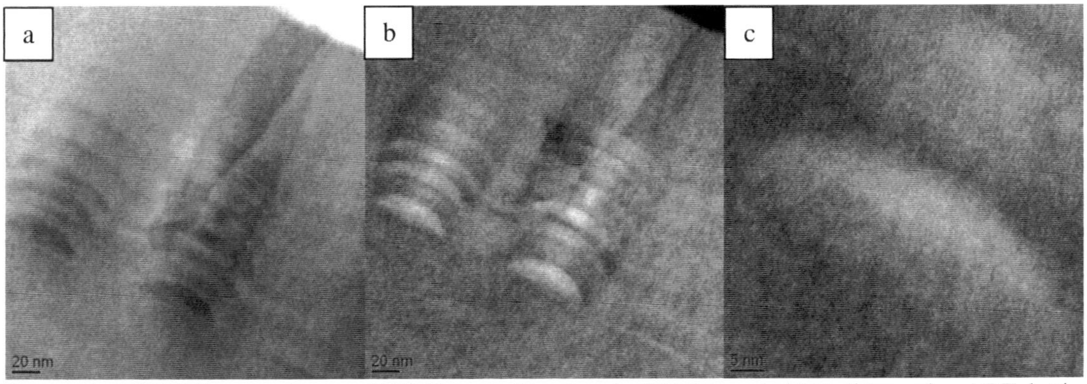

Figure 1. MR2629: overview at 250kX magnification in BF (a) and ADF (b) mode; ADF lattice images of QDs at 1.2MX magnification (c).

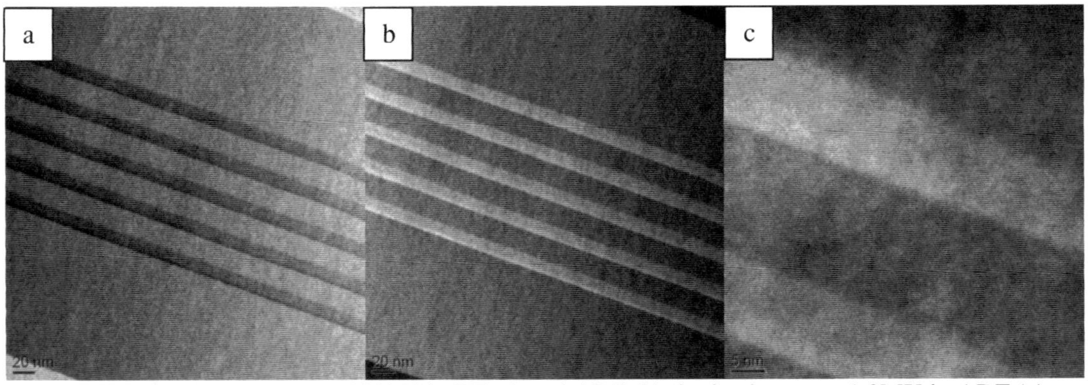

Figure 2. MR2620: overview at 200kX in BF (a) and ADF (b); lattice image at 1.2MX in ADF (c).

Figure 2 shows that the quaternary AlGaInP/InP/GaInP structures (MR2620) are almost planar, and a well-defined In-rich layer ~2.5nm wide with lateral compositional variations is formed above the lower barriers. These may be considered QDs that are still embedded in the wetting layer formed typically during strained layer epitaxy.

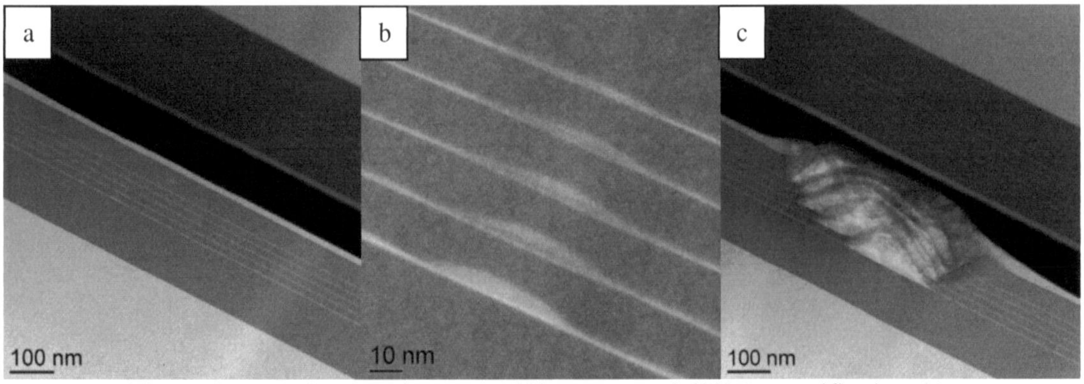

Figure 3. MR2725: ADF images at 80kX (a, c) and 600kX magnification (b).

Figure 3 presents the QDs formed in MR2725, which are mostly very small and difficult to resolve (cf. Figure 3a). Only occasionally do they form proper QDs under 10nm in height (Figure 3b) or agglomerate into very large islands that protrude to the surface (cf. Figure 3c). Sample MR2727 shown in Figure 4, however, exhibits a rather broad and continuous size distribution of QDs.

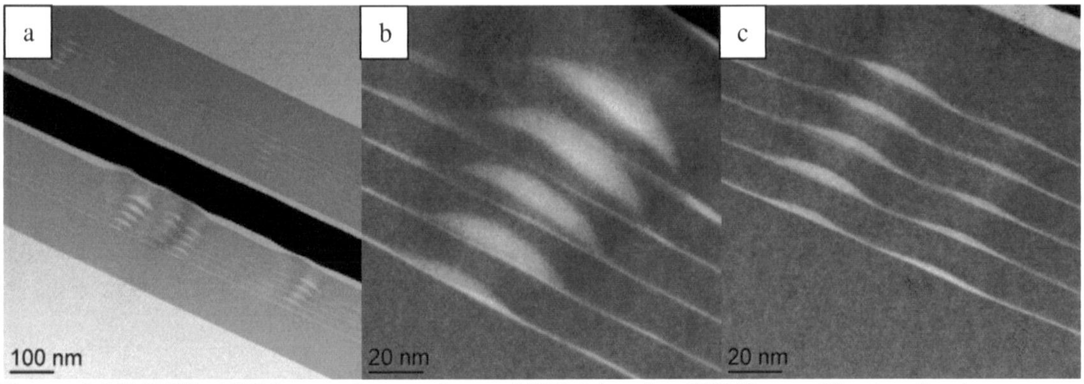

Figure 4. MR2727: ADF images in overview at 80kX (a) and at 500kX magnification (b, c).

4. Discussion

From the ADF STEM images acquired from individual sample regions 10-20μm in total length we produce a table comparing the samples qualitatively.

sample code	height of individual QDs				
	S <2nm	M 2-5nm	L 5-10nm	XL 10-15nm	XXL >50nm
MR2629	-	-	14	96	-
MR2620	180	-	-	-	-
MR2725	1210	3	2	-	65
MR2727	1107	56	34	13	-

Table 2. Comparison of the height of the InP QD in the various samples studied

According to the STEM images, all samples produce InP QDs stacked along the growth direction (only in MR2727 the stacking is slightly oblique). For the lowest growth temperature and growth rate, InP QDs 5-15nm in height are formed between GaInP. Those QDs formed first are very pronounced while the upper QDs grown later are flatter and have been pushed upwards by the underlying ones.

When the growth temperature is increased to 730°C and the growth rate is doubled, the InP layers generally become more planar and it is difficult to observe obvious QDs. An In-rich layer is formed between AlGaInP and GaInP barriers, with intensity fluctuations indicating the onset of formation of

QDs embedded within wetting layer. In order to improve the contrast in the ADF images and study the formation of InP QDs in more detail, the barrier materials around the QDs have been changed to ternary AlInP, retaining the same growth temperature of 730ºC, while the growth rate for one sample the was kept at 0.76nm/sec and for another one it was halved and a 10sec growth interrupt was applied as well. The size and density of the QDs in these samples have been investigated further.

MR2725, which grew fast and without interrupts, yielded many small QDs near to each other, while larger QDs as shown in Figure 3b and 3b only appeared once over ~10µm field of view. Sample MR2727, with reduced growth rate and growth interrupts, has QDs of many different sizes, with heights ranging from ~2nm to 13nm and widths from 17 to 56nm. Figure 5 presents histograms after investigating ~100 QDs. Neglecting contrast reduction due to finite collection angles [4] we can in this system approximate a compositional map of In, the heaviest element, by calculation the square root of the ADF image intensity. The heights and widths of the QDs have been measured as full widths at half maximum (FWHM) of intensity line profiles across the square root of the ADF image intensity map.

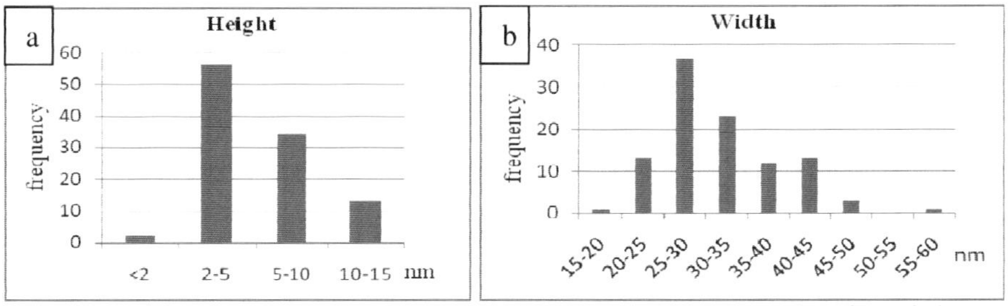

Figure 5. Sample MR2727: height distribution (a) and width distribution (b) of ~100 QDs resolved.

Figure 6 shows a histogram of the distances between adjacent QD stacks acquired from images taken at magnifications of 60kX or 80kX. With reduced growth rate plus growth interrupt the average distance between two nearby QDs is nearly doubled and the standard deviation increased. Moreover, the sample grown slower and with growth interrupt shows a bimodal distribution of the QD spacing, indicating slow growth leads to QD formation while rapid growth inhibits their formation kinetically.

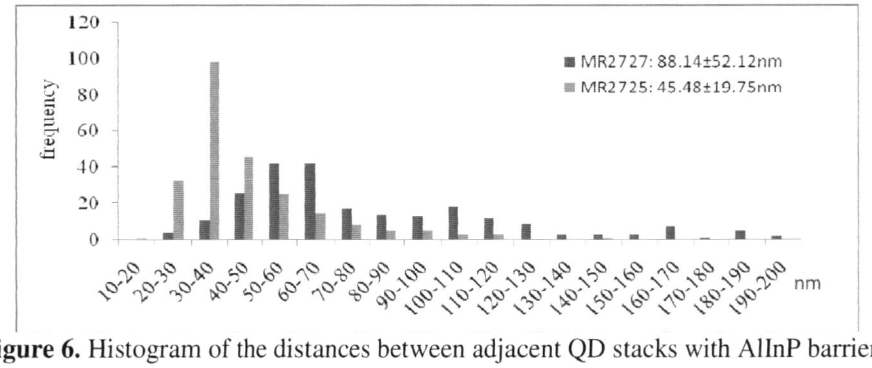

Figure 6. Histogram of the distances between adjacent QD stacks with AlInP barriers.

5. Conclusion

InP QDs can be produced by using MOVPE at relatively low growth temperature of 640ºC or, at higher growth temperature of 730 °C, if slow growth rates are used and growth interrupts are applied. Rapid growth favours epitaxy of quantum wells and suppresses the QD formation.

References

[1] Zundel MK, Jin-Phillipp NY, Phillipp F and Eberl K 1998 *Appl. Phys. Lett.* **73**, 1784
[2] Savitski VG et al. 2010 *IEEE Photon. Technol. Lett.* **22**, 209
[3] McDevitt TL et al. 1992 *Phys. Rev. B* **45**, 6614
[4] Walther T 2006 *J.Microsc.* **221**, 137

Neutralization of positively charged excitonic state in single InAs quantum dot by Si delta doping

N Kumagai[1], S Ohkouchi[1], M Shirane[1, 2], Y Igarashi[1, 2], M Nomura[1], Y Ota[1, 3], S Yorozu[1, 2], S Iwamoto[1, 3], and Y Arakawa[1, 3]

[1]*Institute for Nano Quantum Information Electronics, the Univ. of Tokyo*, 4-6-1 Komaba, Meguro-ku, Tokyo, 153-8505, Japan

[2]*Nano Electronics Research Laboratories, NEC corp.*, 34 Miyukigaoka, Tsukuba, Ibaraki 305-8501, Japan

[3]*Institute of Industrial Science, the Univ. of Tokyo*, 4-6-1 Komaba, Meguro-ku, Tokyo 153-8505, Japan

e-mail address: nkumagai@iis.u-tokyo.ac.jp

Abstract. Positively charged excitonic state in single InAs quantum dot (QD) has been neutralized by Si delta doping and the optical properties of the fabricated samples are characterized by micro-photoluminescence measurements (μ-PL) at low temperature. An InAs QD sample having dominant emissions from positively charged excitonic states is prepared for a reference. By optimizing Si delta doping conditions, positively charged QDs are neutralized successfully and distinctive emissions from neutral exciton and biexciton from single QDs are achieved. For excessive doping condition, μ-PL spectra show optical degradation of lowering of peak intensity and broadening of peak width as doping concentration increases.

1. Introduction

InAs quantum dot (QD) becomes more important key issue due to its high potential in applications for optoelectronics, photonics, spintronics, and quantum information technology. Recently, attractive applications using single InAs QD such as single photon emitter, generating entangled photon pair by cascade emission from neutral biexciton, and single spin manipulation are reported. [1-3]. In order to realize these applications, control of charged state in single InAs QD is significantly important.

Delta doping (sometimes called by sheet doping or spike doping) method is effective for this purpose. It provides extremely narrow profile of dopant and spatial separation between doped layer and dopant. In this study, delta doping method as a means of supplying electron to positively charged QDs is applied to neutralize positively charged excitonic state in single InAs QD.

2. Experiment

InAs QD samples are grown on (001) GaAs substrates by molecular beam epitaxy. Growth temperature (T_g), rate, and nominal coverage of InAs QDs are 485 °C, 0.0042 ML/sec, and 2.1 ML, respectively. Typical QD density is ~$5x10^8$ cm^{-2} and emission wavelength is tuned to around 1μm at room temperature by In-flush method [4]. T_g of GaAs partial cap in In-flush process is 423 °C. A reference sample grown by these conditions has dominant emission from positively charged excitonic states although the QDs are not intentionally doped. Typical μ-PL spectrum of the reference QD is shown in figure 1. μ-PL peaks from excitonic states are assigned by power dependence, photon correlation measurement, and energy diagram [5, 6]. Power dependence is measured for assignment of excitonic or biexcitonic state, and photon correlation measurement is done for identification of cascade emission from biexcitonic state to excitonic one in a same QD. Relation of emission energy from excitonic and biexcitonic states are considered by energy diagram with taking account of charged state and its total spin. The detail of assignments will be published elsewhere [6]. According to these results, PL peaks as shown in figure 1 are assigned to positively charged exciton (X^+) and biexciton (XX^+). Typical feature of positively charged excitonic state is distinct peak (X^+) accompanied with two peaks (XX^+) as shown in figure 1. Accompanied two peaks come from energy branches in triplet state with different total spin [5, 6]. Experimental parameters of delta doping structures, Si sheet density and spacer thickness between InAs QDs and Si sheet layer are varied as shown in table 1. Si sheet layer is sited below InAs QDs. Delta doped InAs QD samples are processed to mesa structure with diameter of ~ 1μm; so that one or two active dots are included in the fabricated mesa. μ-PL measurements are performed at 10K with Ti:Al_2O_3 laser excitation of 780 nm.

Table 1. Delta doping parameters for InAs QDs samples

Sample	**n: Si sheet density**	**d: spacer**
Reference	undoped	-
A	$5x10^{10}cm^{-2}$	40 nm
B	$5x10^{10}cm^{-2}$	20 nm
C	$5x10^{11}cm^{-2}$	40 nm
D	$5x10^{11}cm^{-2}$	20 nm

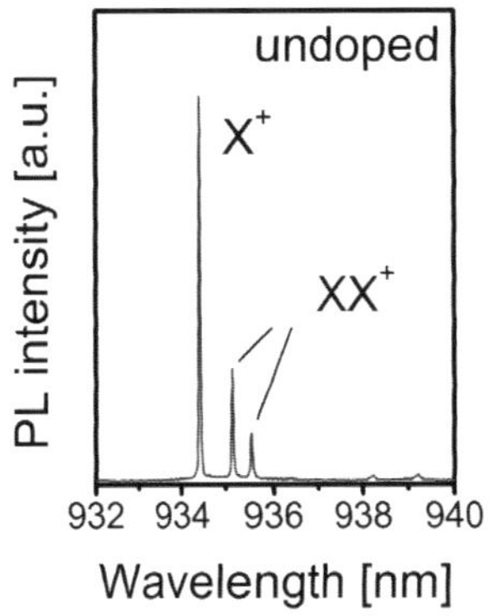

Figure 1. Typical μ-PL spectrum of undoped InAs QDs as a reference. X^+ and XX^+ indicates positively charged exciton and biexciton, respectively.

3. Result and discussions

Figure.2 (a)~(d) show typical μ-PL spectra of sample A~D, respectively. For sample A, the dominant peaks are similar to those shown in figure 1, indicating that neutralization is not achieved and doping concentration may be insufficient for neutralization. For sample B, The feature of spectrum is different from that of the reference. That is, one main peak accompanied with another peak appears. These emissions are assigned to those from neutral excitonic states (X^0 and XX^0) by power dependence and photon correlation measurements. According to the PL results in a wide area, about a half of the charged QDs are neutralized successfully in the case of sample B. For sample C and D, charge states are not assigned since optical degradation is obviously appeared as lowering of peak intensity and broadening of its width. Comparing with sample C, sample D shows broader peak width. Since sample D has higher doping concentration for QDs due to thinner spacer layer than that of sample C, we conclude that this broadening is originated from spectral diffusion with increasing of doping concentration. This kind of serious degradation is not preferred for optical application.

Figure 3 shows typical power dependence of μ-PL spectra of the neutralized sample (Sample B). As pumping power increases, emissions from XX^0 that is important for entangled photon source as mentioned in the introduction appear remarkably. Furthermore, distinctive peak of XX^0 is observed without extra emission as background with increasing of excitation power. These results indicate the successful neutralization of positively charged excitonic states without optical degradation.

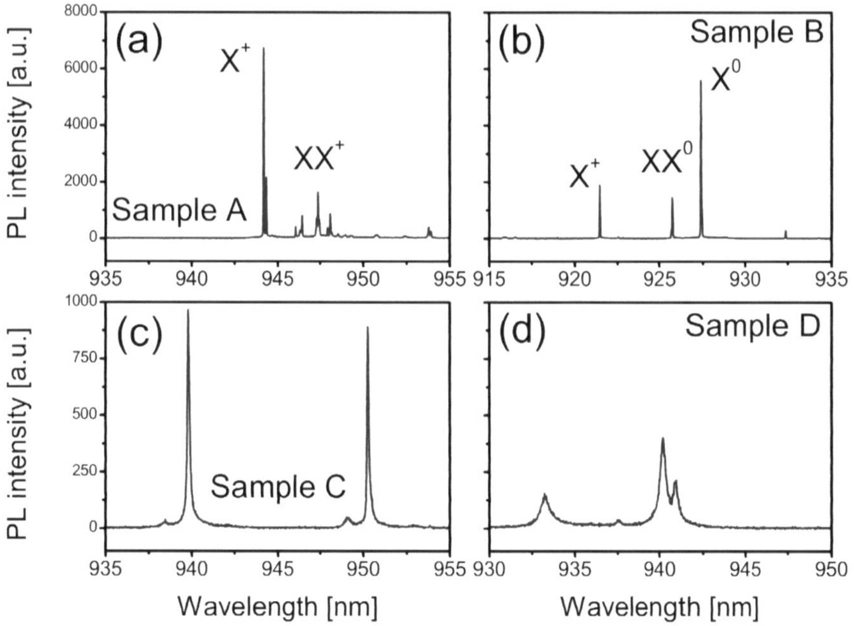

Figure 2. Typical μ-PL spectra of delta doped samples A~D. Charged states in (a) and (b) are assigned by power dependence and photon correlation measurements.

In the case of neutralized sample (sample B), doping concentration is ~100 Si atoms per a dot. Comparing with QD density, this is relatively high doping concentration. In the simple case, one electron is required for neutralization of one QD (X^+), so that doping concentration should be comparable to that of QD density. Although precise mechanism of high doping concentration is not clear, it is probably due to faster capture rate of hole than that of electron in a QD, low activity of Si atoms as dopant due to low temperature, and trap states of electron around QDs. In the present, this remains further subject.

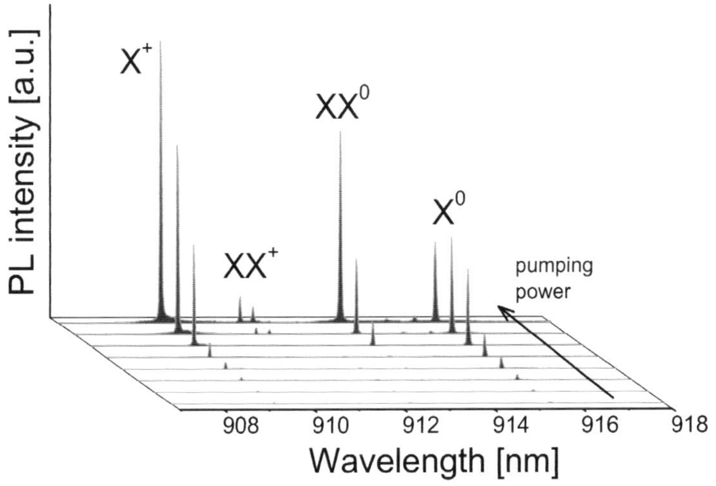

Figure 3. Typical power dependence of μ-PL spectrum of delta doped sample C.

4. Summary

Si delta doping is applied to neutralize InAs QDs with dominant emission from positively charged excitonic states. The charged QDs are neutralized successfully by optimizing the doping condition, although excessive doping leads optical degradation as lowering of peak intensity and its broadening. We have achieved distinctive emissions of XX^0 without extra emission as background that is applicable to entangled photon source.

Acknowledgement
This work was accomplished by the Special Coordination Funds for Promoting Science and Technology. We thank Dr. Watanabe (NanoQuine, The University of Tokyo) for his support on MBE system operation.

References

[1] Stevenson R M , Young R J, Atkinson P, Cooper K, Ritchie D A, Sheilds A J 2006 Nature 439 179.
[2] Kroutvar M, Ducommun Y, Heiss D, Bichler M, Schuh D, Abstreiter G, Finley J J 2004 Nature 432 81.
[3] David P, Ladd T D, Zhang B, Yamamoto Y 2008 Nature 456 218.
[4] Kumagai N, Ohkouchi S, Nakagawa S, Nomura M, Ota Y, Shirane M, Igarashi Y, Yorozu S, Iwamoto S, Arakawa Y Physica E (available on line).
[5] Shirane M, Igarashi Y, Ota Y, Nomura M. Kumagai N, Ohkouchi S, Kirihara A, Ishida S, Iwamoto S, Yorozu S, Arakawa Y Physica E (available on line).
[6] Igarashi Y, M Shirane, Ota Y, Nomura M. Kumagai N, Ohkouchi S, Kirihara A, Ishida S, Iwamoto S, Yorozu S, Arakawa (to be submitted).

Journal of Physics: Conference Series **245** (2010) 012089 doi:10.1088/1742-6596/245/1/012089

Study of structural and electronic properties of metallic nanowires: Bi, Na, Cu, Pb

Sarika Shrivastava[1], Pankaj Srivastava[2], A K Shrivastava[3], Gajanan Pandey[4] and C S Malvi[5]

[1]Gwalior Institute of Information Technology, Gwalior 474015, India
[2]ABV-Indian Institute of Information Tech. & Management, Gwalior 474010, India
[3]SOS in Physics, Jiwaji University, Gwalior 474011, India
[4]Madhav Institute of Technology and Science, Gwalior 474005, India
[5]SPEME, University of Leeds, UK

E-mail: nanosarika@gmail.com

Abstract. The properties of electrons become more striking when progressing from the three- dimensional to lower dimensions; thus there is a need to investigate and recognize the properties of metallic structures at the nanoscale. Devices made from nanowires have several advantages over those made by photolithography. A variety of approaches have been devised to organize nanowires via self–assembly, thus eliminating the need for the expansive lithographic techniques normally required to produce devices of the size of typical nanowires. In the present work, an ab-initio self-consistent density functional method in the local density approximation is employed to study structural and electronic properties of some metallic nanowires such as Bismuth ($Bi_{n)}$), Sodium ($Na_{n)}$), Copper ($Cu_{n)}$), Lead (Pb_n) where n=5. We explored the lowest energy structure and investigate the various physical properties of Bi_n, Na_n, Cu_n, Pb_n (n=5) nanowires. Calculations of lattice parameter, bond length, bond angle, binding energy (BE), internal energy, pressure, band structure and the density of states (DOS) have been carried out in a large energy interval for different isomeric forms like linear chain, zigzag, equilateral triangle, dumbbell, pyramidal, pentagonal, tetrahedral etc. and thereby analyzing the size effects on nanowires.

1. Introduction

The most fundamental properties of materials get changed at the nanoscale with respect to any other scale, the behaviour of a material like its physical, chemical, biological, electrical, magnetic properties is modified when compared to the bulk. Generally, there are two distinct approaches for creating small objects, one by top-down as lithography, etching and deposition techniques and microelectronic devices and other by a bottom-up approach which aims to start with the smallest possible building material- atom, to create a desired product, and this was very well proven by two IBM researchers by creating STM (Scanning tunnelling microscope), to place every atom at the right place. Nanotechnology promises putting essentially every atom in the right place, making almost any structure consistent with the laws of physics and chemistry. The present work is focused on nanowires. Nanowires are playing an integral part in the design and construction of both electronic and optoelectronic nanodevices [1]. Many analyses have been performed in various materials, for example Cu, Si, Ag, Pb, Ni, Al etc. nanowires [2-7] and successful fabrication has been achieved [8]. The increased surface area, very high density of electronic states and joint density of states near the energies of their van Hove singularities, enhanced exciton binding energy, diameter-dependent bandgap, and increased surface scattering for electrons and phonons are just some of the ways in which nanowires differ from their corresponding bulk materials.

2. Methodology

Although a lot of work has been done, the study of metallic nanowire is still at an early stage; hence an extensive investigation is yet to be done. We analyzed the structural and electronic properties of some isomeric structure of metallic nanowires (Bi, Na, Cu, Pb) using computational methods. We have investigated various geometries for different metallic nanowires and then predict the most stable geometry for those particular metallic nanowires in terms of their physical properties. For our computational work ABINIT code [9] has been used to determine the total energy of an assembly of nuclei and electrons placed in a repeated cell according to Density Function

Theory. The computation can be done using plane waves and norm-conserving pseudopotentials. In this code, self consistent calculations will generate the DFT ground- state, with associated energy and density. Afterwards, a non-consistant self calculation might generate Eigen energies at a large number of k-points, for band structures. The electronic density of states can be computed either within the tetrahedron method, or using a smearing technique. There are many advantages of ABINIT code, such as the computation is done using plane waves and pseudopotentials, admits various types of pseudopotentials, the total energy computation is done according to the DFT and LDA, metallic as well as insulating systems can be treated, cell may be orthogonal or non-orthogonal and the code can provide an automatic analysis of bond lengths and angles, and the atomic coordinates.

3. Results and Discussion

3.1 Bi_N (N=1-5) nanowires: Bismuth is really an interesting material to investigate on the nanoscale because bulk Bismuth has very small effective mass carriers, highly anisotropic carrier pockets and a very long mean free path. Thus, these bulk properties lead to a quantum nature that can be observed in nanowires of relatively large diameter. In fact, Bismuth in bulk form is a semimetal and therefore it is not a good thermoelectric material. However, with the introduction of quantum confinement it can be given small diameter nanowire.

Theoretically, *Schmidt* and *Springborg* [10] studied a one dimensional chain of metal atoms including Bi chains but considering only three structures. *Lin et al* [11] studied the transport properties of Bi nanowires arrays by employing a semi classical transport model. However, our current knowledge of the structural and electronic properties of various technologically important metallic nanowires is still quite limited. We are also not aware of any ab-initio theoretical work on the structural and electronic properties of Bi nanowires in various isomeric forms, since the study of metallic nanowires is at early stage. Extensive investigation of semi-metallic Bi nanowires is required. Figure 1 represents the obtained results for Bi nanowires. It has been observed that the equilateral triangle shape is more stable for the Bi nanowire metallic character. The maximum DOS is seen for a three atom triangular wire. A summary of the results is presented in Table 1.

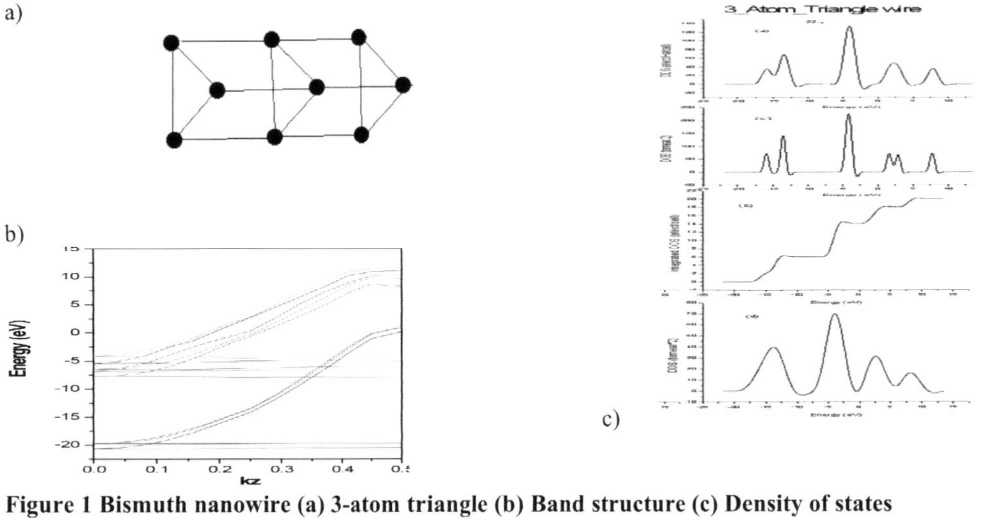

Figure 1 Bismuth nanowire (a) 3-atom triangle (b) Band structure (c) Density of states

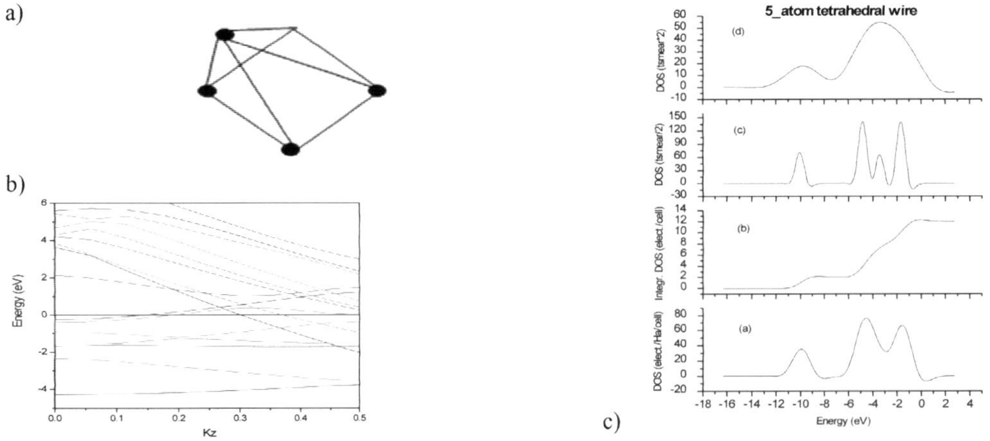

Figure 2 Sodium nanowire (a) 5-atom tetrahedral shape (b) Band structure (c) Density of states

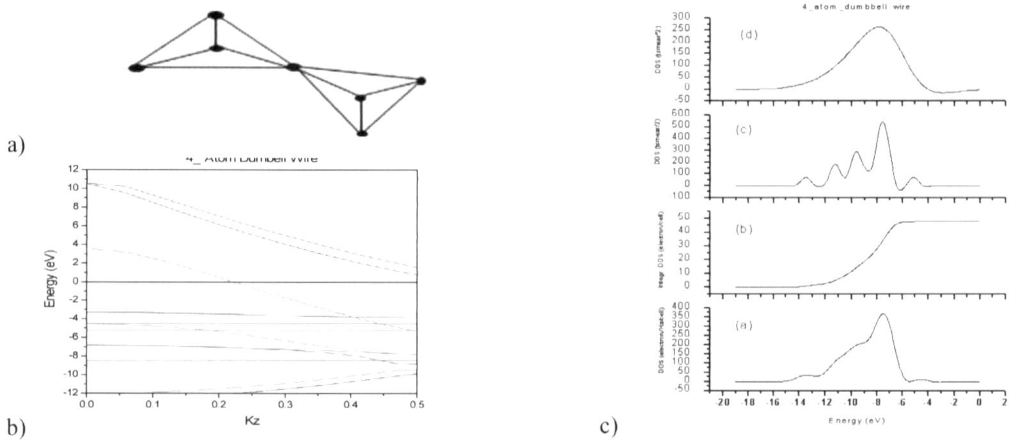

Figure 3 Copper nanowire (a) 4-atom dumbbell shape (b) Band structure (c) Density of states

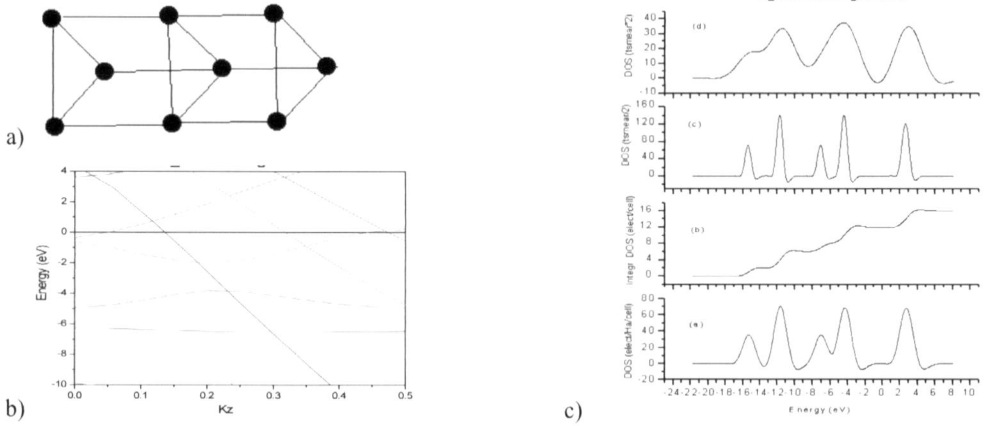

Figure 4 Lead nanowire (a) 3-atom triangle shape (b) Band structure (c) Density of states

Table 1 Summary of results

Metal	Stable shape	Binding Energy (eV)	Lattice constant (Å)	Bond length (Å)	Fig. No.
Bi_N (N=1-5) nanowires	3 atom equilateral triangle	12.59	2.09	2.15	2
Na_N (N=1-5) nanowires	5-atom tetrahedral	2.85	4.02	3.07	3
Cu_N (N=1-5) nanowires	4-atom dumbbell	4.55	2.28	2.29	4
Pb_N (N=1-5) nanowires	3 atom equilateral triangle	12.62	2.70	2.06	5

3.2 Na_N (N=1-5) nanowires: In order to understand the size effect and stability of alkali metals for quantum quantization we have chosen sodium nanowires for our analysis. Although a great deal of work has been done for transition metallic nanowires, so far as alkali metals are concerned more research is needed. *Yannouleas et al* [12] studied the energetics, forces and quantized conductance in jellium-modelled metallic nanowires theoretically. The effect of atomic and electronic structure on mechanical properties of a piece of sodium nanowire was studied with a total energy method based on the Density Functional Theory by *Hakkinen and Manninen* [13]. *Kobayashi et al* [14] employed first-principles method to study the electron transport through monatomic Al and Na wires. The stability of sodium nanowires was studied by *Yanson et al* [15] by modelling them as infinite uniform jellium cylinders. *Tsukamoto and Hirose* [16] studied the electron-transport properties of Na nanowires under applied bias voltages, by using first principles calculations. They predict that a voltage drop locally occurs on the negatively biased end of the nanowire. *Sen et al* [17] studied the atomic and electronic structure of pentagonal nanowires of some alkali, simple transition, noble and inert gas atoms. *Qi et al* [18] analyzed the size effect on the cohesive energy and the lattice parameter of nanoparticle. They revealed that the cohesive energy and the lattice parameter of nanoparticles depend on its size i.e. both of them are size dependent.

Even odd oscillation in conductance of single-row sodium nanowire suspended between semi-infinite electrodes is studied by *Egami et al* [19]. Figure 2 represents the obtained results for Na nanowires. For Na nanowires we found that a five atom tetrahedral wire is more stable; the maximum DOS is seen for a five atom tetrahedral wire. The lattice parameter and bond length goes on increasing with the number of atoms. We have also analyzed the size effect which is important for understanding the actual structural and electronic properties. It is observed that binding energy increases as the number of atoms increases.

3.3 Cu_N (N=1-5) nanowires: *Sen et al* [17] have used first principles calculations to study two pentagonal Cu nanowires along the direction and concluded that these wires should be stable and metallic. The energetics of nanowire of several materials including Cu is studied by quenched molecular dynamics by *Tommei et al* [20] with the aim of comparing wires i.e. fcc and pentagonal which are expected to be among the most favourable ones for wires of intermediate thickness. They revealed that the best pentagonal wires are always those including marks truncations, but the fcc wires are more favourable than the pentagonal ones in any size range. Figure 3 represents the obtained results for Cu nanowires. The highest binding energy is seen in four atom dumbbell shape and hence more stable. Therefore, the maximum DOS is seen for a five atom pyramidal wire followed by four atom dumbbell shape wire

3.4 Pb_N (N=1-5) nanowires: Lead is unique among all metals. Pure lead has little strength and is very soft, because its inter-atomic bonding forces are weak and its melting point is very low. Its resistance to corrosion in aggressive environments is well known. In fact, a very reactive metal, therefore it is commonly used in corrosive environments. The understanding of the conductive properties of extremely small metallic structure is of utmost importance for the potential use in nanoelectronic devices and as interconnects. Theoretically, *Sen et al* [17] performed a first principles study of lead nanowires only in various pentagonal structures and concluded that staggered pentagons have a stable structure. Figure 4 represents the obtained results for Pb nanowires using our ABINIT code. Pb nanowires show more stability in an equilateral triangle shape; thus the maximum DOS is seen for three atom triangular wires. The DOS of three atom triangular wires reveals that the electron are highly delocalized near the Fermi level in comparison to other structures. All the band structure calculations show that large number of channels is available for the electronic conduction hence ballistic conductance is possible.

4. Conclusion

The results of our calculations for various nanowires giving the observed shape and number of atoms are summarized in Table 1. As we found that an equilateral triangle shape is more stable for Bi nanowire metallic character, the maximum DOS is seen for a three atom triangular wire. For Na nanowires, a five atom tetrahedral wire is the most stable; the maximum DOS is seen for a five atom tetrahedral wire. In the case of Cu nanowires, the highest binding energy is seen in a four atom dumbbell shape and is hence the most stable. Therefore, the maximum DOS is seen for a five atoms pyramidal wire followed by a four atom dumbbell shaped wire. Whereas Pb nanowires show more stability in the equilateral triangle shape; thus the maximum DOS is seen for a three atom triangular wire. The DOS of three atom triangular wires reveals that the electrons are highly delocalized near the Fermi level in comparison to other structures. All the band structure calculations show that a large number of channels are available for electronic conduction, hence ballistic conductance is possible.

References:

1. D Appell, *Nature,* (London) 419, 553 (2002).
2. Y- K Kim, S J Park, J P K, G T Kim, S Hong and J S Ha, *Nanotechnology,* 18 015304 (2007).
3. Y-L Chueh, C-H Hsieh, M-T Chang, L-J Chou, C S Lao, J-Y Gan and Z LWang, *Adv. Mater.* (2006).
4. H Ohnishi, Y Kondo and K Takayanagi, *Nature* 395, 780 (1998).
5. Y Kondo and K Takayanagi, *Phys. Rev. Lett.*79, 3455 (1997).
6. Y Kondo and K Takayanagi, *Science* 289, 606 (2000).
7. X Duan, Y Huang, Y Cui, J Wang and C M Lieber, *Nature*, 409:66(2001).
8. X Duan, Y Huang and C M Lieber, *Nano Lett.*, 2:487 (2002).
9. S Goedecker, SIAM *J. Sci. Comput.* 18, 11605 (1997).
10. K Schmidt and M Springborg, *Solid State Communication,* 104(7) 413(1997).
11. Y-M Lin, S B Cronin, J Y Ying, M S Dresselhaus and J P Heremans, *App. Phy Letters* 76(26)3944(2000).
12. C Yannouleas, N E Bogachek and U Landman, *Physical Review* B57, 4872 (1998).
13. H Hakkinen and M Manninen, *Europhysics Letters* 44(1), 80 (1998).
14. N Kobayashi, M Brandbyge and Masaru Tsukada, *Physical Review, B* 62, 8430(2000).
15. A I Yanson, I K Yanson and J M van Ruitenbeek, *Physical Review letters*, 84, 5832(2000).
16. S Tsukamoto and K Hirose, *Physical Review, B 66*, 161402 (2002).
17. P Sen, O Gulseren, T Yildirim, I P Batra and S Ciraci, *Phys Rev. B* 65, 235433 (2002).
18. W H Qi, M P Wang, G Y Xu, Z Li and J Y Chem, Beijing, *P. R. China*, 86 (2002).
19. Y Egami, T Ono and K Hirose, *Physical Review B* 72, 125318(2005).
20. G E Tommei, F Baletto, R Ferrando, R Spadacini and A Danani, *Phys. Rev. B* 69, 115426 (2004).

Journal of Physics: Conference Series **245** (2010) 012090 doi:10.1088/1742-6596/245/1/012090

Absorption Coefficient in Periodic InAs/GaAs Nanostructures

S. Rodríguez-Bolívar, F. M. Gómez-Campos, A. Luque-Rodríguez, J. A. López-Villanueva, J. E. Carceller

Departamento de Electrónica y Tecnología de los Computadores, Facultad de Ciencias, Av. Fuentenueva s/n, C. P. 18071, Universidad de Granada

fmgomez@ugr.es

Abstract. Periodic nanostructure manufacture has been proposed as a procedure for obtaining new materials with tunable physical properties, such as the photon absorption coefficient. In this work we have theoretically investigated this quantity in ordered InAs/GaAs cubic quantum dot systems. We solved the Schrödinger equation associated with these structures, using a set of 13×13×13 plane waves at 12,167 equally spaced points of the Q space. We focused on the transitions between minibands arising from the conduction band. We took into account the different effective masses in each material. We included the effects of the strain by taking a conduction band offset of 0.5 eV, corresponding to strained InAs in GaAs

1. Introduction

During the last few decades, investigations into one-dimensional superlattices have led to interesting applications in optoelectronics [1][2]. Two- and three-dimensional regimented nanostructures would also make available new applications because of the possibility of creating artificial crystalline structures with tunable properties. Therefore, technological efforts have been made to obtain this kind of material.

The physics of two- and three-dimensional periodic nanostructures is not yet well understood. Some works [3][4] have tried to gain an insight into the miniband structure arising from coupling between regimented quantum dots. They used a simple modeling of the systems and obtained some interesting results. Furthermore, other researchers [5] have studied quantization in two-dimensional structures consisting of arrays of crossed quantum wires, and also expect promising physical phenomena.

In this work we present a study of regimented three-dimensional quantum dot arrays. We evaluate the absorption coefficient for optical transitions between states in minibands arising from the conduction bands of InAs/GaAs. Finally we draw our conclusions and suggest future applications of these systems.

2. Theoretical Formalism

We proposed to study the optical absorption coefficient in regimented cubic quantum dot arrays. The quantum dot material is InAs embedded in a GaAs matrix. We included the effects of the strain in the conduction band offset and in the effective masses using the $\boldsymbol{k}\cdot\boldsymbol{p}$ approach of Bir-Pikus [6]. Our investigation focused on cubic quantum dots with an interdot separation equal to the side length of the cube. We took three different cases: array #1 is made up of 3 nm side length cubic quantum dots; array

#2 has quantum dots of 4 nm side length; and array #3 is formed by quantum dots of 5 nm side length. The array alignment is in the [100] direction. The period of the structure is hereafter called "supercell".

We solved the Schrödinger equation of the system using the plane wave method. Because of the periodicity of the system, the wave functions satisfy the Bloch theorem:

$$\Psi_{\vec{q},n}(\vec{r}) = \eta_{\vec{q},n}(\vec{r})e^{iq\cdot r} \quad (1)$$

where $\vec{q}$ is the wave vector associated with the supercell size and $\eta_{\vec{q},n}(\vec{r})$ is a function with the same periodicity as the supercell. Since each $\vec{q}$ is related to several eigenenergies, the subindex n stands for each of the energy minibands arising from the quantization. It is useful to define a Q-space to locate each of the eigenstates of the system in the same way as the K-space is conceived for regular crystalline systems.

3. Miniband Structure

The solution of the Schrödinger equation provides a miniband structure for electrons related not only to the materials used but also to the geometry of the array. Figure 1 shows a part of the miniband structure of array #2. The existence of an electronic energy gap of about 63 meV is noticeable between the lowest energy miniband (from point M in the lowest to point X in the second). Taking into account that optical transitions are vertical in Q-space, the optical energy gap is about 150 meV (transitions at the X point). This could lead to interesting applications in far-infrared optoelectronics.

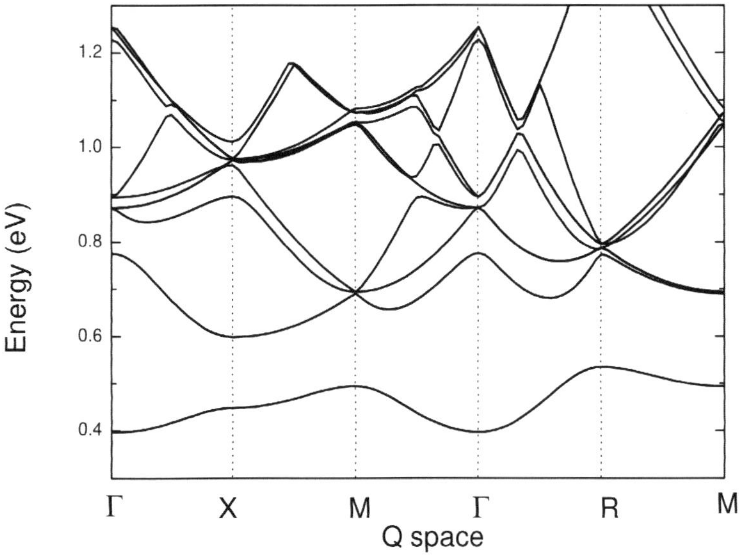

Figure 1. Ten lowest energy minibands of the miniband structure of array #2. These are states arising from energy quantization in the bulk conduction band. The reference of energies is taken at the minimum of the bulk conduction band of InAs.

4. Absorption Coefficient

In order to understand the optical behavior of the system, we carried out calculations of the absorption coefficient for transitions within the miniband structure arising from states in the conduction band. Starting from Fermi's Golden Rule, we developed the theoretical expression of the absorption coefficient in regimented nanostructures:

$$\alpha(\hbar\omega_{op}) = \sum_{n_i}\sum_{n_f>n_i} \frac{w_{n_i\to n_f}(\hbar\omega_{op})}{V} \tag{2}$$

where $w_{n_i\to n_f}(\hbar\omega_{op})$ is the absorption probability for transitions going from the n_i to the n_f minibands, $\hbar\omega_{op}$ is the photon energy and V is the volume of the periodic nanostructure. $w_{n_i\to n_f}(\hbar\omega_{op})$ is obtained from the following expression [7]:

$$w_{n_i\to n_f}(\hbar\omega_{op}) = \frac{\hbar^2\pi e^2}{n_r c_{light}\varepsilon_0 m_0^2\omega_{op}} \times$$
$$\times \int_{Q\,\text{space}} f(E_{n_i})\left[1-f(E_{n_f})\right]\left|\hat{\xi}\cdot\int_{\text{supercell}}\eta^*_{\vec{q},n_f}(\vec{r})\nabla\eta_{\vec{q},n_i}(\vec{r})\right|^2 \delta(E_{n_f}-E_{n_i}-\hbar\omega_{op})\rho_Q d\vec{q} \tag{3}$$

$|e|$ is the absolute value of the electron charge, n_r is the mean refraction index in the system, c_{light} is the speed of light in vacuum, ε_0 is the vacuum dielectric constant, m_0 is the electron rest mass, ω_{op} is the photon frequency, $f(E)$ is the Fermi-Dirac statistic, $\hat{\xi}$ is the polarization of the potential vector of the photon and ρ_Q is the density of states in the Q-space ($\frac{V}{8\pi^3}$ for the system studied in this paper).

For a given array with a miniband structure, equation (3) provides a result depending on light polarization ($\hat{\xi}$) and the position of the Fermi level (in $f(E)$). In order to increase the absorption, it is desirable to place the Fermi level between the two lowest minibands. Nevertheless, our calculations indicate that this implies very high levels of doping, and it could lead to undesired effects because of impurities. Therefore, we placed the Fermi level just at the bottom of the bulk conduction band in InAs. This might be related to a medium or high level of doping, maybe close to those doping results reported elsewhere [8].

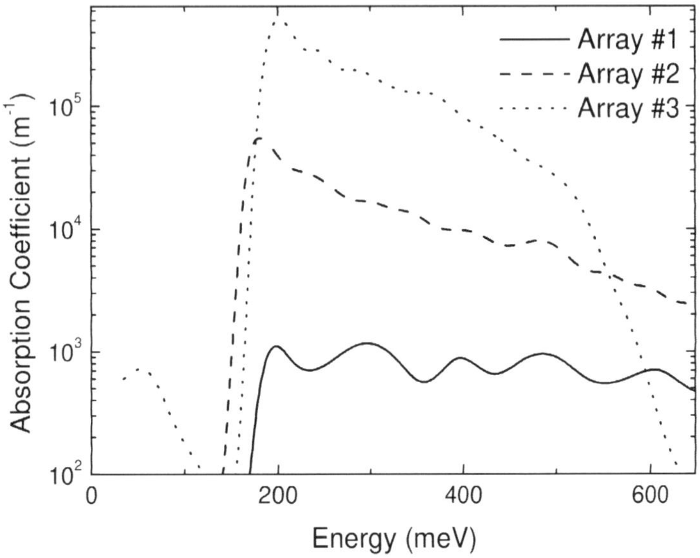

Figure 2. Absorption coefficient for the arrays studied for light polarization in [100].

The absorption coefficients for arrays #1, #2 and #3 are shown in figure 2 for light polarization in the [100] direction. The results correspond to optical transitions between the ten lowest energy minibands. We observed the existence of thresholds for photon absorption determined by the miniband structure of the system. For the cases studied, these threshold are about energies corresponding to far infrared (~200 meV), thus making it possible to propose these systems for use in sensor applications. For the arrays studied, the results provide a greater absorption coefficient for larger dots and greater interdot separations. The reason for this behavior might be the lower width of the lowest energy miniband when decreasing the dot coupling with neighbors, providing an almost constant energy level spread about 20-30 meV and enhancing the absorption for a certain photon energy. On the other hand, array #1 has the lowest energy miniband spread, about 100 meV (see figure 1), and therefore there are many photon energies able to pump carriers from the first to the second miniband, thus decreasing the absorption at the threshold.

A feature of these systems that could be useful for optoelectronic applications is the possibility of removing the carriers after photon absorption through the second miniband of the structure. Carrier transport along the minibands is possible, in contrast with photon absorption by single impurities or single dots. Thus, these systems could be useful for manufacturing QDIPs [9].

In our investigation we used three different arrays, but theoretical investigations could throw light on the influence of dot size and shape or interdot separations in the absorption coefficient, and therefore suggest new periodic nanostructures fitting the desired absorption coefficient profile.

Further, we believe that this type of structure could play an important role in designing intermediate band solar cells. It would be important to achieve the regimentation in order to obtain the miniband structure and to investigate the implications of using different materials in the dot and matrix to find the absorption thresholds of interest for these applications [8].

5. Conclusions

We have investigated the miniband structure formation arising from states of the conduction band in quantum dot arrays of several sizes. Quantum dots are made of InAs embedded in a matrix of GaAs. We took into consideration the stress in the structure, and calculated the absorption coefficient for three different array structures. The results suggest the existence of absorption thresholds at energies close to the optical band gap between the first and second minibands of the structure.

To summarize, we have shown that periodic semiconductor nanostructures could be an interesting topic in optoelectronics, and future studies may clarify their capabilities to develop new applications in this field. We have also suggested future technological applications for them as far-infrared sensors and proposed future theoretical investigations.

References

[1] M. Helm, W. Hilber, T. Fromherz, F. M. Peeters, K. Alavi, R. N. Pathak 1993 *Phys. Rev. B* **48** 1601
[2] S. H. Kwok, H. T. Grahn, M. Ramsteiner, K. Ploog, F. Prengel, A. Wacker, E. Schöll, S. Murugkar, R. Merlin 1995 *Phys. Rev. B* **51** 9943
[3] O. L. Lazarenkova, A. A. Balandin 2001 *J. Appl. Phys.* **89** 5509
[4] C. W. Jiang, M. A. Green, 2006 *J. Appl. Phys.* **99** 114902
[5] J. C. Yi 2008 *Microelect. Journal* **39** 369
[6] G. L. Bir, G. E. Pikus 1960 *Fizika Tverdogo Tela* **2** 2287
[7] F. M. Gómez-Campos, S. Rodríguez-Bolívar, A. Luque-Rodríguez, J. A. López-Villanueva, J. E. Carceller 2010 *Submitted to J. Appl. Phys.*
[8] A. Martí, E. Antolín, C. R. Stanley, C. D. Farmer, N. López, P. Díaz, E. Cánovas, P. G. Linares, A. Luque 2006 *Phys. Rev. Lett.* **97** 247701
[9] P. Martyniuk, A. Rogalski 2008 *Progress in Quantum Electronics* **32** 8

Journal of Physics: Conference Series **245** (2010) 012091 doi:10.1088/1742-6596/245/1/012091

Kinetic Monte Carlo simulation of quantum-dot nucleation and growth in PbSe/PbEuTe multilayers

M Mixa[1], V Holý[1], G Springholz[2] and G Bauer[2]

[1] Department of Condensed Matter Physics, Charles University, 12116 Prague, Czech Republic
[2] Institut für Halbleiterphysik, Johannes Kepler Universität, A-4040 Linz, Austria

E-mail: martin.mixa@mff.cuni.cz

Abstract. We study the influence of different growth conditions on the structural properties of self-organized quantum-dot multilayers PbSe/PbEuTe(111). The kinetic Monte Carlo method is employed and the simulations show that mainly the quality of the dot ordering is affected when the growth conditions are slightly changed.

1. Introduction

Self-organization during a strain-driven heteroepitaxy of semiconductor nanostructures is a promising technique for fabrication of devices based on specific electrical and optical properties of quantum dots (QDs). Mechanisms of spontaneous formation of three-dimensional (3D) quantum dots in the Stranski-Krastanow growth mode as well as the self-assembling principles at their growth have been a field of intensive research in the last few years (for reviews see, e.g., Refs. [1, 2, 3]. For a prospective technological use the achievement of a narrow size distribution and a regular arrangement of the QDs is important. To this purpose the strain-mediated self-organization is advantageously applied at preparation of 3D superlattices of QDs, where the already grown islands influence positions and sizes of the islands in the subsequent layers [3, 4, 5].

Whereas the vertical correlations of dot positions are usually sufficiently ensured by the self-organization effect, the lateral ordering is considerably weaker. A unique example of a semiconductor system (grown on planar substrates) exhibiting good quality of vertical and lateral dot correlations are PbSe/PbEuTe(111) multilayers [6]. Depending on the thickness of the PbEuTe spacer layer, three different arrangements of a PbSe dot superlattice are observed (vertical and trigonal arrangements of the dots and a non-correlated mode), when also a pronounced hexagonal lateral ordering is formed in the both ordered cases [7, 8].

In the paper [9] we introduced a semi-atomistic kinetic Monte Carlo (KMC) model that we have developed in order to simulate the strain-mediated epitaxial growth of self-assembled multilayers. For the first time, this model enabled us to perform simulations of the dot nucleation and growth in multilayers consisting of tens of periods with hundreds of 3D quantum dots. We reproduced all three experimentally observed dot arrangements in the PbSe/PbEuTe(111) multilayers, moreover, with a good quantitative agreement. In this work we want to demonstrate a stability of the results in case the growth conditions are slightly altered. This is important since the experimental samples are also investigated within a range of growth conditions and, further, the conditions are adjusted (measured) only with a limited precision.

2. KMC model description

We shortly summarize the main features of our model (for details see Ref. [9]). The KMC method models a growth simulation as a sequence of random events taking place on a surface [10, 11, 12]. The basic included events are the deposition of adatoms onto the surface and their subsequent surface diffusion. Since the adatom diffusion is a thermally activated process, the hopping rate k of an adatom onto a nearest-neighbor site is given by the Arrhenius-like formula

$$k_{i\to f} = k_0 \exp(-E_{i\to f}/k_B T), \quad (1)$$

where $k_0 = 2k_B T/h \sim 10^{13}\,\mathrm{s}^{-1}$ is a characteristic vibrational frequency and T is the substrate temperature. $E_{i\to f}$ denotes the energy barrier which an adatom must overcome at a hop from the initial site i to the final site f. In our approach based on incorporation of the elastic strain to the adatom diffusion we model $E_{i\to f}$ as a sum of two terms: a substrate term E_S and a strain-field contribution induced by buried dots,

$$E_{i\to f} = E_S + K(w_f - w_i)\, V_0. \quad (2)$$

w_i, w_f are the elastic energy densities at the sites i and f calculated from the strain tensor ϵ_{ij} and the elastic constants C_{ijkl} as $w = 1/2\, C_{ijkl}\, \epsilon_{ij}\, \epsilon_{kl}$. The strain tensor on the surface of wetting layer, calculated in the approach of elastic continuum, comprises the pseudomorphic deformation of the wetting layer and the individual strain contributions from all QDs *buried* under the just growing layer. V_0 is the volume of one adatom (one PbSe molecule) and the coefficient $K > 0$ is a correction factor which absorbs all simplifications and neglects made in the model; above all, the omission of strain field caused by growing dots in Eq. (2).

The key simplification of our model is that the shape of growing dots is *not* atomistically simulated but it is taken *ad hoc* from the experiment. Thus the dots are modeled as trilateral pyramids with $\{100\}$ side facets [6] and only their size changes during the simulation. The main advantage of this approximation is that the simulations are substantially faster, so a larger area and more multilayer periods can be simulated, which is more important for the self-organization effect study. Since the dots are not treated atomistically and thus the described model is not bond-counting, a critical-nucleus concept is introduced in a special way. We consider a nonzero probability of an adatom escape from a dot smaller than the critical nucleus consisting of N_c adatoms. All dots of size equal to or larger than the critical nucleus are taken as absolutely stable, i.e. the escape probability is set to zero for them. (For more details see Ref. [9].)

3. Simulation results

As we adumbrated in the introduction we performed simulations under slightly different conditions (with respect to those used in [9]) to test the stability of our results. A fundamental influence on the epitaxial growth is possessed by the substrate temperature T and the deposition flux F. From this reason we altered just them, whereas the other model parameters remained constant (namely: $E_S = 0.93\,\mathrm{eV}$, $N_c = 100$ PbSe molecules, $K = 7$, total PbSe coverage of 5 ML and its part used for the formation of dots of 1.2 ML, and simulation grid size of 2048×2048 cells). For lucidity, let us label as A the growth conditions with $T = 370\,^\circ\mathrm{C}$ and $F = 0.12\,\mathrm{ML/s}$ used in [9], and as B the changed ones with $T = 360\,^\circ\mathrm{C}$ and $F = 0.08\,\mathrm{ML/s}$.

Figure 1 shows simulation results of three multilayers consisting from 20 periods grown under the conditions B for three different spacer thicknesses d_{SP} of 26, 45, and 60 nm. The first layer in all these multilayers is identical and contains completely disordered array of dots. The dot density in the layer $n = 512\,\mu\mathrm{m}^{-2}$ distinctly differs from that obtained at the growth conditions A where it amounted to $442\,\mu\mathrm{m}^{-2}$. Regardless this fact, the final arrangement of the whole multilayers is *qualitatively* the same for both growth conditions A and B (to compare see [9]). For $d_{SP} = 26$ nm we obtained again a vertically aligned multilayer [panels (a), (b) in Fig. 1], for

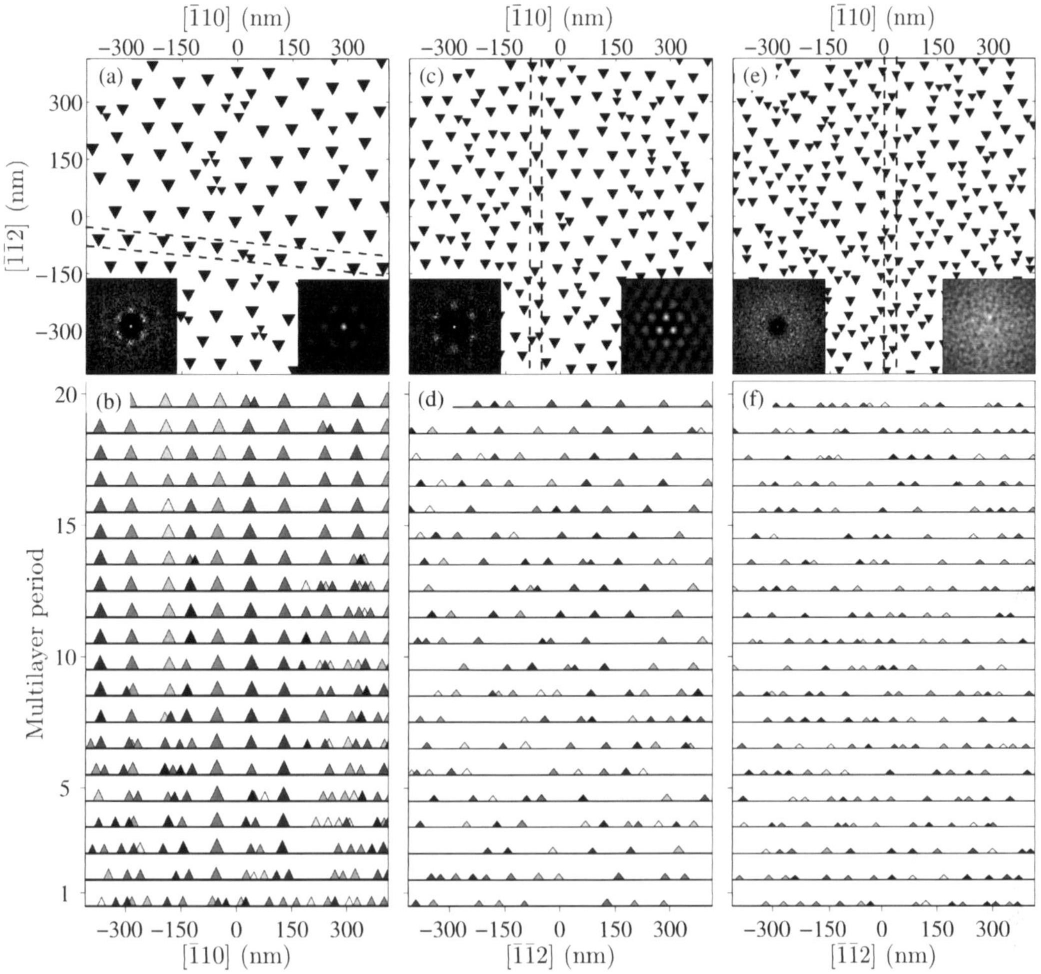

Figure 1. Panels (a), (c), and (e) display the top views on positions and sizes of the dots in the 20th periods of three multilayers simulated with different spacer thicknesses of 26, 45, and 60 nm, respectively. The two dashed lines always mark out the location and orientation of a vertical cross-section through the related multilayer. The cross-sections are displayed in panels (b), (d), and (f), respectively. The gray level of the QD-representing triangles indicates their complementary lateral coordinate in the given cross-section band. The left-bottom-corner insets in the upper panels show the 2D FFT power spectra of the given surface morphology, and the right-bottom-corner insets show the cross-correlation functions of the 19th and 20th dot-layer morphologies.

$d_{SP} = 45$ nm an $ABCABC$ stacking of dot layers is produced [panels (c), (d) in Fig. 1], and for $d_{SP} = 60$ nm an uncorrelated multilayer grew [panels (e), (f) in Fig. 1].

A quantitative comparison between the multilayers grown at the conditions A and B is given in table 1. The initial dot densities n_i in first layers and the dot densities n, mean lateral distances L and correlation lengths L^{corr} in the 20th layers (determined from the FFT power spectra) are compared. The values of L remained constant (within the measurement error) both for $d_{SP} = 26$ nm and $d_{SP} = 45$ nm. It is due to the fact, that the lateral dot distance is primarily determined by the strain field distribution, i.e. by the spacer layer thickness and

Table 1. Comparison of dot-arrangement parameters (dot density n, lateral dot distance L, and lateral correlation length L^{corr}) at the topmost (20th) period of multilayers simulated under two different growth conditions A and B (see the text) for three different spacer thicknesses d_{SP}. The quantity n_i denotes the initial dot density in the first period.

d_{SP} (nm)	growth cond.	n_i (μm^{-2})	n (μm^{-2})	L (nm)	L^{corr} (nm)
26	A	442 ± 10	176 ± 10	87 ± 8	224 ± 30
26	B	512 ± 10	197 ± 10	89 ± 8	219 ± 30
45	A	442 ± 10	295 ± 10	65 ± 3	243 ± 30
45	B	512 ± 10	337 ± 10	65 ± 3	223 ± 30
60	A	442 ± 10	440 ± 10	*disordered*	
60	B	512 ± 10	507 ± 10	*disordered*	

the elastic constants which stayed identical. The dot densities n_i and n for $d_{SP} = 60$ nm are different, however, they do not change (out of the error) during the multilayer growth which confirms the uncorrelated mode. Thus the only one significant difference is in the lateral ordering quality (described by n and L^{corr}) for $d_{SP} = 45$ nm. The more gradual ordering is caused by smaller modulation of the strain field in comparison with the thin spacer of 26 nm.

4. Conclusion

The presented results verified that the self-organization effect is primarily caused by the strain field distribution, namely by the strain field induced by buried dots, since no other mechanism is included in our model. Moreover, they further confirmed that the mean lateral dot distance in ordered multilayers does not depend on the growth conditions if they are varied in a small range, since the distance is also predetermined by the strain field. On the other hand, different growth conditions can influence mainly the quality of a QD superlattice ordering.

Acknowledgments

This work is a part of the research program MSM 0021620834 that is financed by the Ministry of Education of the Czech Republic; the work has been also supported by the EC SANDiE European Network of Excellence and by the project KAN400100652 of the Grant Agency of the Academy of Sciences of the Czech Republic.

References

[1] D. Bimberg, M. Grundmann, and N. N. Ledentsov, *Quantum Dot Heterostructures* (Wiley, New York, 1999).
[2] V. A. Shchukin and D. Bimberg, Rev. Mod. Phys. **71**, 1125 (1999).
[3] J. Stangl, V. Holý, and G. Bauer, Rev. Mod. Phys. **76**, 725 (2004).
[4] J. Tersoff, C. Teichert, and M. G. Lagally, Phys. Rev. Lett. **76**, 1675 (1996).
[5] V. A. Shchukin, D. Bimberg, V. G. Malyshkin, and N. N. Ledentsov, Phys. Rev. B **57**, 12262 (1998).
[6] G. Springholz, V. Holý, M. Pinczolits, and G. Bauer, Science **282**, 734 (1998).
[7] G. Springholz, M. Pinczolits, P. Mayer, V. Holý, G. Bauer, H. H. Kang, and L. Salamanca-Riba, Phys. Rev. Lett. **84**, 4669 (2000).
[8] A. Raab, R. T. Lechner, and G. Springholz, Appl. Phys. Lett. **80**, 1273 (2002).
[9] M. Mixa, V. Holý, G. Springholz, and G. Bauer, Phys. Rev. B **80**, 045325 (2009).
[10] P. Šmilauer and D. D. Vvedensky, Phys. Rev. B **52**, 14263 (1995).
[11] E. Schöll and S. Bose, Solid-State Electron. **42**, 1587 (1998).
[12] M. Meixner and E. Schöll, Phys. Rev. B **67**, 121202(R) (2003).

Bistability and opto-thermal-pulsations in a quantum-dot edge-emitting laser diode

A. Tierno, T. Ackemann

SUPA and Department of Physics, University of Strathclyde, Glasgow G4 ONG, Scotland, UK

E-mail: thorsten.ackemann@strath.ac.uk

Abstract. Self-sustained pulsations in the output of an InAs quantum dot laser diode in the MHz range are reported for the first time. The characteristics (shape, range and frequency) are presented for the free running laser and when optical feedback in the Littrow configuration is applied. The time scale suggest that these are opto-thermal pulsation similar to those reported in quantum well amplifiers. Bistability in the light-current characteristics is observed for wavelengths smaller the gain peak (λ = 1225 nm), but it is not present for wavelength above the gain peak and for the free running lasers.

1. Introduction

Quantum dots (QD) lasers and amplifiers are emerging as an attractive light source in the wavelength range between 1.2-1.3 μm and it is important to assess their performance and stability. One subject of interest are self-pulsing lasers where self-Q-switching due to saturable absorption was observed in many devices [1, 2]. Pulsations are in the GHz region as can be expected for passive Q-switching with the lifetime of the excited state in the nanosecond range. Self-pulsations in edge-emitting QD lasers without an intentionally introduced absorber section were also observed and explained in terms of the inhomogeneous nature of the QD gain, the existence of several confined QD states and the resulting saturable absorption by energy states lower than the laser photon energy [3]. Nevertheless, these oscillations are still in the GHz range.

Bistability is also reported in quantum dot lasers with saturable absorber [1], which is not surprising because saturable absorption is known to promote bistability and self Q-switching.

We are reporting here on self-pulsations in QD edge-emitting lasers without a saturable absorber section taking place on the MHz scale. The oscillation frequency as well as the square-wave shape hint to a thermal origin of the dynamics. Indeed, opto-thermal pulsations with a very similar phenomenology were studied in quantum well amplifiers [4, 5] and are typical for a variety of other nonlinear optical systems [6, 7]. It was shown in [4] that the self-oscillations follow van der Pol-Fitzhugh-Nagumo dynamics, a fairly general scenario of relaxation oscillations, which are characterized by the competition of very different time scales (in [4, 5] for example carrier dynamics and thermal relaxation). Interestingly, the self-pulsing dynamics are found to persist if frequency-selective feedback is applied.

2. Devices and experimental setup

The laser is an edge-emitting diode (QDL) from Innolume GmbH with a length of L = 3.5 mm. It contains InAs QD in a GaAs matrix. It is designed to be single spatial mode with only a shallow edged waveguide and a stripe width of $w = 5\ \mu$m. One facet of the QDL diode is anti-reflection (AR) coated while the other is high-reflection (HR) coated. The laser is mounted on a C-mount and the temperature is controlled by a Peltier element. The emission wavelength is then centered at $\lambda = 1225$ nm.

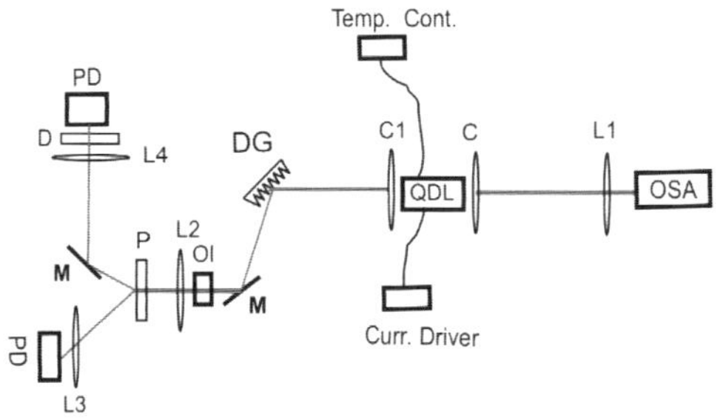

Figure 1. Experimental setup: mirror (M), diffraction grating (DG), aspheric collimator (C1) (f = 3.1 mm), aspheric collimator (C) (f = 8 mm), optical isolator (OI), L1 (f = 30 mm), L2 (f = 200 mm), L3 (f = 50 mm), L4 (f = 35 mm), Wollaston polarizer (P), filter (D), Photodiode (PD), Optical Spectrum Analyzer (OSA).

The experimental setup is illustrated in Fig. 1. The emission in the fast axis is nearly collimated using an aspherical lens (C1) of 3.1 mm focal length and numerical aperture NA = 0.68. Feedback is provided in a Littrow scheme with a diffraction grating with 1450 lines/mm arranged at an angle $\Theta = 75°$ with respect to the incoming beam. The collimator is positioned in a way to optimize threshold reduction by focusing on the grating (external cavity length 123 mm). After the external cavity, there is a lens, L2, for beam shaping and an optical isolator (OI) to prevent feedback from the detection part. A Wollaston polarizer splits the beam in the two polarization components, horizontal and vertical and sent them to two InGaAs detector (Thorlabs PDA255, bandwidth of 50 MHz) to monitor the dynamics. The optical spectrum is also monitored from the HR facet of the diode by a fiber coupled optical spectrum analyzer with a nominal resolution of 0.07 nm. For the characterization of the free running laser we simply replace the grating by a high reflectivity mirror. The threshold current of the free running laser is I = 760 − 800 mA depending on temperature ($T = 6 - 10°$C). With feedback at the same wavelength ($\lambda = 1225$ nm, gain peak) it is reduced to I = 550 mA, a threshold reduction of about 27%.

3. Experimental results

The free running laser shows strong pulsations according to Fig. 2 in a wide range of drive currents. The main polarization component is the horizontal, which is the one shown in Fig. 2, while the vertical polarization has about 10 times lower intensity but exhibits qualitatively the same behavior. The pulsations start around the threshold of the laser. They are large amplitude pulses on a small background (Fig. 2(a)). With increasing drive current their frequency increases and reaches a maximum characterized by approximately 50% duty cycle (Fig. 2(b)). At higher currents, the on-state dominates (Fig. 2(c)) until the pulses are better described as short drop-outs from a high-amplitude state (Fig. 2(d)). The pulsations disappear above I = 850 mA (Fig. 2(d)) and the laser emission is stable afterwards. Quantitatively, the frequency of this pulsations varies between 2.75 MHz and 5.1 MHz (Fig. 2(e)). It is evident from the figures that there is some variation in pulse duration (e.g. Fig. 2(c), (d)), i.e. there is a jitter in the width of the pulses present at all drive currents.

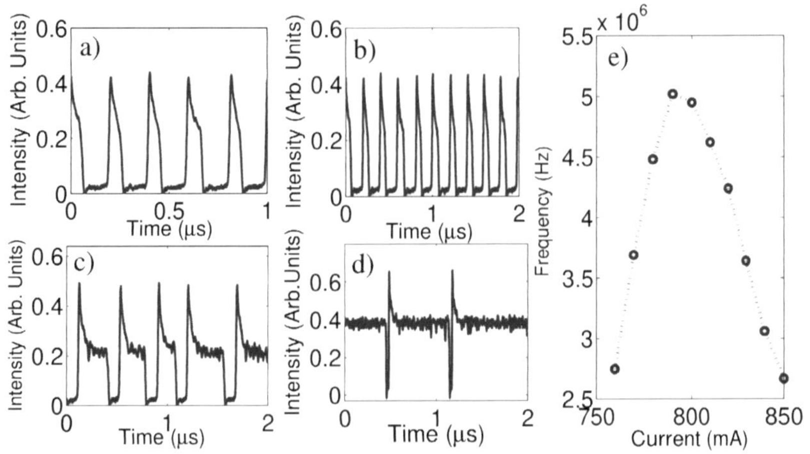

Figure 2. Behavior of the oscillation for the free running laser for increasing current: (a) 795 mA, (b) 810 mA, (c) 850 mA, (d) 870 mA. (e) Frequency behavior versus current in a different realization (laser threshold $I = 760$ mA).

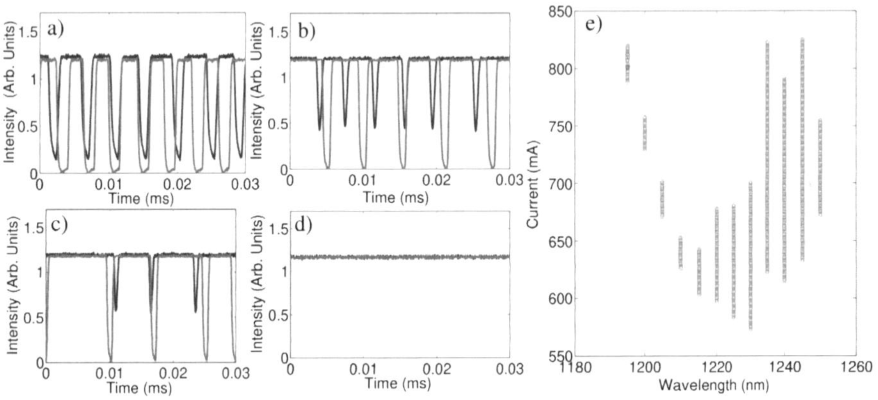

Figure 3. Evolution of oscillations with feedback for increasing current (red curve vertical polarization, blue curve horizontal) at $\lambda = 1195$ nm: (a) 790 mA, (b) 810 mA, (c) 820 mA, and (d) 830 mA. (e) Oscillation range in dependence of wavelength.

The laser with feedback also shows oscillations as evidenced in Fig. 3. In this case the oscillations are more of a square-shape type. They appear at frequencies somewhat lower than the ones in the free running laser and are centered around 1 MHz. Figs. 3(a)-(d) illustrate the development of pulsing in dependence of current. It is taken for lower emission wavelengths ($\lambda = 1195$ nm) than the gain maximum but the qualitative behavior does not depend on wavelength and is quite the same as in the free-running laser.

Fig. 3(e) shows the range of the oscillation for different wavelengths available by tuning the diffraction grating from 1195 nm to 1250 nm. Between 1195 nm and 1220 nm, the range where the oscillation are present is quite small, around 40 mA, then it increases to around 100 mA

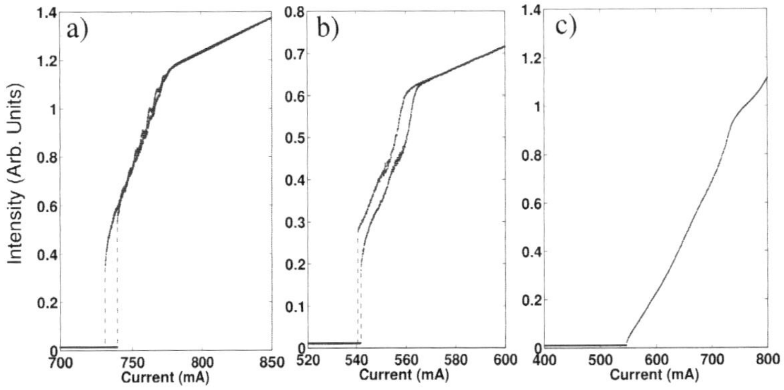

Figure 4. Bistability curve for main polarization component at: (a) $\lambda = 1195$ nm, (b) gain peak, $\lambda = 1221$ nm, (c) $\lambda = 1240$ nm.

around the central wavelength. This range is for both polarizations of the diode.

The time-averaged optical light output versus current characteristics (LI-curves) differ strongly below the wavelength of minimal threshold of 1230 nm and above. At $\lambda = 1195$ nm, Fig. 4(a), there is bistability at laser threshold with a pronounced hysteresis loop. Afterwards, the average intensity increases linearly. At a kink at about 780 mA the slope of the LI-curve changes and the laser becomes stable. Increasing the emission wavelength, the width of the hysteresis loop becomes smaller but an abrupt transition is still present up to $\lambda = 1221$ nm, Fig. 4(b). Fig. 4(c) refers to $\lambda = 1240$ nm where there is no sign of bistability and the laser threshold is continuous. Also the free running laser doesn't present bistability. Frequency-selective feedback is known to support bistability at lasing onset in bulk and quantum well lasers [8]. The dependence on the sign of detuning might indicate a nonlinear index effect.

4. Conclusions

Self-sustained pulsations of the output have been observed in a InAs quantum dot laser diode in the MHz range for the first time. The time scale and other observations suggest that these are opto-thermal pulsations similar to the ones reported in quantum well amplifiers [4]. We are currently investigating the hypothesis of a dynamical change of thermal waveguiding properties. Bistability is present below the gain peak.

Acknowledgments: This work was supported by EPSRC project EP/E025021. We are grateful for useful discussions with Daniil Livshits from Innolume GmbH.

References

[1] Qasaimeh O, Zhou W D, Phillips J, Krishna S, Bhattacharyaa P and Dutta M 1999 *Appl.Phys. Lett.* **74** 1654–1657
[2] Summers H D, Matthews D R, Smowton P M, Rees P and Hopkinson M 2004 *J. Appl. Phys.* **95** 1036–1042
[3] Mokkapati S, Tan H H, Jagadish C and Buda M 2008 *Appl. Phys. Lett.* **92** 021104
[4] Barland S, Piro O, Giudici M, Tredicce J R and Balle S 2003 *Phys. Rev. E* **68** 036209
[5] Marino F, Catalán G, Sánchez P, Balle S and Piro O 2004 *Phys. Rev. Lett.* **92** 073901
[6] Rzhanov Y A, Richardson H, Hagberg A A and Moloney J V 1993 *Phys. Rev. A* **47** 1480–1491
[7] Suret P, Derozier D, Lefranc M, Zemmouri J and Bielawski S 2000 *Phys. Rev. A* **61** 021805(R)
[8] Giudici M, Giuggioli L, Green C and Tredicce J R 1999 *Chaos, Solitons & Fractals* **10** 811–818

Optimization of low density InP/GaInP quantum dots for single-dot studies

C. J. Elliott[1], E. A. Chekhovich[1], M. S. Skolnick[1], A. I. Tartakovskii[1] and A. B. Krysa[2]

[1]Department of Physics and Astronomy, University of Sheffield, Sheffield S3 7RH, UK;
[2]Department of Electronic and Electrical Engineering, University of Sheffield, Sheffield S1 3JD, UK

E-mail: c.elliott@sheffield.ac.uk

Abstract. We achieve well-controlled and reproducible growth by low-pressure metalorganic vapor phase epitaxy (MOVPE) of low density InP/GaInP quantum dots optimized for single-dot physics and applications. We overcome the common occurrence of multi-modal distributions of quantum dot sizes by optimizing the growth on (100) GaAs substrates with a 3° misorientation towards ⟨111⟩. In contrast to other epitaxial techniques for quantum dot growth, very controllable dependence of the quantum dot sizes and densities on the nominal thickness of the InP layer is observed, enabling highly reproducible growth.

1. Introduction

In the last 10 years, experiments on individual quantum dots (QDs) have revealed a wealth of effects due to strong confinement and the resulting isolation of charge-carriers in these nano-structures from the surrounding bulk material [1]. Of particular interest are self-assembled QDs, including InGaAs/GaAs and InP/GaInP QDs, which have been widely researched for applications ranging from single photon emitters [1, 2, 3, 4, 5] to spin qubits for quantum information processing [1, 6, 7, 8]. QDs in both material systems have recently revealed novel spin phenomena of current high interest in semiconductor research and exhibit properties important for spin control on the nano-scale in solid state quantum devices [1, 6]. In optically pumped InP/GaInP QDs, grown by metalorganic vapor phase epitaxy (MOVPE), record high degrees of nuclear spin polarization $\approx 65\%$ as well as ultra-long nuclear depolarization times up to 5000s have been observed[9, 10, 11].

However, samples used in these experiments commonly contain multi-modal distributions of QD sizes, consisting of short wavelength (660 – 730nm), small QDs and longer wavelength (730 – 770nm), large QDs [11, 5, 12, 13, 14, 15]. Although these reproducibly grown samples allow access to individual QDs in the short wavelength range, their properties are influenced uncontrollably by interactions with high density, large QDs. These large QDs have been shown to accumulate high numbers of charges at low temperatures[13], leading to charge instability and additional spin relaxation pathways in the neighboring small QDs. This makes the growth of single-modal distributions essential for single dot studies. In addition, the ability to reproducibly grow interacting double InP/GaInP QDs is enhanced by reproducible single-mode growth, thus avoiding the wide height variation of multi-modal growth, where heights from 1 to 15nm have been observed[5, 12, 13, 14, 15].

In this letter, we report on a MOVPE crystal growth optimization study which enabled us to avoid formation of high densities of large QDs, leading to optimized samples containing only small QDs with densities lower than 10^9cm^{-2}. We investigate the growth conditions required to reproducibly and accurately control both the size and the spatial distribution of single InP/GaInP QDs. This is shown to be achievable by selecting misorientated GaAs substrates and an appropriate InP deposition thickness (d_{InP}). We show that on (100) GaAs substrates with a 3° misorientation towards ⟨111⟩, a distinct range of d_{InP} exists ≈ 0.11 nm, where the natural tendency to develop from a high density of small QDs into several distinguishable distributions of larger QDs is observed. This transition is markedly more gradual with deposition thickness than the one observed in the molecular beam epitaxy growth of widely studied InGaAs/GaAs QDs[16, 1]. Thus the MOVPE method discussed here offers a robust well controlled method for fabrication of a suitable alternative to InGaAs structures in research of quantum effects on the nano-scale. The effects of a gradient in growth reactant concentration, CuPt-type ordering and changing the degree of misorientation of the GaAs substrate in the applied MOVPE method are also discussed.

2. Crystal growth and micro-photoluminescence studies

Growth was performed using low-pressure MOVPE in a horizontal flow quartz reactor, with two wafers exposed for each growth run. The growth temperature of the GaAs buffer and bottom GaInP layer was 700°C. Before proceeding to the deposition of InP and the GaInP capping layer, the wafer was cooled to 650°C. For the samples grown with $d_{InP} = 1.1$ nm, the temperature for the capping GaInP layer was raised to 700°C. The grown GaInP layers were nominally lattice matched to GaAs. A high InP growth rate of 0.11 nm/s was chosen.

All samples were measured at a temperature of 15K using a micro-photoluminescence (μPL) set-up. A HeNe laser with a wavelength of 543nm or 633nm was used to excite carriers in all spectra presented.

Fig.1(a) shows typical μPL spectra recorded at relatively high excitation power ($P = 15\mu$W), for QD samples grown on 3° misorientated substrates with a relatively large InP deposition thickness of $d_{InP} = 1.1$ nm. As seen for both upstream (black line) and downstream (gray line) positions, a pronounced multi-modal size distribution is observed with two broad peaks centered at 705nm and 750nm, corresponding to small and large QDs, respectively. Sharp lines corresponding to PL of individual QDs are observed in the range 725−700nm at a lower excitation power, $P = 1\mu$W, for which population of excited states in QDs is suppressed [see Fig.1(b)]. However, the peak at 750nm remains broad and featureless implying markedly higher densities of large QDs.

One method to reduce the density of large QDs is to increase the angle of misorientation on the substrate surface [12]. This reduces surface diffusion and thus inhibits QD development. Fig.1(c) shows μPL spectra recorded at low excitation power ($P = 1\mu$W) for two samples with $d_{InP} = 1.1$ nm and 0.22 nm (black and gray lines respectively) grown on (110) GaAs substrates with a 10° misorientation towards ⟨111⟩. As expected, it is observed that in the sample with $d_{InP} = 1.1$ nm, there is a marked suppression of the long wavelength μPL peak, but this sample also exhibits an increased density of small QDs which make this method unsuitable for this purpose.

Using the decreased misorientation angle of 3°, we show that growth may be successfully optimized to obtain a low spectral density of small QDs, combined with the tendency not to develop large QDs. As such, all further characterization discussed here was carried out on samples grown on 3° misorientated substrates.

The effect of reducing d_{InP} is shown in Fig.2. The high power spectra shown in Fig.2(a)

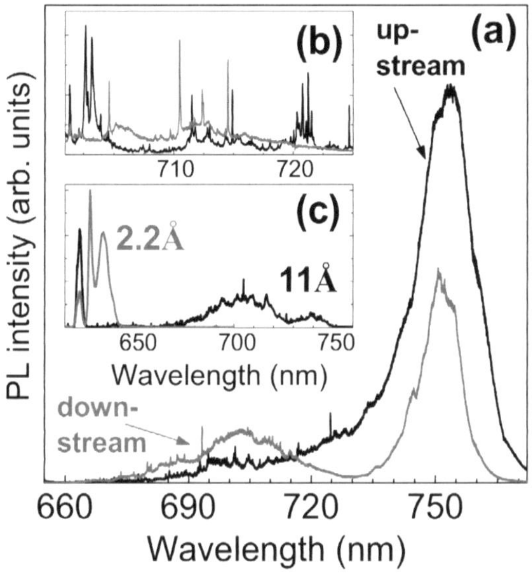

Figure 1. Low temperature μPL spectra of InP/GaInP QDs. (a) d_{InP} = 1.1 nm on a 3° off substrate, P = 15μW, showing the markedly high density of large QDs present. (b) d_{InP} = 1.1 nm on a 3° off substrate, P = 1μW, showing single PL lines of small QDs. (c) Compares d_{InP} = 1.1 nm and 0.22 nm on 10° off substrates, P = 1μW, showing inhibited large QD formation and increased densities of small QDs.

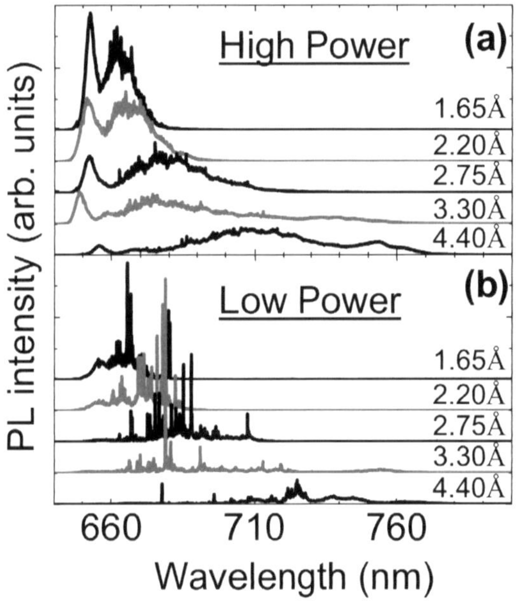

Figure 2. Low temperature μPL spectra of InP/GaInP QDs. Here d_{InP} is varied from 0.165 nm to 0.44 nm, on 3°off substrate. This shows how the samples change from containing a single distribution of small QDs (upper lines) through the transition point where development into large QDs begins (lower lines). Spectra shown are for the same position in the growth chamber and with (a) P = 5μW and (b) P = 0.15μW.

are measured with P = 5μW. They show a sharp peak around 653nm, corresponding to the GaInP barrier emission. At d_{InP} = 0.165 nm, we measure emission from only small QDs with a narrow distribution peaked at 663nm, indicating uniform formation of very small QDs. As d_{InP} is increased the QD μPL peak shifts to longer wavelength and becomes broader. Eventually at d_{InP} = 0.33 nm, a weak peak at 750nm is observed, corresponding to the formation of large QDs which is further enhanced for d_{InP} = 0.44 nm, where the emission exhibits two distinct, broad spectral QD μPL distributions.

Fig.2(b) shows μPL spectra recorded at a low excitation power (P = 0.15μW) for samples as in Fig.2(a). The optimum conditions for the growth of low density of small QDs, with a single distribution, are reached at d_{InP} = 0.275 nm and 0.33 nm. Over this range of d_{InP}, we observe a relatively small number of individual QD PL emission lines distributed in a wide spectral range, allowing clear observation of optical properties of individual QDs. Further spatial selectivity on these samples can be obtained using a standard method by fabrication of opaque metal masks with clear apertures of $\approx$ 1μm diameter[11]. The QD density in samples with d_{InP} = 0.275 nm and 0.33 nm is 1x10^{9}cm^{-2} and 8x10^{8}cm^{-2} respectively (estimated by counting the number of lines in the spectrum at low powers). Under optimized conditions we achieve high spectral isolation of single QD PL lines with typical narrow FWHM of 0.05nm.

We note, that the observed range of d_{InP} from $\approx$ 0.27 nm to $\approx$ 0.33 nm equivalent to 0.2 atomic monolayers (ML), which leads to growth of suitable samples at our InP deposition rate of 0.11 nm/s, is large in comparison to the mechanical growth-control time. This results in very reproducible growth confirmed in our further growth experiments. The variation of the dot density by a factor 1.25 in this range of d_{InP} is up to a factor of 2 (30) smaller than for a similar range of deposition thicknesses in the dot density region of $\approx 10^9$ cm^-2 in molecular beam epitaxial growth of InP/GaInP (InGaAs/GaAs) QDs[16, 17].

3. Conclusions

We have realized MOVPE growth of low density InP/GaInP QDs suitable for single dot applications. The major hurdle of the presence of high densities of large QDs in the previously studied samples has been overcome. We find a reproducible, smooth transition in QD size distribution and density for varying nominal InP deposition thickness and position along the growth chamber. QD development on 3° misorientated substrates provides a wide range of nominal InP deposition thicknesses ($\simeq$ 0.3ML) where conditions are suitable for single dot applications.

Acknowledgments

We thank M. Hugues and N. Babazadeh for fruitful discussions. This work has been supported by the EPSRC Programme Grant EP/G601642/1 and the Royal Society. AIT is grateful for support by an EPSRC ARF.

References

[1] See a collection of reviews on QDs: 2009 *Semiconductor Quantum Bits* ed F Henneberger and O Benson (Singapore: Pan Stanford Publishing).
[2] C. Santori, M. Pelton, G. Solomon, Y. Dale and Y. Yamamoto, Phys. Rev. Lett. **86**(8), 1502 (2001).
[3] P. Michler, A. Kiraz, C. Becher, W. V. Schoenfeld, P. M. Petroff, L. Zhang, E. Hu and A. Imamoğlu, Science **290**, 2282 (2000).
[4] J. Persson, T. Aichele, V. Zwiller, L. Samuelson, O. Benson, Phys. Rev. B **69**, 233314 (2004).
[5] W. -M. Schulz, R. Roßbach, M. Reischle, G. J. Beirne, M. Bommer, M. Jetter and P. Michler, Phys. Rev. B **79**, 035329 (2009).
[6] M. Atature, J. Dreiser, A. Badolato, A. Hogele, K. Karrai, A. Imamoglu, Science **312**, 551 (2006).
[7] R. M. Stevenson, A. J. Hudson, A. J. Bennett, R. J. Young, C. A. Nicoll, D. A. Ritchie and A. J. Shields, Phys. Rev. Lett. **101**, 170501 (2008).
[8] M. Scholz, T. Aichele, S. Ramelow and O. Benson, Phys. Rev. Lett. **96**, 180501 (2006).
[9] E. A. Chekhovich, M. N. Makhonin, K. V. Kavokin, A. B. Krysa, M. S. Skolnick and A. I. Tartakovskii, Phys. Rev. Lett. **104**, 066804 (2010).
[10] E. A. Chekhovich, M. N. Makhonin, J. Skiba-Szymanska, A. B. Krysa, V. D. Kulakovskii, M. S. Skolnick and A. I. Tartakovskii, Phys. Rev. B **81**, 245308 (2010).
[11] J. Skiba-Szymanska, E. A. Chekhovich, A. E. Nikolaenko, A. I. Tartakovskii, M. N. Makhonin, I. Drouzas, M. S. Skolnick, A. B. Krysa, Phy. Rev. **B**, 77, 165338 (2008).
[12] J. Porsche, A. Ruf, M. Geiger, F. Scholz, J. of Cryst. Growth **195** 591 (1998).
[13] D. Hessman, J. Persson, M. -E. Pistol, C. Pryor and L. Samuelson, Phys. Rev. B **64**, 233308 (2001).
[14] M. -E. Pistol, J Phys.: Condens Matter **16**, S3737 (2004).
[15] J. Persson, M. Holm, C. Pryor, D. Hessman, W. Seifert, L. Samuelson and M. -E. Pistol, Phys. Rev. B **67**, 035320 (2003).
[16] Q. Xie, N. P. Kobayashi, T. R. Ramachandran, A. Kalburge, P. Chen and A. Madhaker, J. Vac. Sci. Technol. B **14(3)**, 2203 (1995).
[17] A. Ugur, F. Hatami, W. T. Messelink, A. N. Vamivakas, L. Lombez and M. Atatüre, Appl. Phys. Lett. **93** 143111 (2008).

AUTHOR INDEX

AUTHOR INDEX

AUTHOR INDEX

AUTHOR INDEX

Institute of Physics
Dirac House, Temple Back
Bristol BS1 6BE UK

proceedings.com